Brain-Body-Mind in the
System: A Holistic Approach by Oscillations

Erol Başar

Brain-Body-Mind in the Nebulous Cartesian System: A Holistic Approach by Oscillations

Springer

Erol Başar
Istanbul Kültür University
Istanbul
Turkey
e.basar@iku.edu.tr

ISBN 978-1-4419-6134-1 (hardcover) e-ISBN 978-1-4419-6136-5
ISBN 978-1-4419-6135-8 (softcover)
DOI 10.1007/978-1-4419-6136-5
Springer New York Dordrecht Heidelberg London

Printed on acid-free paper

Springer is part of Springer Science+Business Media (www.springer.com)

In Memory of
The Renowned Physicist and Natural Philosopher

CARL FRIEDRICH VON WEIZSÄCKER

Whose Advice Illuminated My Multidisciplinary Research Pathway

Prologue

Everything in the universe could be explained in terms of a few intelligible systems and simple approaches, upon which the stars and the earth and all visible worlds may have been produced.

René Descartes

All matter in the universe is made up of elementary particles of only few kinds. It is like seeing in one town buildings of different sizes, construction and architecture, but from shack to skyscraper only few different kind of bricks were used, the same in all the buildings. So all known elements of our material world, from hydrogen the lightest, to uranium the heaviest, are built of the same kinds of bricks, that is, the same kinds of elementary particles.

Albert Einstein

Brain-Body-Mind Syncytium

Are the philosophical views of Descartes and Einstein suggestive of the search for elementary principles of brain-body-mind Integration?

The core idea of the book is reflected in the following: EEG oscillations, ultraslow oscillations, and neurotransmitters are quasi-invariants in brain-body-mind function and during the evolution of species.

The brain, spinal cord, overall myogenic system, brain-body rhythms, and neurotransmitters form a functional syncytium. Accordingly, the concept of syncytium brain-body-mind replaces the concept of mind.

In this syncytium, neurons, smooth muscles, neurotransmitters, EEG-, and ultraslow-oscillations that govern all functional processes, are possibly the few intelligible systems or elementary bricks in the sense implied by Descartes and Einstein, respectively. In future, genetic codes that should be jointly analyzed with cognitive processes, will belong to the few intelligible systems

The multiplicity of functional links in brain-body integration does not allow a deterministic description of sensory/cognitive events. The observed uncertainty, and/or deterministic chaos impose the proposition of a nebulous Cartesian system, which should integrate analyses so as to understand brain-body-mind.

Essentials of the Book

Rationale

This is a research book with didactical features on the mechanisms of the mind, encompassing a wide spectrum of results and analyses. Its goal is to develop an empirical-analytical construct denoted as "reasonings to approach the mind." It includes analyses from several disciplines, such as neurophysiology, psychophysiology, neuropsychiatry, vegetative physiology, Darwinian biology, physics, and philosophy; therefore, it should appeal to scientists, graduate students, and students from these fields.

To achieve this task, the book brings together results from brain, vegetative system, brains in evolution of species, emotional processes, neurological, and psychiatric data (specifically, Alzheimer's disease, schizophrenia, and bipolar disorders), which are comparatively evaluated from the perspective of brain oscillations and neurotransmitters. In addition, data from the central nervous system are linked to cardiovascular and overall myogenic coordination of the vegetative system.

In furtherance of a large number of empirical psychophysiological observations, the data were re-evaluated to establish common fundaments and principles in brain-body-mind mechanisms that are governed by the web of multiple large-scale oscillations and neurotransmitters. The probabilistic nature of findings within the scope of Heisenberg's uncertainty principle led to the concept of the quantum brain. Emerging interpretations are discussed in the light of philosophy, especially guided by the principles of René Descartes and dominated by the time and intuition concepts of Henri Bergson. Based on interpretation of considerable empirical evidence, it is tentatively proposed to extend Descartes' Cartesian system with a new framework, denoted as the nebulous Cartesian system,[1] which is also based on the uncertainty principle.

[1]According to F. Capra (1984) the Cartesian model needs a major revolution. In his words:[2]"Transcending the Cartesian model will amount to a major revolution in medical science, and since current medical research is closely linked to research in biology—both conceptually and in its organization—such a revolution is bound to have a strong impact on further development of biology. To see where this development may lead, it is useful to review the evolution of the Cartesian model in the history of biology. Such a historical perspective also shows that the association of biology with medicine is not something new but goes back to ancient times and has been an important factor throughout its history."

Certainly this is not unique in representing a possible framework; one person's work is not sufficient to achieve a necessary breakthrough in neuroscience comparable with those in physics at the turn of the twentieth century. Accordingly, the author considers the present book a crude workshop, by hoping that multidisciplinary neuroscientists may try to establish a trend similar to that of the Copenhagen School in the 1920s. It is also hoped that young scientists reading the results presented from multidisciplinary fields and several theoretical conclusions may amalgamate all of these and develop new thoughts or theories, as I did after reading *Cybernetics* by Norbert Wiener 45 years ago.

Seven Major Factors Encouraged Me to Write the Present Monograph

The writing of a monograph on brain-body-mind is extremely ambitious. This is because I think that Friedrich von Hayek was correct when he stated (1952), "The mind must remain forever in a realm of its own, which we shall never be able fully to explain." Although I am aware of such difficulties, my education in the fields of quantum physics, conventional physiology, and philosophy was the major impetus to begin this project. Seven factors finally influenced my decision to embark on this endeavor, of which the first two have provided the greatest encouragement:

1. *Brain oscillations* has evolved to become a hot and fundamental branch of neuroscience, although during the 1970s Walter Freeman in Berkeley, California and myself were almost alone in publishing a series of studies in this field.[2,3] Presently, the key theme, event-related oscillations, has become a fundamental trademark in studies of memory and cognitive processes.
2. *Quantum Mechanics* and *Fundamentals of Natural Philosophy* were taught to me by Werner Heisenberg and Carl Friedrich von Weizsäcker, who accepted me into their circle first as a student and then as a graduate student – events that were to initiate a tide of good fortune.[4] Quantum mechanics introduced an unavoidable element of unpredictability or randomness into science and also, over the last 3 decades, into brain research as well.[5]

---

[2]Freeman (1975), and Başar (1976, 1980). Measurements of the gamma activity at the cellular level by the groups of Eckhorn et al. (1988) and Singer (1989) were made approximately 12 years later and added important value to research on brain oscillations.

[3]Mountcastle (1992, 1998) stated that the paradigm shift introduced by using brain oscillations has become one of the most important conceptual and analytical tools for the understanding of cognitive processes. He proposes that a major task for neuroscience is to devise ways to study and analyze the activity of distributed systems in waking brains, in particular, human brains.

[4]How did this tide of fortune occur? See the Prelude to Part I.

[5]Başar 1983, Eccles 1986, Hammerhof and Penrose, 1996, etc. For more up to date reviews refer to the March 2010 issue of the online journal, NeuroQuantology.

The other five factors also have been vital:

3. *Nervous Systems.* Theodore Holmes Bullock's opinion is that comparative neuroscience promises important insights into the structure and function of nervous systems. Achieving this potential requires pursuit on a wide front, both in terms of interdisciplinary research and with respect to the species, stages, and states compared (Bullock 1984a,1984b). Starting in 1982 in California, and later in Lübeck, Germany, during a period of almost 25 years of joint research, publishing, and strong friendship with T.H. Bullock, I studied Charles Darwin's theory from an experimental viewpoint.
4. *Research on Cardiovascular System and Vegetative Physiology* provided me experience and knowledge to bridge brain and body. Although after publishing the monograph, *Vasculature and Circulation* (Elsevier 1981), I did not carry out experimental research on this topic, my 20 years of teaching physiology to medical students at The Medical University Lübeck (until 2000) allowed me to keep track of developments in vegetative physiology. In turn, this indicated the necessity of including the vegetative system within the understanding of mind. The concept of *brain-body-mind* replaced the word *mind*.
5. *Clinical Research on Neurology and Psychiatry* started in Lübeck, Germany with analysis of patients with multiple sclerosis.[6] This research currently continues in Istanbul, Izmir, and Bremen with patients with Alzheimer's disease, bipolar disorders, and schizophrenia.[7] Results on pathology bring an important dimension to the understanding of mind and breakdown of mind.[8]
6. *Our investigations into the recognition of faces and emotional events related to episodic memory* – among the most complex forms of brain processing, during which the brain travels back in time within a fraction of a second – fascinated me.[9]
7. Last, but not least, my strong scholastic *education in philosophy* enabled the emerging interpretations of empirical results in the light of philosophy, especially guided by the principles of René Descartes and dominated by the *time and intuition* concepts of Henri Bergson.

Besides the seven factors explained in the preceding, during the years of my experimental and theoretical work I had the privilege to meet and engage in discussion with a number of prominent colleagues, including Hermann Haken,[10] Joaquín Fuster, Robert Galambos, Henri Begleiter, John Polich, and Roy John. Joaquín Fuster spent hours surveying this monograph's outline during a world congress in Istanbul. Following a symposium in Lübeck in 1994,[11] Riita Hari and F. Lopes da

[6]With Prof. K. Wareczka in Lübeck and Prof. Başar-Eroğlu in Lübeck and Bremen.

[7]With Prof. Canan Başar Eroğlu and coworkers.

[8]With Prof. Canan Başar Eroğlu (Bremen), Görsev Yener (Izmir), and Ayşegül Özerdem (Izmir).

[9]With Asist. Prof. Dr. Bahar Güntekin.

[10]In 1983, I worked with Herman Haken to organize a symposium on the Synergetics of the Brain (Başar et al., 1983), and thus began an interesting and fruitful scientific friendship.

[11]E. Başar et al., 1997.

Silva worked as co-editors of a special issue on the functional correlates of alpha activity, which has been very influential in emphasizing the functional importance of alpha activity.

The Importance of Holistic View of Rhythms

To approach the brain-mind, we also have to observe the machineries of invertebrate ganglia and brains during the evolution of species (Fig. P.1). Furthermore, physiological processes and anatomical changes need to be analyzed during maturation of the brain from infancy through adulthood to old age (see Chaps. 10 and 11).

In pathologic brains, the release of transmitters and, accordingly, oscillatory processes and control of cognitive processes, are highly altered (see Chap. 13). Therefore, the analysis steps in Fig. P.1, which also includes a loop indicating the influences of pathology (Alzheimer's disease, schizophrenia, and bipolar disorders) constitute a minimal analysis prerequisite approach to the integration of brain-body-mind.

The importance of rhythms of the brain is explained by Buzsáky (2006) in a brilliant manner:

> Neuroscience has provided us some astonishing breakthroughs, from noninvasive imaging of the human brain to uncovering the molecular mechanisms of some complex processes and disease states. Nevertheless, what makes the brain so special and fundamentally different from all other living tissue is its organized action in time. This temporal domain is where the importance of research on neural oscillators is indispensable.

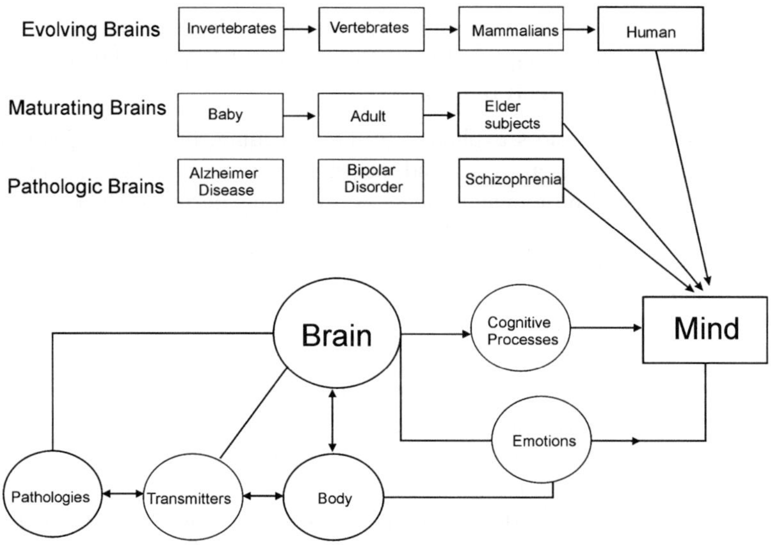

Fig. P.1 Schematic explanation of the steps proposed for an approach to brain-body-mind

However, the present book demonstrates that the brain is not the unique living tissue showing oscillatory organization; and for breakthroughs we need an adequate Cartesian system taking account the whole body and not only the brain. Accordingly, research on brain-body-mind is essential.

Influences of Descartes, Heisenberg, Einstein, and Wiener

From the time of the Renaissance philosophers, the study of mind required a multidisciplinary analysis, as clearly indicated in the quote from René Descartes at the top of this discussion. As a young student in Munich between 1959 and 1962, I had the good fortune to learn from Werner Heisenberg, about a basic idea related to Newton's theory, quantum mechanics, and the S-matrix. Further, Prof. Heisenberg advised me to finish my studies in quantum theory and elementary particles in Hamburg at the German Electron Synchrotron DESY. Following my studies in Hamburg, I completed a thesis entitled "Calculation of cross sections in an experimental setup for measuring scattered elementary particles as strong resonances resulting from interaction of gamma rays on protons." My research started with some fundamental bricks of matter, as explained in a most didactical manner by René Descartes and Albert Einstein, outlined at the very beginning of the Prologue. In neuroscience, the view of Descartes is reflected by the work of Alfred Fessard (1961), which constitutes a leitmotif within the book.

In the field of brain-body-mind, these bricks (or building blocks) are oscillations – EEG-oscillations and ultraslow oscillations – in the brain and organs of the vegetative system, which occur in different frequency channels, and interact with various neurotransmitters, such as acetylcholine, GABA, dopamine, etc. The buildings made from these bricks are sensory processes, memory, emotions, and possibly also the processes of traveling back in time as well as creativity and intuition. These bricks also constitute the basis for the brain's string theory, which is included in the book as a model of thought.

I also mention the important controlling function of clocks in all control systems and devices, ranging from locomotives to space ships; this serves as a reminder that there are special types of clocks, denoted as "Wiener regulators."

Einstein's concepts about bad clocks and good clocks (synchronized clocks) also establish core concepts of the brain's oscillatory dynamics and control in the body. Synchronized clocks are vital elements controlling brain-body mind functioning; the breakdown of synchronized clocks is observed in pathological brains, and accordingly the breakdown of the mind.

Norbert Wieners' *Cybernetics* (1948) was conceived as a common ground on which engineers, biologists, mathematicians, psychologists, and others could meet and discuss, in a common scientific language, the problems of control and communication that appear in various forms in their scientific fields. The work of Wiener provided an inspiring framework for thinking broadly in parallel within multidisciplinary fields. One thing is absolutely clear: The research leading to the foundation of brain dynamics[12] was anchored to the idea of signal processing and communication within the brain.

After the publication of the 1976 book by Başar, a reviewer commented that passages related to Norbert Wiener were somewhat historical and not up to date. I do not share this opinion. I re-read parts of Norbert Wiener's book several months ago; it remains inspiring for young scientists in our new century. Such multidisciplinary and conceptual works are needed more than ever.

Advice of Carl Friedrich Von Weizsäcker

In the period between 1962 and 1968 I had the privilege of participating in the "Thursday colloquia" of Prof. von Weizsäcker. The colloquia took almost 3 h every Thursday and were attended by a maximum of 10–12 participants. During my study years in Munich, and later in Hamburg, Heisenberg and Weizsäcker conveyed many anecdotes, discussions, and Gedankenexperiments[13] from Einstein, Bohr, Schrödinger, Pauli, and Dirac, as well as the Solvay conferences. The time concept of Einstein was also frequently explained and discussed by Prof. von Weizsäcker. In my physiology and biophysics career, I was continually influenced by these eminent natural philosophers by always thinking about elementary mechanisms and unifying processes, as explained in the previous section.

In 1965, Professor Weizsäcker strongly encouraged me to switch my research from high-energy physics to physiology; furthermore, he convinced me to start with conventional cardiovascular research before brain research. It is mostly because of this background, enriched with the advice of these great scientists that during the 1970s I was able to initiate work on brain oscillations in parallel with Walter Freeman at Berkeley.

At the time I started my research in physiology I posed a question to Prof. von Weizsäcker in his home, "What should I do first when planning or starting an experiment?" He simply answered, "Erol, firstly you have to think on the concept; if you plan an experiment to understand the structure of the atom, you have to strongly focus your thoughts on the concept of the atom." This means that you have to think about the atom, and in doing so you will also consider the relationships among electrons, protons, and the fields around the atom, i.e., gravitation.

I applied this approach to the understanding of mind, and took account of the concepts and conceptual frameworks pertinent to neurodynamics and physiology in the years that followed. This particular advice was certainly the most important.

[12] After the publication of the book Başar (1976) a reviewer commented that passages related to Norbert Wiener was somewhat historical and not up to date. I do not share this opinion.

[13] Experiment of thought simpler.

Undoubtedly, the mind has to do with the brain. I do not study the brain as a control organ, independent of our body. The brain is continually in interaction with the autonomous vegetative system; furthermore, the body is neither independent of our environment nor of archetypes (accordingly, genetics and the unconscient dominated by genetic constitution). Therefore, my experiments were planned according to the basic concepts of such scientists as Galileo, Descartes, Newton, Darwin, Bergson, Heisenberg, Berger, von Hayek, and von Weizsäcker. The unifying concept reached in this book was developed during years on that pathway.

The Endeavor with Theodor Holmes Bullock[14] in Evolution of Species

One of the most revolutionary developments in biological sciences was Charles Darwin's introduction of the evolution of the species. Darwin worked within a "transformist" framework of the living world, initiated earlier by Jean-Baptiste de Lamarck (see also Changeux 2004).

Joint experimental research, steady and strong interactions with Ted Bullock took place over a period of more than 20 years. We first started experimenting in La Jolla; then I developed a laboratory to perform measurements with Helix pomatia in Lübeck. The long distance between California and north Germany, then later relocation to Turkey did not create an important obstacle. Ted Bullock's opinion that comparative neuroscience promises important insights into the understanding of structures and functioning of nervous systems, including studies of species, stages, and states coincided with my multidisciplinary brain dynamics research program, published in 1980. Mostly, we began from opposite positions when starting experimental research. Ted always preferred to start with Newton's slogan "*hypotheses non fingo*," whereas I preferred to start with a concept relying on earlier experiments. Ted Bullock and I published several papers and edited two books. Among these was the 1992 book, *Induced Rhythms in the Brain*, produced as a result of a Workshop in New York, which attracted authors from a broad spectrum of neurosciences; and which became a standard reference work at that time. The concepts of the development of intuition in evolution and the bridging of neuroethology with Bergson's understanding of mind and the increase of coherence in various brain structures during evolution are fruits of this longstanding collegial friendship.

[14]Theodore Holmes Bullock, who was one of the most distinguished scientists in the American Neuroscience Community, is known as the father of Neuroethology. See also Prelude to Part III.

Pathologic Brain: Breakdown of Oscillations and of Mind in Dementia

The present book also deals with cognitive impairment in Alzheimer's disease, bipolar disorders, and schizophrenia, conditions often leading to a breakdown of the mind or changes of brain-body-mind. As stated in a recent review on cognitive impairment (Başar and Güntekin 2008), consideration of brain oscillations within clinical studies is rare, although oscillations in cognitive processes have begun to emerge within the neuroscience literature.

As indicated, our group's contribution to developments in the study of the impairment of cognition has been one of the factors encouraging the writing of the present book.[15] The comparison of changes of cognitive responses in three types of cognitive impairment and their treatment with neurotransmitters opens a new avenue for research on brain-body-mind integration. This is explained in Chap. 13. However, the results are as yet only the tip of an iceberg, and the study model may open new methods of interpretation in the coming years. For example, we have observed enhanced gamma oscillations in an isolated ganglion of the invertebrate ganglion of the snail species Helix pomatia after the increase of dopamine in the organ bath. What happens in schizophrenia after dopamine medication?

In Alzheimer's disease, with severe impairment of mind, the gamma coherence between frontal and parietal locations was not shown to decrease, whereas in bipolar patients, a large decrease of gamma coherence was observed. What are the functional implications to be learned from those comparisons, and what are the changes in mind in such cases? These questions will be investigated in the future.

Emotion, Episodic Memory, and Intuition Form One Combined Entity: Darwin and Reappraisal of Bergson's Philosophy

According to Solms and Turnbull (2002), emotion is akin to a sensory modality that provides information about the current state of body self, as opposed to the state of the object world. "Emotion" is the aspect of consciousness that is left if all

[15]Our cooperation group of Başar-Eroğlu in Bremen is one of only a few groups analyzing schizophrenia in a frequency window with multiple oscillations.

Görsev Yener, who strongly influenced my return to Turkey 10 years ago, then joined my laboratories as a young associate professor of neurology in order to analyze oscillations upon cognitive load in Alzheimer patients. Additionally, she was my graduate student, completing her thesis in biophysics in Izmir.

Several years later, Ayşegül Özerdem, also a young associate professor in psychiatry clinics, joined to teach our graduate students; however, she also decided to study biophysics as a graduate student. Accordingly, we had a unique chance to perform comparative studies in neuropsychiatric diseases. After my coming to Istanbul, we continued our comparative clinical research with both colleagues, this time complemented by joint programs with the Bakirköy Neuropsychiatry Hospital (the traditional and largest Hospital in Turkey) and also with Maltepe University.

externally derived contents are removed. Le Doux (1999) states that memory is generally understood to be the process by which we bring back to mind some earlier conscious experience. The original learning and remembering in this case are both conscious events. Further, this author is of the opinion that emotional and declarative memories are stored and retrieved in parallel and their activities are joined seamlessly in our conscious experience. "Emotion is not just unconscious memory: It exerts a powerful influence on declarative memory and other thought processes." Thus, emotions or feelings are conscious products of unconscious processes. In this context, we show the importance of episodic memory by means of experiments. The thesis is that, according to knowledge obtained by means of electrophysiological measurements, it may be possible to explain phenomena that are at the boundary of conscious and unconscious processes. In that context, this work describes ways of analyzing what Eric Kandel termed "traveling back to the past." Intuition is the ability of humans to perform creative methods of thinking and apply these processes to produce inventions or discoveries. Further, it is tentatively stated that creativity and intuition form one combined entity.

In the late twentieth century, simple measurement strategies overwhelmed fundamental conceptual questions. The efficiency of multi-disciplinary analysis on brain-body-mind encompasses an attempt to reconcile the relevant natural philosophy of Bergson with contemporary knowledge, in the hope of adding value to the progress of neuroscience. Henri Bergson (1907), who studied the work of Charles Darwin, came to the conclusion that the superiority of the human brain in comparison with lower species is defined by the capacity for intuitive and creative thinking. In our opinion, the existence of two types of memories, the concept of duration and the description of human intuition inspired by Darwin's theory on the evolution of species (1859), constitute a solid chain of ideas that were described in an era of research not undertaken with electronic instrumentation and diverse methods of analyzing anatomy and electrophysiology.

The chapters of the book related to metaphysics explore strategies and/or methods for understanding the mechanisms of traveling back in time, and also the understanding of creative episodes; accordingly, to find way to analyze the questions posed by Kandel (2006), Andreasen (2005), and Penrose (1989).

As stated, I consider this book to be a type of workshop or workbench leading to reasonings to understand brain-body-mind. Ted Bullock and I started to talk about synchrony and coherences in the 1980s and published a joint review paper in 1988.[16] Later, Bullock published most relevant papers on coherence in the brains of various species, ranging from Aplysia to the human brain, showing the increase of coherence during evolution. The most important idea here is this: It is possible to compare Einstein's bad clocks (asynchronous clocks) in Aplysia ganglia with the good clocks (clocks in synchrony) in the healthy human brain. Further, it is possible to compare bad clocks of pathologic brains with the good clocks of the healthy brain. The minds of Aplysia (if any) and the minds of Alzheimer's patients and healthy subjects are certainly different.

---

[16]Bullock and Başar.

Are the concepts of time described by Einstein's synchronous clocks and Bergson's inhomogeneous time (Durée) comparable? To understand episodic memory, in which the brain is effectively traveling back in time, we need the concept of duration. In brains with dementia, neither Einsteinian good clocks nor Bergsonian clocks are working! The episodic memory zips 20 years into 500 ms. Alzheimer's patients in the late stages of the disease do not remember; both types of clocks are gone.

Several examples of that kind are presented within this book, and I think these types of multidisciplinary steps may lead to significant developments in the understanding of mind, possibly within few decades. Presently, we can measure electrical responses that demonstrate that the patient's brain's differentiates known and unknown faces. In other words, progress and refinements have already penetrated the difficult barrier to the realm of biological sciences. Such progress means that meta-processes in the brain become testable processes.

Neuron Populations, Overall Myogenic System, EEG-, and Ultraslow Oscillations Embedded in Neurotransmitters Are Co-acting by Forming a Syncytium

According to research, even the lymphatic system (residual drainage system of the body) works in oscillatory accordance with the frontal brain (master of the body), and is also a constituent of a syncytial functioning. Even heartburn, a moderate loud tone, or excess of lymphatic fluid in the legs disturb this syncytial work, what we call mind; not only dementia, and not only aging.

Part VI presents several possible models of brain-body-mind. Especially in Chap. 25, an ensemble of constituents or essence of the brain-body-mind problem is described. This is called "Reasonings in Search of Brain-Body-Mind" and resumes the core features of the present book.

These reasonings will not be repeated in here. Nevertheless, we include here two representative paragraphs that should convey the core idea of the book:

1. The concept of the syncytium of brain-body-mind replaces the concept of mind because of the fundamental findings manifested in EEG oscillations, ultraslow oscillations, and neurotransmitters that are quasi-invariants in brain functioning and also the link between the brain and the vegetative system. To understand the mind requires connaissance of these quasi-invariants and also their changes in pathology, especially in dementia, i.e., impairment of mind. A unifying trend proposes the brain's string theory, similar to the string theory in physics.
2. According to the preceding statements, the mind cannot be defined with a unique sentence, as often stated in the literature. We can try to learn and investigate the question, "How does the mind work?" The answer to this question involves multifold functional implications in the brain and body. The brain, body, and mind form a combined entity. The influence of the unconscious and intuition are considerable, being yet only partially understood.

3. Genetics is not a major topic in the book. However, combined with the study of neurotransmitters, oscillations, and paradigms of the cognitive processes, this area will possibly provide the most important window in the study of brain-body-mind. This is already reflected in the pathway opened by Begleiter and Porjesz, which is briefly explained in several chapters, especially Chap. 13.

Bullock et al. (2005) discussed recent evidence suggesting that the Neuron Doctrine, conceived nearly a century ago by Ramon y Cajal, cannot encompass important aspects of information processing in the brain. Intercellular communication by gap junctions, slow electrical potentials, and action potentials initiated in dendrites, neuromodulatory effects, extrasynaptic release of neurotransmitters, and information flow between neurons and glia all contribute to information processing. Revisiting the Neuron Doctrine, these authors suggest that future research outside its limits may lead to new insights into the unique capabilities of the human brain. In extension to these views of Bullock et al. (2005) and also empirical evidence of the present book, we tentatively formulate that not only neural populations, but also the ensemble of brain and body processes could be explained as a syncytial work.

Although a full understanding of the mind will possibly remain an enigma, it should not be considered as a mission completely impossible.

Newly refined measurements and interdisciplinary analyses can lead to greater insights into the mechanisms of the mind. As Bullock would say, *Open your eyes, new discoveries are awaiting us!*

At the time of Descartes, Pascal, and Locke it was inconceivable to measure the efforts of memory or any type of cognitive processes. Within just the last 30 years there has been considerable progress in measuring thoughts or approaching brain-body-mind integration; therefore, I think that the sighting of new land within a few decades is conceivable. Logically, The New Land will potentially accompany a more complicated enigma. We need new and more refined measurements, employing concepts derived from interdisciplinary thinking.

According to Lord Kelvin, "Science is all measurement; but all measurement is not science." Extending this through the wisdom of von Weizsäcker, I merely tend to say, "Only measurements based on a concept form valuable science."

References

Andreasen N (2005) The creating brain. The neuroscience of genius. Dana Press, New York

Başar E (1976) Biophysical and physiological systems analysis. Addison-Wesley, Amsterdam

Başar E (1980) EEG–brain dynamics. relation between EEG and brain evoked potentials. Elsevier, Amsterdam

Başar E (1983a) Toward a physical approach to integrative physiology: I. Brain dynamics and physical causality. Am J Physiol 14:R510–R533

Başar E (1983b) Synergetics of neuronal populations. A survey on experiments. In: Başar E, Flohr H, Haken H, Mandell A (eds) Synergetics of the brain. Springer, Berlin, pp. 183–200

Başar E, Güntekin B (2008) A review of brain oscillations in cognitive disorders and the role of neurotransmitters. Brain Res 1235:172–193

Başar E, Weiss C (1981) Vasculature and circulation. Elsevier, Amsterdam

Başar E, Flohr H, Haken H, Mandell AJ (eds) (1983) Synergetics of the brain (Proceedings of the International Symposium on Synergetics at Schloss Elmau, Bavaria, May 2–7). Springer, Berlin.

Başar E, Hari R, Lopes da Silva FH, Schürmann M (eds) (1997a) Brain alpha activity: new aspects and functional correlates. Int J Psychophysiol 26:1–482

Bergson H (1907) L'évolution Créatrice. Presse Universitaires de France, Paris.

Bullock TH (1984a) Physiology of the tectum mesencephali in elasmobranchs. In: Vanegas H (ed) Comparative neurology of the optic tectum. Plenum Press, New York pp. 47–68

Bullock TH (1984b) Ongoing compound field potentials from octopus brain are labile and vertebrate-like. Electroencephalogr Clin Neurophysiol 57(5):473–483

Bullock TH Başar E (1988) Comparison of ongoing compound field potentials in the brains of invertebrates and vertebrates. Brain Res Rev 13:57–75

Bullock TH, Bennett MV, Johnston D, Josephson R, Marder E, Fields RD (2005) Neuroscience. The neuron doctrine, redux. Science 310(5749):791–793

Buszaky G (2006) Rhythms of the brain. Oxford University Press, New York

Capra F (1984) The turning point. Flamingo, London, p. 97.

Changeux J.-P (2004) The physiology of truth: neuroscience and human knowledge. Harvard University Press, Cambridge, MA.

Darwin C (1859) The origin of species by means of natural selection or the preservation of favoured races in the struggle for life. John Murray, London.

Eccles JC (1986) Do mental events cause neural events analogously to the probability fields of quantum mechanics? Proc R Soc Lond B Biol Sci 227(1249):411–428

Eckhorn R, Bauer R, Jordan R, Brosch W, Kruse M, Munk M, Reitboeck HJ (1988) Coherent oscillations: a mechanism of feature linking in the visual cortex. Biol Cybern 60:121–130

Fessard A (1961) The role of neuronal networks in sensory communications within the brain. In: Rosenblith WA (ed) Sensory communication. MIT Press, Cambridge, MA, pp. 585–606

Freeman WJ (ed) (1975) Mass action in the nervous system. Academic Press, New York

Hameroff S, Penrose R (1996) Orchestral reduction of quantum coherence in brain microtubules: a model for consciousness. Math Comput Simul 40:453–480

Hayek FA (ed) (1952) The sensory order. University of Chicago Press, Chicago

Kandel ER (2006) In search of memory: the emergence of a new science of mind. W. W. Norton, New York

Le Doux JE (1999) Emotion, memory, and the brain. In: Damasio A (ed) The scientific American book of the brain. The Lyons Press, Guilford, CT, pp. 105–117

Mountcastle VB (1992) Preface In: Başar E, Bullock TH (eds) Induced rhythms in the brain. Birkhäuser, Boston, pp. 217–231

Mountcastle VB (1998) Perceptual neuroscience: the cerebral cortex. Harvard University Press, Cambridge, MA.

Penrose R (1989) The emperor's new mind: concerning computers, minds and the laws of physics. Oxford University Press, New York.

Singer W (1989) The brain: a self-organizing system. In: Klivington KA (ed) The science of mind. MIT Press, Cambridge, MA, pp. 174–179

Solms M, Turnbull O (2002) Emotion and motivation. In: Solms M, Turnbull O (eds) The brain and the inner world. Other Press, New York, pp. 105–137

Wiener N (1948) Cybernetics or control and communication in the animal and the machine. Massachusetts Institute of Technology, Cambridge, MA

How Can This Book Be Read?

Types of Chapters

The monograph, which begins with a Prologue and an Essentials, is divided into six parts; it includes 26 chapters, and a closing Epilogue. The prologue mirrors the core concept of the book. A more detailed version following the Prologue describes the essence of the book, including narrative developments. The presentation of these developments emphasizes the causalities behind the rational of the monograph.

The Preludes provide useful orienting material making the context of the parts transparent before reading the Parts; and also provides bridges between the Parts that have different contexts.

There are three different types of chapters:

1. Data-analyzing chapters contain large amount of experimental results accompanied mostly with interim syntheses and conclusions. Chapters 1–6 and 10–13 belong to the category of data analysis chapters and build the empirical foundation of the book; in turn, they open the way for the understanding of the conceptual foundation of the book.
2. A second type of chapters includes syntheses and lead to conceptual frameworks. In Part II, Chaps. 7–9 lead to the holistic approaches of memory, whole brain work and the holistic view of the brain-body, further preparing the notion of brain-body-mind syncytium.

   The chapters of Part III, 14–16, introduce a physics/biology interface and include methodological proposals for new Cartesian systems to approach brain-body-mind. Part V includes chapters on metaphysics and philosophy.
3. The third kind of chapters is mostly pure theoretical, conceptual, and/or modeling. All the chapters of Part VI belong to this category, and are interrelated by forming a new type of trend or interdisciplinary pathway in bridging physics biology, cutting edges of philosophy, and metaphysics. Chapters 21–26 aim further to open possibilities for new syntheses for mathematically oriented and also clinically interested research scientists and graduate students.
4. The Appendices provide useful material related to mathematical methods and the resonance concept. They are detached from the appropriate chapter providing a possible discontinuity within the chapters.

How should one start and continue reading?

Readers who are already oriented with methodological and physiological knowledge may omit at the first reading the first and second types of chapters and appendices. They may start by reading the second or third types of chapters and then jump back to data chapters when necessary.

Usefulness of Pivot Chapters, The Prologue, and Essentials of the Book

The book contains material from biological sciences, physical sciences, and philosophy; therefore, the reader can get the impression that the reading may be difficult. However, difficulties can be avoided by means of a selective reading, as explained in the following.

The Prologue, Essentials, Preludes, and Chaps. 8, 9, and 21–26 can be considered "pivot reading material" for a reader in possession of enough prerequisite or good knowledge. At the beginning of the reading the essential and the Preludes can give a good idea about the ensemble of the heavy package consisting of experimental material and conceptual development. All pivot readings can provide the reader with the basic messages that the monograph tries to give. Often, there are redundant seeming materials in such chapters. However, often the same material is explained with different viewpoints. The field of oscillatory neuroscience is still New Land for a number of neuroscientists. Accordingly, such redundancy may improve a better didactical style.

Chapters 12, 13, and 16–18 are co-authored by Bahar Güntekin, who was also responsible for co-writing from the beginning.

Acknowledgements

A great number of people have contributed to the achievement of this project. The data acquisition goes back to the 1970s; therefore, it is not possible to mention the names of all those who contributed to this volume.

Dr. Bahar Güntekin, PhD, merits a special vote of thanks for the realization of this project. About 5 years ago she received a stipend from the Turkish Scientific Research Council in my research center in Izmir and started to work with me on the manuscript three full days a week. Then she came to Istanbul when I moved to the Istanbul Kultur University and continued to work day in and day out on the difficult manuscript. She helped me to integrate the basic ideas, especially the concept of the Nebulous Cartesian System, continuously raising questions in the style of the Ancient Greek disciples. Further, she offered all types of necessary assistance, such as reading my hand writing and taking long dictations. The writing of this book would have been very difficult without her disciplined engagement.

Mrs. Melis Diktaş, our communication specialist, provided important correspondence and all the necessary secretarial work for the book with great expertise and engagement. Miss Elif Tülay, MS, was in charge of the reference list and overall error finding throughout. She implemented impeccable assistance.

Mrs. Bilge Turp, BS, was highly competent in solving engineering problems and accomplished several illustrations.

Besides the names of our mentioned staff members, three neuroscientists provided enormous contributions to the book.

Professor Görsev Yener (neurologist), MD, PhD, joined my laboratories at DEU University in Izmir 10 years ago. She was teaching graduate students; then she chose to become a graduate student herself in my laboratories and started her first clinical projects in Izmir, with the electrophysiological analysis of Alzheimer's patients as her doctoral dissertation. She is currently a visiting member at the Kultur University in Istanbul and we continue our joint research interest in dementia.

Professor Ay egül Özerdem (psychiatrist), MD, PhD, joined my laboratories at DEU University in Izmir 6 years ago. She was teaching our graduate students; then she also chose to become a graduate student herself in my laboratories and started her first clinical projects in Izmir, with the electrophysiological analysis of bipolar patients as her doctoral dissertation. She is also currently a visiting member at the

Kultur University in Istanbul and we continue our joint research interest in psychiatric disorders.

I am highly indebted to Görsev and Ayşegül for our steady discussions and growing joint research. I am gaining concrete knowledge from their clinical experiences.

Last but not least I am very thankful to Prof. Dr. Canan Başar Eroğlu (neurobiologist and psychologist), who joined my Institute at the Hacettepe University in Ankara exactly 40 years ago as an undergraduate student. She performed all the experiments on smooth muscle organs and obtained results that opened the way to the concept of the overall myogenic system. Later, at the Medical University of Lübeck, she was responsible for all the intracranial measurements during sensory-cognitive processes.

Institutional Acknowledgments

The most important institutional and generous financial support for the achievement of this book came from Fahamettin Akıngüç, the honorary chairman of the board of trustees of the Istanbul Kultur University. Later, Dr. Bahar Akıngüç Günver, the present chairman of the board of trustees, took over all responsibilities for the promotion and further development of the Brain Dynamics and Cognition Research Center.

The former rector of our university, Prof. Dr. Tamer Koçel, took major steps to establish The Research Center on Brain Dynamics, Cognition and Complex Systems. Our present rector, Prof. Dr. Dursun Koçer, is highly supportive of the research at our center. I am very thankful to these members of the Istanbul Kultur University.

In the last 20 years several granting agencies financially supported the research presented in this book: Deutsche Forschungs Gemeinschaft, Bundes Ministerium für Forschung und Bildung, Ministry of Education Schleswig-Holstein in Kiel and Volkswagen-Stiftung in Germany, Turkish Scientific and Technical Research Council, and Department of Planning of Turkey.

Contents

Part I Foundations: History and Fundamental Processes in Nature

Prelude to Part I

1 Brain-Body-Mind Problem: A Short Historical and Interdisciplinary Survey 5
1.1 Introduction 5
1.1.1 What Is the Mind? 5
1.1.2 What Is Thought? 6
1.1.3 The Brain-Body-Mind Problem 7
1.2 Earlier and New Thoughts on the Mind 8
1.2.1 Introductory Remarks 8
1.3 Rene Descartes' Essential Work 8
1.3.1 Two-Dimensional Coordinate System 9
1.3.2 Three-Dimensional Cartesian System 10
1.4 Cardinal Questions of René Descartes and Alfred Fessard Constitute the Core Philosophical Framework of This Book 11
1.5 Galileo Galilee 11
1.6 Isaac Newton 12
1.7 Thoughts on the Mathematical and Intuitive Mind By Blaise Pascal 13
1.8 David Hume 15
1.9 John Locke: Sensations and Ideas 15
1.10 Gottfried Leibniz 16
1.11 Immanuel Kant 17
1.12 Henri Bergson 17
1.12.1 Intuition 18
1.13 A Comparative Treatise of the Conceptual Frameworks of Pascal, Locke and Bergson 20

2 **Frameworks in the Integration of the Sciences** 23
2.1 Introduction 23
2.2 Charles Darwin and the Voyage of the Beagle 24
2.3 Norbert Wiener and Cybernetics 25
2.4 Hermann Haken: Synergetics and Laser Theory 26
2.5 René Thom: Catastrophe Theory and Forced Oscillations in the Brain 28
2.6 Prigogine: Dissipative Structures 29
2.7 The Importance of Einstein's Three Concepts in Brain Research: (1) Synchrony of Clocks, (2) Brownian Motion, and (3) Unconscious Problem Solving 30
2.7.1 Synchronization of Clocks in the Brain (Synchronization of Oscillations of Neurons and of Neural Populations) 31
2.7.2 Brownian Motion 32
2.7.3 Unconscious Problem Solving 33
2.8 Werner Heisenberg 33
2.8.1 Microscope Model of Werner Heisenberg 33
2.9 Boltzmann's Statistical Mechanics 35
2.9.1 Statistical Mechanics in Biology and Physics from Griffith's Perspective (1971) 35
2.9.2 Global Neurodynamics: The View of Rosen (1969) 36
2.10 Santiago Ramon Y Cajal 36
2.11 Hans Berger and Electroencephalography 37
2.12 Hebb, Hayek, and Helmholtz 38
2.13 Jacques Monod: "The Chance and the Necessity" (1971) 40
2.14 Otto Loewi and the Discovery of Acetylcholine 41
2.15 A Synthesis from the Concepts of Wiener, Prigogine, Thom, and Haken 42

Part II Whole Brain Work and a Holistic Approach to Brain-Body Integration

Prelude to Part II

3 **Brain Structures, Transmitters, and Analyzing Strategies** 47
3.1 Signaling in the Brain 47
3.1.1 The Cellular Hypothesis 48
3.2 Functional Anatomy of the Auditory Pathway 49
3.3 Visual Pathway 52
3.4 Cerebral Cortex Anatomy and Global Function 53
3.4.1 Association Cortex and Frontal Lobe 55
3.5 Neurotransmitters 60
3.6 Why the Analysis of EEG Is Important 64
3.7 Some Principles of Biological System Analysis Applied to Brain Research 66
3.7.1 Why Establish a Program for Brain Research? 66
3.7.2 Steps of the Program 68
3.7.3 Mathematical Methods of the Program 69

4 Autonomous Nervous System, Cardiovascular System, and Smooth Muscles 71
4.1 Autonomous Nervous System and the Web of Overall Myogenic System 71
4.2 The Cardiovascular System 72
4.3 The Lymphatic System 76
4.4 Dynamics of Smooth Muscle Contractions 78
4.5 Dynamic of Blood Flow in the Cardiovascular Organs 80
4.5.1 Dynamics of the Arterial Impedance of the Ascending Aorta 81
4.5.2 Kidney In Situ 84
4.5.3 Intestinal Vasculature In Situ 85
4.5.4 Dynamics of Arterial Impedance and Flow 86
4.5.5 Dynamics of the Neural Control of the Vascular Bed 87
4.5.6 Renal Vascular Resistance 88
4.5.7 Coronary Vascular Resistance 93
4.5.8 Dynamics of the Microcirculation of the Kidney 93

5 Overall Myogenic-Coordination: Building Stones in the Whole-Body Integration and Tuning 95
5.1 Interim Synthesis of Overall Myogenic Control of Flow and Vascular Resistance 95
5.2 The Overall Myogenic System 95
5.3 Effects of Overall Myogenic Coordination on Local Circulatory Control 98
5.3.1 Interaction with the Peristalsis of Visceral Organs 98
5.3.2 Is There Any Interaction Between the Peristalsis of Visceral Organs and Auto-oscillations of Blood? 100
5.4 The Possible Role of Overall Myogenic Coordination in the Brain-Body-Mind Incorporation 101
5.5 Response Susceptibility of the Peristalsis Organs 103
5.6 Respiratory Coordination 103
5.7 The Integration of the Overall Myogenic System in Brain-Body-Mind 105

6 Dynamics of Sensory and Cognitive Processing 107
6.1 Introductory Issues 107
6.1.1 From Single Neuron to Neuron-Populations and the Brain 107
6.1.2 From Ramon Ý Cajal to Vernon Mountcastle 108
6.2 Neural Coding 109
6.3 EEG and Event-Related Oscillations as Information Codes in the Brain 110
6.3.1 Frequency Coding at Different Levels of Coding in the Brain 110
6.3.2 Do General Transfer Functions Exist in the Brain? 110
6.3.3 Natural Frequencies of the Brain 111

6.4 Emphasis of Multiple Oscillations in Brain Functioning ... 112
6.4.1 The Views of Fuster and Klimesch on the Role of Oscillations in Memory Processing ... 112
6.5 Selectively Distributed Oscillatory Systems in Brain Function: Distributed Multiple Oscillations in Brains ... 113
6.5.1 Concept, Definition and Methods ... 113
6.6 Remarks on Physiology of Selectively Distributed Oscillatory Processes ... 119
6.6.1 Connections of the Sensory-Cognitive Systems in the Brain ... 120
6.7 A Survey of Work on EEG-Oscillations ... 122
6.7.1 Alpha Activity ... 122
6.7.2 Earlier Experiments on Induced or Evoked Theta Oscillations ... 123
6.7.3 Alpha Oscillations in Perception and Cognition: The Alphas ... 124
6.7.4 Theta Oscillations in Perception and Cognition ... 130
6.7.5 Delta Oscillations in Cognition ... 131
6.7.6 Activation of Alpha System with Light ... 131
6.7.7 Activation of the Alpha System with Auditory Stibmulation ... 132
6.8 Superposition Principle and Theta and Delta Frequency Windows Shown with Examples in Cognitive Processes ... 132
6.8.1 Activation of Theta and Delta Systems Following Cognitive Inputs ... 137
6.9 The Importance of Gamma Oscillations in Sensory, Cognitive and Motor Processes ... 137
6.9.1 Historical Survey ... 137
6.10 Selectively Distributed and Selectively Coherent Oscillatory Networks ... 141
6.11 Interim Conclusions ... 144
6.11.1 A Brain-Body-Mind Interpretation Needs the Concept of Oscillatory Dynamics ... 145

7 Dynamic Memory ... 147
7.1 Different Levels of Memory ... 147
7.1.1 Fuster's View on Memory Networks ... 147
7.1.2 A Tentative Model Related to EEG Activation ... 147
7.1.3 Inborn (Built-in) Networks (Level I) ... 149
7.1.4 What Is Physiological Memory? What Is Fundamental Memory? ... 151
7.1.5 Living System Settings Incorporated in Physiological Memory ... 152
7.1.6 Genetic Factors Are Fundamental in Living System Settings and Physiological Memory ... 152

7.1.7 Working Memory, Dynamic Memory (Level II) 154
7.1.8 Perceptual Memory 154
7.1.9 Incorporation of Oscillatory Codes in Physiological Memory Consisting of Phyletic, Sensory, and Perceptual Memory 155
7.1.10 What Is Motor Memory and Procedural Memory Developed During Life? 158
7.2 Dynamic Memory in Whole Brain: Memory States Instead of Memories 159
7.2.1 Alpha, Theta, and Delta Oscillatory Processes During APLR 159
7.2.2 Are Dynamic EEG-Templates Created During Processing of the Alliance of APLR? Do Such Templates Build a (Virtual) Short-Term Storage of the New Learned Material? 160
7.2.3 Recent Examples of Brain Oscillations in the Cognitive Processes of Healthy Subjects 161
7.2.4 Are All Functions of the Brain Linked with Memory? 164
7.3 Complex or Multiple-Matching Evolving Memory, and the APLR-Alliance 164
7.3.1 Multiple and Complex Matching Processes: Reciprocal Activation of Alpha, Delta, Theta, and Gamma Circuits in the Whole Brain. Reentry 166
7.3.2 Prolonged Oscillations, Delays, and Coherent States During Complex Matching in the Whole Brain 167
7.4 Matching of Multiple Oscillations in the Whole Brain 169
7.5 Longer-Acting-Memory and Transition to "Persistent Memory" in the Whole Brain 172
7.5.1 Evolving Memory Is Identified as Multiple-Level Functioning in CNS 172

8 Whole-Brain Work 179
8.1 The Theory of the Whole-Brain Work: An Approach to Brain Function by Means of Brain Dynamics 179
8.1.1 Level A: From Single Neurons to Oscillatory Dynamics of Neural Populations 179
8.1.2 Level B: Super-Synergy of Neural Assemblies 181
8.1.3 Level C: Integration of Attention, Perception, Learning, and Remembering 182
8.1.4 Level D: Causality in Brain Responsiveness 182

9 Does the Brain-Body-Mind Work as a Dynamic Syncytium? 185
9.1 General Principles: Descartes' Suggestion and Fessard's Extended Question 185
9.1.1 How to Approach General Transfer Functions? 185
9.1.2 Cranial Nerves 186

9.2 Overall Myogenic System Revisited 186
9.3 Sympathetic Nerves of the Heart and the Spinal Cord 188
9.3.1 Dynamics of Ultra-Slow Potentials in the Auditory Cortex 191
9.4 Rhythmic Coordination in the Brain, Overall Myogenic System, and Spinal Cord 192
9.5 Globally Coupled Oscillators in Brain-Body-Mind Integration 193

Part III The Brain in Different States: Evolution, Maturing, Emotion, and Pathology
Prelude to Part III

10 The Brain in Evolution of Species and Darwin's Theory 199
10.1 A Unifying Step in Brain Function: The General Transfer Functions in the Brain According to Fessard 199
10.1.1 Do Some General Transfer Functions in Electrical Activity of Nervous System Exist During the Evolution of the Species? 200
10.2 Dynamics of Potentials from the Brain of Invertebrates 200
10.2.1 Introduction 200
10.2.2 Anatomy and Physiology of the Invertebrate (Gastropod) Nervous System 201
10.2.3 The Relationship Between the EEG of Vertebrates and Field Potential Fluctuations of Invertebrates 203
10.3 Neurochemical Modulation 208
10.3.1 Neurotransmitters in Evolution? An Important Invariant for Understanding the Mind 212
10.4 Dynamics of Potentials from the Brain of Anamniotes (Low Vertebrates): Goldfish and Ray 213
10.4.1 The Oscillatory Processes in the Ray and Goldfish 214
10.4.2 Similarities and Differences During the Evolution of Species 218
10.4.3 Importance of Synchrony: The Work of T.H. Bullock 220
10.5 Concluding Remarks on the Evolution of the Brain 221
10.5.1 A Global Scheme of Alpha Response in the Evolution of the Species and Maturation of the Brain 223
10.5.2 Remarks on Darwin's Theory 224

11 The Maturing Brain 225
11.1 Introduction 225
11.2 Changes in Structure and Synaptic Organization of the Human Brain 226
11.2.1 The Aim of This Chapter 227
11.2.2 Spontaneous and Evoked Alpha Activity at Occipital Sites in Three Age Groups 228

11.2.3 A Comparative Analysis of Frontal Vs. Occipital 10 Hz Activity in Young and Middle-Aged Adults 229
11.2.4 Single-Sweep Analysis of Visual EPs in Young and Middle-Aged Adults 230
11.3 Brain Response Susceptibility 232
11.3.1 Excitability of the Brain: Spontaneous Electroencephalogram Rhythms and Evoked Responses 232
11.3.2 Electroencephalogram in Children Might Provide a Useful Natural Model for Testing the Hypothesis of Brain Response Susceptibility 233
11.3.3 Aging and Topology-Related Changes in Alpha Activity and Brain Response Susceptibility 233
11.4 Conclusion: Importance of Maturation in Brain-Mind 234
11.4.1 Important Comment to the Parallelism of Alpha Activity During Maturation of the Human Brain and the Evolution of the Species 235

12 Oscillatory Dynamics of the Emotional Brain: Links of Emotion to Episodic Memory 237
12.1 Emotions: Introduction 237
12.1.1 What Is Emotion? Definition and General Philosophy 238
12.2 Oscillatory Dynamics of Emotion 238
12.2.1 Grandmother Experiments, Experimental Strategy, and Procedure for Recognition of Known and Unknown Faces 239
12.2.2 The Efficiency of the Grandmother Paradigm for Differentiation of Memory Components or States 240
12.2.3 Does Activation of Larger Neural Populations Indicate the Reactivation of Episodic Memory Components? 241
12.3 Oscillatory Dynamics of Facial Expressions 243
12.4 Oscillatory Dynamics of a "Loved Person" Versus Unknown Faces and Simple Light Stimuli 247
12.4.1 fMRI Studies 248
12.4.2 Electrophysiology Studies 248
12.4.3 Interim Discussion 250
12.4.4 Dynamic Localization 252
12.5 Integration of Episodic Memory and Emotion 253
12.5.1 Links Between Emotion and Persistent Memory 253
12.5.2 Links Between Emotion and Dynamic Memory 254
12.5.3 Links Between Emotion and Earlier Episodes and Longer-Acting Memory 254
12.6 Future of Emotion Experiments, Link with the Intuitive Brain, and Metaphysical Thoughts 255

13 Pathologic Brain: Impairment of Mind Based on Break of Oscillations and Modulation of Neurotransmitter Release 259
13.1 The Importance of Pathology in Understanding Brain Function and Mind 259
13.2 Importance of Brain Oscillations and Neurotransmitters in Impairment of Mind 260
13.3 The Basic Properties of Neurotransmitters 261
13.4 Some Relevant Experiments Related to Oscillations and Transmitters 263
13.4.1 Theta Oscillations 263
13.4.2 Gamma Oscillations 264
13.5 Alzheimer's Disease and Mild Cognitive Impairment 265
13.5.1 Oscillatory Responses in Delta, Theta, and Alpha Bands 265
13.5.2 Comparison of Sensory-Evoked and Event-Related Oscillations 269
13.5.3 Sensory-Evoked and Event-Related Coherences in Alzheimer's Disease 269
13.5.4 Comparison of Sensory-Evokedand Event-Related Oscillations 273
13.6 Schizophrenia 276
13.7 Bipolar Disorders 281
13.8 A Comparative Analysis of Alzheimer's Disease and Bipolar Disorder in the Gamma Frequency Range 284
13.9 Pathologies with Special Analysis of Genetic Methods: Genetic Disorders 289
13.10 Essential Changes in Brain Oscillation Lead to Essential Changes of the Mind in Pathological Brains 290
13.10.1 Delta 291
13.10.2 Theta 291
13.10.3 Alpha 291
13.10.4 Beta 291
13.10.5 Gamma 291

Part IV The Physics-Biology Interface: A Cartesian System for the Twenty-First Century
Prelude to Part IV

14 Chaos and Quantum Approach: Gateway to a Twenty-First Century Cartesian System 301
14.1 Introduction 301
14.2 Some Definitions Related to Brain Dynamics 302
14.3 Chaos in Brain Function 302
14.3.1 Deterministic Chaos 302
14.3.2 The EEG Has Strange Attractors: The EEG Is Not Noise 304
14.3.3 Typical Examples of Chaotic EEG Behavior 305

14.4 Remarks on Quantum Dynamics ... 312
14.4.1 Quantum Dynamics and Brain Oscillations ... 313
14.4.2 How May the Brain Show Chaotic Behavior and a Quantum Type of Uncertainty? ... 314
14.5 The Link Between Two Uncertainty States: Chaotic Behavior and Quantum-Like Behavior ... 316
14.5.1 Two Types of Uncertainties: Web of Chaotic Brain/Quantum Brain ... 316
14.5.2 String Theory as a Unifying Brain Concept ... 317

15 The Brain in Probabilistic Hyperspace ... 319
15.1 A Unifying Step in Brain Function: The Most General Transfer Functions in the Brain According to Fessard (1961) ... 319
15.2 Generalization of Questions from Descartes and Fessard Concerning Brain-Body Interaction ... 320
15.2.1 The Sympathetic System and EEG Oscillations ... 320
15.3 Mutual Excitation and Overall Tuning in the Brain and the Overall Myogenic System ... 321
15.4 What Are Multiple Causalities? What Is a "Hyper-Probabilistic Cartesian System?" ... 321
15.5 A Commentary on Some Philosophical Thoughts Related to the Brain in Probabilistic Hyperspace ... 324
15.6 From the Cartesian System in Probabilistic Space to a Nebulous Cartesian System ... 325

16 Quantum Brain and the Nebulous Cartesian System ... 329
16.1 Possible Ways to Approach Functioning of the Brain-Body-Mind Incorporation in the Framework of a "Nebulous Cartesian System" ... 329
16.1.1 S-matrix Formulation of Heisenberg, Brain Dynamics, and Physical Causality ... 329
16.2 Feynman Diagrams ... 331
16.2.1 Brain-Body Feynman Diagrams ... 332
16.2.2 Computing of Brain-Body Feynman Diagrams ... 336
16.2.3 Possible Advantages of "Brain-Body Feynman" Diagrams ... 337
16.3 Does the Language of the Brain-Body-Mind Need the Evolution of a New Discipline? Parallels to Quantum Theory, String Theory, and Chaos ... 339

Part V Metaphysics of the Brain: The Cutting Edge of Philosophy
Prelude to Part V

17 Darwinism, Bergsonism, Entropy, and Creative Thinking ... 345
17.1 Darwinism and L'Evolution Créatrice ... 345
17.1.1 Darwin's Theory ... 345

17.2 Electrical Activity from Aplysia Ganglion to Human Frontal Cortex: Possible Role of the "Alpha Oscillation" During Evolution of Species, in "Creative Evolution" and Maturing Brain 347
17.2.1 The Evolution of Alpha Activity During Cognitive Loading, and Its Role in the Maturing Brain 348
17.3 The Role of Coherence in Brain Evolution 349
17.4 Maxwell's Demon in Cognitive Process Entropy 350
17.4.1 What is Maxwell's Demon? 350
17.4.2 Does a Maxwell's Demon Exist During Some Cognitive Processes? 351
17.5 Hebb, Kandel, and Edelman: Entropy Changes 353
17.5.1 Hebb's Theory: Growth of Neural Assemblies 353
17.5.2 Fundamental Results by Kandel Support Hebb's Theory 353
17.5.3 Re-entrant Signaling: A Theory of Higher Brain Function (Scope of G.M. Edelman) 354
17.6 Is Hawking's Scope on Entropy Also Needed in Brain Research? 355
17.7 Conclusions and a Tentative Synthesis 356
17.7.1 From Body–Brain to Mind 358
17.7.2 What Is the Place of Bergson's Work in Memory and Quantum Brain? 358

18 Bergson's Intuition Memory and Episodic Memory 359
18.1 The Importance of Bergson's Philosophy in the Era of New Physics and Contemporary Biology 359
18.2 Essentials of Intuition 360
18.3 S-matrix 360
18.4 Bergsonism: Material and Memory and Creative Evolution 361
18.5 What Is Time? What Is Duration? 362
18.6 A New Interpretation of Intuition and Duration in Relation to Creative Processes 364
18.7 Lessons from Bergsonism 365

19 Towards Metaphysics: Conscient and Unconcient States 367
19.1 A Short Physiological and Psychological Classification of Unconscious States 367
19.2 Physiologic Unconsciousness 368
19.2.1 Anesthesia 368
19.2.2 Dream States 369
19.2.3 Autonomous System 370
19.3 Psychological Unconsciousness 370
19.3.1 Habitual Behavior 370

19.4 Sigmund Freud and Gustav Jung 371
19.4.1 Sigmund Freud and the Unconscient 371
19.4.2 Gustav Jung and Archetypes 372
19.5 A General Scheme Jointly Analyzing Unconscious Conscious and Preconscious States 372
19.6 Do New Integrative Trends of Mind and Consciousness Exist? 373

20 Mysteries of the Mind: Conscient and Unconscient States in Creativity and Sleep 375
20.1 Thoughts on the General Scope of Metaphysics 375
20.2 Web of Metaphysics and Unconsciousness: Consciousness Versus Unconsciousness 376
20.2.1 Is the "Esprit De Finesse"[2] of Blaise Pascal Now Measurable? 377
20.2.2 Is "Le Temps Perdu" of Marcel Proust a Pioneering Description of Episodic Memory? 377
20.3 Unconscious Problem Solving 378
20.3.1 Poincaré 378
20.3.2 Conan Doyle and Einstein 379
20.3.3 The Dream of Otto Loewi 379
20.4 How Is Creativity Elicited? 379
20.4.1 Mozart 379
20.4.2 Balzac's Description of Stefan Zweig (1932) 380
20.4.3 Hermann Hesse's Essay: How to Start Writing a Book 380
20.5 The Important View of Eric Kandel on the "New Science of Mind" 381
20.5.1 A Remark on Hawking's Arrow of Time 382
20.6 Episodic Memory and Search for "Temps Perdu:" Traveling Back to our Apartment in Istanbul In 1944 383
20.6.1 Measurement of Episodic Memories and Emotions Can Be Crucial to Interpret the Creative Mind 384
20.7 What Is the Nature of Creativity in the View of N. Andreasen (2005)? 385

Part VI Essentials and Unifying Trends in Brain Body Mind
Prelude to Part VI

21 Leitmotifs and Common Concepts: An Interim Description 395
21.1 Introduction 395
21.2 General Principles of the Brain-Body-Mind in the Web of Biology/Physics 396
21.2.1 Overall Frequency Tuning in Brain-Body Functioning 396

21.2.2 Synchrony of Clocks According to Einstein: Good and Bad Clocks in the Brain 397
21.2.3 The Concept of Resonance in the Brain and in Nature 398
21.2.4 The Fundamental Role of Causality in Classical Mechanics, Quantum Mechanics, Brain, and Evolution Theory 398
21.2.5 Statistical Mechanics, and the Quantum Brain 399
21.2.6 Chaos 400
21.2.7 String Theory 400
21.2.8 Brownian Motion and Einstein's Concept in Search of Hidden (or Invisible) Processes 400
21.3 Common Concepts 401
21.3.1 Entropy in Brain Structures 401
21.3.2 Entropy During Evolution and Maturation 402
21.3.3 Going Out of the System and Darwinism 402
21.3.4 Micro-Darwinism in Brain-Body Interaction 403
21.4 Interaction of Schools in Search of Brain-Body-Mind 404
21.5 Common Codes, Principles, and Rules in Brain-Body Integration 405
21.6 Various Overlapping Principles, Concepts, and Methods in Biological and Physical Systems 406
21.7 The Importance of Brain Metaphysics 407
21.8 Frameworks 408

22 Oscillations and Transmitters Are Quasi-invariants in Brain-Body-Mind Integration 409
22.1 Parallel Analysis of "Whole Brain Body and Brains" as Minimal Study Prerequisite 409
22.2 EEG Oscillations as Quasi-Invariants and Importance of Coherences 410
22.3 Ultraslow Oscillations Are also Quasi-Invariants 411
22.4 Web of Oscillations and Transmitters as Quasi-Invariants 412
22.5 A Synopsis on the Relation of Neuropathologies to Brain-Body-Mind 413

23 Unifying Trends: Globally Coupled Oscillators in Brain-Body 417
23.1 New Possible Models as Integration of the Brain with the Body 417
23.2 A Model of Globally Coupled Oscillators in Brain-Body-Mind Integration 417
23.2.1 Overall and Mutual Excitability 418

24 Unifying Concepts: Brain-Body's String Theory 419
24.1 "Brain's String Theory" in the Brain-Body Syncytium 419
24.2 What Can We Attain with Such Modeling Concepts? 422

25 Unifying Concepts: Dynamic Syncytium of Brain-Body-Mind and Intuitive Processes 423
25.1 Is Hayek's View Related to New Psychology a Precursor of the Integrative Mind? 423
25.2 Intuition Revisited: Intuition and Emotion Highly Influence the Machineries of the Mind 424
25.3 Intuition, Episodic Memory, and Emotion Form a Functional Syncytium 426
25.4 Reasonings on the Web of Brain-Body-Mind: Joint Interpretation of the Brain-Body's String Theory and Intuition 426

26 The Need for a Paradigm Shift and a New Cartesian System 431
26.1 The Importance of Philosophy 431
26.2 A Synthesis from the Concepts of Wiener, Prigogine, Thom, Hayek, and Haken 433
26.3 Possible Shift to a New Concept Based on Results of the Present Book 436

Epilogue 439

Appendix A 443

Appendix B 455

Appendix C 461

References 475

Author Index 505

Subject Index 519

Part I
Foundations: History and Fundamental Processes in Nature

Prelude to Part I

The first chapter of the book briefly describes the work of a few philosophers who opened the way to reasoning on cognitive processes. The second chapter explains essential concepts of other relevant scientists/philosophers who enormously influenced "interdisciplinary science" in the last four centuries.

Certainly, the list of prominent scholars is not complete. My narrative description encompasses the work of scientists most pertinent to the rationale and structure of the book. Why did I select those men of science for the description of history and important frameworks? Those scientists heavily influenced my scientific path and, accordingly, the structure of the present monograph and the proposed unifying concepts of the book.

Moreover, in the last 50 years I had the opportunity to personally meet some of the path-breaking scientists of the last century, some of whose work I studied deeply. Accordingly, it is not by accident that I have embarked on a multidisciplinary research pathway, which facilitated the writing of the present multidisciplinary book.

The leitmotifs included in the book were developed day by day during the last 50 years, according to frameworks launched by these scientists. My experiments were planed according basic concepts of men such as G. Galileo, R. Descartes, I. Newton, Ch. Darwin, H. Bergson, W. Heisenberg, F. von Hayek, and C.F. von Weizsäcker (Fig. 1).

As stated in the Prologue, the scientific style of one of my outstanding mentors, von Weizsäcker, has heavily influenced my way of thinking and this allowed me, years later to bring together ideas, concepts of scholar from multidisciplinary areas.

Influential Books and Readings

Several books have influenced my thinking, but three in particular changed my entire scientific life. During a vacation in the spring of 1965, I was at home in Istanbul and I read Norbert Wiener's famous book, *Cybernetics*. Possibly, the book

Fig. 1 Carl Friedrich von Weizsäcker (June 28, 1912–April 28, 2007)

triggered the memories of an accumulation of knowledge from earlier years and thus, in this unexpected way, my scientific interests completely changed. To bring this event into perspective, I want to outline my scientific career up to that point. This will explain my enthusiasm for Norbert Wiener's book.

In the last 3 years of my Lycée studies (1956–1958) I had the opportunity to acquire a wide range of knowledge from diverse areas of science and philosophy. From Niels Bohr I tried to understand experiments on *atomic spectra*[1] detailed in his famous book. Parallel to this, from the interesting lessons of our philosophy teachers at school, I was able to deeply discuss the philosophy of René Descartes and to learn about the transcendent view of Henri Bergson concerning his concepts of time, not measurable with physical clocks (duration) and memory.[2] René Descartes and Henri Bergson became my heroes.

In 1959, a year after leaving school, I started studying physics at the University of Munich in Germany.[3] In the first 3 years of physics education, fortune smiled on me and I was accepted as a young student in the family circle of the esteemed Professor Werner Heisenberg. Although I was not able to attend Heisenberg's lectures, during several visits to his home I had the opportunity to learn the essences of quantum theory directly from him. Three years later, Heisenberg suggested that I should attend the University of Hamburg, where I was invited this time into the

[1]The famous book of N. Bohr on atomic spectra.

[2]Our philosophy teacher was Monsignor Pierre Dubois, who was later nominated as "The Cardinal of the Catholic Church in Istanbul from French government."

[3]As I came to Munich as physics student at the end of 1959, by a tide of fortune, I made the acquaintance of Adelheid Graefin zu Eulenburg as a result of my interest in French philosophers. Graefin Eulenburg, C.F. von Weizsäcker's sister, immediately introduced me to Prof. Heisenberg and his family.

circle of the eminent physicist and philosopher Carl Friedrich von Weizsäcker.[4] For many years, I participated in his colloquia for graduate students.

In April 1965, after finishing my Master Thesis on elementary particle physics in Hamburg, I returned home to Istanbul for a 4-week vacation. I had planned to return to my studies at the German Electron Synchrotron and work on a doctoral dissertation on "high-energy physics."

My decision to study physics at the universities of Munich and Hamburg (1959–1965) had been based on my interest in the structure of the atom, and especially of elementary particles. After picking up Wiener's book, my focus suddenly turned to the phenomena of brain mechanisms. What was this sudden quantum jump? Was it an *intuitive decision*? In fact, it was probably based on my intellectual pursuits starting around 1956; because the first chapter of Norbert Wiener's book is entitled "Newtonian and Bergsonian Time" and you can imagine what this web of physics, biology, and philosophy induced in my young scientific mind. Only years later did I realize that this sojourn in Istanbul was the turning point in my whole scientific career. I read and re-read Norbert Wiener's "Cybernetics" and Henri Bergson's work. Brain, memory, and brain research had me in its thrall.

During the years of my experimental and theoretical work I had the privilege to meet and discuss with a number of prominent colleagues, including T.H. Bullock, Hermann Haken, Joachin Fuster, Henri Begleiter, and Roy John. The work of these important men of science will be cited in several chapters of the book.

I think that the rationale for choosing a limited number of scientists in this introductory Part I could be explained with this prelude.

[4]So what happened on my return to Hamburg after the vacation? I explained my intention to work on brain research to Prof. Dr. Ernst von Weizsäcker, who was a graduate student in biology at that time. Through his contacts with the Institute of Physiology in Hamburg, I was offered a position of Scientific Assistant at that institute. However, at these laboratories I was only able to work on conventional physiology, particularly on the central problems of the circulatory system, but not in brain research. However, heavily influenced by Wiener's book and Henri Bergson's *Material and Memory*, I was very much attracted to work in the realm of brain research. Finally, due to the advice of my mentor, Carl Friedrich von Weizsäcker I decided to start on a doctoral dissertation at the Institute of Physiology in Hamburg, and later began my journey in the exploration of the brain in New York.

Chapter 1
Brain-Body-Mind Problem: A Short Historical and Interdisciplinary Survey

René Descartes (1596–1650) made the daring suggestion that:

> *Everything in the universe could be explained in terms of a few intelligible systems and simple approaches on which the stars, and the earth and all visible world may have been produced.*

1.1 Introduction

1.1.1 What Is the Mind?

According to the encyclopedia, *mind* refers to the collective aspects of intellect and consciousness that are manifest in some combination of *thought, perception, emotion, will*, and *imagination.*

There are many theories concerning the mind and how it works, dating back to the ancient Greeks, Plato and Aristotle. Modern theories, based on a scientific understanding of the brain, see the mind as a phenomenon of psychology and often psychiatry, and the term is frequently used more or less synonymously with consciousness.

Which human attributes make up the mind is a much-debated question. Many scientists argue that only the "higher" intellectual functions constitute mind, particularly reason and memory. In this view the emotions – love, hate, fear, joy – are more "primitive" or subjective, and should be seen as different in nature or origin to the mind. Other scientists argue that the rational and emotional sides of a human being cannot be separated, as they are of the same nature and origin and should all be considered to be part of the individual mind. This book takes the view that *mind* is inseparable from the psychological and physiological functions of the body. Most strongly, it is argued that physiology of the brain-body and psychology are strongly interwoven and inseparable.

In popular usage the word *mind* is frequently synonymous with the word *thought*: It is that private conversation with ourselves that we carry on inside our heads. Thus, we "make up our minds," "change our minds," or are "of two minds"

E. Başar, *Brain-Body-Mind in the Nebulous Cartesian System: A Holistic Approach by Oscillations*, DOI 10.1007/978-1-4419-6136-5_1,

about something. One of the key attributes of the mind in this sense is that it is a private sphere to which no one but the owner has access. No one else can know our mind; they can only know what we communicate.

According to Eric Kandel (2006), a new biology of mind has gradually emerged over the last 50 years. In the 1960s the new discipline attempted to find common elements in the complex mental processes of animals, ranging from mice to monkeys to humans. The approach was later extended to simpler invertebrate animals, such as snails, honeybees, and flies. This modern cognitive psychology was at once experimentally rigorous and broad based. It focused on the arrangement of behavior from simple reflexes in invertebrate animals to the highest mental processes in humans, such as attention, consciousness, and free will, which are traditionally the concern of psychoanalysis. Kandel further says that cognitive psychology merged with neuroscience in the 1970s, and the result was cognitive neuroscience. This new discipline introduced biological methods of exploring mental processes into modern cognitive psychology. In the 1980s cognitive neuroscience was enhanced by methods of brain imaging such as fMRI, PET, brain mapping, and not least, oscillatory brain dynamics. In 1980s molecular biology added great value to our understanding of the biological mind.

Chapter 3 globally outlines the methods of this new discipline of cognitive neuroscience (Fig. 3.11). The present chapter brings together several of the mentioned methods. The question, What is mind? cannot be answered with a unique discipline; it should encompass various disciplines, methods, and systems of thought.

Physiology, biochemistry, philosophy, physical concepts, psychology, Darwin's evolution of species, maturation of the brain, and pathology, should provide deeper insight for reasoning related to mind. This view is schematically shown in Fig. 22.1.

1.1.2 What Is Thought?

According to the encyclopedia, *thought* is a mental process that allows human beings to model the world, so as to deal with it effectively according to their goals, plans, ends, and desires. Words referring to similar concepts and processes include cognition, sentience, consciousness, idea, and imagination. Thinking involves the cerebral manipulation of information, as when we form concepts, engage in problem solving, reason, and make decisions. Thinking is a higher cognitive function, and the analysis of the thinking process is part of cognitive psychology.

Memory is an organism's ability to store, retain, and subsequently recall information. Although traditional studies of memory began in the realms of philosophy, in the late nineteenth and early twentieth century memory was placed within the paradigms of cognitive psychology.

Imagination is accepted as the innate ability and process to invent, partial or complete, personal realms within the mind from elements derived from the sense perceptions of the shared world. The term is technically used in psychology for the process of reviving in the mind percepts of objects formerly obtained from perception.

Consciousness is a quality of the mind generally regarded to comprise elements such as subjectivity, self-awareness, sentience, sapience, and the ability to perceive the relationship between oneself and one's environment. It is a subject of much research in philosophy of mind, psychology, neuroscience, and cognitive science.

1.1.3 The Brain-Body-Mind Problem

The *brain-body-mind problem*, i.e., the relationship of the mind to the body, is commonly seen as the central issue in the philosophy of the mind, although there are other issues concerning the nature of the mind that do not involve its relation to the physical body. This book argues that thought processes are strongly anchored to physiological processes and thus also with the mechanisms of transmitter releases, as analyses of Alzheimer's and bipolar patients clearly show.

Furthermore, this may be the first time that the necessity is introduced of analyzing the mind problem during the evolution of species by means of electrophysiological evolution. The maturation of the mind from child to adult and to elder brains is also a crucial step that is addressed in Chaps. 10 and 11.

At the beginning of scientific development, philosophers had no knowledge of certain given mechanisms. First, fundamental questions were raised. These fundamental philosophical questions opened the way to essential measurements within the limits of the instrumental progress of the time. Accordingly, philosophy as an integration of the multi-disciplines of basic science merits important attention. This was the starting point toward the establishment of positive sciences. Accordingly, it is necessary to start with fundamental philosophical questions and related concepts.

This chapter gives a brief and concise survey of the work and concepts of selected scientists, starting from the early days of the Renaissance. Each scientist made immense contributions by asking questions, and added to the grand avenue of science and philosophy.

It is useful to review the trends of earlier centuries by trying to find common principles. From time to time science makes great jumps, although at other times there is stagnation for decades or even centuries in a particular branch of science. Sometimes a branch of science or a theory is considered to be "dead;" then, after several years a renaissance of this "dead" branch is observed. In fact, the author of this book began by considering the "dead" branch of the electroencephalogram, which had been declared a "smoke" by several neuroscientists, especially Sir John Eccles. However, the validity of this instrument was revived. Therefore, it is necessary to re-examine the past, and those theories that have been seriously criticized. For example, some authors (Damasio 1994) declared that Descartes' theory related to the mind was unsuccessful, labeling it "Descartes' error." However, Descartes was not in error. Some of his approaches are illuminating, although others are limited. Every scientist achieves a limited path. The conglomerations or "the Holon" of essays of several scientists can partly illuminate future problems. There are fewer than 15 scientists and philosophers who made considerable contributions to the evolution of science. The author has read approximately 70% of the cited works in the original books and articles. As Ramon y Cajal (1911) stated, this is important to avoid disagreeable surprises.

In this chapter some descriptions are long and some very short, as this is not a book on the history of the sciences. Only the essential features from the work of scientists that are the prerequisite to understanding complex phenomena in diverse chapters of the book are given. For example, the causality principle described by David Hume is presented, but not his work on morality.

1.2 Earlier and New Thoughts on the Mind

1.2.1 Introductory Remarks

This section analyzes and evaluates the conceptual frames of René Descartes, Blaise Pascal, John Locke, and Henri Bergson – four philosophers who were geometricians who attempted to develop fundamental ideas for the evolution of the sciences. Descartes' ideas dominated thought from the seventeenth century. Most important discoveries were made between the seventeenth and the beginning of the twentieth century. As new discoveries in physics and biology were established, the importance of using new techniques became clear. In this way, new discoveries opened the way to the development of new machines; and these in turn opened the way to new types of observations and accordingly to new discoveries. A description of Renaissance philosophers should also embrace the work of Galileo Galilee and Isaac Newton.

At the beginning of the twentieth century the development of statistical thermodynamics and, later, quantum theory led to a coordinate system that included uncertain probabilities. Physicists, therefore, were able to make progress following the development of quantum dynamics and the theory of relativity. The creation of the branch of psychology starting with James Stuart and Hermann Helmholtz opened the way to the inclusion of cognitive processes. Although Pascal and Descartes had already mentioned the relevance of cognitive phenomena, measurements of this area were not feasible during the twentieth century. Norbert Wiener (1948) in his book, *Cybernetics*, made predictions about the use of computers and the relevance of their use in the twentieth-century world. Now in the twenty-first century it is possible to measure brain-body processes and make predictions. The following sections describe the few philosophers' work so as to clarify the gap between the time of Descartes-Pascal and Henri Bergson.

1.3 Rene Descartes' Essential Work

Descartes believed that science should be grounded in absolute certainty rather that observation and prediction. Three of the important principles that describe his philosophy follow.

1. To employ the procedure of complete doubt to eliminate every belief that does not pass the test of undeniability (skepticism).
2. To accept no idea as certain that is not clear, distinct, and free of contradiction (mathematicism).
3. To found all knowledge on the bedrock of the certainty of self-consciousness, so that "*I think, therefore I am*" becomes the only innate idea unshakable by doubt (subjectivism).

Descartes' first principle says, in a nutshell, that everything is untrue until proved true. He attempts to install a great wall of doubt between truth and unproved statements. Unlike the American philosophy "innocent until proved guilty," Descartes pushes for the view "guilty until proved innocent." It is these rigorous standards of proof that can filter many of the half-truths that the scientific community chooses to believe.

Descartes' second principle is related to the first, but only in the rigorous standard that truth needs to be infallible in all aspects of fact. This follows modern philosophy, comparing it again with an aspect of law. It is similar to the philosophy "guilty beyond a reasonable doubt." If there is even the slightest doubt that proof is not totally true, it cannot be true.

Descartes' third principle is the strongest in its implications. In my interpretation, Descartes states that the one thing that cannot be questioned is consciousness. His famous "*Cogito ergo sum*" (I think; therefore, I am) is the only unquestionable truth. A person exists because he thinks. If you think, "Do I exist?" the simple act of thinking is enough to prove existence. *Cogito ergo sum* is the base, the bedrock, of all truth and knowledge. It stands as the foundation of the building of knowledge, the strong taproot of the tree of knowledge, the keystone in the arch of knowledge.

The essentials of the Cartesian system are described in the following. Part IV of this volume contains the proposal, a new Cartesian system, tailored to the needs of a twenty-first century approach to brain-mind.

The word *Cartesian* means relating to the French mathematician and philosopher Descartes, who, among other things, worked to merge algebra and Euclidean geometry. This work was influential in the development of analytic geometry, calculus, and cartography

The idea of this system was developed in 1637 in two writings by Descartes. In Part Two of his *Discourse on Method*, he introduces the new idea of specifying the position of a point or object on a surface, using two intersecting axes as measuring guides. In *La Géométrie*, he further explores these concepts.

1.3.1 Two-Dimensional Coordinate System

The modern Cartesian coordinate system in two dimensions (also called a rectangular coordinate system) is commonly defined by two axes, at right angles to each other, forming a plane (xy-plane). The horizontal axis is labeled x, and the vertical axis is labeled y. In a three-dimensional coordinate system, another axis, normally labeled z, is added, providing a sense of a third dimension of space measurement.

1.3.2 Three-Dimensional Cartesian System

The coordinates in a three-dimensional system are of the form (*x*,*y*,*z*). An example of two points plotted in this system are shown in Fig. 1.1, points *P*(5, 0, 2) and *Q*(-5, -5, 10). Notice that the axes are depicted in a world-coordinates orientation with the *z*-axis pointing up.

The *x*, *y*, and *z* coordinates of a point (say *P*) can also be taken as the distances from the *yz*-plane, *xz*-plane, and *xy*-plane, respectively. Figure 1.1 shows the distances of point *P* from the planes.

The *xy*-, *yz*-, and *xz*-planes divide the three-dimensional space into eight subdivisions known as octants, similar to the quadrants of two-dimensional space. Although conventions have been established for the labeling of the four quadrants of the *x*′-y *plane, only the first octant of three-dimensional space is labeled. It contains all of the points whose* x, y, *and* z *coordinates are positive. That is, no point in the first octant has a negative coordinate.* The three-dimensional coordinate system provides the physical dimensions of space – height, width, and length – often referred to as the three dimensions. It is important to note that a dimension is simply a measure of something, and that another dimension can be added for each class of features to be measured. Attachment to visualizing the dimensions precludes understanding the many different dimensions that can be measured (time, mass, color, cost, etc.). It is the powerful insight of Descartes that allows us to manipulate multi-dimensional objects algebraically, avoiding compass and protractor for analyzing in more than three dimensions (Fig. 1.2).

Although Descartes' method had its advocates, it was also criticized by his contemporaries, such as the mathematician Pierre de Fermat, and ultimately dismissed. Leibniz states that Descartes' rules amount to saying "take what you need, and do what you should, and you will get what you want."

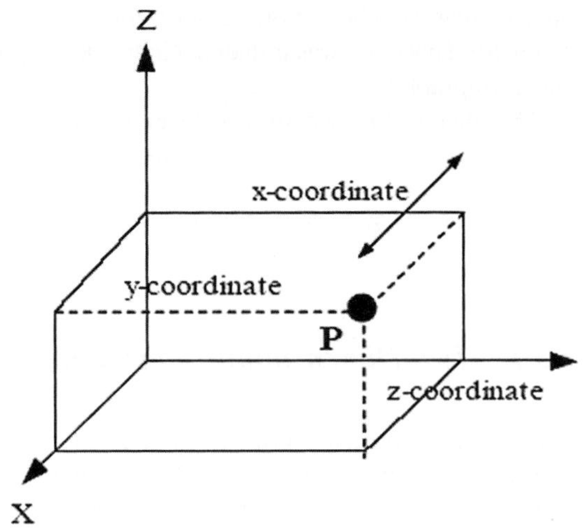

Fig. 1.1 Three-dimensional coordinate system

Fig. 1.2 René Descartes (March 31, 1596–February 11, 1650)

1.4 Cardinal Questions of René Descartes and Alfred Fessard Constitute the Core Philosophical Framework of This Book

René Descartes posed the fundamental suggestion related to general systems in the universe: "Everything in the universe can be explained in terms of a few intelligible systems and simple approaches."

Alfred Fessard (1961) emphasized that the brain must not be considered simply as a juxtaposition of private lines leading to a mosaic of independent cortical territories, one for each sense modality, with internal subdivisions corresponding to topical differentiations. The fundamental question of Fessard is the following: *What are the principles dominating the operations of hetero-sensory communications in the brain?* Further, Fessard (1961) indicated the necessity of discovering the principles that govern *the most general, or transfer functions*, of multi-unit homogeneous messages through neuronal networks.

The questions of R. Descartes and A. Fessard are cardinal inquiries that govern the leitmotivs and core conceptual framework that led to a unifying trend in the understanding of brain-body-mind (see Chaps. 22–26).

1.5 Galileo Galilee

In this book, oscillations, rhythms, neurotransmitters, and resonance phenomena have a fundamental role, which is true of all processes in nature. The physics of the *harmonic oscillator;* that is, *Galileo's pendulum*, made it possible to measure the flow of time and leads far beyond a device for making accurate clocks. These oscillators have been found to be the basis not only of what we hear as music and see as the color of light but, via quantum theory, of what we understand as the fabric of the universe. An interesting book by Roger G. Newton (2004) describes events from the rhythm of time to the making of matter. *Without oscillators there would be no particles: any air to breath, no fluids to sustain life, and no solid matter to form the earth.* Although biological mechanisms do not work with the accuracy or stability of modern clocks, a sense of time and its rhythms is built into the functioning of the human body. The autonomy of biological clocks is now a well-established fact. In addition to the heartbeat, some internal time keepers have shorter periods, called ultraslow oscillations. The electrical activity of the brain is one of the most important rhythms in the human body.

Galileo's first important scientific discovery was the property of the simple pendulum of a heavy bob suspended by a cord long enough not to swing so widely that its period is independent of the amplitude of its oscillation. In Appendix C the basic principles of the harmonic oscillator as well as the resonance phenomena in physical systems and nature are described. The observation of *synchrony* and *asynchrony* in clocks as described by Albert Einstein was important in the development of the relativity theory (see Chap. 2). The atom as a harmonic oscillator, neurons as basic oscillators in the brain, heartbeat, and rhythms of smooth muscles in the vasculature and peristaltic organs all obey the general principle of oscillators (see Appendix C).

Galileo's physics is accordingly the basic element in several physical and biological systems. However, the principles of oscillations or oscillatory phenomena are a basis on which to build a conceptual framework binding several sciences.

1.6 Isaac Newton

Although Galileo discovered the isosychronism of the pendulum as a fact of nature, he did not offer an underlying reason for his seminal observation. Newton – regarded by many as the greatest figure in the history of science – was the scientist who found the explanation. His treatise, *Philosophiae Naturalis Principia Mathematica*, described universal gravitation and three laws of motion, laying the groundwork for classical mechanics. The unifying and predictive power of his laws was central to the scientific revolution. Newton carried out fundamental work on gravitation theory, optics, and decomposition of white light. However, possibly his most important contribution was the use of analytical geometry created by Descartes. The union of the concepts of Galileo, Descartes, and Newton opened the way for the huge development of science from the sixteenth century to the present day.

In Newtonian mechanics, all physical phenomena are reduced to the motion of material particles, caused by their mutual attraction; that is, the force of gravity. The effect of this force on a particle or any other material object is described mathematically by Newton's equations of motion, which form the basis of classical mechanics. They were considered fixed laws according to which material objects moved, and were thought to account for all changes observed in the physical world. In the eighteenth and nineteenth centuries Newtonian mechanics were used with tremendous success. Newtonian theory explained the motion of planets,

Fig. 1.3 Isaac Newton (January 4, 1643–March 31, 1727)

moons, and comets down to the smallest details, as well as the flow of the tides and various other phenomena related to gravity. Newton's mathematical system established itself quickly as the correct theory of reality and generated enormous enthusiasm among scientists and laypersons alike.

At this point it is important to refer again to the fundamental evolution in physical and engineering sciences that had been reached by the union of Descartes' concept and Newton's work. At the beginning of the twentieth century the rigid Cartesian system was no longer efficient. A great jump or a type of bifurcation happened after the discovery of the new physics by Einstein and later by the Copenhagen school pioneered by Niels Bohr. At the same time Sigmund Freud and Bergson also opened a new area, which was not deterministic. Bergson introduced the concept of duration, and Freud's explanation of dreams explicitly changed the notion of time. These concepts ushered in a completely new understanding of the brain by introducing the non-deterministic view. The deterministic time of Newton had become the good old days (Fig. 1.3).

1.7 Thoughts on the Mathematical and Intuitive Mind By Blaise Pascal

The difference between the mathematical and the intuitive mind are explained by Pascal (Fig. 1.4) as follows:

> "In the '*mathematical mind*,' the principles are palpable, but removed from ordinary use; so that for want of habit it is difficult to turn one's mind in that direction: but if one turns

Fig. 1.4 Blaise Pascal (June 19, 1623–August 19, 1662)

> it thither ever so little, one sees the principles fully, and one must have a quite inaccurate mind who reasons wrongly from principles so plain that it is almost impossible they should escape notice. But in the '*intuitive mind*' the principles are found in common use and are before the eyes of everybody. One has only to look, and no effort is necessary; it is only a question of good eyesight, but it must be good, for the principles are so subtle and so numerous that it is almost impossible but that some escape notice. Now the omission of one principle leads to error; thus one must have very clear sight to see all the principles and, in the next place, an accurate mind not to draw false deductions from known principles."[1,2]

Therefore, the reason that some intuitive minds are not mathematical is that they cannot at all turn their attention to the principles of mathematics. But the reason that mathematicians are not intuitive is that they do not see what is before them, and that, accustomed to the exact and plain principles of mathematics, and not reasoning till they have well inspected and arranged their principles, they are lost in matters of intuition where the principles do not allow for such arrangement (Pascal, *Pensées*, 1660).

According to Pascal, mathematicians are only exact provided all things are explained to them by means of definitions and axioms. Otherwise they are inaccurate and insufferable, for they are only right when the principles are quite clear. Further, people of intuition, who are only intuitive, do not have the patience to first reach the principles of things that are speculative and conceptual.

For explanations the reader is referred to Chaps. 17–20, which relate to unconsciousness and intuition.

[1]Pensées by Blaise Pascal translated by W.F. Trotter.

[2]In mathematics there are also intuitive solutions. See the story of H. Poincarre in Chap. 20.

1.8 David Hume

Hume's positive, naturalistic approach has much in common with contemporary cognitive science, and his concept was a revolutionary one in physical sciences at the turn of the twenty-first century. Hume outlined a strategy that concerns human understanding. The first view looks at humans as active creatures that are driven by desire and feelings, painting a flattering picture of human nature. Philosophers make us feel that what they say about feelings is useful and agreeable. In this way simple people confronted with these views are readily inclined to accept them.

According to Hume, causation is the only principle that takes us "beyond the evidence of our memory and senses." It establishes a link or connection between past and present experiences with events that we predict or explain so that all reasoning concerning a matter of fact seems to be founded on the relation of cause and effect. Certainly, Hume's philosophy is not limited to description or comprehension of the causation. Here it is only necessary to take his original view on the cause and effect; in later chapters related to memory and those on common principles, the cause-and-effect concept is one of the dominating leitmotivs so as to understand brain-body-mind integration. As Hume postulated, there are simple and complex ideas. This view can be extended by saying that there are simple and complex causes, or *multiple causalities*, in brain-body-mind integration (analyzed in Chaps. 15 and 16).

One of the primary aims of this book is to develop a new Cartesian system, which emerges from the existence of multiple causalities in brain-body functioning. In theoretical physics Newton is a philosopher working with simple causes. Accordingly, the causality principle developed by Hume opened the way to modern science and Heisenberg's indeterminism injected the probabilistic behavior in the physics of atoms. C.F. von Weizsäcker defined causation as *nebulous wave packets* that are not precisely located in the atomar space (micro-space). It is seen in brain research that the chaotic nature of brain oscillations implies the probabilistic nature of the causality principle. Another very important aspect is Bergsonian intuition and duration, which completely destroys the importance of causality in human reactions. In our sentiments and emotions we also present a model of multiple causalities related to our emotions (see Chaps. 12, 15, 17, and 18).

1.9 John Locke: Sensations and Ideas

Locke's greatest philosophical contribution was his book, *Essay*, and he recorded his own account of the origin of that work. Locke's work is dominated by the concept of *sensations*. Understanding, like the eye, although it makes us see and perceive all other things, takes no notice of itself; and it requires art and pains to set it at a distance and make it its own object. Locke does not "meddle with the physical consideration of the mind"; he has no theory about its essence or relation to the body. At the same time, he has no doubt that, if due pains are taken, understanding can be studied like anything else. We can observe its object and the ways in which it operates on them.

Furthermore, *ideas* in general plays a major role in Locke's philosophy. All the objects of the understanding are described as ideas, and ideas are spoken of as being in the "mind." Locke's first problem was to trace the origin and history of ideas and ways in which the understanding operates on them. This wide use of the term *idea* is inherited from Descartes. Locke pointed to the variety of human experience, and the difficulty of forming general and abstract ideas.

Locke thought that "everyone is conscious of them in himself, and men's words and actions will satisfy him that they are in others." His first inquiry was "how they come into the mind." The next task was to show that they constitute the whole material of our knowledge. All our ideas, he said, come from experience.

There is another perception of the mind concerning *the particular existence of finite beings without us*, which, although it goes beyond bare probability, does not yet reach either of the foregoing degrees of certainty. This is called *knowledge*. This view has some similarity to the intuition of Bergson, which is elicited on accumulation of knowledge. There can be nothing more certain than that the idea we receive from an external object is in our minds: This is intuitive knowledge. The closing Chapters of Book IV of the *Essay* are devoted to a consideration of that kind of apprehension of reality that Locke calls "judgment," as distinguished from "knowledge." Reason must be our last judge and guide in everything.

1.10 Gottfried Leibniz

Leibniz's best known contribution to metaphysics is his theory of monads, as exposited in *Monadologie*. Monads are to the metaphysical realm what atoms are to the physical/phenomenal. Monads are the ultimate elements of the universe; they are "substantial forms of being" and are eternal, indecomposable, individual, subject to their own laws, uninteracting. Monads are centers of force. Substance is force, whereas space, matter, and motion are merely phenomenal. In Leibniz's philosophy, the essential features are similar to Descartes' view. Unlike atoms, monads possess no material or spatial character. They also differ from atoms by their complete mutual independence, so that interactions among monads are only apparent. Each monad follows a preprogrammed set of "instructions" peculiar to itself, so that a monad "knows" what to do at each moment. By virtue of these intrinsic instructions, each monad is like a little mirror of the universe. Monads need not be small; for example, each human being can be considered to be a monad. In this case free will is problematic.

Monads present the continuation of ideas of important philosophical approaches:

- Interaction between mind and matter arising in the system of Descartes.
- Lack of individuation inherent to the system of Spinoza, which represents individual creatures as merely accidental.

1.11 Immanuel Kant

The German philosopher Immanuel Kant wrote the core of his philosophy in his work, *Critic of Pure Reason* (1787). In this book Kant attempted to set up a contrast between things that exist in the outside world and actions of the human mind. For Kant, pure reason without reference to the outside world was impossible. He has borrowed this idea from the empiricist David Hume. What one knows, according to the empiricists is the result of what has gathered up with one's senses. The essential part of Kant's philosophy is that human beings only have access to the phenomenal world. They can have no knowledge of the true nature of *things-in-themselves*.

1.12 Henri Bergson

Henri Bergson was one of the leading philosophers of the twentieth century. *Duration, memory*, and *Élan Vital* mark the major stages of Bergson's philosophy. *Intuition* is the method of Bergson. Intuition is not a feeling, an inspiration, nor a disorderly sympathy, but one of the most fully developed methods in philosophy. It has strict rules constituting that which Bergson calls *precision* in philosophy. Bergson emphasized that intuition, in his understanding, methodologically, already presupposes *duration*. "These conclusions on the subject of duration were decisive. Step by step they led me to raise intuition to the level of philosophical method" (Deleuze 1966).

Bergson relied on the intuitive method to establish philosophy as an absolutely "precise" discipline; as precise in its field, as capable of being prolonged and transmitted as science itself. Further, without the methodical thread of intuition, the relationships among *duration, memory*, and *Élan Vital* would remain indeterminate from the point of view of knowledge.

The most general methodological question is this: How is intuition, which primarily denotes an immediate knowledge (connaissance), capable of forming a method, once it is accepted that method essentially involves one or several mediations? Bergson frequently presents intuition as a simple act. But, in this view, simplicity does not exclude a qualitative and virtual multiplicity, the various directions in which it comes to be actualized.

Bergson considered that he had made metaphysics a rigorous discipline, one capable of being continued along new paths that constantly appear in the world. The following is a short commentary on the history of Bergsonian philosophy.

As a mathematic genius and concrete scientist Bergson introduced considerable important concepts to the cutting edges of natural philosophy. His popularity declined for various reasons, one being the criticisms made by the British philosopher Bertrand Russell, who, contrary to Bergson, was an elegant politically oriented social philosopher. The French philosopher Deleuze wrote an important book on Bergsonism and started a Renaissance of Bergson's ideas. Following is a short summary of the important points. Chapter 18 contains further information on the essential development of Bergsonism philosophy and his most relevant contributions to the metaphysics of brain function and Darwin's theory.

1.12.1 Intuition

Bergson saw intuition not as an appeal to the ineffable, a participation in a feeling, or a lived identification, but as a true method. This method sets out, first, to determine the conditions of problems; that is to say, to expose false problems or wrongly posed questions, and discover the variables under which a given problem must be stated as such.

Bergson defined duration as a type of multiplicity. This is strange word, because it makes *multiple* a noun rather than an adjective. Intuition is seen as method, philosophy as rigorous science, and the new logic as a theory of multiplicities. Bergson invokes metaphysics to show how a memory is not constituted after present perception but is strictly contemporaneous with it, since at each instant duration divides into two simultaneous tendencies, one of which goes toward the future as the other falls back into the past.

According to Bergson, new ideas in science always appear strange at first, but these are precisely the ideas that may be the most fruitful; they may well be ideas engendered by philosophical intuition. Accordingly, he stated:

> I take the view that several of the great discoveries, of those, at least, which have transformed the positive sciences or created new ones, have been so many soundings in the depths of pure duration. The more living was the reality touched, the more profound had been the sounding.

Really important intuitions are rare events by nature. The point is that Galileo's, Newton's, and Leibniz's treatments of motion are the absolutely essential turning points in the history of science. Modern science could not be realized without them. Bergson believed the intuitions leading to such discoveries were achieved only haphazardly, although it is now possible to search for them methodically.

In his *Évolution créatrice* (1907) (*Creative Evolution*), Henri Bergson declared that the most lasting and fruitful of all philosophical systems are those that originate in intuition. If one believes these words, it appears immediately with regard to Bergson's system how he has made fruitful the intuitive discovery that opens the gate to the world of his thought.

What we usually call time, which is measured by the movement of a clock or the revolutions of the sun, is something quite different. It is only a form created by and for the mind and action. At the end of a most subtle analysis, Bergson concluded that it is nothing but an application of the form of space. Mathematical precision, certitude, and limitation prevail in its domain; cause is distinguished from effect and hence, raises that edifice, a creation of the mind, whose intelligence has encircled the world, raising a wall around the most intimate aspirations of our minds toward freedom. These aspirations find satisfaction in living time. Here, cause and effect are fused; nothing can be foreseen with certainty, because certainty resides in the act, simple in itself, and can be established only by this act. (See Chaps. 14–16 for the Cartesian system in probabilistic hyperspace.) Living time is the realm of free choice and new creations, the realm in which something is produced only once and is never repeated in quite the same manner.

According to Bergson, imagination and intuition are sometimes capable of flights in which intelligence lags behind. It is not always possible to decide whether the imagination is seduced or the intuition recognizes itself and allows itself to be convinced. In any event, reading Bergson is always highly rewarding. In the account of his doctrine, *Évolution créatrice*, Bergson created a concept of striking grandeur, a cosmogony of great scope and unflagging power, without sacrificing a strictly scientific terminology. It may be difficult at times to profit from its penetrating analysis or the profundity of its thought, but one always derives a strong aesthetic impression from it without any difficulty (Deleuze 1966).

Possibly, one can apply to this intuition, the central point of the Bergsonian doctrine, the brilliant expression that he uses about intelligence and instinct – the perilous way toward vaster possibilities. Within the limits of its knowledge, intelligence possesses logical certainty, but intuition, dynamic like everything that belongs to living time, must without doubt content itself with the intensity of its certainty. This is the drama: Creative evolution is disclosed, and humans find themselves thrust on stage by the *Élan Vital* of universal life that pushes them irresistibly to act, once they have come to the knowledge of their own freedom, capable of divining and glimpsing the endless route that has been traveled with the perspective of a boundless field opening onto other paths. Which of these paths is humanity going to follow? Chapter 18 returns to the Bergsonism concept with its transcendent nature.

Einstein's theories of relativity and quantum dynamics are completely different from the classical physics of Newton. Things are completely changed in the physics of the twentieth century. In the search of brain-mind we are still far from including this transcendent view. However, new steps can be seen on the horizon. The recent work of Kelso and Engstrom (2006) offers an ambitious and much-needed analysis of the "complementarity" concept of Niels Bohr within an extended physical-psychological-philosophical framework. The necessity of new frameworks, including a Cartesian one, is contained within a very interesting book by Fritjof Capra, *The Turning Point* (1982). In Chap. 18 a semi-empirical approach is introduced to search for the biological causation of evolution. Charles Darwin's evolution of species will be studied using electrophysiological tools to reconcile this theory with the creative evolution and *Élan Vital* of Bergson (see also Başar and Güntekin 2008, 2009). Accordingly, it is possible to concretize Bergson's view on Darwin's work with modern methods that were not available in Bergson's lifetime. For this reason a detailed chapter is included in Chap. 17 on the electrophysiology in the evolution of species to provide a prerequisite for Chap. 18, which is related to the creative evolution of *Élan Vital* and intuition. The philosophical views of Freud and Jung are described in Chaps. 19–20. The concept of time in dreams and in the views of Freud, Jung, and Bergson might provide new clues in the search for the boundaries of the metaphysics of the brain.

1.13 A Comparative Treatise of the Conceptual Frameworks of Pascal, Locke and Bergson

According to Locke *sensations* (phyletic memory) and *ideas* are almost equivalent. Sensations are the base of knowledge leading to complex ideas and a type of intuition. Ideas in general play a major role in the philosophy of Locke. All the objects of the understanding are described as ideas, and ideas are spoken of as residing in the mind. Locke's first problem is to trace the origin and history of ideas, and the ways in which the understanding operates on them. This wide use of the term *idea* is inherited from Descartes. Further, judgments are different from sensations. Table 1.1 outlines the evolution of concepts and ideas by four major philosophers.

Reviewing Bergson's concept combined with those of Pascal, Descartes, and Locke leads to the conclusion that the most developed thought function is Bergsonian intuition, which consists of instinct, associations of ideas, judgment, and reasoning. These entities converge in changes of substances, that are, in turn, a consequence of complex ideas. The existence of intuitive developments possibly enhances new connections in Hebb's (1951) sense. The invasion ability of such Hebbian networks can help brain growth during its evolution and maturation.

This chapter has presented the revolutionary ideas of prominent philosophers and scientists, who established fundamentals of philosophy and science. In the beginning science-philosophers created new and substantial theories; some were also able to perform experiments that were based on their essential thinking. Philosophy and science (including empirical science) were inseparable until the beginning of the twentieth century.

This begs the question, What were the most important contributions of the philosophers to science? The answer is that leading philosophers ask leading questions. Finding the right path to answer the questions can, at the very least, lead to

Table 1.1 Summary of the evolution of concepts and ideas by Descartes, Pascal, Locke, and Bergson

René Descartes	Blaise Pascal	John Locke	Henri Bergson
Cogito ergo sum	Mathematical Mind	Cogito ergo sum	Élan Vital
Mathematical Mind	Intuitive Mind	Skepticism	Substance; complex ideas
Skepticism		Sensations; phyletic memory	Instinct phyletic memory
		Substance; complex ideas	Judgments are different from knowledge
		Judgments are different from knowledge	

partial answers, and this is what scientists did from the seventeenth to the end of the twentieth century. By following this process, the approach to the mind gained new concrete territories, as stated in the "Prologue".

This volume presents empirical evidence about new systems to approach brain-body-mind integration; and the "Conclusion" poses new questions, based on this evidence. Some of these questions have metaphysical structures. It is hoped that these metaphysical questions can one day find empirical foundations and become reality. This is why this chapter has presented the first steps performed in the "metaphysics of the brain-mind."

Chapter 2
Frameworks in the Integration of the Sciences

2.1 Introduction

Chapter 1 describes the essential concepts of a few Renaissance philosophers, who were also mathematicians and/or physicists. Galileo Galilee and Isaac Newton were included in this category because their basic works in physics form the fundamental framework of scientific philosophy. These thinkers opened the way to modern sciences and were the predominant philosophers up to the beginning of the twentieth century. Although Henri Bergson does not belong to the Renaissance philosophers, he was an interdisciplinary working mathematician and academic who bridged concepts of mind from the Renaissance to modern science by forging essential steps in the theory of memory. He also provided an ultimate approach to both Albert Einstein's concept of time and Charles Darwin's theory of evolution.

In fact, Chapters 1 and 2 could have been presented as a single chapter, because the topics under evaluation also have a relevant philosophical impact. However, it seemed better to split the fundamental scientific-philosophical approaches into two chapters, with this one describing the important and fundamental discoveries at the end of the nineteenth and beginning of the twentieth centuries that opened the way to modern biology and contemporary physics. René Descartes and Newton's mechanical viewpoint gave way to probabilistic and statistical approaches in the twentieth century. There is a separation between the Newtonian mechanical viewpoint and statistical approaches. Accordingly, what is described in the present chapter leads to a new conceptual framework that is explained in Part IV.

The reader may wonder about the choice of 12–13 important developments by the most recognized scientists. The answer to this question can be found in Chapters 15–20.

E. Başar, *Brain-Body-Mind in the Nebulous Cartesian System: A Holistic Approach by Oscillations*, DOI 10.1007/978-1-4419-6136-5_2,

2.2 Charles Darwin and the Voyage of the Beagle

In the 1830s, Charles Darwin started his famous voyage on the Beagle from England to South America, and thus opened the way to modern biology. Darwin studied medicine and theology, and could be considered a zoologist as well. For two thirds of this 5-year journey, Darwin was on land carefully noting a rich variety of geological features, fossils, and living organisms and collecting an enormous number of specimens, which were sent to Cambridge together with reports about his findings. The voyage of the Beagle summarizes Darwin's findings and provides social, political, and anthropological insights into the wide range of people he met. One of the most important books by Darwin was entitled, *On the Origin of Species by Means of Natural Selection or the Preservation of Favored Races in the Struggle of Life* (usually abbreviated to *The Origin of Species*). Darwin's theory rests on two fundamental ideas:

1. The concept of *heritable variation*, which appears spontaneously and at random in individual members of a population and is immediately transmitted through descent.
2. The idea of *natural selection*, which results from a "struggle for life." Only individuals whose heredity endowment enables them to survive and reproduce in a particular environment can multiply and perpetuate the species.

In the 1970s, Jacques Monod suggested an extension of Darwin's model to cultural evolution and the progression of ideas. Related arguments were also developed by Karl Popper. The proposition separately advanced by these authors is that cultural evolution arises from the externalization of the inner representations of the brain, the sharing of these representations between the brains of individual members of a social group, and ultimately their storage in extracellular memories (Changeux 2004).

The monograph by Monod (also mentioned in Chapter 17) is a brilliant piece of scientific conceptual work. However, Monod did not adequately take into account the work by Bergson, which he criticized as somewhat impulsive and not scientifically grounded. On the contrary, Bergson gave a superb intimation of how to understand the evolution of ideas (see Chapter 17 and 18). Nevertheless, Monod's *The Chance and Necessity* (1970) opened the way to bridge molecular biology and evolution (see also Sect.2.1).

As strongly emphasized in several chapters, Darwinism in the general sense provides an important conceptual method in all sciences. Darwin used comparative physiology to develop his theory on the evolution of species; thus, he went out of the system to become able to understand the system.

Einstein took a similar step by studying gravitation in various galaxies to understand the Earth's gravitational fields. Başar (1976) proposed "going out of the system," a conceptual method to approach physiological functions. The starting point was that to understand the causes of the autoregulation of blood flow in the kidney one has to go out from the specific system under investigation and look into other systems. For example, the coronary system of the heart and the contractile behavior of smooth muscles in the vasculature and peristaltic organs also should be analyzed to learn about the circulation of the kidney. (This is explained in Chapters 4 and 5.) In the brain one cannot perfectly understand the functioning of the hippocampus without establishing comparisons and links to physiology of the reticular formation. In other words, a holistic view is needed, which is why attention has been drawn to the holistic approaches of Darwin and Einstein, who provided magnificent examples and lessons to scientists trying to solve core problems in the natural sciences and physics.

2.3 Norbert Wiener and Cybernetics

Cybernetics is the science of control and communication – the transmission, exchange, and processing of signals in animals and machines (Fig. 2.1). Although Wiener's formulation was based on very little experience in biology, it predicted the course of the development of research in the field of cybernetics. His definition includes everything that the term *cybernetics* encompasses today. Although several research scientists find Wiener's approach somewhat old-fashioned, the intellectual impact of his work had a strong influence on several disciplines and provided a turning point for the establishment of the schools of Ilya Prigogine on dissipative structures and Herman Haken on synergetics. The view of René Thom in catastrophe theory and the general nonlinear approach to sciences have immensely profited from Wiener's vision. The book provided an inspiring framework for thinking broadly in parallel in multidisciplinary fields. One thing is absolutely clear:

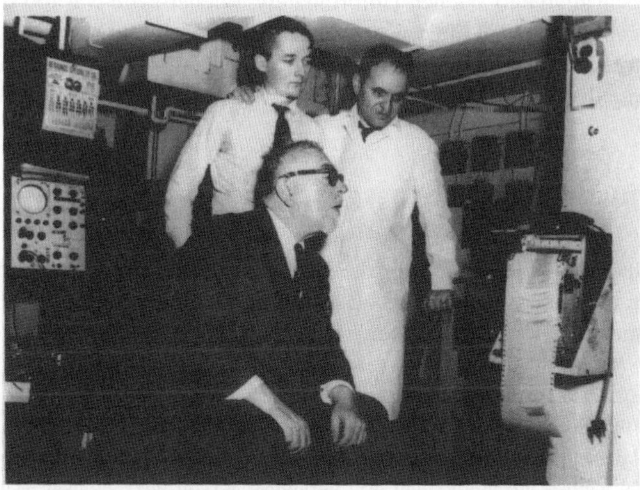

Fig. 2.1 Norbert Wiener (November 26, 1894–March 18, 1964) at work

The research leading to the foundation of *Brain Dynamics*[1] was anchored in the idea of signal processing and communication in the brain.

In the Introduction to *Cybernetics*, Wiener gives a detailed description of the experiences and thoughts that preceded the founding of *cybernetics*: On the basis of reflections and conversations with scientists in many specialties, especially physicians, it became clear to him, as a mathematician, that control processes take place and information is transmitted and stored in the human organism as well as machines (Hassenstein 1971). Wiener advised scientists working in multidisciplinary areas that a physiologist working with a mathematician would never be able to develop as powerful mathematical techniques as would a mathematician; but a physiologist would at least be able to understand the mathematical tools being jointly applied. Similarly, a mathematician or physicist would never be able to develop a physiological preparation with the skill of a biologist. However, he or she can understand what is going on in the physiological system as well as the main propose of investigating a given function.[2]

Wiener discovered functional similarities between technical processes and living organisms. The new science of cybernetics was intended to create a scientific framework considering it as a separate branch of science with the idea of conforming to the functional principles in technology and biology. Cybernetics was conceived as a common ground on which engineers, biologists, mathematicians, psychologists, etc., could meet and discuss in a common scientific language the problems of control and communication that appear in various forms in their scientific fields.

This means that the concepts of cybernetics should be neutral and abstract; they should contain no specifically technological or biological characteristics that would make them inapplicable to another field. By considering these entire essential proposals it can be seen that the science created by Wiener was a school of thought that was unique at the beginning of the twentieth century and resembled the ancient Greek Academy of Athens. However, the philosophers of Athens and later of Ionia did not have the tools available to Wiener. Unfortunately, Wiener's life was too short for him to realize his applications in biological sciences. However, despite this his predictions have come true about the future governing role of computers. The applications of cybernetics are detailed elsewhere in this book (see especially Chapters 3 and 6).

2.4 Hermann Haken: Synergetics and Laser Theory

A recent important development in the physics-like theories applied to biological systems is framework synergetics. The word *synergetics* is composed of two Greek words and means "working together." Haken (1977) states, "In many disciplines,

[1]Başar 1976, 1980; Freeman 1975.

[2]The author of present book took this advice very strongly into consideration. Although educated as physicist, he learned physiology and also developed several physiological techniques in his laboratories. In this way it was possible for him to become one of the neuroscientists who launched the field of brain dynamics and oscillations.

ranging from astrophysics over biology to sociology, we observe very often that cooperation of many individual parts of a system leads to macroscopic structures of functionings." In its present state, synergetics focuses its attention on those situations in which the functioning structures of the systems undergo changes on a macroscopic scale. In particular, synergetics investigates how the subsystems produce these changes in an entirely self-organized manner. The subsystems are usually discrete, e.g., atoms, cells, or human beings. An important group of phenomena are oscillations (temporal structures) that occur in a self-organized manner. Here a rod of laser-active material with two mirrors at its end faces is pumped energetically from the outside, and the atoms emit light (Fig. 2.2).

The essential feature to be understood is this: If the laser atoms are pumped only weakly by external sources, the laser acts as an ordinary lamp. The atoms, independently of each other, emit wave tracks with random phases. The coherence time of about 10^{-11} s is evident on a microscopic scale. The atoms, visualized as oscillating dipoles, are oscillating at random. If the pump is further increased, suddenly within a very sharp transition region the line width of laser light may become on the order of 1 cycle/s so that the laser is evidently in a new, highly ordered state on a macroscopic scale. The atomic dipoles now oscillate in phase, although they are excited by the pump completely at random. Thus, the atoms show the phenomenon of self-organization. Evidently the macroscopic properties of the laser have changed dramatically in a way reminiscent of the phase transition of the Ferro magnet, for example. The laser analogy and cooperative phenomena at the atomic level are presented here to provide an additional metaphor for the phenomenon of frequency stabilization, i.e., the transition to a highly ordered state on a macroscopic scale as seen in the brain responses (see also Chapter 6).

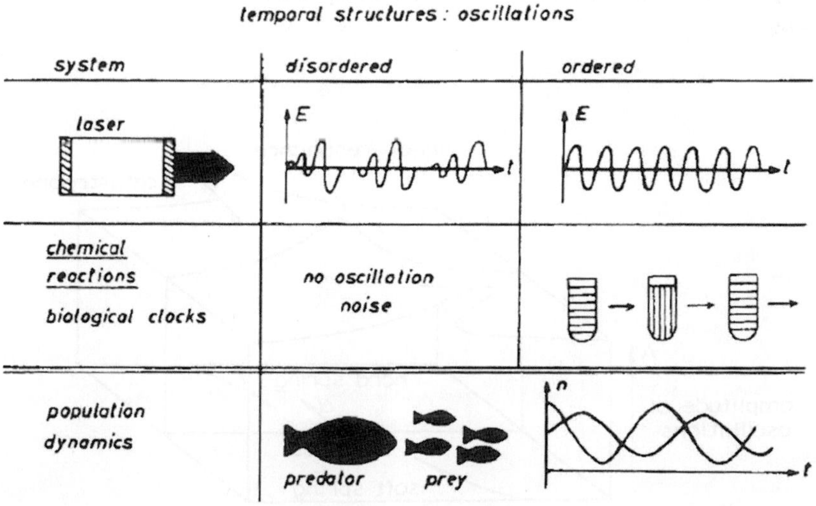

Fig. 2.2 Self-organized oscillations from physics, chemistry, and population dynamics (from Haken 1977)

2.5 René Thom: Catastrophe Theory and Forced Oscillations in the Brain

Eric Christopher Zeeman (1977) discussed the classical oscillators, the Van der Pol oscillator, and especially Duffing's equation in brain modeling in terms of catastrophe theory. Forced oscillations can x + ax = Fc be modeled by Duffing's equation: where $k > 0$ is a small damping term, a small nonlinear term (a = -1/6 for a simple pendulum), and F cos Qt is a small periodic forcing term with frequency fi close to 1, the frequency of the linear oscillator. The amplitude A of the resulting oscillation depends on the parameters, and Fig. 2.3 shows graph A as a function of a and ft (keeping k and F fixed). There are two cusp-catastrophes with a, 0 as conflicting factors. At each cusp the upper and lower sheets represent attractors (stable periodic solutions) whereas the middle sheets represent saddles (unstable periodic solutions). If the frequency of the forcing term is gradually changed to cross one of the cusp lines, starting from the inside and going to the outside of the cusp, then the amplitude A will exhibit a catastrophic jump. There will also be a sudden phase shift at the same time (Zeeman 1977).

The description of the mathematics of the catastrophe theory (Thom 1975) is beyond the scope of this book; therefore, a detailed explanation of the graph technique or terminology used by Zeeman is not attempted here. However, it is important to note that according to Zeeman's theory the brain's activity can be modeled by forced nonlinear oscillations. Furthermore, Zeeman's catastrophe model exhibits catastrophic jumps in amplitude at resonant frequencies and sudden phase shifts at the same time. These theoretical statements are most pertinent to the analysis presented in this book because sudden jumps of amplitude and phase shifts are also obtained at resonant frequencies of the brain response.

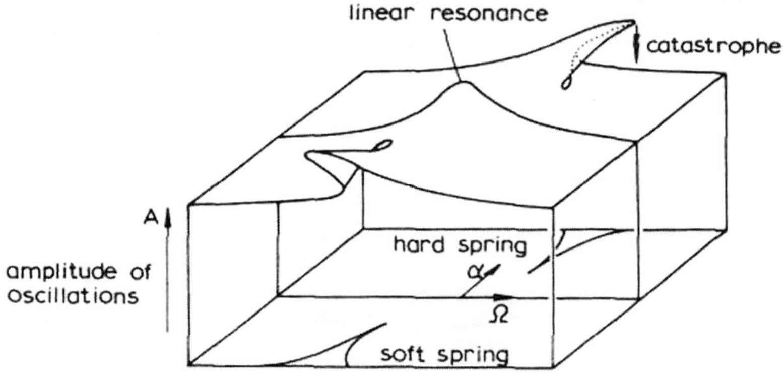

Fig. 2.3 The oscillation of a forced non-linear oscillator bifurcates according to the cusp-catastrophe (from Zeeman 1977)

2.6 Prigogine: Dissipative Structures

The mechanisms of self-organization in the genesis of oscillations through various kinds of interaction in physical, chemical, biological, psychological, and social systems has been deeply explored by Aharon Katzir-Katchalsky et al. (1974) and Ilya Prigogine (1980) in studies of dissipative structures and chaotic state transitions. According to Prigogine's theory, no system is structurally stable; fluctuations lead to instabilities and new types of function and structure. The evolution of a dissipative structure is a self-determining sequence according to Fig. 2.4.

This approach combines both deterministic and probabilistic elements in the time evolution of the macroscopic system.

Freeman's viewpoint (1999) is that complex biochemical feedback pathways within cells support the emergence of oscillations at cycle durations of minutes, hours, and days, and they underline the recurrence patterns of normal cyclical behavior as well as epileptic fits, mood disorders, and other pathologies. Further, large numbers of neurons form macroscopic population under the influence of external and internal stimuli and endogenous neurohormones. Freeman's opinion is that these populations are more closely related to the nerve cell assemblies conceived by Hebb (1949). In these assemblies, relationships of neurons to the mass are explained by Haken's synergetic theory (1977), whereby the microscopic neurons contribute to the macroscopic order and then are "enslaved" by that order, similar to particles in lasers and soap bubbles.

According to Prigogine, "living processes" were in some sense pushed outside nature and physical laws. One was tempted to ascribe an accidental character to a living organism and imagine the origin of life as being the result of a series of highly improbable events. A sharp distinction is made between events and regularities in classical dynamics.

At most, we could use Boltzmann's probabilistic interpretation of the second law of thermodynamics to ascribe a probability to each possible condition. One an initial condition is specified, the system will be led to its most probable state through an irreversible process.

Life, considered to be a result of "improbable" initial conditions is, therefore, compatible with the laws of physics (initial conditions can be arbitrarily chosen), but does not follow from the laws of physics (which do not prescribe the initial conditions). This is the outlook supported, for example, by Monod's well-known book.[3]

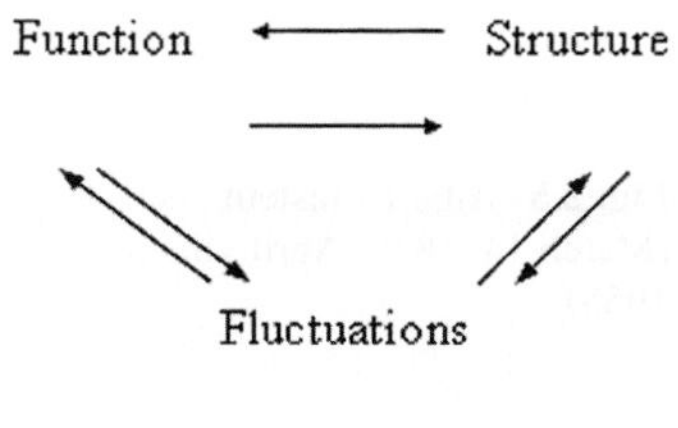

Fig. 2.4 Dissipative structures

[3]Monod (1970), Le Hasard et la Nécessité, Seuil, Paris.

Moreover, the maintenance of life appear, in this view, to correspond to an ongoing struggle of an army of Maxwell demons[4] against the laws of physics to maintain the highly improbable conditions that permit its existence. The results summarized by Prigogine support a different point of view. Far from being outside nature, biological processes follow the laws of physics, appropriate to specific nonlinear interactions and conditions far from equilibrium. Thanks to these specific features, the flow of energy and matter may be used to build and maintain functional and structural order.

The reader is referred also to Chapter 17, in which the possible role of a Maxwell's demon in cognitive processes and creative evolution is detailed.

2.7 The Importance of Einstein's Three Concepts in Brain Research: (1) Synchrony of Clocks, (2) Brownian Motion, and (3) Unconscious Problem Solving

What is a clock? Any physical phenomenon may be used as a clock, provided it exactly repeats as many times as desired. Taking the interval between the beginning and the end of such an event as one unit of time, arbitrary time intervals may be measured by the repetition of this physical process. All clocks, from the simple hourglass to the most refined instruments, are based on this idea. It is, therefore, inconvenient to have only one clock; therefore, if we know how to judge whether two or more clocks show the same time simultaneously and run in the same way, we can imagine as many clocks as we like in a given coordinating system (Einstein and Infeld 1938) (Fig. 2.5). Provided

Fig. 2.5 Albert Einstein (March 14, 1879–April 18, 1955)

[4]For the second law of thermodynamics see Chapter 17.

Fig. 2.6 Clocks (from the collection of the Başar family)

the clocks are all at rest relative to the coordinating system, they are "good" clocks and are *synchronized*, meaning that they show the same time simultaneously.

2.7.1 Synchronization of Clocks in the Brain (Synchronization of Oscillations of Neurons and of Neural Populations)

There are two classes of synchronized clocks in the brain: First, synchronous neural oscillators in a given special brain structure (Eckhorn et al. 1988; Singer 1989), and second, large-scale synchrony between distant structures (Başar 2004; Bressler and Tognoli 2006; Varela et al. 2001; von Stein and Sarnthein 2000). The electroencephalogram (EEG) consists of the activity of an ensemble of generators producing oscillatory activity in several frequency ranges. These "brain oscillators" are active in a random way, usually. However, with application of sensory-cognitive stimulation, these generators become coupled and synchronized; they start acting in a coherent way. This synchronization and enhancement of EEG activity produces the "evoked" or "event-related" oscillations that may be phase-locked to the stimulus; or they may be non-phase-locked to the stimulus and thus have an "induced" character (Fig. 2.6).

The compound event-related potential (ERP), which includes the responses of ensembles of neural populations, represents a transition in the brain from a disordered state to an ordered one. The morphology of the ERP waveform is an outcome of the superposition of evoked/event-related oscillations. The "natural frequencies" of the brain that compose these oscillations range from the delta band (0.5–3.5 Hz) to theta (3.5–7 Hz), alpha (8–13 Hz), beta (15–30 Hz), and gamma bands (30–70 Hz). That the oscillations are the basic responses of the brain nowadays finds strong support from a large number of neuroscientists who endeavor to understand the brain and the way it functions in cognition (Bressler and Tognoli 2006; Freeman 2006; Yordanova and Kolev 1998b).

In Haken's *Synergetics* (1977, 2004), the synchrony of oscillators plays a major role in the laser effects used in many applications. In biological systems and especially the brain, on the other hand, the synchronization of clocks plays a crucial role in the realization and control of the integrative functions. Although it is a technical phenomenon in physics, it is an explanatory model in biological systems. Electrocorticograms (EcoG) have a broad-band spectrum; within it, all frequencies are simultaneously present and are separately waxing, waning, and shifting phase (Bullock 1988a, b; Bullock et al. 1990).

There are also clocks that are not synchronized; according to Einstein there are *bad clocks*. Bad clocks are observed in case of pathologies, presented in Chapter 13.

2.7.2 Brownian Motion

Einstein and Infeld (1938) described the tracks of molecules in Brownian motion. However, they did not only describe the tracks, but also analyzed the causes of Brownian motion. In searching for causes of gravitation, Einstein wished to understand the causes of dissipating energy. To establish what is happening in the galactic system, he predicted black holes. Thus he not only used descriptions of the astrophysical events; he also combined the existing knowledge on the motion of stars and considered the laws of physics. With such an approach, he described the nature of stars and the galaxy; thereafter he arrived at the concept of black holes, an existence invisible to conventional observation techniques.

What is Brownian motion? A suspended particle is constantly and randomly bombarded from all sides by the molecules in the liquid. If the particle is very small, the hits it takes from one side will be stronger than the bumps from the other side, which will cause it to jump. These small random jumps make up Brownian motion. The first mathematical theory of Brownian motion was developed by Einstein in 1905 (Einstein and Infeld 1938). Einstein showed that the overall visible motion, averaged over many observations, exactly matches that expected if the little particles were atoms or molecules. Brownian movement exists if the bombarded particles are sufficiently small. It exists because this bombardment, owing to its irregular and haphazard character, is not uniform from all sides and cannot be averaged out. The observed motion is, thus, the result of the unobservable one.

One of the aims of EEG research is try to discover brain functions. Accordingly, the analysis of Brownian motion trajectories initiated by Einstein (Einstein and Infeld 1938) is an excellent theoretical model or metaphor for the brain functions, which is latently present in the puzzling engrams that the EEG-oscillations form. In a number of explanatory formulations (Başar 2006; Begleiter and Porjesz 2006; Bressler and Tognoli 2006; Bullock 2006; Freeman 2006; Galambos 2006), the trajectories of EEG-oscillations are used for discovering their hidden sources (origins). These formulations show the immense usefulness of function-oriented investigation of brain signals for understanding the way the system functions. As Einstein's fundamental model shows, signal analysis alone will never be sufficient.

2.7.3 Unconscious Problem Solving

In describing the way Sir Arthur Conan Doyle's detective Sherlock Holmes solves problems, Einstein pointed out the following:

> The great detective, however, realizes that no further investigation is needed at the moment, and that only pure thinking will show the pattern of relation between the collected facts. So he plays his violin, or lounges in his armchair enjoying a pipe, when suddenly, by Jove, he has it! Not only does he have an explanation for the clues at hand, but he knows that certain other events must have happened. Since he now knows exactly where to look for it, he may go out, if he likes, to collect further confirmation on his theory.

This very important viewpoint is presented in Chapter 20 on unconscious states. Although in the present chapter only the empirically founded facts are analyzed, it is important to emphasize here that Einstein too was interested in the metaphysics of the brain.

2.8 Werner Heisenberg

2.8.1 Microscope Model of Werner Heisenberg

The uncertainty principle in quantum physics was formulated by Werner Heisenberg during the period of the Copenhagen School at the beginning of the twentieth century. To justify the philosophical framework of this principle, Heisenberg developed a model of thought.[5] If one day a microscope with very high resolution could be used, the experimenter would be able to observe the interaction of a gamma ray with an electron in the aperture of the microscope. Heisenberg assumes that at the time the gamma ray, which is used for the illumination of the electrode, would undergo an interaction with the electron, meaning that supplying energy to the electron should change the position of the electron according to the laws of physical motion. When the observer aims to localize the position of the electrode, he or she will certainly fail. The observer would then discern not the exact position of the electron at the moment of collision with the X-ray light, but only the position of the electron following the displacement (Fig. 2.7). No observation is possible without using a gamma light; the exact localization of the electron is impossible by using the light. This model of thought was the subject of discussions after the development of quantum mechanics. Finally, the experimental requirements of Heisenberg were fulfilled and the microscope theory was supported by the experiments of Christopher Foot (1994) and in this way Heisenberg's dream was realized.

[5]Works of Bohr, Schrödinger, Pauli, Dirac, Born, and Weizsäcker.

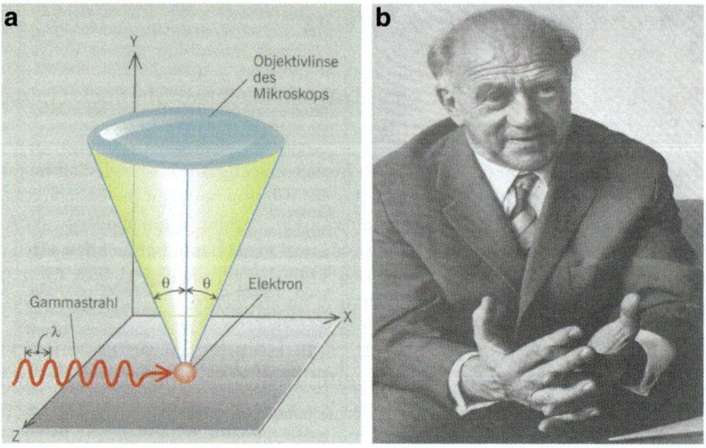

Fig. 2.7 The Gedanken Experiment by Werner Heisenberg: microscope theory

Fig. 2.8 Development of alpha activity on sensory stimulation

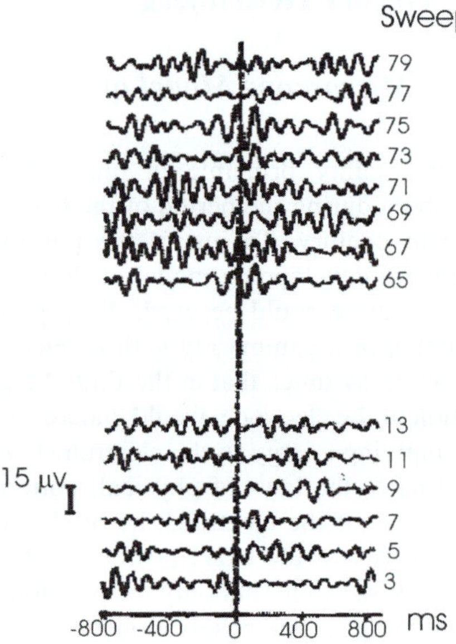

Is it possible to translate the uncertainty principle manifested by the microscope thought experiment to brain research? Consider the experimental recording in Fig. 2.8, in which the brain is stimulated by a sequence of peripheral stimulations. The spontaneous activity of the brain incessantly changes. The development of alpha activity with increasing amplitudes has, in turn, an important influence on the alpha responses. The brain is learning and goes from a preliminary state to a learned state. The same situation occurs with the microscope analogy. At the moment of application of the cognitive input, the brain state is changed. Accordingly, it is not possible to determine the exact cognitive response to cognitive inputs or cognitive inputs with emotional components.

The laws of quantum physics are statistical. This means that they are valid not for a single system, but for an aggregation of identical systems. They cannot be confirmed by measurements on one individual, but by a series of repeated measurements from that individual. In Einstein's words, "*Quantum physics formulates laws governing crowds and not individuals. Not properties but probabilities are described.*" (Einstein and Infeld 1938). Laws do not disclose the future of systems, but govern the temporal changes in these probabilities. In quantum physics, laws are valid for a great congregation of individuals. Similarly, laws concerning the brain specifically in cognitive processing are valid not for single neurons, but for neural populations. What applies to quantum mechanics also applies to the dynamics of chaotic systems. In such systems also, not properties but probabilities are described, laws disclose the change of the probabilities over time, and they are valid for congregations of individuals (see also Chapter 16 and 24).

2.9 Boltzmann's Statistical Mechanics

2.9.1 Statistical Mechanics in Biology and Physics from Griffith's Perspective (1971)

Griffith (1971) discussed concepts of *statistical neuron-dynamics* and tried to formulate the similarity between *statistical mechanics* and *neurodynamics* as follows:

The situation is superficially very similar to that which is obtained in statistical mechanics, as it applies to the relation between macroscopic thermodynamic quantities and the underlying microscopic description in terms of the complete specification of the states of all the individual atoms or molecules,... These are, firstly, that we could not, even if we knew all the necessary parameters, actually solve in detail the 10^{10} or more coupled neuronal "equations of motion" necessary to follow the state of the system in detail as a function of time. Secondly, that there exists a simpler "macroscopic" level of description which is really our main ultimate object of interest so that we do not wish, even if we could, to follow the "microscopic" state in detail but merely wish to use it to understand the time development of the macroscopic state. One most important aspect of this is that we only wish to specify, at the macroscopic level, the initial conditions of any calculation we may make. This leads immediately to the problem of whether the fundamental assumptions of equal a priori probabilities and random a priori phases hold for nerve cell aggregates, and, if not, whether we can find anything to replace them (Griffith 1971).

Griffith's remarks are more important today than they were 30 years ago because new trends or avenues in brain research clearly indicated the need to introduce new frameworks to analyze the integrative brain function by introducing *cell aggregates instead of single cells.*

2.9.2 *Global Neurodynamics: The View of Rosen (1969)*

A similar problem statement was created by Rosen (1969) asking the following question: "*What is the role of statistical mechanics in gas dynamics?*" The gas laws that describe gas dynamics are based on the ensemble of molecules in an isolated system. One does not describe gas dynamics with the dynamics of single molecules in an isolated system. However, after the laws are experimentally determined, one tries to correlate the macro-system laws with dynamics in the micro-level, i.e., with gas molecules. In other words, the laws of gas dynamics were determined before these laws were exactly correlated with molecular properties. This is a complementary explanation to Griffith's problem.

Başar (1980, 1998) commented on the questions of Rosen and Griffith as follows:

> In the analysis of brain waves we are certainly interested to discover the particular properties of individual neurons and their relation to the gross activity. To further examine the problem of the correlation between single unit activity (micro-activity) and gross activity (macro-activity).

Rosen (1969) explained the concepts of statistical mechanics and physics and their relation to Neurobiology as follows:

> What is the micro-description? We know, that here, the fundamental state variables are the displacements and momenta of the individual particles which make up our system. According to Newtonian dynamics, the kinetic properties of the system are given by the equations of motion of the system, which express the momenta as functions of the state variables.

The basic postulates of "Newtonian Dynamics" are the following point: Knowing the state variables at one instant and the equations of motion, we are supposed to be able to answer any meaningful question that can be asked about the system at any level. Statistical mechanics however, identifies a macro-state with a class of underlying microstates, and then expresses the global state variables as averages of appropriately chosen micro-observables over the corresponding class of microstates.

2.10 Santiago Ramon Y Cajal

In the twentieth century a great amount of research suggests that it is possible to understand the functioning of the brain once there is sufficient explanation for the specific functions of individual nerve cells and their connections. The transformation of neural information and its storage as memory involve only nerve cells and their interconnections.

However, at the end of the nineteenth century it was generally believed that the brain is made up of a continuous net of nerve tissue, a "reticular network" or "syncytium." The first morphological studies of the nervous system were done by the Spanish anatomist Santiago Ramon y Cajal. He proposed that the functions of the brain could be understood by analyzing the functional architecture of the nervous system. Applying Golgi's silver staining technique to the study of nerve tissue, he observed that only some cells are stained in their entirety. This led to his formulation of the "neuron doctrine," which states that the brain is made up of discrete units rather than a continuous net of nerve tissue or "syncytium," as was originally thought. He proposed that nerve impulses travel from the dendrites of a neuron to its cell body and then along the axon to the dendrites of the neighboring neuron. This flow of information would be a finite process.

The neuron is a transmitter, because it converts the conducted electrical signals into chemical messages and then conveys or "transmits" them from one neuron to a neighboring neuron. Neurons are connected at specialized contact points called *synapses*. English physiologist Charles Sherrington (1861–1952) worked out the details of the reflex arc in the spinal cord of mammals (*The Integrative Action of the Nervous System* 1906). Although the book of Sherrington was republished in 1948, it is noteworthy that he did not include memory and cognitive functions in the integrated action of the nervous system.

2.11 Hans Berger and Electroencephalography

Hans Berger's discovery of EEG dominates several parts of the book. Here, we add only the handwriting of Hans Berger related to encephalography. Figure 2.10 is self-explanatory (Fig. 2.9).

Fig. 2.9 Ramon y Cajal

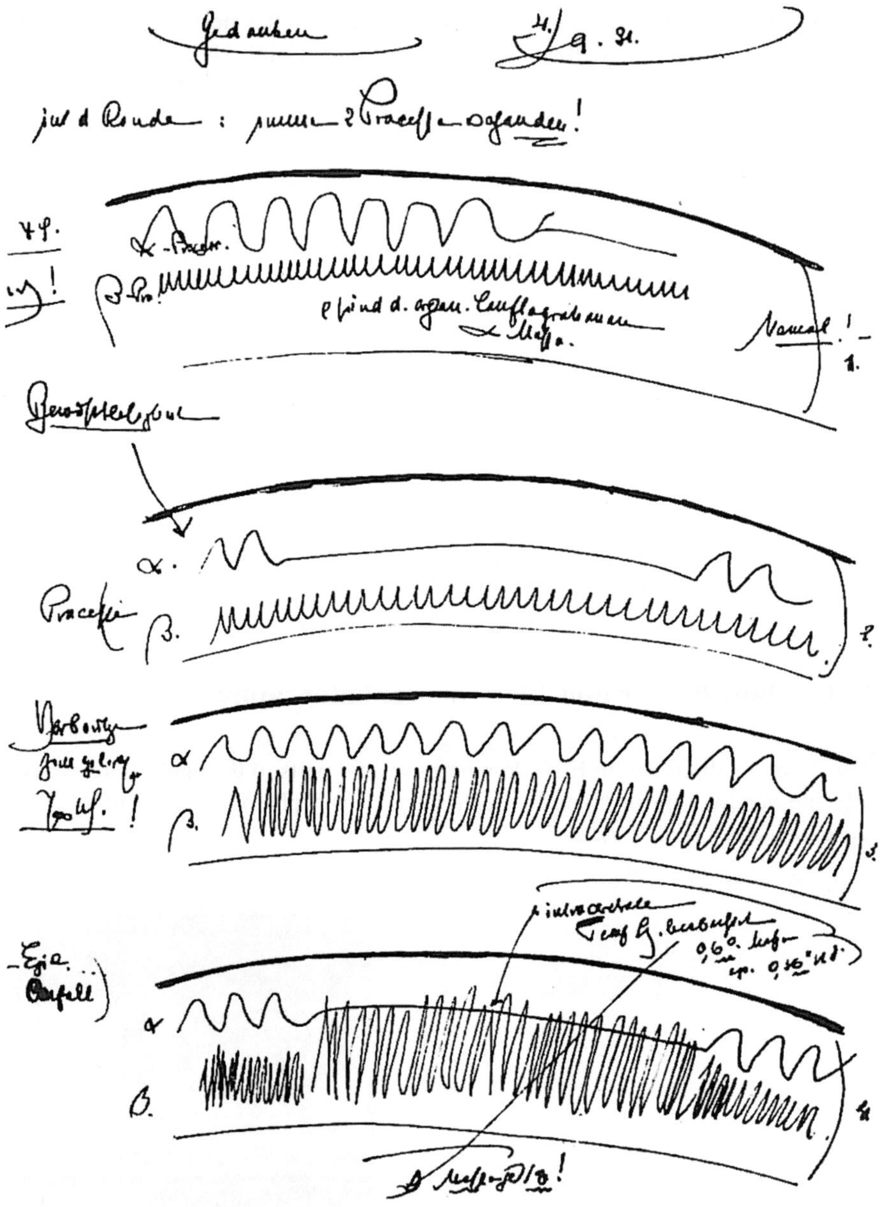

Fig. 2.10 Facsimile page from Berger's protocol giving his concept of the alpha- and beta-wave processes in normal and certain pathological conditions. Berger's handwriting is a mixture of normal German and a special form of shorthand. In English translation: "Thoughts 21/9/31. In the cortex: Always 2 processes present! (1) ψϕ. Psychophysical, Alpha-process. Nutrition! Beta-process. That is the organ. Conflagration of Mosso. Normal! (2). Unconsciousness. Process Alpha. Beta. (3). Preparation for epileptic seizure. Aura! Alpha. Beta. (4). Epileptic seizure. Alpha. Beta. Intracerebral temperature increase measured 0.6°, Mosso 0.36° in the human. According to Mosso, not always, however." (from Jung. Jenenser EEG symposium, 30 Jahre Elektro-enzephalografie, p. 47, 1963. Courtesy of VEB Verlag Volk und Gesundheit)

2.12 Hebb, Hayek, and Helmholtz

In the first half of the twentieth century two important books introduced outstanding holistic and dynamic approaches to brain functioning, Donald Hebb's book (1949) related to the organization of behavior inspired several neuroscientists in search of the "Hebb neuron." Speculations on the existence of the Hebb neuron and Hebb's theory are explained in Chapters 7 and 8. According to Hebb, the functioning of the brain after learning is a different brain compared with the same brain before the learning process.

Although Friedrich Hayek developed his theory of "theoretical psychology" almost 20 years before the publication of Hebb's book, Hayek's book was published much later (1952). The chain of ideas developed in this theory is highly pertinent to the dynamic nature of the living brain. Hayek states:

> We shall see that the mental and the physical word are in the sense two different orders in which the same element can be arranged; though ultimately we shall recognize the mental order as part of the physical order.

Hayek argues that it is the whole history of the organism that determines its action. New factors contribute to this determination on later occasions that were not present at first. This idea is much better explained in the following sentence: "We shall find out that the same set of external stimuli will not always produce the same responses, but also that altogether new responses will occur" (compare also Fig. 2.8). Here is a dynamic interpretation of brain responsiveness similar to the statement made by the Ionian philosopher Heraclites: "One never can step twice into the same river." One of Hayek's most important statements is related to perception and memory in that they are inseparable functions. This view later received excellent support from Fuster, Baddeley, Desimone, and Başar. Therefore, perception is always an interpretation, the placing of something into one of several classes of objects. An event of an entirely new kind, which has never occurred before and sets up impulses that arrive in the brain for the first time could not be perceived at all.

Here it is important to emphasize the parallels with the theories of Hebb and Hayek. The brain that is learning or is targeted by several stimuli; accordingly it will be changed both physiologically and anatomically. According to Hebb, there are changes in the connectivity of neurons in the learning brain, thus changing both the anatomical structure as well as the electrical activity. Hebb and Hayek both discussed the dynamic brain.

Although neither of these scientists mentioned structural and entropy changes during learning, it is clear that the concept of altered entropy exists in both scientists' theories. This central question is discussed in Chapter 17. Although theoreticians such as Prigogine and Wiener took advantage of Hayek and Hebb's biological models, they did not find an important bridge between neural connectivity and changes in the entropy of the learning brain. Chapter 7 attempts to create this bridge.

Hayek asks, "What is mind?" and he discusses the relation between mind and body or mental and physical events. The difficulty of any fruitful discussion of the body-mind problem consists largely in differentiating what part of our knowledge can properly be described as knowledge of mental events, as distinguished from our knowledge of physical events. After discussing physical events, the physiological responses to physical events, Hayek comes to the following definition: What we call "mind" is a particular order of a set of events taking place in some organism and some manner related but not identical to the physical order of events in the environment.

Hayek considers the nervous system an instrument of classification. He classifies emotion as a special type of disposition for a type of action, which in the first instance is not necessitated by a primary change in the state of the organism, but which consists of complexes of responses appropriate to a variety of environmental conditions. Fear, anger, sorrow, and joy are attitudes toward the environment, and particularly toward fellow members of the same species. This means that a great variety of external events, and also some condition of the organism itself, may evoke one of several patterns of attitudes or dispositions that will affect the perception of, and the responses to, any external event. Emotions may thus be described as "affective qualities similar to the sensory qualities and forming part of the same comprehensive order of mental qualities."

According to Hayek, the term *experience* is related to memory; however, it is a plastic memory. If stimuli are applied to the central nervous system, then this system gains a type of experience. However, when the same stimuli occur again, they have special significance for the organism, even though not having any meaning for the individual. Hayek proposes that we must distinguish between two different kinds of physiological "memory" or traces left behind by the action of any stimulus. One is the semi-permanent change in the structure of connections or paths, which determines the courses through which any change of impulses can run (similar to Hebb's principle). The other is the pattern of active impulses proceeding at any moment as results of a stimulus received in the recent past, and perceived also as merely part of continuous flow of impulses of central origin, which never cease altogether, even when no external stimuli are received. At this point the reader is referred to Chapters 7 and 8, which discuss memory.

Hayek's most important conclusion on the evaluation of impulses from the organism is that it is the whole story of the organism that determines its action. New factors contribute to this determination on the later occasion that were not present on the first: "We shall find not only the same set of external stimuli will not always produce the same responses, but also that altogether new responses will occur." This is similar to the coordinated movement of the organism, which is not determined by the movement of an individual muscle, but to the whole complex of body muscles.

Chapters 15 and 16 introduce the Brain S-matrix, which takes into account the whole history of the organism. Hayek does not comment on the S-matrix, but this concept includes the application of the S-matrix, which includes the history of whole brain-body organism.

Hayek explains perception as an interpretation or the placing of something into one or several classes of objects. An entirely new kind of event, which has never

Fig. 2.11 Hayek (May 8, 1899–March 23, 1992)

occurred before, and which sets up impulses that arrive in the brain for the first time could not be perceived at all. This explanation is in accordance with Helmholtz's opinion with regard to perception.

Helmholtz puts the emphasis on the effect of experience in determining sensory qualities, and he goes far beyond ascribing to experience the creation of their spatial order. It is today widely recognized that "the manner in which we see things of the external world is sometimes affected by experience to an overwhelming extent" and that "it is often difficult to decide, which of our visual experiences are determined immediately by sensation and which, on the contrary, are determined by experience and practice." His conception of the "unconscious inference" by which stimuli that do not lead to conscious experience and yet are utilized in the perception of a complex position comes very close to the theory developed here.

Chapters 7 and 8 contain descriptions of the phyletic memory, which is very well described by Fuster (1995a) and later by Başar (2004): every sensation, even the "purest" must therefore be regarded as an interpretation of an event in the light of the past experience of the individual or the species.

Hayek's conclusion is that the mind must remain forever in a realm of its own, in which we can now only directly experience it, but which we shall never be able fully to explain or "reduce" to something else. Even though we may indicate that the mental event of the kind that we experience can be produced by the same forces that operate in the rest of nature, we shall never be able to say which particular physical events "correspond" to a particular mental event (In Fig. 3 is a picture of Hayek).

2.13 Jacques Monod: "The Chance and the Necessity" (1971)

Monod actually begins by showing that the difference between natural and artificial things is illusory, as natural things are also built for a purpose. Living beings are characterized by three properties: teleonomy (organisms are endowed with a purpose that is inherent in their structure and determines their behavior); autonomous morphogenesis (the structure of a living organism is a result of interactions within the organism itself); and reproductive invariance (the source of information expressed in a living organism is another structurally identical object; it is the information corresponding to its own structure).

From his analysis of how DNA and proteins work, Monod concludes that humans are the product of *chance, an accident in the universe*. The paradox of DNA is that a mono-dimensional structure such as the *genome* could specify the function of a three-dimensional structure such as the body. The function of a protein is underspecified in the code; it is the environment that determines a unique interpretation. There is no causal connection between the syntactic (genetic) information and the semantic (phenotypic) information that results from it. Then the growth of our body, the spontaneous and autonomous morphogenesis, rests on the properties of proteins. Monod concludes that life was born by accident, and then evolved by natural selection, as discovered by Darwin. Biological information is inherently determined by chance. The concept developed by Monod is discussed in detail in Chapter 17.

2.14 Otto Loewi and the Discovery of Acetylcholine

One of the most important developments at the beginning of twentieth century is the experiment of Loewi leading to discovery of acetylcholine. The role of transmitters in the understanding of the mind is crucial, and what Loewi has achieved is one of the most important discoveries in brain research (Fig. 2.12).

In his most famous experiment, Otto Loewi took fluid from one frog heart and applied it to another, slowing the second heart and showing that synaptic signaling used chemical messengers. He called the chemical Vagusstoff. It was later found that this chemical corresponded to acetylcholine. We return to this important discovery in Chapters 3, 13, 22, and 24.

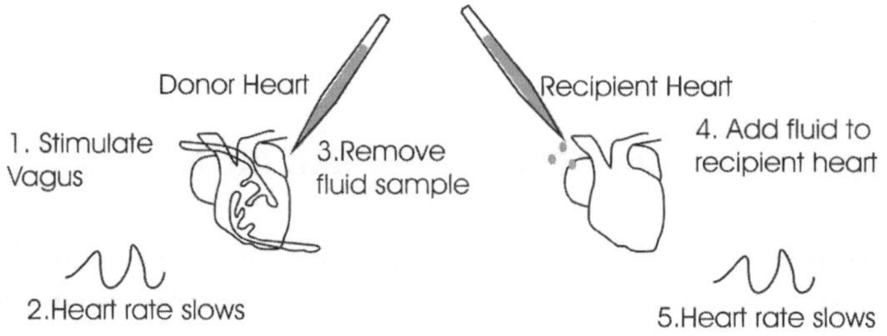

Fig. 2.12 Experiment of Otto Loewi

2.15 A Synthesis from the Concepts of Wiener, Prigogine, Thom, and Haken

These four philosopher-scientists strongly emphasized some common features in interdisciplinary sciences. They carefully analyzed the following ideas: the concepts of order and disorder, the second law of thermodynamics, entropy, and nonlinear phenomena. The energy input in lasers induces the transition of oscillating atoms from a disordered to an ordered state similar to the brain oscillations on sensory stimulation. Prigogine stated that according to the second law of thermodynamics, the emergence of ordered states in the creation of life is improbable. Wiener already mentioned the role of a Maxwell Demon in living processes. As seen in Chapter 17, Monod also emphasized the importance of the Maxwell Demon in the creation of life. These frameworks also have a common general frame. All of these scientists started from physical, mathematical, or chemical metaphors by trying to identify common abstract mechanisms or symbols to create new interdisciplinary sciences. Unfortunately, none of these frameworks have their origin in biological empiricism. One essential biological framework is Darwin's evolution theory, in which "natural selection" plays a major role. However, in turn a selection needs a type of transition for moving a new order. Wiener, Prigogine, Thom, and Haken also mention the importance of non-linear phenomena; and deterministic chaos is related to this. According to these philosopher-scientists, new branches of sciences must deal with the second law of thermodynamics, equilibrium, feedback mechanisms, and communication and information processes.

With a profound approach to the phenomena analyzed in these frameworks and by amalgamating these trends with Bergson's concept of creative evolution processes, we will develop the idea of launching a new framework, or a new Cartesian system. The scope of the present book and the aim in launching a new Cartesian system consists of a synthesis of the excellent ideas governing these described frameworks in the twentieth century. The present author has worked with these four frameworks, so as to launch the EEG-brain dynamics[6] concept, which is now the prevailing approach in publications from a number of neuroscience laboratories. Therefore, the aim here is not to deny the importance of existing frameworks, but to enlarge them and also incorporate the philosophical schools following Renaissance, quantum physics, and the new results in chaotic brain dynamics. The Cartesian system of the twenty-first century is not intended to discover final solutions, but to raise questions that could be answered with the help of many experiments and scientists. This system will provide a working branch, as in Wiener's cybernetics, and have the additional possible advantage of collecting experiences from existing frameworks.

Figure 2.13 illustrates, globally, the evolution of philosophy and sciences from old Athens and the Renaissance to the development of physics and the new contemporary unifying schools. In the twentieth century cybernetics, quantum

[6]Definition of dynamics.

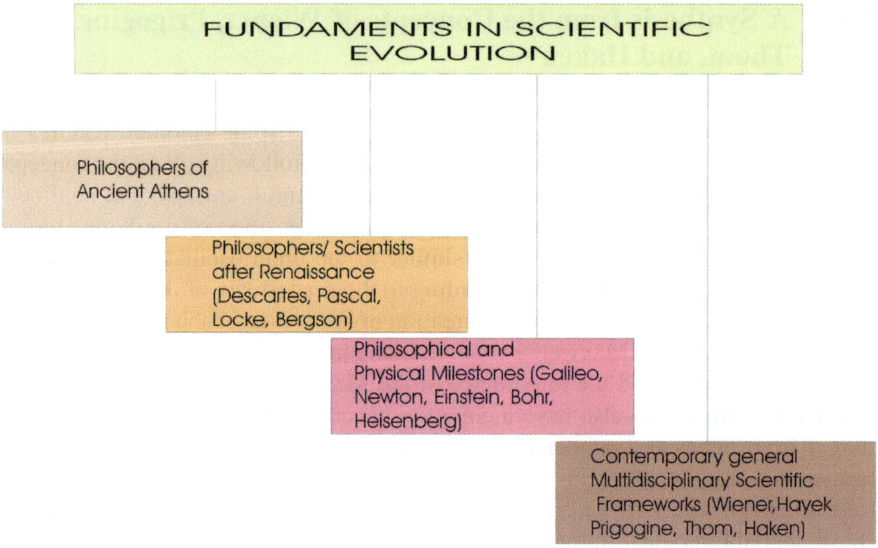

Fig. 2.13 Some fundamental approaches during the evolution of science (compare with Fig. 26.1)

theory, chaos theory, dissipative structures, and synergetics provided essential steps along the way to the branch that we call brain dynamics. By application of the concepts and methods of the mentioned schools, scientists have collected vast empirical data to approach brain-body-mind integration. Furthermore, the application of these various concepts and the rich amount of data collected should serve to find new types of evaluations closer to the language of the brain and the understanding of the brain-mind by developing new approaches. These possibilities are outlined in Chapters 14, 15, 16, 23, 24, 25 and 26.

Part II
Whole Brain Work and a Holistic Approach to Brain-Body Integration

Prelude to Part II

Day-Zero in New York

One day towards the end of July 1969, in my home in Piermont, New York, I was looking at the beautiful sunset on the Hudson River. That day I had, for the first time, demonstrated the alpha response as a component of a simple light-evoked potential. I was a postdoctoral fellow at the Research Center of the Rockland State Hospital (now the Nathan Kline Research Institute) and I was already aware of the functional role of oscillations in physiological systems from my experience with cardiovascular organs and with smooth muscles. The importance of oscillatory processes in quantum dynamics was also clear to me, because of my earlier career in physics. That day was quite moving, because the possibility of predicting brain-body interactions relying on oscillations was very exciting: A new opportunity was emerging, to improve our understanding of EEG in relation to the vegetative system and neurotransmitters. However, in contemporary textbooks the EEG was described as a noise signal and alpha was considered as the idling of the brain.

During the Hudson River sunset I had two thoughts: (1) alpha oscillation is not a noise signal (or cannot be a simple noise) and (2) showing the fundamental role of EEG-oscillations in brain-body function will be a difficult and protracted problem. In fact, during the intervening period, the role of EEG as functional was overshadowed by other themes in the neuroscience literature. In 1981, the Galambos group in San Diego mentioned the importance of gamma response. A review of the references within Chaps. 6 and 7 shows that, between 1980 and 2000, studies of sensory-cognitive brain oscillations invaded the neuroscience literature.

Our experimental results on brain oscillations during the period 1970–1980 opened the way to the publication of a monograph on EEG–brain dynamics in 1980. During this period also, measurements on smooth muscles and the overall myogenic system were accomplished by Canan Başar-Eroğlu and my group. Because of this parallel development, the monograph on Vasculature and Circulation was strongly influenced by the holistic approach of EEG–brain dynamics. Finally, these earlier findings, greatly extended by the new measurements on ultra-slow oscillations, led to the holistic view or the syncytium of brain-body discussed in Chap. 9.

In Chap. 3 some anatomical and physiological methods and strategies are explained. In Chap. 4 core material and results related to the dynamics of the cardiovascular system and the machinery of the autonomous vegetative system are introduced. The joint analysis of general body dynamics leads to the definition of an overall myogenic system (OMS), which has a key role in understanding brain-body integration. The cardiovascular system is certainly fundamentally clearer and less complicated in comparison with the brain. In order to perform initial steps, we first consider five processing systems: (1) the vascular system and smooth muscles; (2) the lymphatic circulation, which is adjacent and interwoven with the circulatory system; (3) co-activation of peristalsis and circulation; (4) local control of blood flow: autoregulation; (5) and last but not least, we mention the link between the central nervous system and the body by means of cranial nerves (see also Chap. 9). Accordingly, at the beginning of Part II, in Chaps. 4 and 5, an important prerequisite to understanding dynamics of processes in the vegetative system is achieved.

Then, in Chap. 6 we explain the dynamics of sensory and cognitive functions, and in Chap. 7 we try to explain the memory function of the brain and different memory states. In turn, this will help in understanding the functional organization within the brain. Further, the integrative aspects of memory and all other brain functions will be brought to a new type of schematically presentation. This analysis leads to a new theory of "whole brain work," discussed in Chap. 8. In Chap. 9 a different type of organization will be explained that considers the whole body: the description of the brain-body link. Therefore we first consider the links within the brain and then describe possible dynamic functional links between the organs of the autonomous system and the brain by means of oscillatory dynamics and coherence functions. As a result of joint consideration of the dynamics of the brain and vegetative system, the possibility is presented for new insights into brain-body-mind integration. To develop empirically based concepts for the mechanisms of brain-body-mind, the whole-brain work and whole body work must be analyzed in parallel. Accordingly, prerequisites for the questions, "What is mind?" or "What is brain-body-mind?" will be ready to be discussed in Part VI.

The vegetative system functions with well-organized oscillatory dynamics: Several organs, including the blood vessels, the respiratory system, and the lymphatic system, are precisely tuned to similar frequency ranges. Besides this, the vegetative system is linked and tuned with the spinal cord, the brain stem through cranial nerves, and possibly with the cerebellum. At the beginning of this book we give considerable emphasis to the cardinal questions of Descartes and Fessard; consequently, in Chap. 9 we will pose a more general question, "Are there some general rules and/or transfer functions in the functioning of the brain-body system?"

This means, do we find some similar frequency characteristics in the dynamics of the brain, spinal cord, peristaltic organs, blood vessels and lymphatic system? We will see important implications of this general question, not only in Chap. 9, but also in the concluding Part VI.

Chapter 3
Brain Structures, Transmitters, and Analyzing Strategies

3.1 Signaling in the Brain

Brain research is one of the last great frontiers in the biological sciences. The unraveling of its mysteries is comparable in complexity and intellectual challenge to the research in elementary particle research or astrophysics.

The functioning nervous system can be studied at a number of levels of organization (Fig. 3.1). Biologists investigate the properties of molecules that perform tasks important for brain function. Physiologists study the characteristics of individual nerve cells or ensembles. Psychologists explore patterns of behavior and its modification – learning. Computational neuroscientists attempt to bring these fields together to model higher brain functions in terms of the known properties of molecules and cells.

This book focuses on the lower levels of the system shown in Fig. 3.1. The major emphasis is on signaling, or information transfer, within and between nerve cells. Such signaling is essential for an organism to (1) sense information about its environment; (2) import this information into its brain, where it can be processed; and (3) generate a behavioral response.

In this book we will focus on functional correlates of neural populations of the human and animal brain. However, we emphasize that cellular and molecular neurobiology is extremely important. All the levels of organization depicted in Fig. 3.1, from behavior of animals to single brain cells and the molecules that regulate cellular activity, are important. No single level is inherently more important than any other, and information from all of them will be necessary for even the most rudimentary understanding of normal and abnormal brain functions. Thus, brain research demands a multidisciplinary approach, one that bridges traditional scientific disciplines and facilitates collaboration among scientists with very different experimental approaches.

E. Başar, *Brain-Body-Mind in the Nebulous Cartesian System: A Holistic Approach by Oscillations*, DOI 10.1007/978-1-4419-6136-5_3,

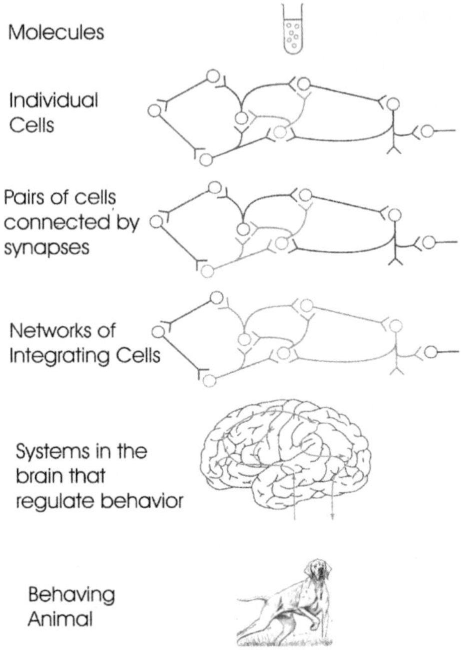

Fig. 3.1 Levels of organization for studying structure and function in nervous systems. Depending on their background and training, different scientists may take different approaches to the study of the nervous system (modified from Levitan and Kaczmarek 2002)

3.1.1 The Cellular Hypothesis

Around the end of the nineteenth century, the great neuroanatomists Santiago Ramon y Cajal and Camillo Golgi argued about whether the brain consists of enormous numbers of discrete cells or is a continuous syncytium of tissue. As a result of this, individual neurons show up clearly in tissue sections that actually contain a large number of neurons. Ramon y Cajal correctly identified these discrete entities as individual nerve cells (Fig. 3.2).

It must be remembered that the essence of nervous system function is signaling, or information transfer, both *intra*cellularly from one of a cell to another, and *inter*cellularly between cells. It is a fundamental premise of cellular neurobiology that a great deal will be learned about how the nervous system works by investigating:

1. Those aspects of neural structure that specialize them for information transfer
2. The mechanisms of intercellular neuronal signaling
3. The patterns of neuronal connectivity and mechanisms of intercellular signaling
4. The relationship of various patterns of neuronal connectivity to different behaviors
5. The ways in which neurons and their connections can be modified by experience

In the following, we will present the functional anatomy of the auditory and visual pathways.

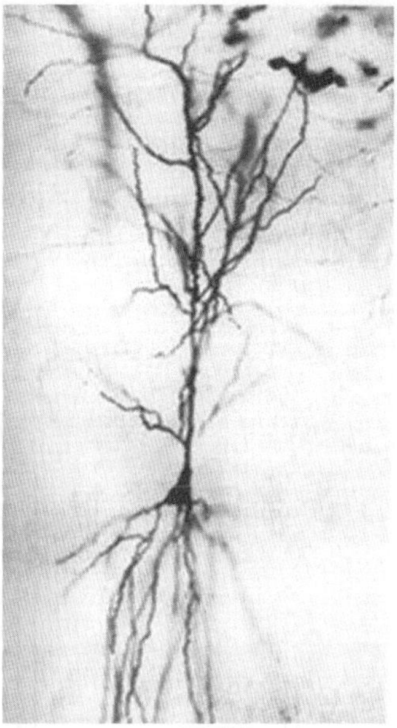

Fig. 3.2 A single Golgi-stained neuron in the hippocampus. The Golgi stain allows the shape of the cell body, and the complex dendritic arborization of this hippocampal neuron to be resolved from the "tangled thickets" seen when all surrounding neurons are also stained

3.2 Functional Anatomy of the Auditory Pathway

The physiological description of the auditory pathway can be divided into two parts: (1) the *peripheral auditory system* (the ear and primary neurons, i.e., the auditory and cochlear nerves); (2) the *central auditory system* from cochlear nucleus to cortex. Figure 3.3 provides a schematic representation of the auditory pathway using systems theory configuration.

According to Regan (1989), the auditory pathway is described, using the terminology of systems theory, as follows: (1) A *signal conditioning apparatus* (the outer ear), followed by (2) a parallel spectral analyzer (the inner ear), and followed by (3) a parallel analyzer (the central auditory pathway). The sensory pathways are parallel analyzers.

For the reader's first orientation, Fig. 3.3 illustrates the major auditory pathway and some of the related nuclei (details concerning the cochlear nucleus and thalamocortical projections follow in Fig. 3.4). Third order neurons send their axons via the lateral lemniscus to the inferior colliculus (IC), where some cross to the opposite side. (Some terminate at a lower level in the nucleus of the lateral lemniscus.) A few fibers cross from the nucleus of the lateral lemniscus through the commissure of Probst to the contralateral nucleus, and still other fibers cross through the inferior collicular commissure, from one inferior colliculus to the other. From the inferior colliculus, the pathway then passes through the peduncle of the inferior colliculus to the medial geniculate nucleus (MG), where all the fibers synapse. From here, the auditory tract spreads by way of the auditory radiation to the auditory cortex (Brazier 1968; Guyton 1971).

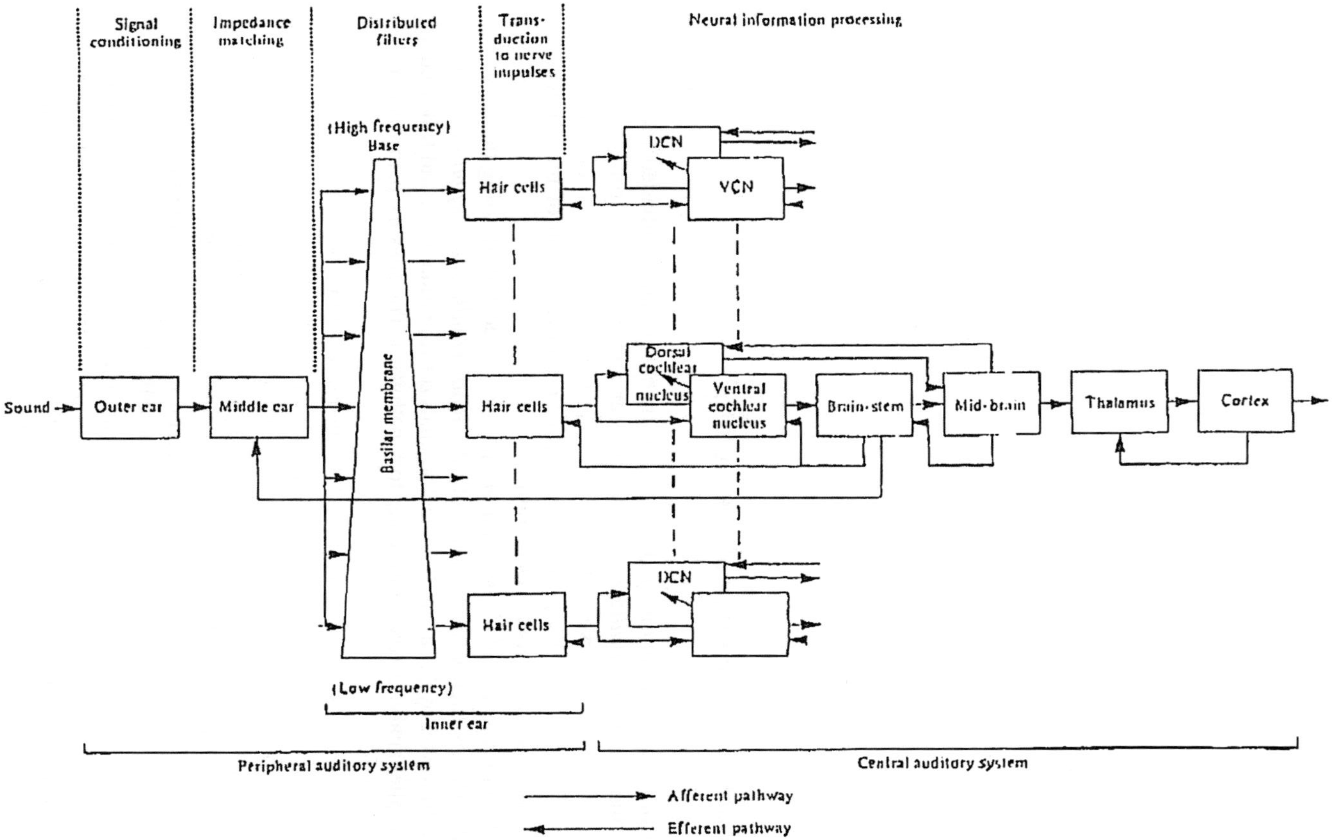

Fig. 3.3 Schematic anatomical and functional map of the auditory system. At the level of the basilar membrane, the incoming auditory wave (time signal) is processed in multiple parallel frequency channels. The vertical axis in this figure represents frequency. This tonotopic map organization is retained at the cochlear nucleus, wherein a further divergence occurs into parallel pathways only two of which, DCN and VCN, are shown. The DCN pathway is thought to be concerned with the nature of the auditory stimulus and the VCN pathway with its location in space (from Evans 1982)

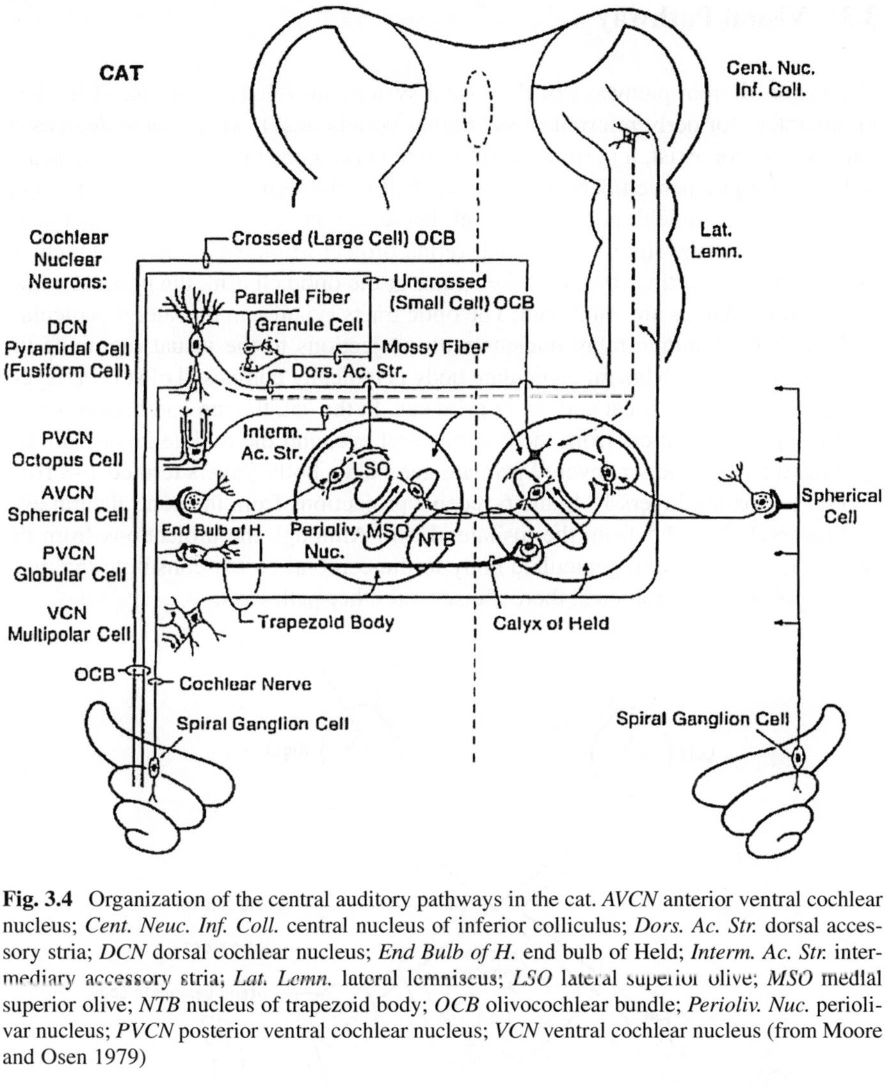

Fig. 3.4 Organization of the central auditory pathways in the cat. *AVCN* anterior ventral cochlear nucleus; *Cent. Neuc. Inf. Coll.* central nucleus of inferior colliculus; *Dors. Ac. Str.* dorsal accessory stria; *DCN* dorsal cochlear nucleus; *End Bulb of H.* end bulb of Held; *Interm. Ac. Str.* intermediary accessory stria; *Lat. Lemn.* lateral lemniscus; *LSO* lateral superior olive; *MSO* medial superior olive; *NTB* nucleus of trapezoid body; *OCB* olivocochlear bundle; *Perioliv. Nuc.* periolivar nucleus; *PVCN* posterior ventral cochlear nucleus; *VCN* ventral cochlear nucleus (from Moore and Osen 1979)

Several points of importance in relation to the auditory pathway should be noted. Collaterals from the main auditory pathway pass into the brainstem reticular formation (RF). Direct connections between the inferior colliculus and the auditory cortex are also established. Several important pathways also exist from the auditory system into the cerebellum: (1) Directly from the cochlear nuclei; (2) from the inferior colliculi; (3) from the reticular formation; and (4) from the cerebral auditory areas.

3.3 Visual Pathway

The major afferent pathways of the visual system are relatively simple (Fig. 3.5). In amniotes (formerly referred to as "higher vertebrates"), where some degrees of binocular vision exist, a portion of each optic nerve goes to each side of the brain. In humans, optic nerve fibers from the left half of the retina (representing the right half of each visual field) project to the left lateral geniculate nucleus of the thalamus, and fibers from the right half of each retina project to the right lateral geniculate nucleus. The re-sorting of fibers takes place at the optic chiasm, the point at which the two optic nerves are combined. The optic tracts synapse in the lateral geniculate body and the thalamic relay nucleus with projections to the visual region of the cerebral cortex. Each lateral geniculate body is typically composed of several layers or regions, three in the cat and six in primates. In the cat, the top and bottom layers receive optic tract fibers from the contralateral eye, and the middle layer from the ipsilateral eye. In the six-layered primate geniculate body (layers termed 1–8 from ventral to dorsal), layers 1, 4, and 6 receive projections from the contralateral eye, and layers 2, 3, and 5 from the ipsilateral eye. Although the projections from the optic tract to the lateral geniculate body to the cerebral cortex constitute the main visual pathways in amniotes, there are several other pathways.

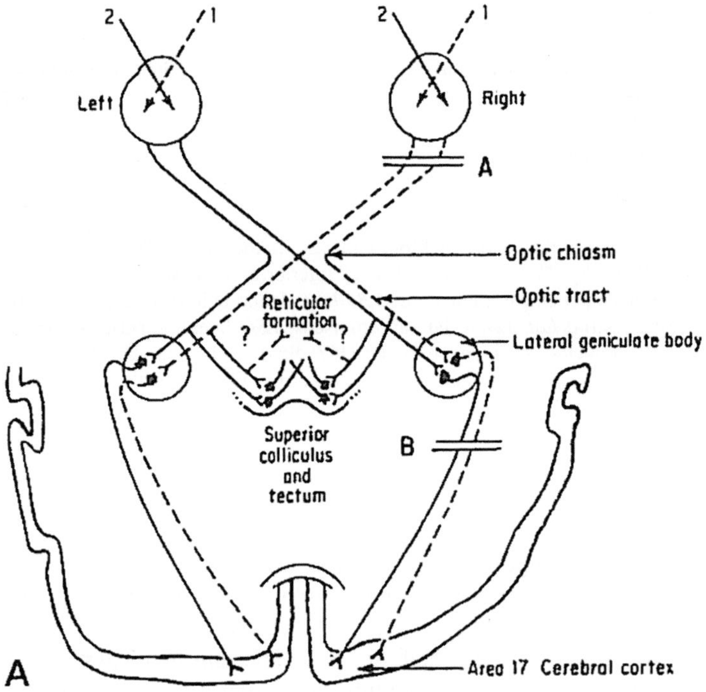

Fig. 3.5 Schematic diagram of the visual system. Section at A eliminates input from the right eye, but section at B eliminates input from the right half of each eye (from Gardner 1952)

1. Some optic fibers project to the superior colliculus in the midbrain.
2. A portion of the optic tract fibers is also connected to the pretectal area (a region in front of the superior colliculus) in the midbrain.
3. Electrophysiological and anatomical evidence indicates the existence of projections from the visual system into the midbrain reticular formation (French et al. 1953).

From Fig. 3.6, it is evident that signal transport in the visual pathway occurs, covering major centers of the brain. In addition to those centers that are directly related to the visual pathway, the hippocampus and reticular formation are also involved in the signal transmission through the visual pathway.

3.4 Cerebral Cortex Anatomy and Global Function

Because the electroencephalographer usually analyzes signals of the human cortex, we will commence with the description of the human cortex. The cerebral cortex is characterized by infolding, i.e., fissures and gyri. Each cerebral hemisphere can be divided into four lobes, which are named for the overlying bones of the skull: frontal, parietal, temporal, and occipital (Fig. 3.7a). Large regions of the cerebral cortex are committed to movement and sensory processing. Areas that are directly committed to such functions are called primary, secondary, and tertiary sensory and motor areas. The vast amount of physiological literature indicates that the primary motor cortex, lying within the precentral gyrus, contains neurons projecting directly to the spinal cord.

Kelly and Dodd (1991) point out that, "the primary sensory areas (i.e., visual, auditory, somatic, sensory) receive information from peripheral receptors with only a few synapses interposed. The primary visual cortex is located at the back of the occipital lobe. The primary auditory cortex lies in the temporal lobe. The primary somatic sensory cortex lies on the postcentral gyrus."

Surrounding the primary areas are high order (secondary and tertiary) sensory and motor areas. These higher order cortical areas process more complex aspects of a single sensory modality or motor function than the primary areas. The purpose of the higher order sensory areas is to achieve more detailed analysis and integration of information originating from the primary sensory areas.

Also classified with the higher order areas is a portion of the posterior parietal lobe called the posterior parietal cortex. This region is somewhat transitional between sensory and motor functions. Not only does it serve as a higher order sensory area for both somatic sensation and vision but, in addition, it interrelates aspects of sensation and movement.

Kelly and Dodd (1991) note that, "three other large regions of cortex, called association areas, lie outside the primary, secondary, and tertiary areas." In primates, the association areas constitute by far the largest area of cortex. Their function is mainly to integrate diverse information for purposeful action, and they are involved to different degrees in the control of three major brain functions: perception, movement, and motivation.

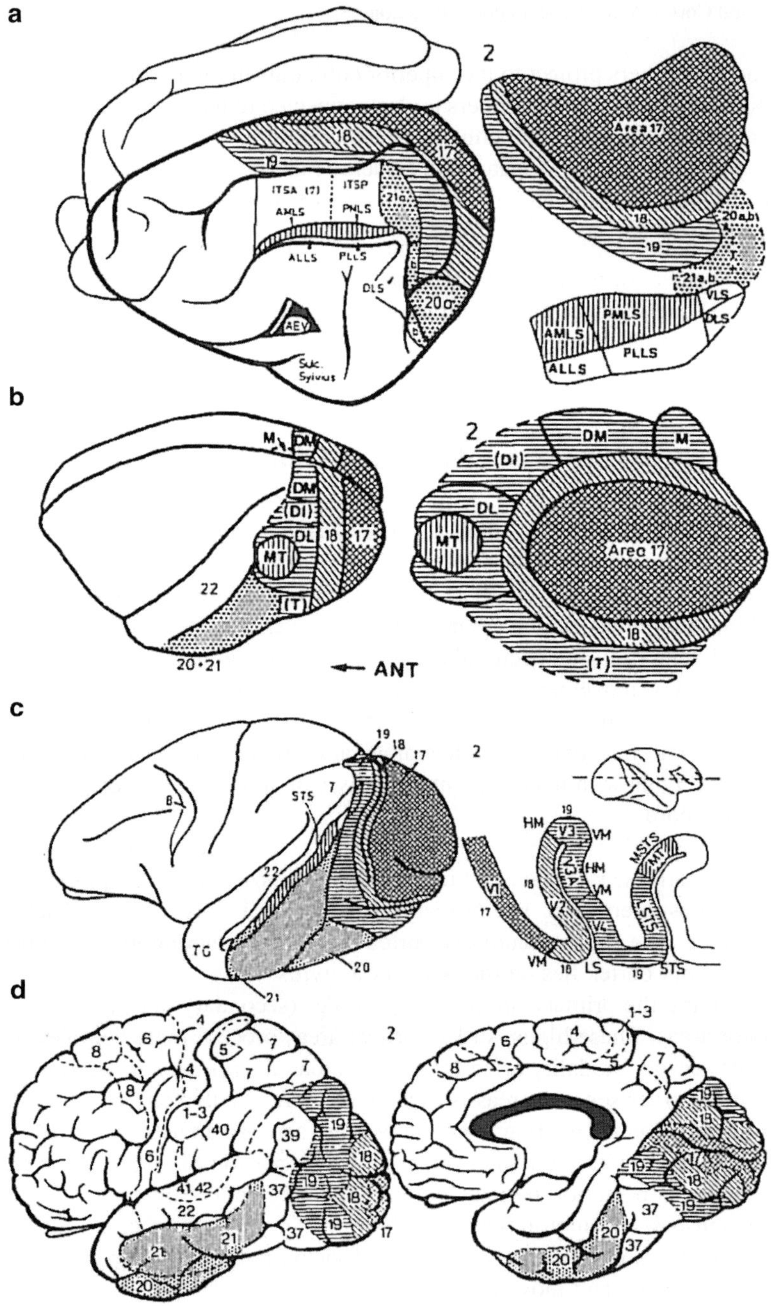

Fig. 3.6 The visual areas in the cat (**a**), in the New World monkey (**b**), in the rhesus monkey (**c**), and in humans (**d**). (**a**l) Lateral view of the cat brain from the top. The areas around the lateral suprasylvian sulcus (LS), which is slightly turned up, are designated in accordance with their topographical position as anterior (A) and posterior (P) as well as medial (M) and lateral (L) (AMLS, PMLS; ALLS; PLLS). The MLS areas are also designated as the Clare-Bishop area. The anterior ectosylvian visual area (AEV) is situated in the bottom of the sylvian sulcus. (**a**2) Extension of the various visual areas including the medial extent of area 17 and folding up of the LS areas.

The parietal-temporal-occipital association cortex occupies the interface between the three lobes for which it is named. It is concerned with higher perceptual functions related to somatic sensation, hearing, and vision, (which are) the primary sensory inputs to these lobes. Information from these different sensory modalities is combined in the association cortex to form complex perceptions. The prefrontal association cortex occupies most of the rostral part of the frontal lobe; one important function of this area is the planning of voluntary movement. The limbic association cortex is located on the medial and inferior surfaces of the cerebral hemispheres, in portions of the parietal, frontal, and temporal lobes, and is devoted mainly to motivation, emotion and memory.

To summarize, the primary sensory areas of the cerebral cortex are devoted to the reception and initial cortical processing of sensory information. The primary areas project to higher sensory areas that further elaborate and process the sensory input. The higher order areas connect to the association areas; these provide the link between sensation and action by making connections with the higher order motor areas.

Figure 3.8 illustrates the 10-20 system of EEG electrode placement, thereby demonstrating the relationship between the cortex and the conventional EEG electrode sites.

3.4.1 Association Cortex and Frontal Lobe

According to Kupfermann (1991), localization of function means that certain areas of the brain are more involved with one kind of function than with others. Most functions require the *integrative action* of neurons in many regions, and therefore localization does not imply that any specific function is exclusively mediated by only one region of the brain. The association areas of the brain are involved in *the integration of more than one sensory modality* and, additionally, with movement planning. This means that the association areas are those regions of the cerebral cortex that perform more *multimodal functions* than the primary sensory and primary motor areas; they are presumed to be involved in complex functions.

With respect to the question, "How does information reach an association cortex?", Kupfermann (1991) points out that each primary sensory area of the cortex is adjacent to, and connected with, a series of higher order sensory regions. The association areas have – contrary to primary sensory cortices – a much less precise map of the

Fig. 3.6 (continued) (**b**) The various visual areas in New World monkeys. *MT* middle temporal; *DL* dorsolateral; *DM* dorsomedial; *M* medial area. The areas are hatched as in the cat. The temporal association areas 20 and 21, which likewise have functions in visual behavior, are *dotted*. It is an open question whether areas 21a and b of the cat are really homologous to area 21 of primates, or whether they should be assigned to one of the numerous fields of visual representations of area 19. (**cl**) Areas 18 and 19 extend only barely to the surface of the convexity and mainly extend deep into the lunate (18) and superotemporal (19) sulci. (**c2**) Horizontal section through the occipital lobe as indicated in the sketch above it. The various peristriate areas V 2–4 are largely concealed in the sulci, as are areas 18 and 19. The *arrows* indicate the boundaries of area 18. MSTS (medial supratemporal sulcus) probably corresponds to area MT in the New World monkey and the Clare-Bishop area in the cat. (**d**) In humans, there are so far no unequivocal findings on the subdivision of area 19 into several visual areas. However, it is to be assumed that they also exist. The temporal association areas 20 and 21 are *dotted* (from Creutzfeld 1995)

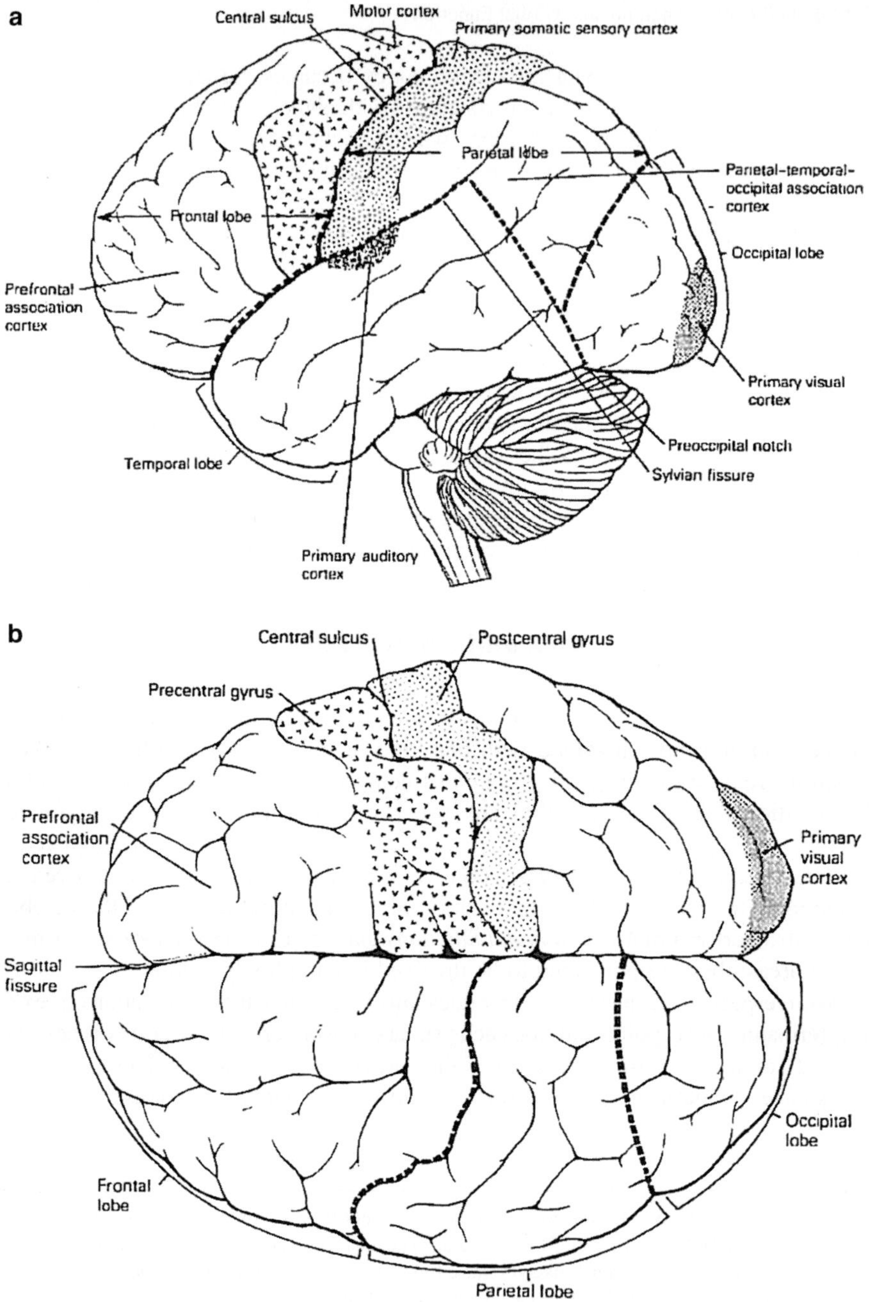

Fig. 3.7 The major divisions of the human cerebral cortex. (**a**) Lateral view of the hemisphere. In this view, it is easier to appreciate both the primary cortical areas and the association areas. The primary auditory cortex lies near the junction of the temporal and parietal lobes. Two large association areas are visible: the prefrontal association cortex and the parietal-temporal-occipital association cortex. The Sylvian fissure is the most prominent cleft visible in a lateral view of the brain.

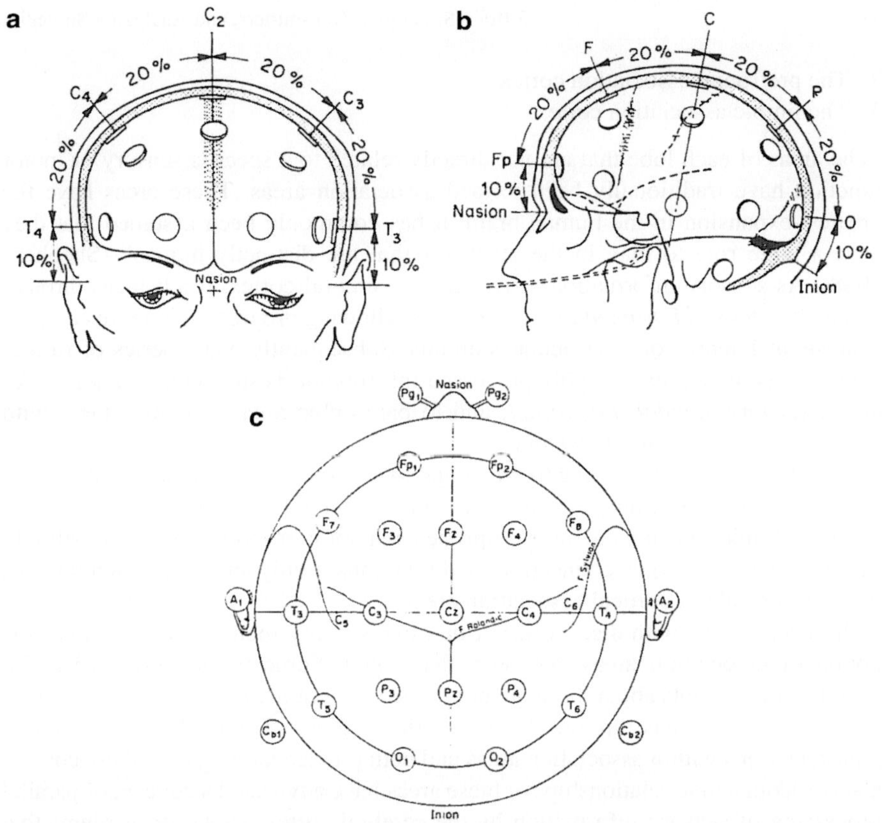

Fig. 3.8 10-20 System of electrode placement. (**a**) Frontal view of the skull showing the 10-20 method for measurement of the central line of electrodes. (**b**) Lateral view of the skull to show the 10-20 method of measurement from nasion to inion at the midline. *Fp* frontal pole position; *F* frontal line of electrodes; *C* central line of electrodes; *P* parietal line of electrodes; *O* occipital line. Percentages represent proportions of the measured distance from the nasion to the inion. Note that the central line is 50% of this distance. The frontal pole and occipital electrodes are 10% from the nasion and inion, respectively. Twice this distance, or 20%, separates the other line of electrodes. (**c**) (**a**, **b** and **c** from Jasper 1958)

peripheral receptive sheet, and are concerned with more *complex aspects of sensory processing*.

According to Kupfermann (1991) three major associations cortices exist:

1. The parietal-temporal-occipital association cortex

Fig. 3.7 (continued) (**b**) Dorsal view, with anterior toward the left. The saggital fissure separates the two hemispheres. The frontal lobe is rostral to the central sulcus. The precentral gyrus of the frontal lobe contains the motor cortex. The postcentral gyrus, which contains the somatic sensory cortex, lies posterior to the central sulcus and is a part of the parietal lobe. The occipital lobe lies at the caudal margin of the hemisphere and contains the visual cortex. The temporal lobe, which lies ventrally, is not visible in this view of the brain (from Kelly and Dodd 1991)

2. The prefrontal association cortex
3. The limbic association cortex

"The areas of each lobe that are not directly related to a specific sensory or motor function have traditionally been termed association areas. These areas have the greatest expansion in the human brain; it has commonly been assumed that they have a large role to play in the attributes that are distinctly human" (Shepherd 1988). As sensory information arrives at the cerebral cortex, it advances through successive steps of *intramodality elaboration*, allowing progressively more complex analysis of features of a particular stimulus. Subsequently, via a series of further connections, this already highly processed information is subjected to *multimodal areas*, for *cross-modal integration*; and to paralimbic areas, for investment with emotion and *placement in memory*.

Finally, connections leading to the frontal lobe association areas allow integration of both sensory and limbic information to provide stability in time and space to an organism while carrying out an appropriate behavior in response to sensory stimuli. This correlation between connections and functions is only tentative; much remains to be learned about cortical association areas.

In recent years, with discoveries of multiple sensory-motor representations, the notion of association areas has been challenged. Some investigators have also raised serious doubts about the sequential processing of incoming information from primary sensory areas through the association regions. The evidence of multiple representations within association areas and better understanding of thalamocortical and corticothalamic relationships of these areas have advocated a concept of parallel processing of sensory information by the cerebral cortex. However, it seems that perhaps both sequential and parallel processes are essential for the analysis of incoming information (Pandya et al. 1979).

As mentioned, each primary sensory cortex is adjoined by parasensory association areas. Although there are a number of such parasensory association areas for each modality, it is useful to categorize each unimodal association area into two broad divisions. The first, or proximal association area, lies adjacent to the primary sensory cortex and receives cortical input directly in this area, whereas the more distal association area is second in line, receiving cortical sensory input by way of the proximal area.

The basic patterns of cortical connections are quite similar in all three modalities. The connections between the primary auditory cortex and association areas are illustrated in Fig. 3.9. The first auditory association area (AAI) receives input from the primary auditory cortex (AI) and projects, in turn, to the second auditory association area (AAII). Whereas the first association area projects to the second auditory area (AII), the second-order parasensory region projects to paralimbic regions: the

Fig. 3.9 (continued) (**c**) Schematic representation of the major association connections of cortical association areas in humans. *Numbers* refer to the cytoarchitectural areas of Brodmann; association areas on the cortical surface are shown by stippling (**a**, **b** from Pandya 1987); **c** from Creutzfeld 1995)

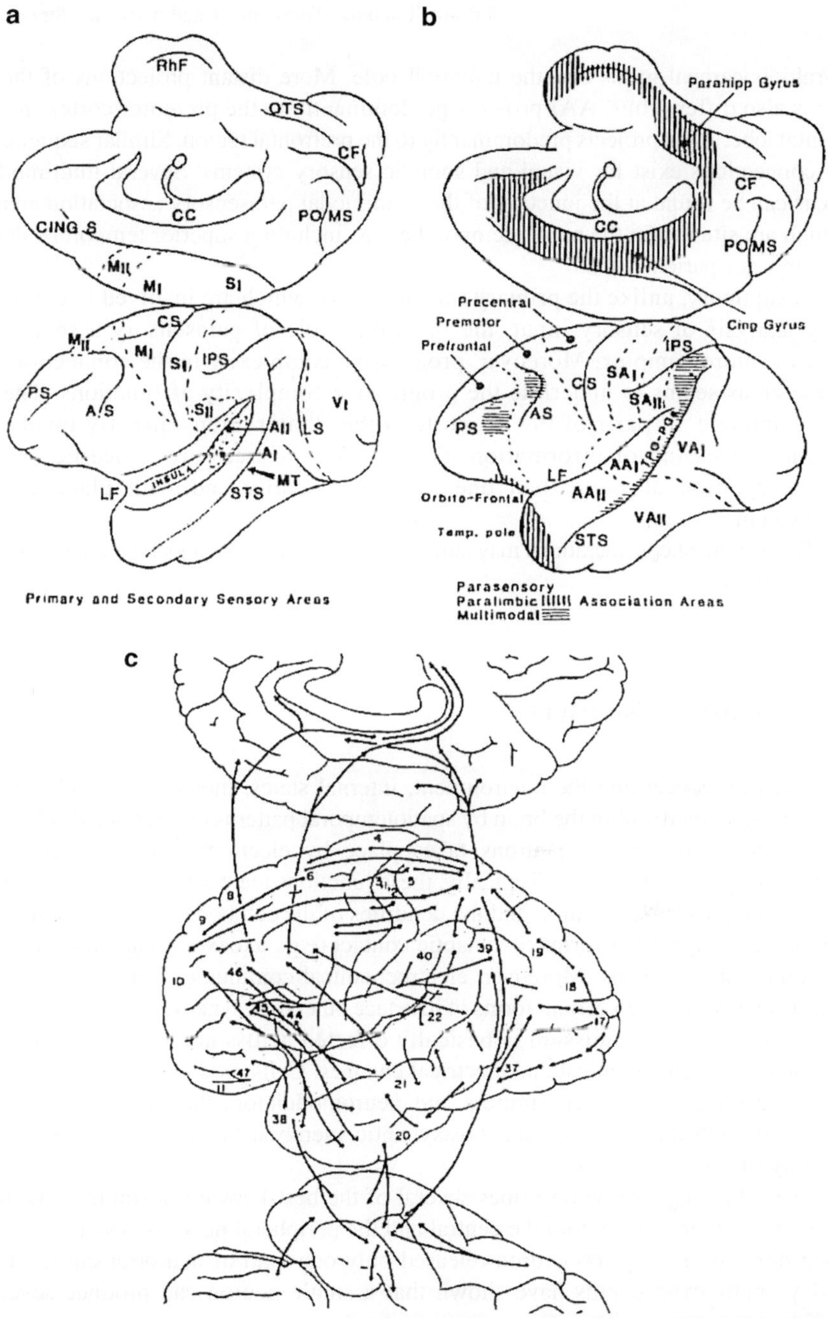

Fig. 3.9 Diagrams of the lateral and medial surfaces of cerebral hemisphere of rhesus monkey. (**a**) The locations of primary and second (supplementary) sensory-motor cortices (auditory areas AI and AII, somatic areas SI and SII, visual areas VI and MT, motor areas MI and MII). (**b**) The three major divisions of association cortex: parasensory association areas (auditory association areas AAI and AII, somatic sensory association areas SAI and SAII, visual association areas VAI and VAII), frontal association areas (premotor and prefrontal areas), and paralimbic association areas (cingulate gyrus, parahippocampal gyrus, temporal pole, and orbitofrontal cortex).

parahippocampal gyrus and the temporal pole. More distant projections of these areas also differ. Thus, AAI projects predominantly to the premotor cortex in the frontal lobe; AAII projects predominantly to the prefrontal region. Similar sequences of connections exist for visual and somatic sensory systems. Several multimodal areas can be found at the junction of these unimodal parasensory association areas, which are situated in the parietotemporal areas, including superior temporal sulcus and inferior parietal lobule.

In summary, unlike the primary sensory areas, which are involved in elementary analysis of sensory input, the functional role of parasensory association areas is more complex. Moreover, progressive complexity in the connection of these areas seems to underline the progressive complexity of functions. These areas include all parts of the neocortex other than regions directly involved in the processing of information: the association areas are concerned with the integration of more than one sensory modality and with planning of movement.

These general considerations may suffice to provide a global understanding of association areas.

3.5 Neurotransmitters

Information concerning the environment, internal states, motives, and behavioral acts is transmitted within the brain by spatiotemporal patterns of neuronal discharge. Communication between neurons takes place by electrochemical transactions, mostly at synaptic junctions. Typically, these junctions are between the axon terminals of the presynaptic neuron and the dendrites of the cell body of the postsynaptic neuron, although other forms of synaptic contact (e.g., axoaxonic) are also present in certain brain regions. Moreover, electrochemical communication between cells may take place through certain forms of interface other than synapses (e.g., ephapses). In any case, the transmission is basically effected across nerve cell membranes by interdependent chemical and electrical changes. Cells produce certain chemical substances called neurotransmitters and neuromodulators that, through specific receptors embedded in pre- and postsynaptic membranes, modify the electrical activity of other nerve cells.

The following section describes several of the best-known transmitters that are involved in functions in both the central and the peripheral nervous systems. It was long thought that a given neuron released only one kind of neurotransmitter. But today, many experiments have shown that a single neuron can produce several different neurotransmitters.

Acetylcholine is a very widely distributed excitatory neurotransmitter that triggers muscle contraction and stimulates the excretion of certain hormones. In the central nervous system, it is involved in wakefulness, attentiveness, anger, aggression, sexuality and thirst, among other things.

Example of Disorder Involving Acetylcholine

Alzheimer's disease is associated with a lack of acetylcholine in certain regions of the brain.

Norepinephrine is a neurotransmitter that is important for attentiveness, emotions, sleeping, dreaming and learning. Norepinephrine is also released as a hormone into the blood, where it causes blood vessels to contract and the heart rate to increase.

Example of Disorder Involving Norepinephrine

Norepinephrine plays a role in mood disorder such as manic depression

Dopamine is an inhibitory neurotransmitter involved in controlling movement and posture. It also modulates mood and plays a central role in positive reinforcement and dependency.

Example of Disorder Involving Dopamine

The loss of dopamine in certain parts of the brain causes the muscle rigidity typical of Parkinson's disease.

GABA (gamma-aminobutyric acid) is an inhibitory neurotransmitter that is very widely distributed in the neurons of the cortex. GABA contributes to motor control, vision and many other cortical functions.

Example of Disorder Involving GABA

Some of the drugs that increase the level of GABA in the brain are used to treat epilepsy and to calm the trembling of people suffering of Huntington disease.

Glutamate is a major excitatory neurotransmitter that is associated with learning and memory.

Example of Disorder Involving Glutamate

It is also thought to be associated with Alzheimer disease, whose first symptoms include memory malfunctions.

Neurons that use GABA and glutamate as neurotransmitters are >80% of the neurons in the brain, and constitute the most important inhibition and excitation systems, respectively, of the substantia nigra and pars compacta.

Neurotransmitters are produced within the body of the nerve cell and transported along the axon to its terminal synaptic vesicles, which in some cases are considerably remote from the soma – for example, catecholamine transmitters generated in certain cells of the lower brainstem are conveyed all the way to the cortex by the axons of those cells. The rate of neurotransmitter synthesis within a given cell is subject to a variety of metabolic factors, but the level of the substance at the terminals is maintained relatively constant.

The designation *neurotransmitter* is used in general for all chemical transmitter substances. According to Fuster (1995a), in recent years, and especially in the context of prefrontal physiology, the distinction has been made between neurotransmitters proper and neuromodulators. Glutamate and GABA are included among the first neurotransmitters, which are characterized by rapid effects on the ion channels that mediate transmission. For example, glutaminergic pyramidal cells of the prefrontal cortex engage in rapid, persistent activity during working memory, while GABA-ergic neurons help to tune the network's firing by inhibiting responses to irrelevant stimuli and memories (Rao et al. 2000). Neuromodulators, on the other hand, include almost all other transmitters (monoamines, acetylcholine, etc., and sometimes also glutamate and GABA when they act on certain receptors). They are relatively slow acting and often involved in changes of general state (e.g., between sleep and wakefulness) or reward value.

In the most typical intercellular transaction, the arrival of an action potential in the axon terminal of a given neuron results in the opening of calcium (Ca^+) channels in the cell's terminal presynaptic membrane, whereby the cation flows into the terminal. The accumulation of intracellular Ca^{2+} promotes the release (exocytosis) of a neurotransmitter substance from the synaptic vesicles of the presynaptic membrane into the extraneuronal synaptic space.

Figure 3.10 schematically illustrates the process for the more common – "classical" – neurotransmitters, as well as the sites or steps in the action of some drugs on their respective synapses.

Neurotransmitter studies are especially relevant to the functions and pathology of the frontal lobes, as explained in Chapters 13 and 22. There is also evidence that the modulation of synaptic transmission in the most plastic brain regions, such as the association areas of the cortex (e.g., the prefrontal cortex), is at the foundation of learning and memory. On the other hand, there is the evidence that certain pathological conditions of neurotransmitter systems are at the foundation of neuropsychiatric disorders. For example, a clear relationship exists between the degeneration of a dopaminergic system in the basal ganglia and Parkinson's disease; and schizophrenia is associated with abnormalities of another dopaminergic system, also involving the basal ganglia and the prefrontal cortex (see references in Chapter 13).

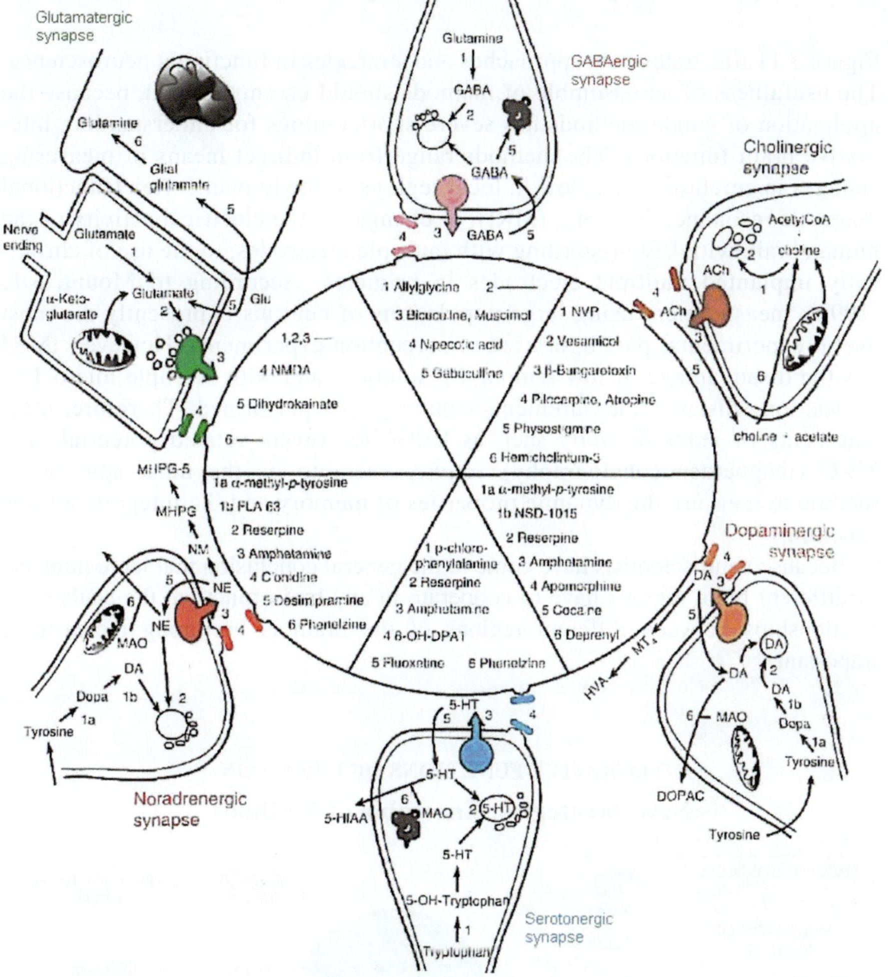

Fig. 3.10 Synaptic processes – depicted in and around synaptic vesicles – for six major neurotransmitters: glutamate, GABA (γ-aminobutyric acid), norepinephrine (NE), dopamine (DA), serotonin (5-HT), and acetylcholine (ACh). Each neurotransmitter undergoes six basic successive processing steps (designated by *number*): (1) synthesis, (2) vesicular uptake, (3) release, (4) receptor binding, (5) reuptake, and (6) inactivation. In the sector of the central circle corresponding to every transmitter, chemical agents are inserted that act at the steps indicated by the *numbers* that precede them. *Abbreviations: DOPAC* 3,4-dihydroxyphenylacetic acid; *5-WIAA* 5-hydroxyindoleacetic acid; *MAO* monoamine oxidase; *NMDA* *N*-methyl-D-aspartate; *NVP* naphthyl vinyl pyridinium ion; *8-OH-DPAT* 8-hydroxydipropylaminotetralin. FLA-63 and NSD-1015 are pharmaceutical code names (from Wilcox and Gonzales (1995), modified, with permission)

3.6 Why the Analysis of EEG Is Important

Figure 3.11 illustrates new approaches and strategies in functional neuroscience. The usefulness of an ensemble of methods should be emphasized, because the application of single methods has severe shortcomings for understanding integrative brain functions. The methods range from indirect means of measuring changes in cerebral blood flow in local regions of the human cortex (functional magnetic resonance imaging, fMRI), or changes in the electrical activity of the human brain with EEG recording with multiple electrodes, to the use of chronically implanted multiple electrodes in primates. According to Mountcastle (1998), measurement using large populations of neurons is presently the most useful experimental paradigm used in perception experiments. However, fMRI has the disadvantage of low temporal resolution, and with multiple microelectrodes, long distance measurements cannot yet be performed. Therefore, measurements of *macro-activity* such as EEG/ERP (event related potential) and MEG (magnetoencephalography) activity seem to be the most appropriate method to measure the dynamic properties of memory and the integrative brain function.

Because neuroscientists have come to the general conclusion that large numbers of different brain regions have to cooperate in any brain function, the analysis of relationships between different regions of the brain is becoming increasingly important.

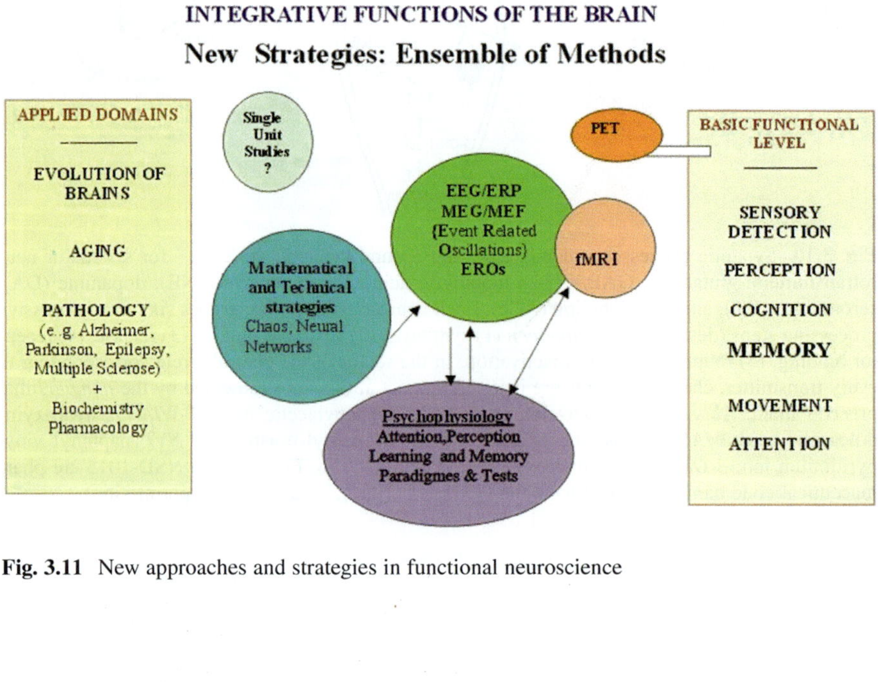

Fig. 3.11 New approaches and strategies in functional neuroscience

In the following section, we will briefly discuss the outcomes of methods and strategies shown in Fig. 3.11. The expression *strategy* refers here to the combined application of several methods, in parallel or sequentially.

1. Studies at the single-cell level have been of great importance in elucidating the basic physiological mechanisms of communication between cells (Eccles 1973; Mountcastle 1998). However, the importance of these studies for understanding integrative brain functions is questionable because, during the integrative processes the whole brain is involved, as Adey (1966, 1989) and Adey et al. (1960) merely underlined, and the new trends in neuroscience clearly emphasize (see also Freeman 1999).
2. Positron emission tomography is an invasive procedure applied to patients. It has large temporal resolution in the range of a half hour and offers no possibility for dynamic measurements at the level of microseconds.
3. The methods incorporating analyses of EEG/ERPs (and especially event related oscillations, EROs) and fMRI provide further excellent strategies to illuminate brain functions, because they cover dynamic changes in the brain and the morphological structure. Magnetoencephalography and study of event related magnetic fields (MEF) greatly increase the spatial resolution in comparison with EEG and ERP. Accordingly, these methods are likely to provide excellent results in future applications.
4. The new strategies are interwoven with the use of relevant mathematical and psychophysiological strategies. These are:
   (a) Mathematical and systems theoretical approaches including in recent decades (1) the concepts of *chaos*, *entropy*; (2) modeling with *neural networks*, interpretation of *frequency domain approach*, new approaches utilizing wavelet analysis and spatial coherence and temporal coherence.
   (b) Psychological strategies with the use of behavioral paradigms and application of neuropsychological tests (Karakaş et al. 2002, 2003).
   (c) An important strategy not included in Fig. 3.11 is recording with chronically implanted intra-cranial electrodes in the animal brain.

To achieve relevant progress in functional neuroscience, it became fundamental to apply several methods together (Freeman 1999). However, the application of all strategies in every laboratory is not yet possible. Figure 3.11 illustrates further the levels of basic CNS-functions (right side) and the applied domains (left side). Functions such as *sensory detection*, *movement*, and *memory* can be successfully analyzed by using individual methods or strategies from several research domains, such as *evolution*, *aging*, *pathology*, and *pharmacology* (use of drugs or pharmacological agents in pathology). The application of combined strategies in all these fields has led to new horizons for understanding the integrative functions of the brain, especially of memory function. The role of memory in the human mind and behavior cannot be overemphasized, because very few aspects of higher nervous function could operate successfully without some memory contribution. Perception, recognition, language, planning, problem solving and decision-making all rely on memory (Damasio and Damasio 1994).

3.7 Some Principles of Biological System Analysis Applied to Brain Research

3.7.1 Why Establish a Program for Brain Research?

In the 1970s, Başar and co-workers tried to determine, first, the dynamics of brain responses in an abstract way, naming this approach "A Program For Biological System Analysis." They then tried to show, based on the existing neurophysiological data, which particular neural responses could give rise to the general transfer functions. The program was extended and modified in 1998, referring to this approach as "Brain Dynamics Research Program" (Başar 1976, 1980, 1998). In the meantime, a large number of laboratories and research groups partially or largely applied some steps, or the global concept, of this program. In addition to the classical analysis tools of general systems theory, this program includes some supplementary experimental methods and *methods of thought* in accordance with the special nature of the living systems. The program has three main classes of methods: (1) abstract methods of general systems theory; (2) specific methods for living systems; (3) methods of thoughts and research principles.

Figure 3.12 illustrates a more advanced version of the Biological Systems Analysis and Brain Dynamics Research Programs, with the methods of thought or research principles being separately displayed. The rationale to develop a research program was based on ways to elucidate the black box (the brain).

The definition of the so-called *black box* indicates that there are three basic quantities involved in biological investigations: input (stimulus), the system, and the output (response). If the stimulus and response are known or are measured variables, it should be possible to estimate the properties of the system (e.g., whole brain). The determination of the abstract frequency characteristics or the transfer function of the system under study usually causes experimental and sometimes also conceptual difficulties. This is partly because of the fact that the parameters measured may change rapidly. On the other hand, the mathematical representations merely help to identify the frequency positions of all components without determining the exact nature of these components. At this stage, the researchers have to elucidate the black box. Because the determination of a system's mathematical characteristics alone does not allow statements about the biophysical nature of the phenomenon, the most difficult challenge was to establish the *biological systems analysis theory*.

The application of this program has been along the same line of thinking as Fessard, Griffith, and Rosen, to develop transfer functions; and of Hebb, to look to long distance coherences in the brain's macroscopic electrical activity.

For the concept of *gray box*, it was noted that investigators within the field of brain studies usually deal with gray boxes (partially elucidated black boxes) and not with completely black boxes: we use the term *gray box* to refer to an apparatus or a system that performs a defined operation, and from which we have some information

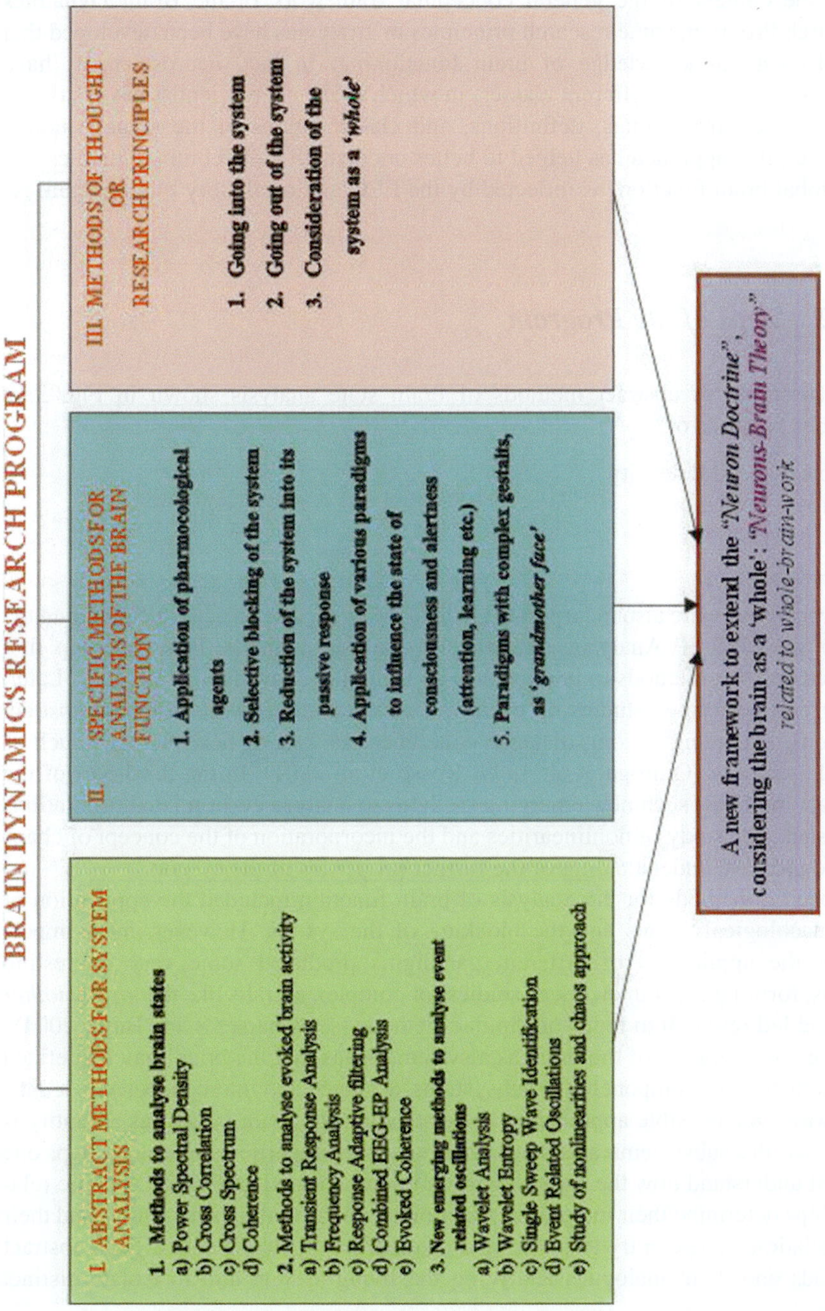

Fig. 3.12 Brain dynamics research program (modified and extended from Başar 1998)

concerning the structure or processes making possible (realizing) the defined operation. In a gray box we have partial information concerning the structure and processes that realize the input-output relationship (Başar 1998).

In the context of the general conceptual framework of the Brain Dynamics Research Program, some research principles or strategies have been developed that can deepen our knowledge of brain functioning. In fact, neuroscientists have divided research into different classes, in which every neuroscientist has his or her own unique surroundings, definitions, and classifications of the signal studied. However, this approach has helped to better understand global brain dynamics and the global brain function, as reflected by the EEG and oscillatory brain responses.

3.7.2 Steps of the Program

The ensemble of abstract methods of brain state analysis shown in Fig. 3.12 includes analyses of:

1. Power spectral density
2. Cross correlation
3. Cross spectrum
4. Coherence

Other methods were also incorporated to analyze evoked brain activity, including the combined EEG-EP Analysis, and wavelet analysis methods. Başar's group first applied abstract methods to brain waves by using the conventional methods. Later, this group performed studies on event related oscillations with all of the abstract methods, extending to long distance coherence, as well as new methods such as wavelet entropy (Quiroga et al. 1999; Rosso et al. 2001). In the third part of the abstract methods, such new emerging methods to analyze event related oscillations are listed. The study of nonlinearities and the incorporation of the concept of chaos aim to increase understanding of the further properties of the system.

Specific methods for the analysis of brain function included the application of pharmacological agents and the blocking of the system. However, most importantly, the application of different paradigms produced some very interesting results, forming the framework of studies for complex gestalts like the *grandmother cell* and led research to focus on similar methods (see Chapter 8 and Başar 2004).

The consideration of the brain as a system means that the brain may be defined as a collection of components or subsystems, arranged and interconnected in a definite way. One possible approach to understanding the brain system as an entity is to isolate the subsystems and study their specific properties. As a next step, one should understand how the subsystems are interconnected and which specific relationships determine their integrative functioning. By revealing subsystems and their interrelations, one can try to model and reconstruct the whole entity. Both abstract methods and their analogues analyzing the living system aim to isolate distinct system components. This approach is informative and defines a research strategy that is generally termed *going into the system*.

The conceptual framework provides us, however, with another, much more important research strategy that cannot be simply realized by any of the analysis methods available, nor by their combined application. This strategy is called *going out of the system*, and is defined as a method of thought. It is well known, for example, that the content of a word is an abstract representation extracting the most essential attributive features from an enormous entity of single concrete objects. In a similar manner, by using the method of thought, one can approach the essential principles in brain functioning by removing specific concrete representations and by extracting from them at the same time common building units. This can be achieved by going out of the system.

Thus, the principle of going out of the system is not only important when comparing two systems or two completely different brains, such as the human brain, octopus brain, or the brain of invertebrates, which includes in fact the comparative physiology and anatomy. This was the essential step undertaken by Darwin when establishing comparative biology. The expression *going out of the system* involves another important comparative level: Comparison of frequency response of the cortex, hippocampus, and other structures in the same brain also gives important information regarding parallel processing, thus contributing to revealing fundamental building blocks. Within this research strategy, one should also consider the interpretation of results obtained at different levels of investigation, such as cellular, structural, and system levels. The concept of Micro Darwinism is described in Chapter 10.

These statements clearly indicate that the brain's macrodynamic function is governed by oscillatory EEG dynamics, providing an important key to understanding brain function.

The grandmother gestalt experiments in Chapter 12 are largely based on the implications of this program. This framework represents an important extension of the neuron doctrine, to considering the brain as a whole.

During studies related to brain function, the Brain Dynamics Research Program's methods provided the conventional tools, as well as continuously developing new principles and applications. The program also reflected the wide spectrum approach of Başar and co-workers, not limiting the research field to a single frequency (e.g., 40 Hz) window, thus facilitating the super-synergy concept. When new methods become available, they are thus applicable within this framework, e.g., the wavelet entropy studies (Quiroga et al. 1999, 2001a; Rosso et al. 2001).

3.7.3 Mathematical Methods of the Program

Mathematical methods can be found in Başar et al. (2002) and Başar (1998). Analysis of chaos, wavelet, and wavelet entropy can be found in the extensive literature. Accordingly, we introduce to the glossary the methods of amplitude frequency characteristics, and coherence, because this method cannot be found in the newest references.

Chapter 4
Autonomous Nervous System, Cardiovascular System, and Smooth Muscles

4.1 Autonomous Nervous System and the Web of Overall Myogenic System

The autonomic nervous system, which regulates what we usually call our innards, is the part of the body linked with the brain. The autonomic nervous system regulates our vital functions without conscious control. We breathe, our heart beats, our stomach digests, and our bladder muscles contract. Further, we secrete saliva, insulin, and digestive enzymes. Our skeletal muscles are able to show vasodilatation and vasoconstriction without our conscious control of them. These functions are operating mainly on structures *hidden* from view. The autonomic system acts on smooth muscle (in the blood vessels and intestines, cardiac muscles and glands). The autonomic system also has afferent pathways, carrying signals from our innards to the brain and spinal cord.

The visceral functions are organized at peripheral level both (1) intrinsically, that is; in the visceral-motor organs or effectors themselves and (2) at neural level, by means of the autonomic visceral ganglia (Fig. 4.1).

As the coming sections will show, the autonomous rhythmic behavior of visceral organs occurs only in certain consistent frequency bands. We call these frequency bands the frequency channels of overall myogenic coordination. According to our results, autonomous visceral activity has an important driving effect on the autonomous control of blood flow in peripheral circulatory organs. The overall myogenic system plays the role of a distributed heart, with its complex pumping action and its control of local blood flow.

Başar and Weiss (1981) introduced the concept of *the Overall Myogenic System*. We want to systemize, quantify, and treat the performance of smooth muscle dynamics of an overall and coordinated system. The overall myogenic system is linked and interwoven with the circulatory system; however, according to its visceral functions, it can also be considered as a separate, parallel system. The overall myogenic system will be schematically explained in Chapter 5.

E. Başar, *Brain-Body-Mind in the Nebulous Cartesian System: A Holistic Approach by Oscillations*, DOI 10.1007/978-1-4419-6136-5_4,

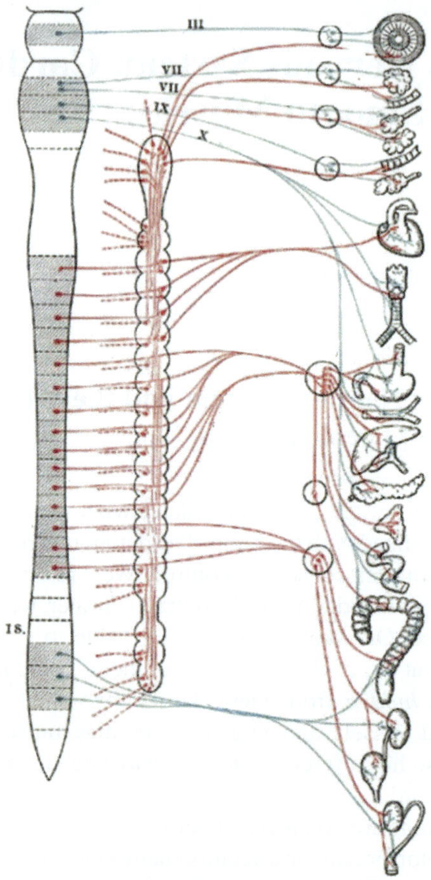

Fig. 4.1 Arrangement of the peripheral autonomic nervous system. *Continuous lines*: preganglionic axons. *Lines dotted at the end*: postganglionic axons. The sympathetic innervation of vessels, sweat glands, and piloerector muscles is not shown

4.2 The Cardiovascular System

The cardiovascular system includes the heart, arteries, capillaries, and veins, each with differentiated functions and structures. It is a transportation system, which delivers to all cells of the living body the materials needed for their proper function. It also carries away the waste products of their metabolism. The circulation provides a means of communication between the cells and the external environment, by brining oxygen and nutritive materials to the cells and relieving them of carbon dioxide and other metabolites (see Fig. 4.2).

It is interesting to note that approximately 79% of the entire blood volume of the body is in the systemic circulation. The heart contains only 9% of the blood and the pulmonary vessels contain 12%. Accordingly, not only the controlled pumping

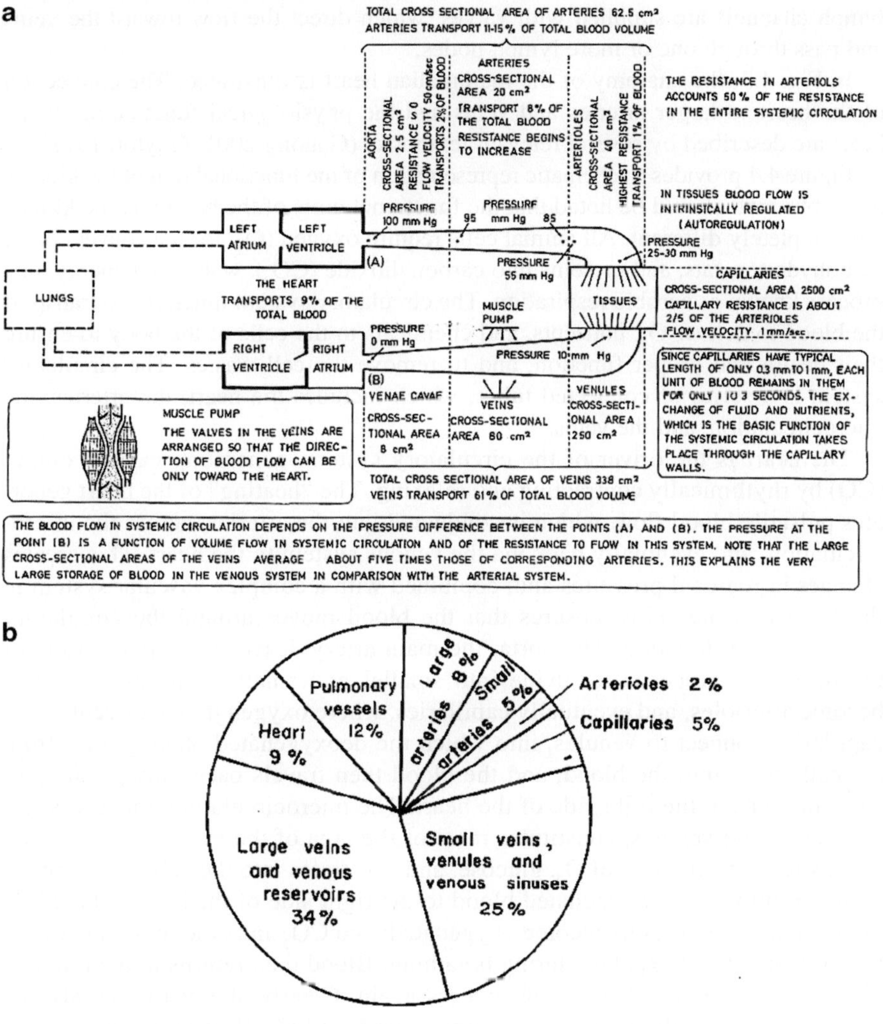

Fig. 4.2 (**a**) A functional scheme of the systemic circulation (from Noyan 1980). (**b**) Distribution of blood flow in the cardiovascular system

action of the heart, but the dynamics of the systemic circulation (or vascular dynamics), merit considerable attention in the study of the regulatory mechanisms of the circulation. Perhaps it would be more appropriate to talk about a second or a "peripheral distributed heart" instead of the vascular system. Guyton (1971) assumed that cardiac output is mostly regulated with intrinsic mechanisms of the regional blood flow.

The vascular system includes not only the arteries, capillaries, and veins of the systemic and pulmonary circulations, but another network called the *lymphatic system*. The fluid contained in this system (the lymph) carries solutes from the extracellular fluid and returns them to the circulating blood. Many of the larger lymph channels are supplied with valves, which direct the flow toward the veins and pass through one or more lymph nodes.

In Fig. 4.3, the anatomy of the mammalian heart is illustrated. The connection of the heart to larger arteries and veins and the physiological functioning of the heart are described by several relevant textbooks (Ganong 2001; Guyton 1971).

Figure 4.4 provides a schematic representation of the functional unit of the kidney, the nephron. It should be noted that the functional units of the heart and the kidney are completely different. All animal cells require oxygen (O_2) for the conversion of carbohydrates, fats, and proteins into carbon dioxide (CO_2), water, and energy, in a process known as aerobic respiration. The circulatory system functions to transport the blood to deliver O_2, nutrients, and chemicals to the cells of the body to ensure their health and proper function, and to remove the cell wastes. The circulatory system is a series of connected tubes, which includes the heart, the arteries, the microcirculation, and the veins.

The heart is the driver of the circulatory system, generating cardiac output (CO) by rhythmically contracting and relaxing. The "beating" of the heart generates pulsatile blood flow, which is conducted into the arteries, across the microcirculation, and eventually back via the venous system to the heart. This creates changes in regional pressures and, combined with a complex vascular system in the heart and the veins, ensures that the blood moves around the circulatory system in one direction. The aorta, the main artery, leaves the left ventricle of the heart and proceeds to divide into smaller and smaller arteries until they become arterioles, and eventually capillaries, where oxygen transfer occurs. The capillaries connect to venules, into which the deoxygenated blood passes from the cells back into the blood, and the blood then travels back through the network of veins to the right side of the heart. The microcirculation, the arterioles, capillaries and venules, constitutes most of the area of the vascular system and is the site of the transfer of O_2, glucose, and substrates into the cells. The venous system returns the deoxygenated blood to the right side of the heart, where it is pumped into the lungs to become oxygenated; and CO_2 and other gaseous wastes are exchanged and expelled during breathing. Blood then returns to the left side of the heart, where it begins the process again. Clearly, the heart, vessels and lungs are all actively involved in maintaining healthy cells and organs, and all influence hemodynamics.

The factors influencing hemodynamics are complex and extensive, but include CO, circulating fluid volume, respiration, vascular diameter and resistance, and blood viscosity.

The blood flow and the pressure in the cardiovascular system are controlled by several submechanisms of the autonomous nervous system.

Autoregulation is a manifestation of local blood flow regulation. It is defined as the intrinsic ability of an organ to maintain a constant blood flow despite changes in perfusion pressure.

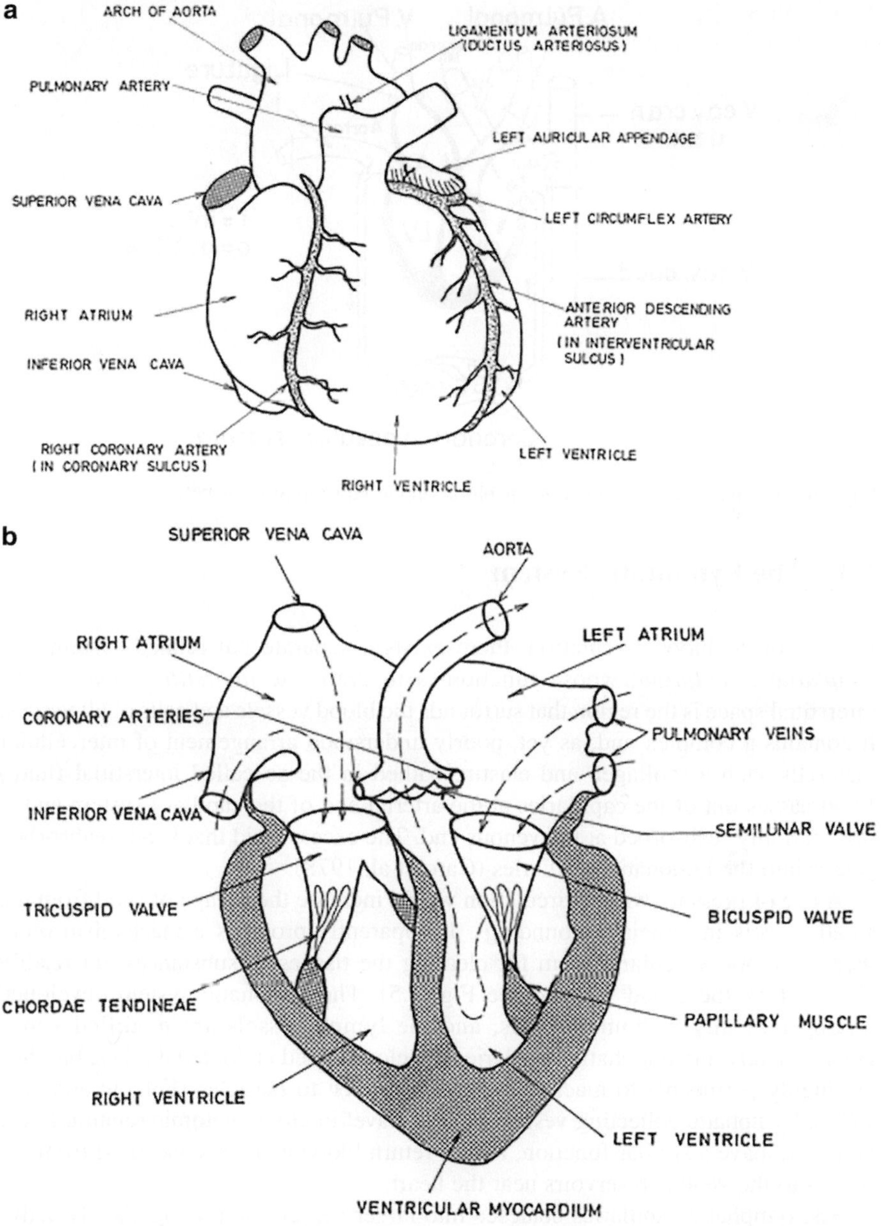

Fig. 4.3 (**a**) Anatomy of the mammalian heart. (**b**) Section through four chambers of heart showing valve structure. *Arrows* indicate flow of blood (except flow from right ventricle through pulmonary artery)

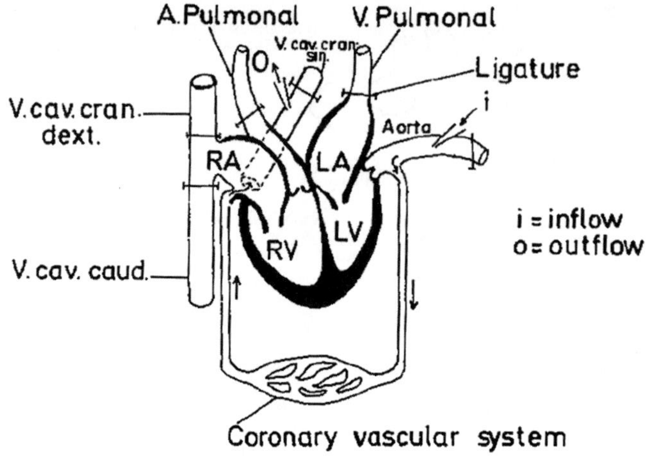

Fig. 4.4 Schematic representation of a nephron (modified from Smith 1959)

4.3 The Lymphatic System

In addition to blood circulation, there exists a separate but related system, the *lymphatic circulation*, whose function is to drain the *interstitial space*. (The interstitial space is the region that surrounds the blood vessels and cells of all tissues. It contains a complex and, as yet, poorly understood arrangement of intercellular materials such as collagen and elastin, bathed in the so called interstitial fluid.) Fluid passes out of the capillaries at the arterial end of the capillary system and is then partially reabsorbed at the venous end. The excess fluid that is not reabsorbed passes into the lymphatic capillaries (Caro et al. 1978).

A rise of pressure within circulation would increase the escape of fluid from the small vessels into their surroundings. It apparently provides a mechanism other then the blood vascular system for clearing the tissues of substances not readily absorbed by the blood vessels (see Fig. 4.5). The lymphatic system developed phylogenetically in homeotherms, and the lymph vessels are modified veins. Histologically, the lymphatic capillaries are close-ended endothelial tubes, but they are highly permeable to macromolecules and even to particles (Folkow and Neil 1971). Lymphatic collecting vessels tend to travel in close anatomic relation to the veins and have a similar function, i.e., to return blood elements and fluid from the tissues to the venous reservoirs near the heart.

The lymphatic capillaries coalesce into larger vessels, and myogenically active smooth muscles begin to appear in the walls of these vessels. Lymph glands are interposed in the course of the larger lymphatics. On reaching the glands, the lymph vessels subdivide into smaller channels, which, entering the gland, open into the sinuses of the lymph node. Fine vessels drain these sinuses and, becoming confluent, reform larger trunks. Figure 4.6 shows a drawing of a lymph node of a dog, which illustrates the afferent and efferent lymphatics. The lymphatic fluid, which is

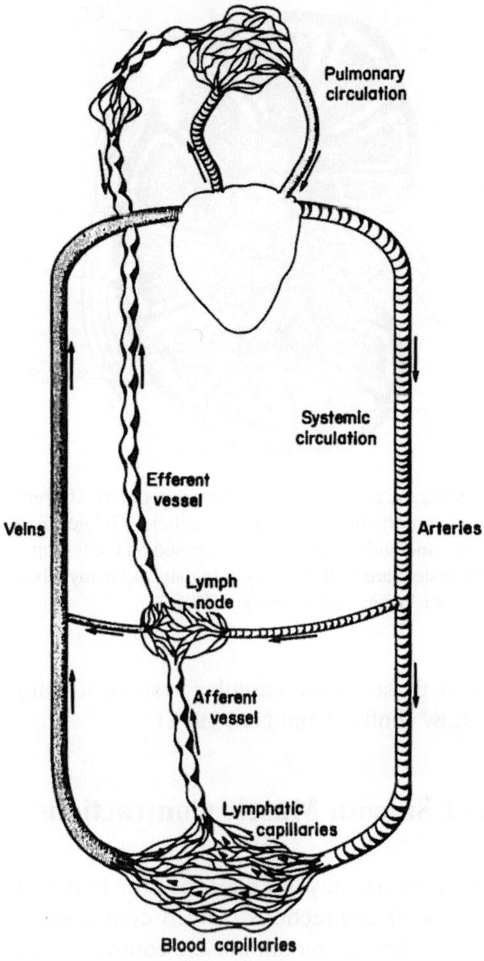

Fig. 4.5 Diagrammatic representation of blood and lymph circulations in the mammal (from Yoffey and Courtice 1970)

conducted to the lymph node with the afferent lymphatic, circulates first into the cortical or subcapsular sinus and, from there, to medullar sinuses into the hilum and into efferent vessels. The lymph nodes, which have a rich blood supply (Lundgren and Wallentin 1964), contain phagocytic cells, which attack and destroy foreign material conveyed to them by the lymph. They also manufacture lymphocytes and plasma cells, and produce antibodies. The lymph nodes form a powerful defense mechanism against invading bacteria.

The two great terminal channels, the right and left thoracic ducts, empty into the right and left subclavian veins, respectively, at their junction with the jugular veins. With a daily lymph flow of 2–4 L, and with lymphatic dimensions that are not

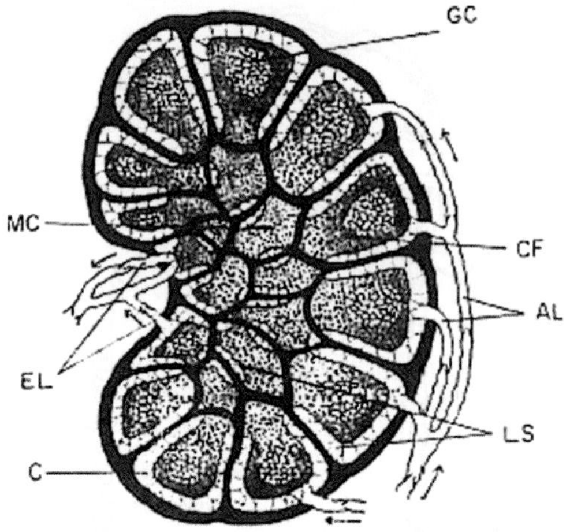

Fig. 4.6 Diagram illustrating the architecture of a lymph node. *AL* afferent lymphatics; *EL* efferent lymphatics; *C* capsule; *LS* lymph sinus; *CF* cortical follicle; *GC* germinal center; *MC* medullary cord. This is a conventional and widely used diagram, based on the lymph node of the dog, in which the capsule and trabecular system are well developed. In man and many laboratory animals, trabeculae are not so conspicuous (from Yoffey and Courtice 1970)

markedly smaller than those of the vascular system, it follows that the flow of lymph is extremely slow (Folkow and Neil 1971).

4.4 Dynamics of Smooth Muscle Contractions

Smooth muscle contractions may be classified in two ways: (1) spontaneous contraction and (2) evoked contractions (forced contractions induced by passive stretch). Accordingly, studies of smooth muscle contractions that give rise to vasomotion in peristalsis and lymphatic flow have necessitated the use of the same conceptual and mathematical tools for both systems.

Spontaneous contractions: Smooth muscles in different organs of the body show intrinsic mechanical activity. These muscles are found in the circulatory system; in the gastrointestinal tract (stomach, small intestine, large intestine); and in the urogenital tract (uterus, urinary bladder, scrotum, penis).

Evoked contractions: Smooth muscles respond to passive mechanical stretching with active contraction (or with an active increase of tension) (Bozler 1947; Burnstock and Prosser 1960; Sparks 1964; Wachholder 1921).

We summarize the basic acting ability of smooth muscle as follows:

1. Smooth muscles show spontaneous mechanical activity.
2. Smooth muscles have spontaneous electrical activity.
3. Smooth muscles respond to a quick stretch with large contraction.
4. Smooth muscles respond to electrical stimulation with contraction.

5. Smooth muscles respond to pharmacological chemical or hormonal stimulation with contraction or relaxation.
6. Smooth muscles play important effector roles in the coordination of the circulation, water regulation, the urogenital mechanism, and the digestive system.

Figure 4.7 shows a conventional experimental setup to measure smooth muscle contractions.

Figure 4.8 shows all possible patterns of the spontaneous tension changes of the guinea pig portal vein. Figure 4.9 shows typical tension changes of the guinea pig stomach.

Figure 4.10 shows typical power spectra of spontaneous contractions (peristalsis) of the guinea pig stomach.

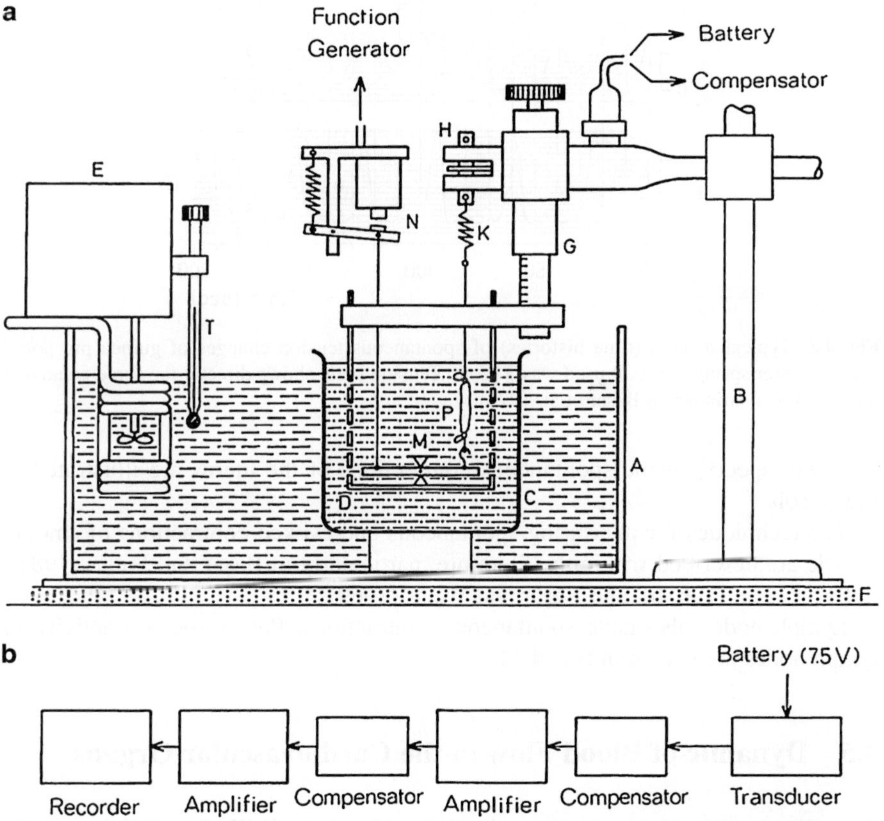

Fig. 4.7 (**a**) Experimental setup to record spontaneous mechanical tension changes of smooth muscle. (A) Aquarium filled with water, (B) support, (C) muscle bath, (D) Krebs solution, (E) thermoregulator, (G) device used to apply passive stretch to muscle, (H) force transducer, (K) spring, (M) device used to apply mechanical tension to muscle, (N) electromagnetic device used to apply mechanical step function to muscle, (P) isolated smooth muscle, (T) thermometer. (**b**) Schematic diagram of the experimental setup in (**a**) (from Eroğlu 1974)

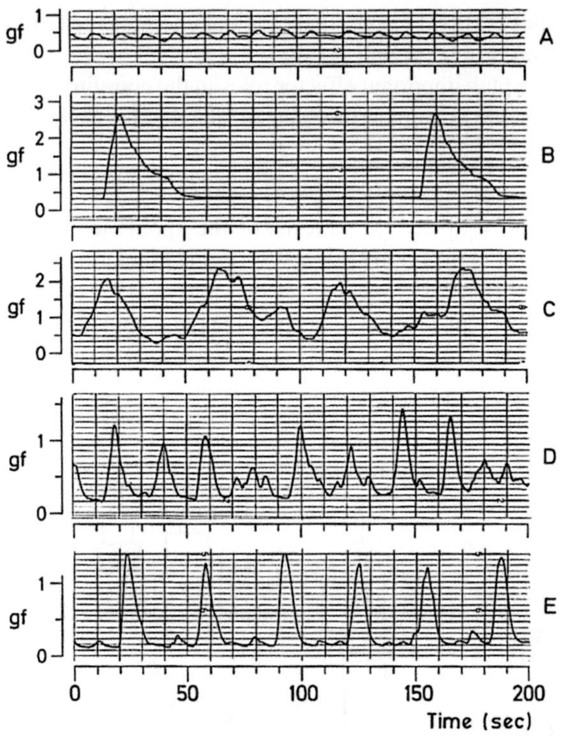

Fig. 4.8 Typical records (time histories) of spontaneous tension changes of guinea pig portal vein. Transfer spring constant used was 3 g/cm. Tension scale shows directly the force created in the portal vein strip (from Başar et al. 1974)

Power spectra of the spontaneous contractions of the uterus are illustrated in Fig. 4.10b.

The techniques for measuring spontaneous and evoked contractions of smooth muscle are described within the literature, particularly in Başar and Weiss (1981), in Başar et al. (1968b), and Başar and Weiss (1969).

Lymph nodes also have spontaneous contractions. Power spectral activity of lymph nodes are shown in Fig. 4.11.

4.5 Dynamic of Blood Flow in the Cardiovascular Organs

An experimental setup to measure circulatory dynamics is illustrated in Fig. 4.12.

Since the first theoretical studies on the analysis of local circulation systems, several experiments have been performed with various organ preparations, including organs in situ and the dynamics of neural control of the vascular bed. All the results depict maximum vascular resistance in the range 0.02–0.01 Hz. These results will be reviewed in the following sections.

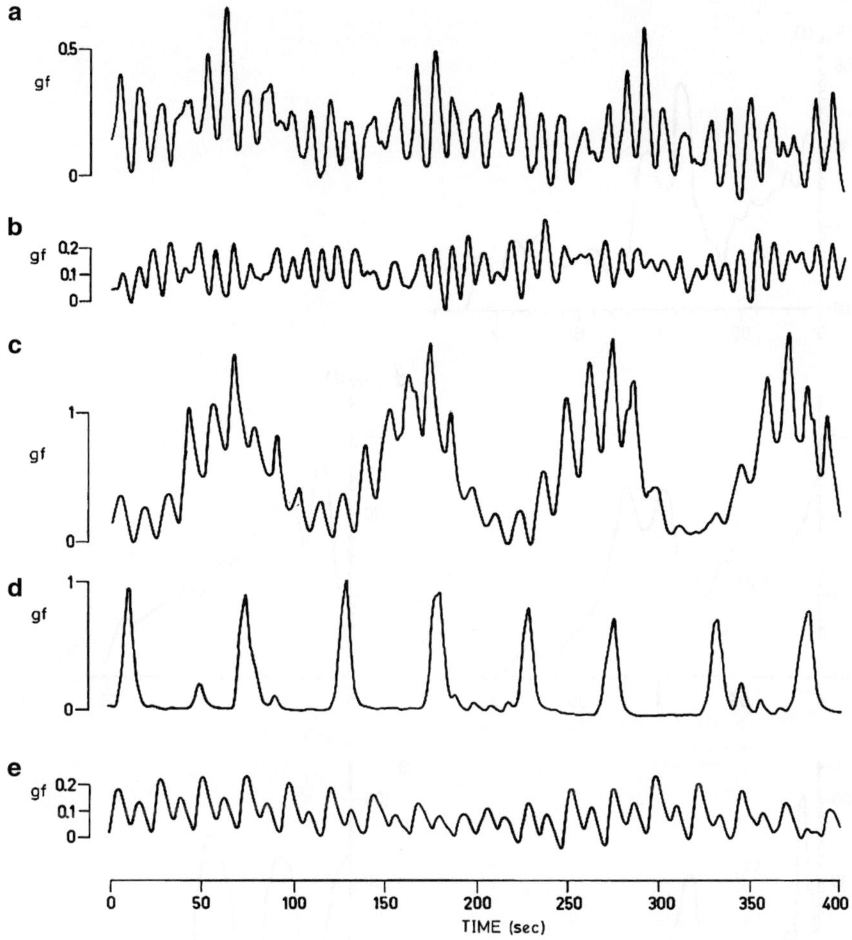

Fig. 4.9 Typical records of spontaneous tension changes of guinea pig stomach. Transfer spring constant used was 3 g/cm. Tension scale shows directly the force created in the stomach strip (from Başar-Eroğlu et al. 1979)

4.5.1 Dynamics of the Arterial Impedance of the Ascending Aorta

Taylor (1966) investigated the arterial impedance in dogs in the low frequency range of 0.00125–0.125 Hz, using vagal stimulation with random bursts of stimulation to produce varying periods of reduced aortic flow. The vascular resistance (the aortic impedance) was greatest in the frequency range between 0.01 and 0.05 Hz (Fig. 4.13).

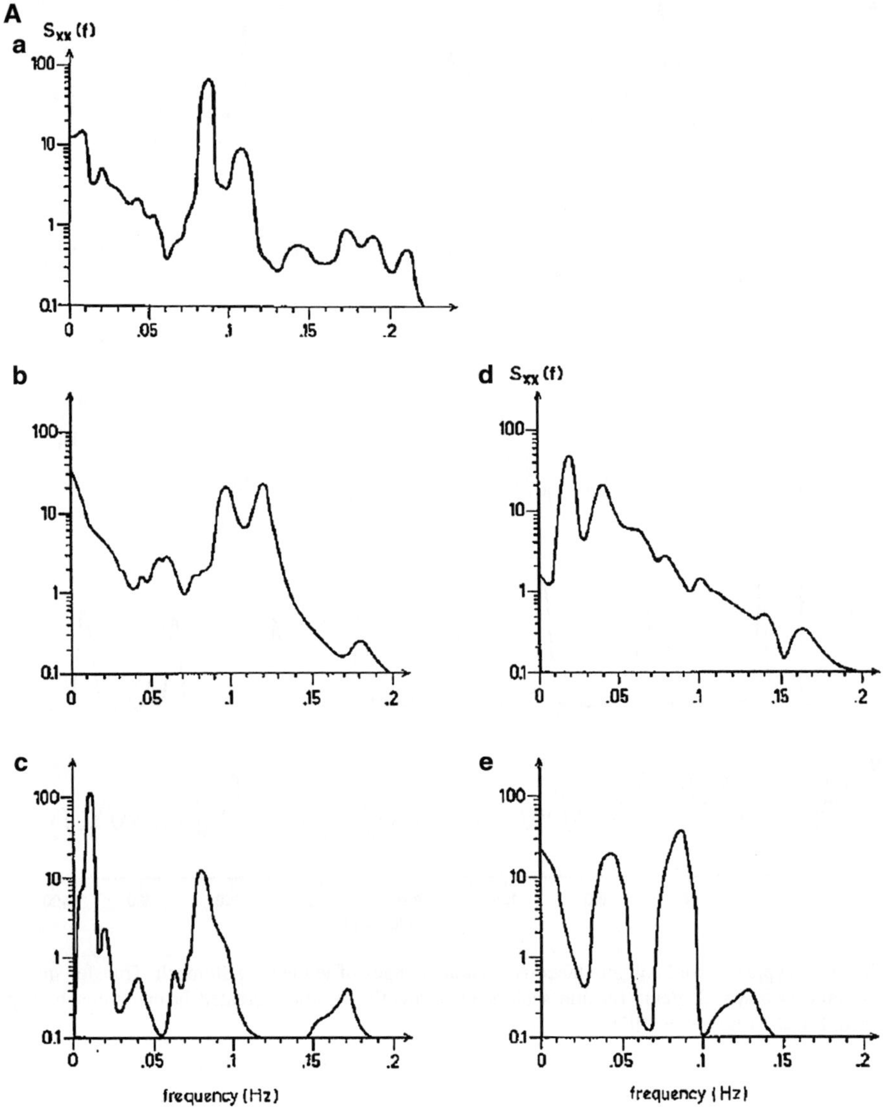

Fig. 4.10 (**A**) Typical power spectra of the spontaneous tension changes of the stomach. Curves a–e were obtained using the spontaneous contraction patterns of Fig. A.5 (a–e), respectively. Along the abscissa is the frequency in Hz; along the ordinate is the power spectral density $S_{xx} f$ in relative units (from Başar-Eroğlu et al. 1981).

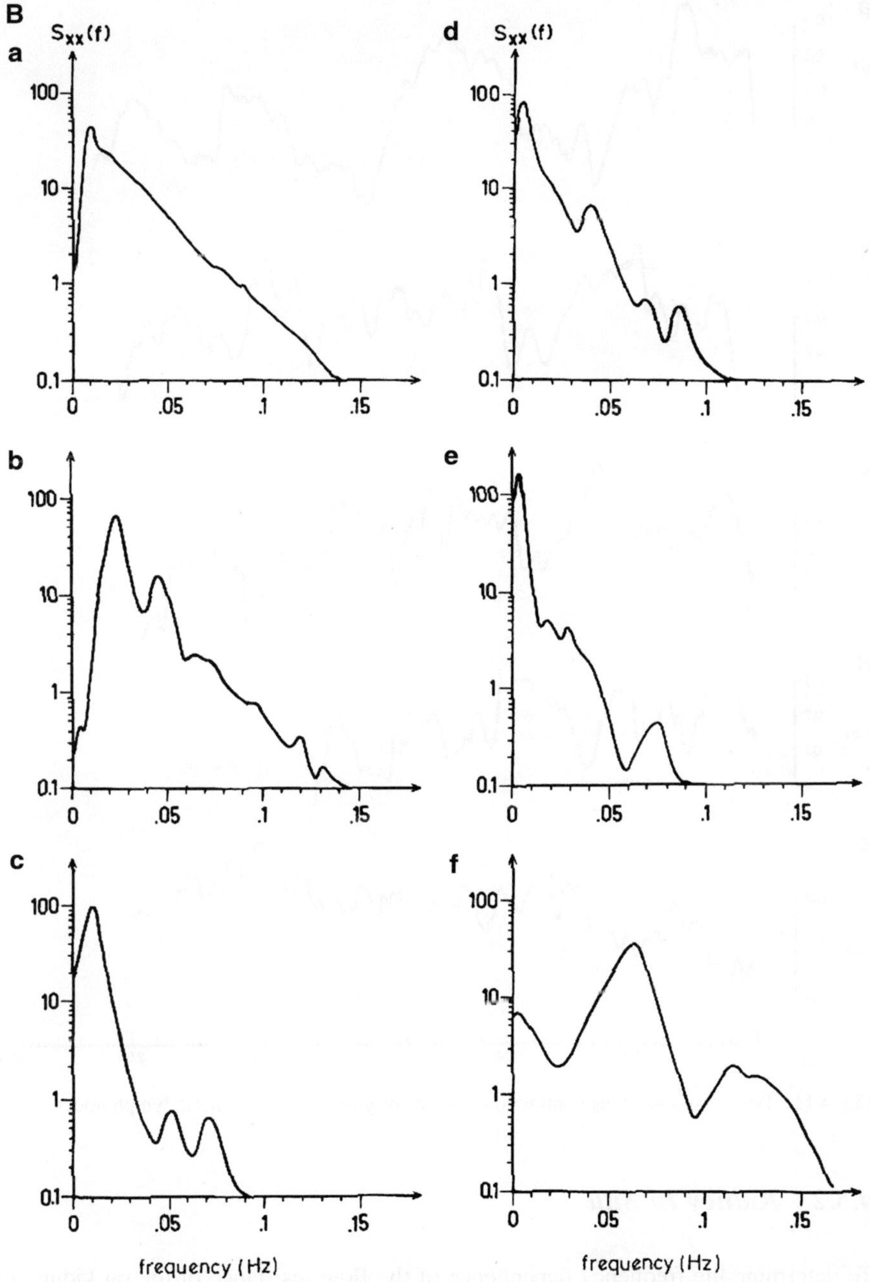

Fig. 4.10 (continued) (**B**) Typical power spectra of the spontaneous tension changes of the uterus. Curves a–f were obtained using the spontaneous contraction patterns of Fig. A.12 (a–f), respectively. Along the abscissa is the frequency in Hz; along the ordinate is the power spectral density $S_{xx}f$ in relative units (from Tümer 1980).

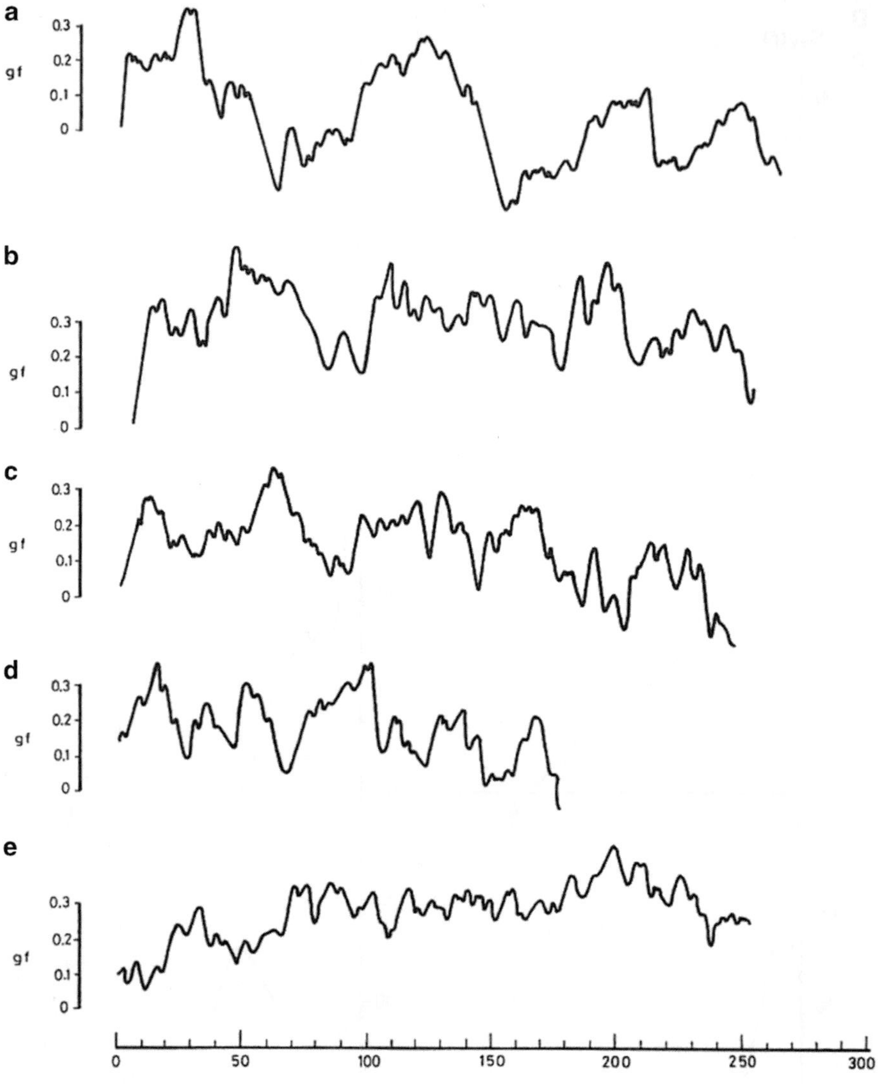

Fig. 4.11 Typical spontaneous contraction patterns of guinea pig mesenteric lymph nodes

4.5.2 *Kidney In Situ*

To determine the frequency dependence of the flow resistance of the rat kidney in the living animal, stochastic pressure changes of low amplitude were superimposed on the perfusion pressure of the renal artery (Eggert et al. 1979). Similar to the experiments of Başar et al. (1968a), these investigators considered perfusion pressure as the input signal and venous blood flow as the output signal. The venous blood

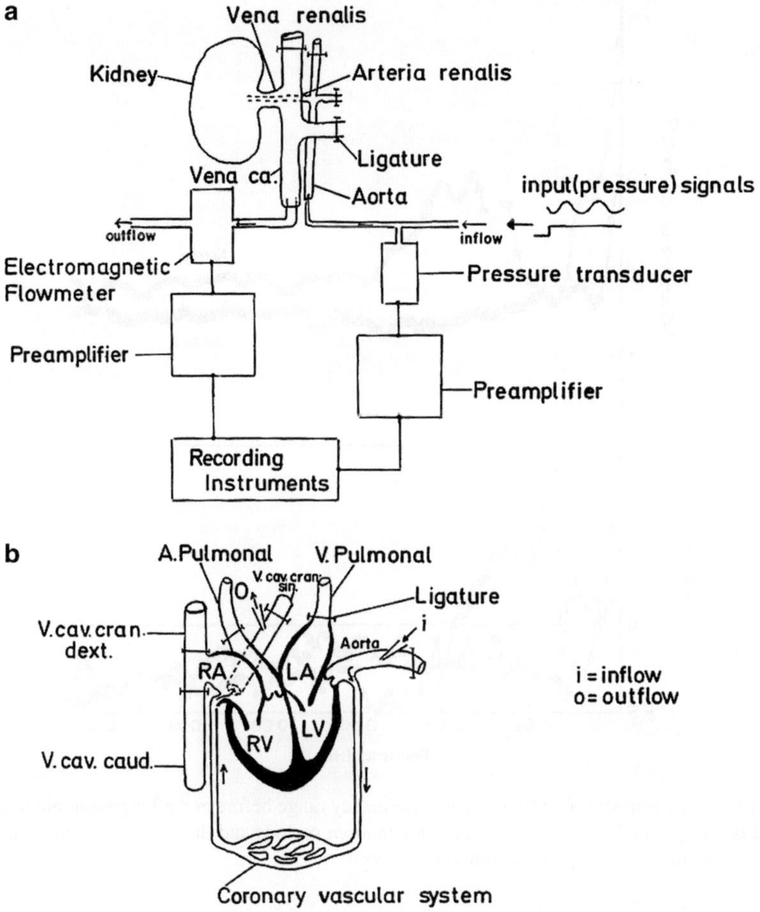

Fig. 4.12 (**a**) Schematic drawing of experimental setup with isolated kidney preparation. (**b**) Schematic drawing of the isolated rat heart. During experiments on the coronary system, the heart preparation replaces the kidney preparation in the experimental setup shown above (from Başar 1974)

flow (output signal) was cross-correlated with the input noise (stochastic signals). Figure 4.12 shows an original registration of blood pressure and volume flow.

4.5.3 Intestinal Vasculature In Situ

Lutz (1966, 1978) studied the frequency response of the vascular resistance of intestinal vasculature in situ to rhythmic venous pressure changes. Under sinusoidal pressure stimulation at the venous side, this author has observed a maximum of vascular resistance around 0.015 Hz.

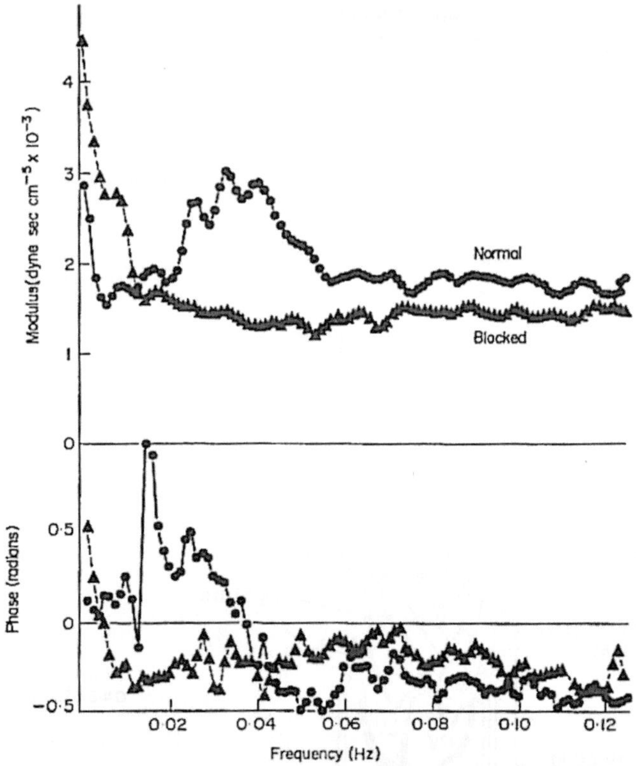

Fig. 4.13 Aortic impedance in the ultra-low frequency range before and after ganglionic blockade, studied in a dog. The heart rate was altered for random periods and the resulting changes in aortic flow and pressure were analyzed (from Taylor 1966)

4.5.4 Dynamics of Arterial Impedance and Flow

Taylor (1966) investigated the arterial impedance in dogs in the low frequency range of 0.00125 Hz to 0.125 Hz, using vagal stimulation with random bursts of stimulation to produce varying periods of aortic flow. The frequency characteristics of the aortic flow resistance were obtained with statistical systems theory methods. The vascular resistance (the aortic impedance) reached its highest values in the frequency range between 0.01 Hz and 0.05 Hz (Fig. 4.13). Note the striking similarity to the curves in Fig. 4.16. Spelman and Pinter (1978) measured the effects of pressure in iliac blood flow and iliac blood pressure in the unanesthetized baboon. The iliac flow-frequency characteristics were computed within the frequency range 0.007–0.1 Hz, under conditions of controlled iliac blood flow and controlled iliac blood pressure. The frequency characteristics of flow showed the following features (Fig. 4.14).

The presence of a resonant frequency at 0.02 Hz is clear, according to these authors, because the phase of the flow was at maximum at that frequency, whereas

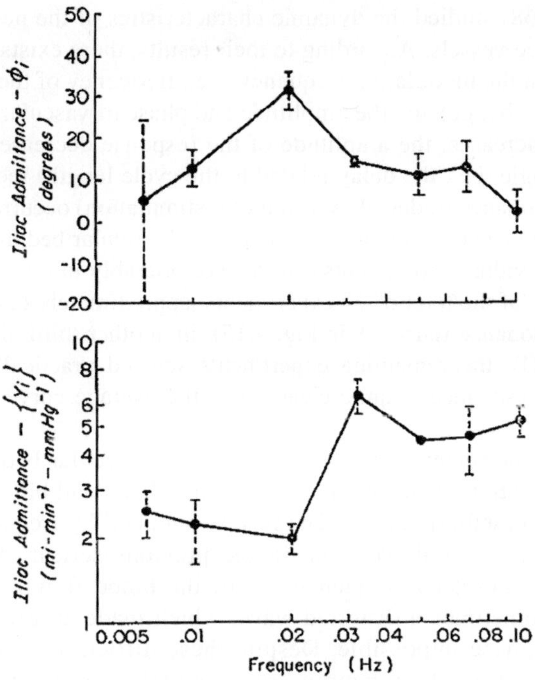

Fig. 4.14 Frequency characteristics of flow in the iliac artery between 0.007 and 0.1 Hz. Flow was controlled. *Upper curve* is the phase characteristic; *lower curve* is the amplitude characteristic. The data are the means of several cycles, and the *bars* are the standard deviations about the means (from Spelman and Pinter 1978)

the amplitude of flow showed a minimum. Because the measurements were performed in the intact and unanesthetized animal, it can be assumed that the central nervous system was able to exercise its full range of control on the peripheral vasculature. However, the data of Spelman and Pinter (1978) argue against the CNS mechanism, and are consistent with changes caused by the local circulation.

According to the discussion by Spelman and Pinter (1978), the minimum vascular flow observed around 0.02 Hz is not because of the influence of the CNS.

4.5.5 Dynamics of the Neural Control of the Vascular Bed

The dynamic aspects of neural control of the vessels can be studied by stimulating the vasomotor nerve innervating a vascular bed with sinusoidally modulated stimulus frequencies and relating the observed changes in resistance or capacitance to the stimulating frequency used (Sagawa 1972). The normal nervous control of vasomotor tone acts through both frequency modulation of the impulse train in individual fibers and recruitment of different fibers. Using this method

Penáz et al. (1968) studied the dynamic characteristics of the nervous control of various resistance vessels. According to their results, there exists a constant relationship between the modulating frequency (i.e., frequency of the stimulated sine wave vasomotor change) and the amplitude and phase of vascular response; when the frequency increases, the amplitude of the response decreases and the phase lag (or phase angle, i.e., the delay related to the cycle length) increases.

A form of resonance (induced by vasomotor stimulation) occurred in most of the amplitude-frequency characteristics of the femoral vascular bed that are illustrated in Fig. 4.15. Individual experiments differed considerably in the magnitude of this resonance peak: Of the total of 20 experiments, approximately one third showed a very marked resonance (curve A in Fig. 4.15); in another third, the peak was less evident (curve B); the remaining experiments showed practically no resonance (curve C). The resonance is quite clear-cut in the average curve plotted from all experiments.

Analogous experiments have been made on the vascular bed of the intestine in cats; the left major splanchnic nerve was stimulated and blood flow recorded in the superior mesenteric artery. The results obtained by frequency-modulated stimulation were less uniform than in the previous series: In some records, irregularities or spontaneous fluctuations of the blood flow occurred, and the response was sometimes a distorted wave, which made an exact estimation of amplitude and phase impossible. Despite these difficulties, after elimination of unsuitable records, the remaining records yield sufficient data to plot the frequency response characteristics (Penáz et al. 1968). The characteristics are shown in Fig. 4.16. The resonance peak is present here again, and is perhaps still more marked than in the previous series. (The incision in the middle of the peak is statistically insignificant.)

It is interesting to note that the vascular resistance in the femoral artery and in the mesenteric artery reach maximal values at vasomotor stimulation frequencies of 0.01–0.02 Hz. The first consequence of this finding can be formulated as follows: The frequency sensitivity of the smooth muscle vascular effector seems to be independent of whether the stimulation is electrical or mechanical vascular resistance

4.5.6 Renal Vascular Resistance

In this chapter we have described the dynamics of organ volume (i.e., of the vascular resistance) for a variety of circulatory organs such as the kidneys, coronary system, mesenterial bed, iliac artery, etc. We have seen that, in all these circulatory systems, the pressure-induced vascular resistance reached maximal values in a frequency range between 0.008 and 0.1 Hz. In most of the experiments, the peak of the vascular resistance (or the maximum of volume flow) was centered at a value around 0.02 Hz. Başar et al. (1968a) previously pointed out the possibility of several peaks of vascular resistance in the frequency range of 0.01–0.5 Hz. However, a rigorous statement could not be made, because the data evaluation was

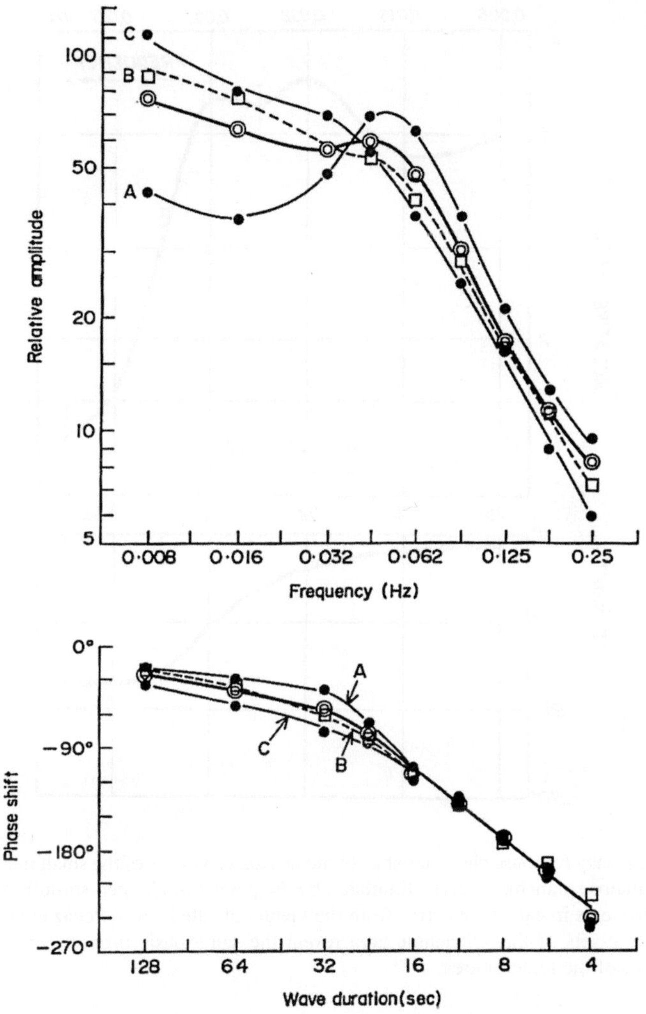

Fig. 4.15 Frequency response characteristics of the efferent vasoconstrictor control of resistance vessels of the femoral vascular bed in rabbits. *Heavy line*, average curve from 20 experiments; A, B, and C, averages of 7, 7, and 6 experiments, respectively, showing a different degree of resonance (from Penáz et al. 1968).

carried out by means of IBM punch cards after visual inspection of the flow results. Accordingly, the number of read (and punched data) points was restricted, thus leading to a limited frequency resolution. The fact that the stretch evoked frequency characteristics of the vascular smooth muscle showed several maxima of force in the frequency range between 0.001 and 0.2 Hz (Başar and Eroğlu 1976) stimulated Başar-Eroğlu et al. (1979) to a new investigation with the help of an on-line computer. They analyzed the pressure-induced frequency characteristics of the isolated rat kidney by evaluating the flow responses to pressure steps above 100 mmHg. In this way, the recording of data was practically unlimited,

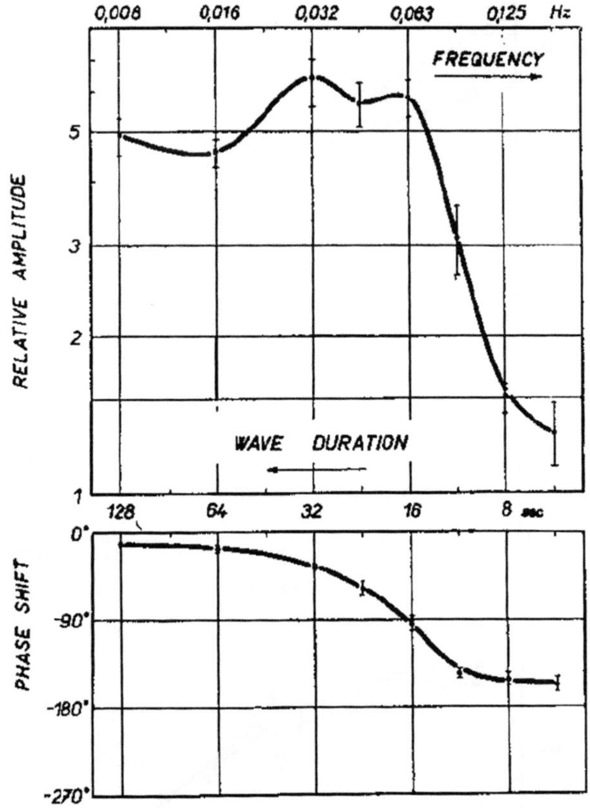

Fig. 4.16 Frequency response characteristics of the resistance vessels of the small intestine; efferent fibers of the major splanchnic nerve stimulated by frequency-modulated stimuli. Average curve from 20 experiments in cats; mean error from the mean indicated (from Penáz et al. 1968). Thus, the resistance vessels of the splanchnic area reveal the same basic property of a second-order system as those of the femoral area.

thus allowing a higher frequency resolution. Figure 4.17 illustrates 4 examples of 13 evaluated experiments. Along the abscissa is the frequency in logarithmic units. Along the ordinate is the vascular resistance (i.e., $P(\acute{\omega})/V(\acute{\omega})$), which is carried out instead of flow, again in relative units and decibels. In this evaluation, the vascular resistances are given directly, to allow comparison with the forced contraction of the vascular smooth muscle. The amplitude-frequency characteristics of the kidney vascular resistance usually depicted 3 or 4 maxima within the frequency range 0.008–0.5 Hz (Fig. 4.17)

The mean value of the frequency-dependent vascular resistance experiments is given in Figs. 4.17 and 4.18, together with mean value amplitude-frequency characteristics of the smooth muscles such as the aorta and portal vein on passive stretch.

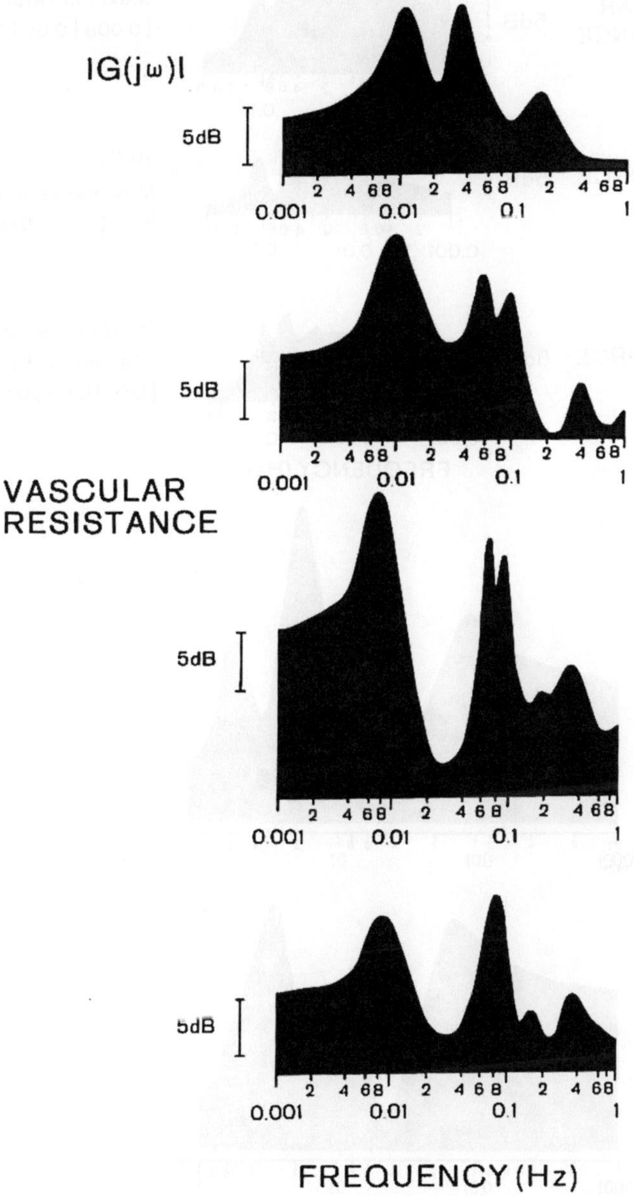

Fig. 4.17 Typical amplitude-frequency characteristics of vascular resistance in the kidney. Along the abscissa is the frequency in logarithmic units. Along the ordinate is the vascular resistance ($P(\acute{\omega})/V(\acute{\omega})$) in relative units and decibels (from Başar and Weiss, 1981)

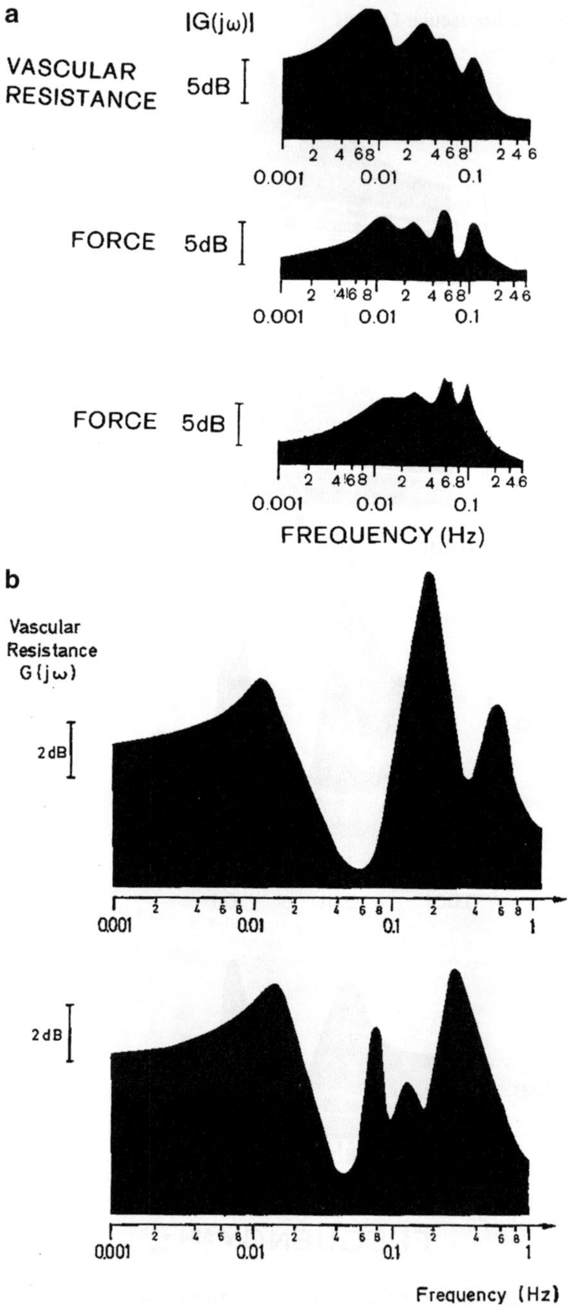

Fig. 4.18 (**a**) Mean value amplitude-frequency characteristics of aorta and portal vein (forced oscillations). Along the abscissa is the frequency in logarithmic scale. Along the ordinate is the amplitude of the tension in relative units and decibels (from Başar-Eroğlu et al. 1969). (**b**) Two typical examples of the amplitude-frequency characteristics of the coronary vascular resistance. Along the ordinate is the vascular resistance ($P(\dot{\omega})/V(\dot{\omega})$) in relative units and decibels. Along the abscissa is the frequency in logarithmic units

The pressure evoked vascular resistance to flow (evaluated from the volume flow in the auto-regulating kidney) has exactly the same selectivities as the contractile forces evoked in the vascular smooth muscle upon stretch. The first interpretation of the results is as follows: The dynamics of local flow regulation are dominated by the dynamics of the smooth muscle effector.

4.5.7 *Coronary Vascular Resistance*

Presented here are preliminary results of studies on the amplitude-frequency characteristics of the coronary system of the rat heart, which were recently started with the help of an on-line computer. Figure 4.18b shows two typical examples of the amplitude-frequency characteristics of the coronary vascular resistance, obtained with the same methodology for isolated kidneys. The first results on the fine structure of the frequency characteristics show that, in fact, the coronary vascular resistance has maximum amplitude at around 0.01, 0.06–0.08, and 0.1–0.4 Hz. The fluctuations are larger, and the peaks less stable, in frequency characteristic curves

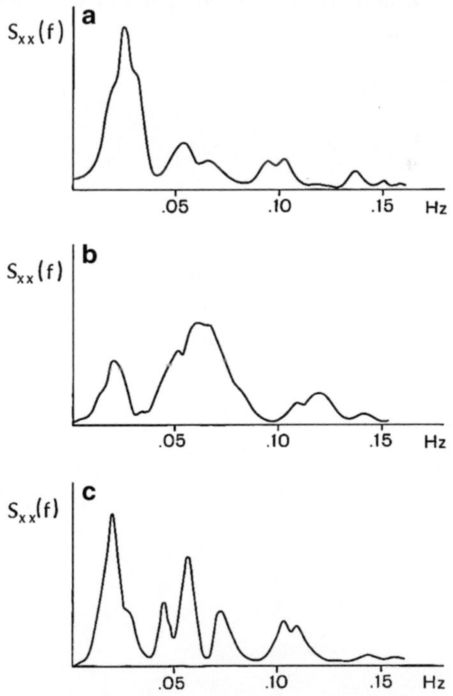

Fig. 4.19 Power spectra of flow auto-oscillations of microflow in the kidney (from Eggert and Weiss 1980)

obtained from the heart experiments in comparison with the kidney experiments. The less favorable experimental conditions caused by spontaneous fluttering of the heart are a serious obstacle to determining the frequency characteristics with the same exactitude as in kidney experiments. However, we can state that the coronary system of the rat heart has similar fine structure to the vascular resistance as in the kidney, i.e., it shows accordance with smooth muscle tension characteristics.

4.5.8 Dynamics of the Microcirculation of the Kidney

At the beginning of the present chapter, we have described the dynamics of flow in various isolated and in situ organ preparations. The results indicated myogenic reactivity components with similar time and frequency characteristics, thus demonstrating the importance of the myogenic structure of circulatory local control mechanism. Having studied the dynamics of flow in macrocirculation, we want to compare the results of studies at the microcirculation level, and determine whether the assumptions and principles derived from the studies of macroflow also hold at the microcirculatory level. Figure 4.19 presents frequency distribution of the microflow in the renal cortex. It is to note that dynamics of the microflow in the kidney shows the same spectral composition in comparison with the macro flow of the isolated kidney preparation.

Chapter 5
Overall Myogenic-Coordination: Building Stones in the Whole-Body Integration and Tuning

5.1 Interim Synthesis of Overall Myogenic Control of Flow and Vascular Resistance

A comparative presentation of amplitude-frequency characteristics in the kidney (macro flow and micro flow) and the coronary vascular system is shown in Fig. 5.1. The right side of the illustration (b) shows the contractions of the aorta, lymph node, and stomach on stretch increase. In the amplitude characteristics of the circulatory organs, the vascular resistance is illustrated, but not the flow characteristics. The comparison of all theses curves indicates that vascular resistance at the microscopic and macroscopic levels, and the contraction properties of the aorta, lymph nodes, and the stomach, show good accordance. In other words, the contractility of the vascular bed, large arteries such as aorta, large veins such as the portal vein, lymph nodes, peristaltic organs such as the stomach, intestines, and uterus all show the same resonant properties.

The pushing of the digested mass from the stomach to the intestine, the pushing of blood flow to the lymphatic system, and the contractility of the coronary system and kidneys all occur in the same frequency band of the overall myogenic system. According to these facts, an integrative myogenic control system is demonstrated in the body. Certainly tuning of frequency bands and coherences are only globally manifested; however, this provides a solid working and communicating framework for functioning of the organs of the vegetative system. This will be tentatively explained in the forthcoming section.

5.2 The Overall Myogenic System

Başar and Weiss (1981) measured and reviewed mechanisms of control, auto-oscillations of blood flow, contractility of the vasculature, forced oscillations in the peripheral circulatory system, spectral activity of peristaltic organs, and dynamics of the lymph nodes and the lymphatic system. They found that all of these subsystems

E. Başar, *Brain-Body-Mind in the Nebulous Cartesian System: A Holistic Approach by Oscillations*, DOI 10.1007/978-1-4419-6136-5_5,

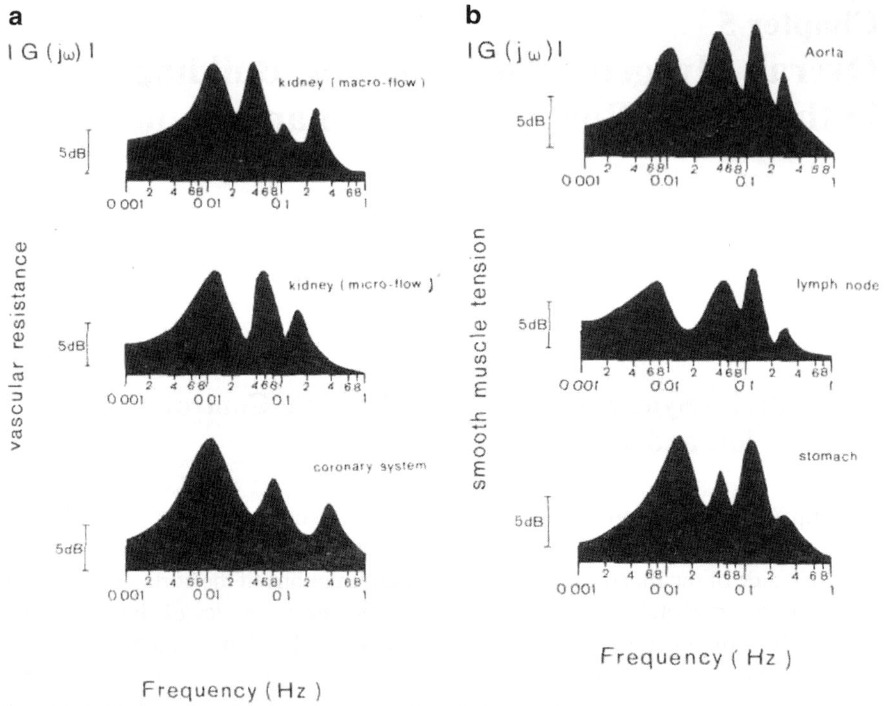

Fig. 5.1 Comparison of amplitude-frequency characteristics in various organ preparations. (**a**) *Left side:* vascular resistance in relative units and decibels is shown along the ordinate *h*. The frequency in logarithmic scale is along the ordinate. (**b**) *Right side:* smooth muscle tension in relative units and decibels is shown along the ordinate. Frequency in logarithmic scale is along the abscissa. Note the striking resemblance (and/or congruence) of flow and resistance characteristics (from Başar et al. 1980)

showed a resonant property of oscillatory activity in the ultraslow frequency range of 0.01, 0.04–0.06, and 0.1 Hz These authors introduced the concept of the *overall myogenic system* after quantifying and treating the performance of smooth muscle dynamics, by considering this as the dynamics of an overall and coordinated dynamic system. The overall myogenic system incorporates:

1. The *vascular system*, with all the arteries, arterioles, etc. in the systemic circulation.
2. The *lymphatic system*, with lymphatic vessels and nodes.
3. The *visceral system*, which performs the visceral functions of the vegetative system and peristalsis in vegetative function (see Fig. 5.2).
4. The smooth muscle cells are building blocks and basic effectors of the overall myogenic system. (See Fig. 5 showing the organization of the overall myogenic system, and schematically describing the effects of overall myogenic control.) According to the measurements of Başar and Weiss (1981) the overall myogenic coordination occurs

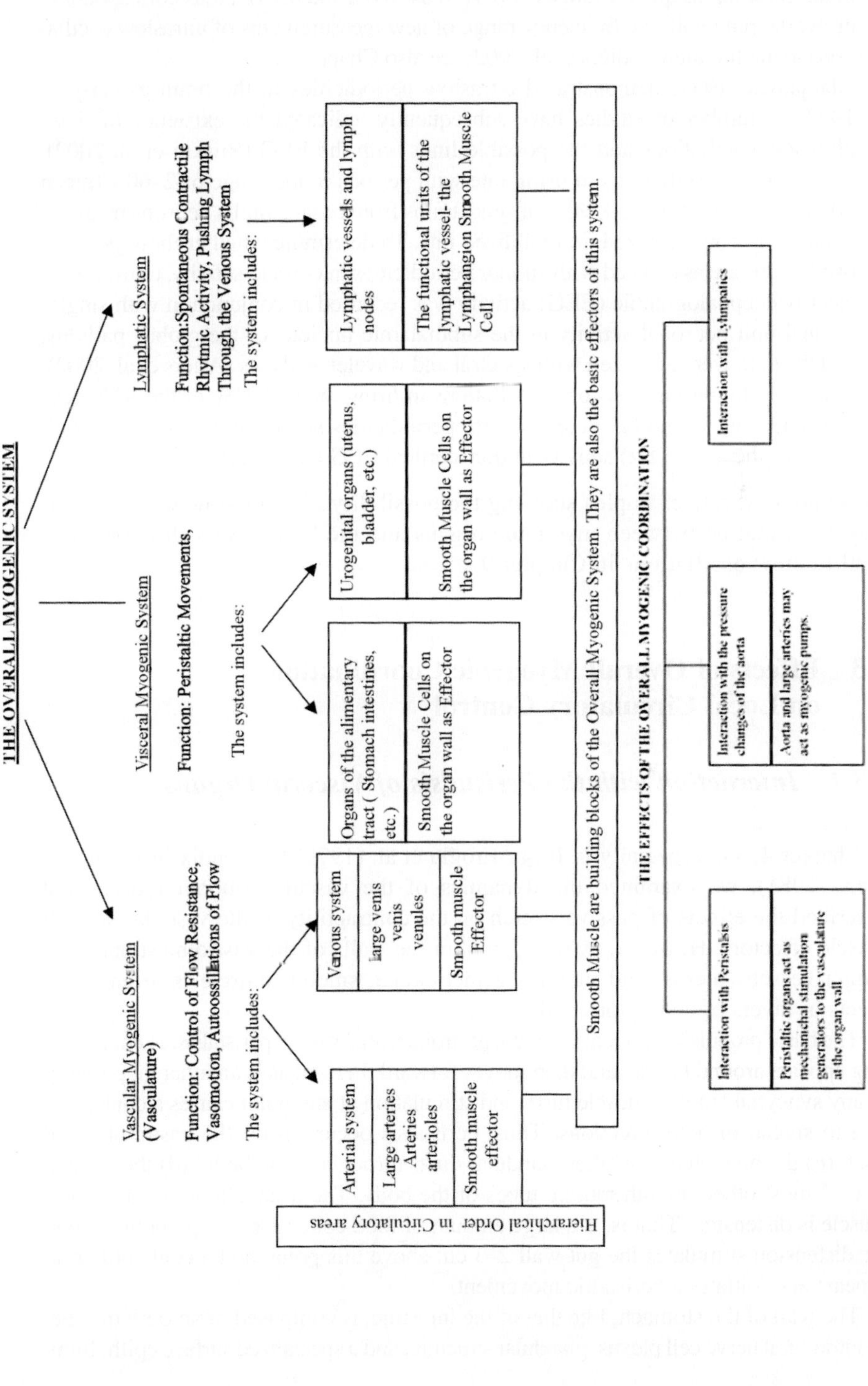

Fig. 5.2 **(a)** Schematic presentation of the overall myogenic system. **(b)** Schematic presentation of the effects of the overall myogenic system

in the common frequency ranges of 0.01, 0.04–0.05, and 0.1 Hz, thus corresponding or overlapping with the frequency range of new measurements of ultraslow oscillations in the literature (Allers et al. 2002; see also Chapter 9).

5. Aladjalova (1957) demonstrated ultraslow periodicities in the brain as early as 1957. A number of studies have subsequently indicated the existence of these ultraslow oscillations and the possible links with the EEG (Ruskin et al. 2002). Multi-second oscillations in firing rate with periods in the range of 2–60 s (mean 20–35 s) are present in 50–90% of spike trains from basal ganglia neurons recorded from locally anesthetized, immobilized rats. To determine whether these periodic oscillations are associated with similar periodicities in cortical activity, transcortical electroencephalographic (EEG) activity was recorded in conjunction with single- or dual-unit neuronal activity in the subthalamic nucleus or the globus pallidus, and the data were analyzed with spectral and wavelet analyses (Allers et al. 2002). The study found multi-second oscillations in firing rates of 31% of the STN neurons and 46% of the GP neurons, with periodicities significantly correlated with bursts of theta (4–7 Hz) activity in transcortical EEG recordings.

These are concrete examples showing the possibility of ultra-slow wave oscillatory coordination between myogenic organs and the brain. We will come back to ultra-slow oscillations in Chapter 9.

5.3 Effects of Overall Myogenic Coordination on Local Circulatory Control

5.3.1 Interaction with the Peristalsis of Visceral Organs

In Chapter 4, and especially in Başar-Eroğlu et al. (1979) (Appendix in Başar and Weiss 1981), we examined the dynamics of the vascular smooth muscle and described the effects of passive stretch on the contractility of the vascular smooth muscle effectors. However, as stated, within the walls of the gastrointestinal tract, lymph vessels, uterus, and nictitating membrane, smooth muscle is arranged in sheets or layers of contiguous cells.

The basic propulsive movement of the gastrointestinal tract is peristalsis. A contractile ring appears around the gut and then moves forward. Peristalsis is an inherent property of any syncytial smooth muscle tube, and stimulation at any point causes a contractile ring to spread in both directions. Thus, peristalsis occurs in (a) the gastrointestinal tract; (b) the bile ducts; (c) other glandular ducts throughout the body; (d) the ureters; and (e) most other smooth muscle tubes of the body. The usual stimulus for smooth muscle is distension. That is, if a large amount of food collects at any point in the gut, the distension stimulates the gut wall 2–3 cm above this point, and a contractile ring appears and initiates a peristaltic movement.

The wall of the stomach, like that of the intestine, is composed of smooth muscle, an intramural nerve cell plexus, glandular structure, and a specialized surface epithelium.

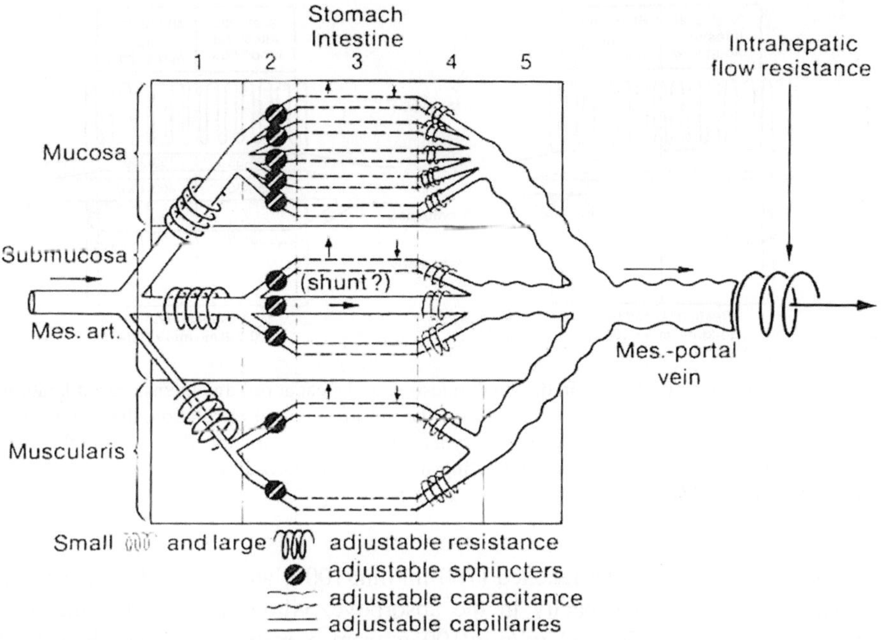

Fig. 5.3 Schematic of the different tissue compartments in the stomach and intestines, with their specialized parallel-coupled vascular circuits and consecutive sections: (1) Precapillary resistance vessels; (2) "sphincter" sections; (3) capillaries; (4) postcapillary resistance vessels; and (5) capacitance vessels (from Folkow and Neil 1971; modified after Folkow 1967)

For this reasons the gastric vascular bed, like that of the intestines, should be considered as comprising a set of parallel-coupled circuits, specialized for the requirements of these different tissue compartments (see Jacobson 1967) (Fig. 5.3). During resting conditions, gastric blood flow is some 20–40 ml/min/100 g and during maximal dilatation it may increase to some 150 ml/min/100 g. The smooth muscle component of the gastric wall (and that of the small and large intestine) may receive blood up to 40 ml/min/100 g tissue during maximal vasodilatation – a figure that is not very different from that of phasic skeletal muscles. However, it seems likely that the difference between the resting and maximal metabolism is far less in smooth muscle than in skeletal muscle, and that the blood supply to smooth muscle suffices to cover metabolic demands aerobically. Mechanical interferences with flow, as produced by intense gastrointestinal contractions, might occasionally interfere with the blood supply. The blood supply of the smooth muscle compartment is 10–15 ml/min/100 g during basal conditions. The blood supply of the intramural ganglionic plexus is likely to be rich, but no measurements exist concerning this small tissue compartment (Folkow and Neil 1971). The blood supply to the smooth muscle portion of the small intestine is of the same order as in the stomach. In the mucosa (Fig. 5.4) there is an especially rich vascularization of the secretory crypts, wherein maximal blood flow figures rivaling those in the

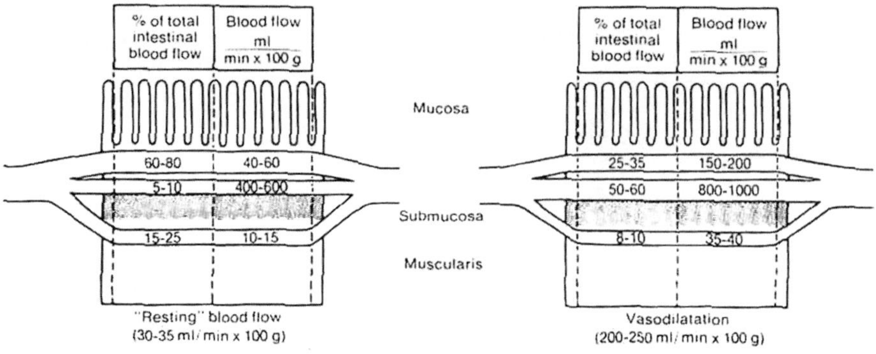

Fig. 5.4 Blood flow distribution in the small intestine of the cat at rest and intense vasodilatation. The three vascular pathways depict, in essence, the villous mucosal circulation, the submucosal circulation, especially that of the secretory crypts, and the muscularis circulation. Note the huge flows in the submucosa, presumably reflecting the rich vascularization of the secretory crypts (from Lundgren 1967)

salivary glands seem to be reached (700 ml/min/100 g and more). In the cat, the maximal blood vessel capacity in the absorptive, villous part of the intestinal mucosa may reach 150–200 ml/min/100 g from a resting value of 40–60 ml (Lundgren 1967).

The important features of the uterine circulation are entirely different, according to whether or not the uterus is gravid. The uterus possesses three layers: The superficial serosal layer, the myometrium, and the endometrium. The scheme of blood supply to the endometrium is shown in Fig. 5.5. As is seen from this figure, the circulatory architecture of the uterus comprises the main arteries, the main venous circulation, and the arteriolar circulation. The circulation in the uterus, like that in all muscular viscera, functions despite contraction and relaxation of the muscular viscera, and the muscular components of the organ. Reynolds (1963) assumed that the myometrial smooth muscle acts in concert with that of blood vessels themselves. The consequence of strong uterine contractions on the systemic circulation is shown by the fact that adulatory changes in arterial blood pressure occur as the post-partum uterus contracts.

5.3.2 Is There Any Interaction Between the Peristalsis of Visceral Organs and Auto-oscillations of Blood?

In the description of the contraction of various part of the stomach, we have seen the contractility dynamics of the stomach (Figs 4.9and 4.10A). These spontaneous contractions have different stages of contractility in the frequency channel of 0.01–0.02, 0.05, and 0.08 Hz. We call $G(j\omega)/(j\omega)$ the transfer function of the vascular elements in the gastric wall. The mechanical activity of the gastric wall possibly

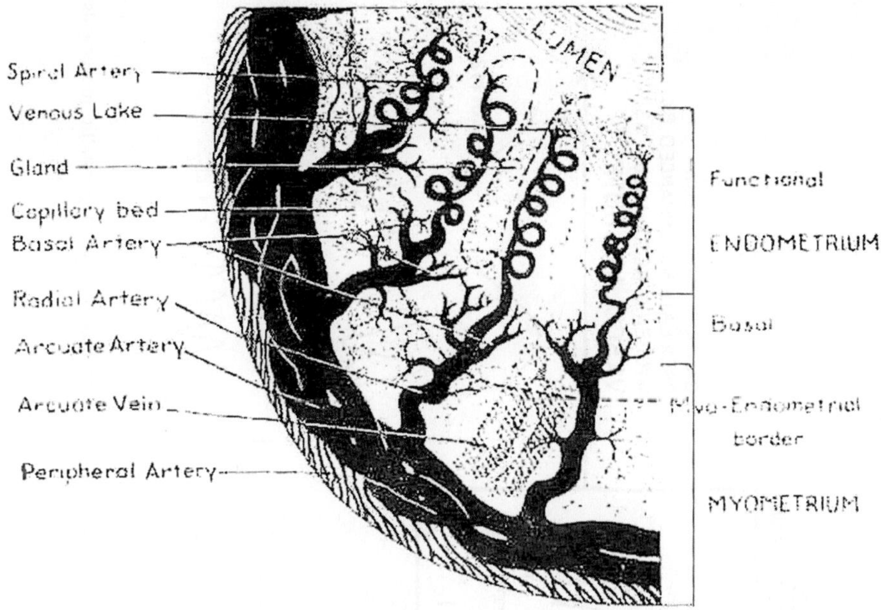

Fig. 5.5 Schematic representation of arterial supply to portions of uterus simplex (monkey, human) (from Reynolds 1947)

serves as adequate mechanical stimulation for the smooth muscle effectors of the vascular elements. In turn, the vascular elements elicit myogenic reactions. Let us suppose that the components of contractions of the stomach wall have nearly sinusoidal shapes. We can then make a prediction of the frequency characteristic of blood flow through the vascular system of the stomach (Fig. 5.6). The mechanical forces developed by stomach smooth muscle contraction (or peristalsis) induce tension changes in the walls of gastric vasculature. These changes have periodic shapes to elicit myogenic resonances and, accordingly, to achieve a steady control of blood flow. Because the peristaltic movements of the gastric wall and of the vascular bed are in good agreement, we call this phenomenon the control of blood flow due to overall myogenic coordination.

5.4 The Possible Role of Overall Myogenic Coordination in the Brain-Body-Mind Incorporation

The elements of the overall myogenic system can be activated by electrical pulses coming from the autonomic nervous system; they can also be activated by means of local causes. For example, a sudden increase of arterial pressure in the kidney or in the brain gives rise to auto-regulation of blood flow, which is performed without

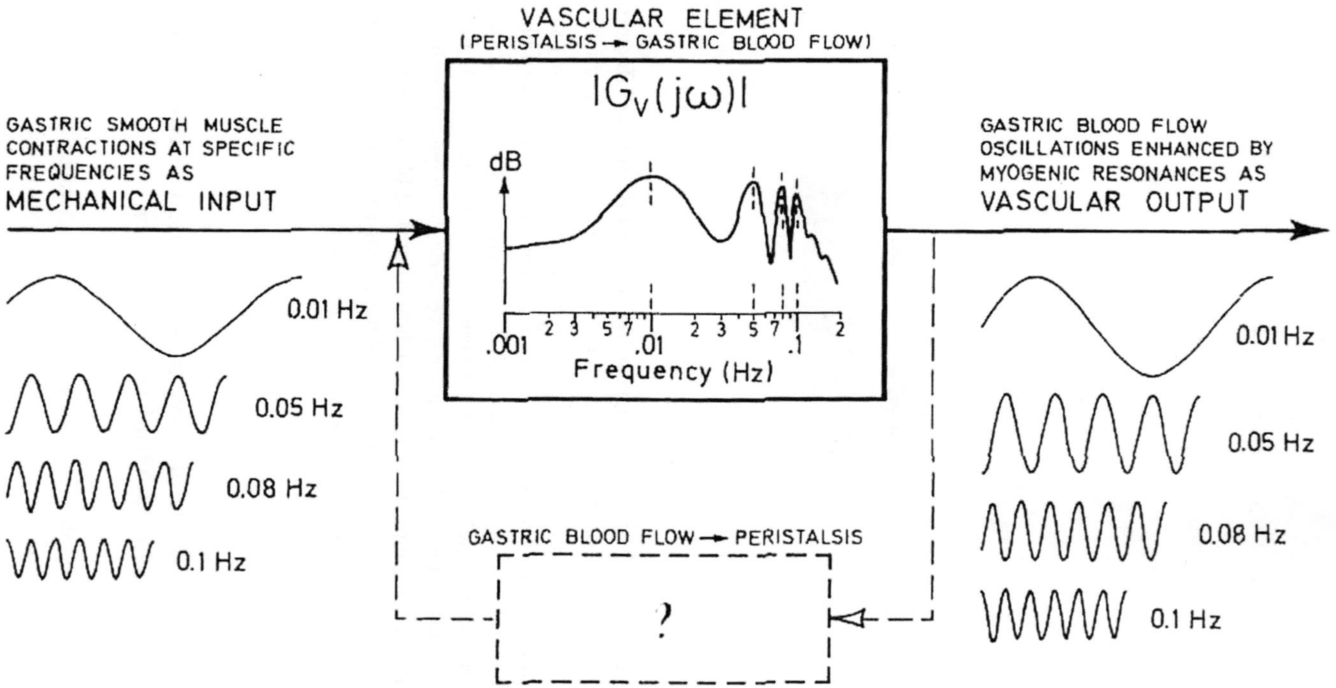

Fig. 5.6 A systems theoretical explanation of the elicitation of enhanced myogenic resonances of a vascular bed because of peristaltic movements (see text)

interaction with the central nervous system or the autonomic nervous system. An increase of heavy nutritive elements in the stomach can also induce rhythmic contractions or peristaltic movements of the stomach that are useful for digesting nutritive elements. The uterus in gravidas is also susceptible to producing rhythmic contractions before the birth of a baby (also monthly). Several other examples are available for the peristaltic motion of several organs, their independent functioning from the central nervous system, and their readiness to transfer impulses coming from the CNS to contractions or dilations.

All of these organs have different types of functions; however, they are functioning or depict contractility in the same frequency range of 0.01–0.1 Hz.

This means that all these organs of the myogenic control system are tuned to contract in the same frequency channels of between 0.01 and 0.1 Hz. In the next chapters, when we study the brain, we will see that similar oscillatory neural activities are also found in the brain stem. We go a step further, and suggest that the central nervous system and the organs of the overall myogenic systems are tuned to the same frequency ranges.

Recent findings, however, indicate that this overall tuning is not limited to these ultraslow frequencies. According to the work of Gebber and also W. Freeman, the EEG frequencies of 10 and 40 Hz are tuned with parasympathetic activity of 10 Hz in the heart and respiratory activity with 40 Hz in the bulbus olfactorius (see Chapter 9).

5.5 Response Susceptibility of the Peristalsis Organs

Another important observation is related to the oscillatory activity of the elements of the overall myogenic system. All of these organs, ranging from the vasculature to the stomach or to the uterus, show spontaneous activity in given frequency ranges. In 1980 we termed this phenomenon the response susceptibility of the system (see also Chapters 7 and 9). There are three general principles in the brain-body incorporation. (1) All these organs have resonance properties. (2) In turn, all these resonances properties have common properties (they occur in similar frequency channels). (3) In these common frequency channels, all these organs show spontaneous oscillatory behavior; they are also susceptible to show resonances in these common frequency channels.

For further discussion of this concept, see also Chapters 22–25 in Part VI.

5.6 Respiratory Coordination

Freeman (1991) states that, in living individuals, EEGs always oscillate, or rise and fall, to some extent, but the oscillations are usually quite irregular. When an animal inhales a familiar scent, what we call a burst can be seen in each EEG tracing.

All the waves from the array of electrodes suddenly become more regular, or ordered, for a few cycles – until the animal exhales. The waves often have a higher amplitude and frequency than they do at other times.

The burst waves are often called 40 Hz waves, meaning that they oscillate at about 40 cycles per second. Because the frequency can actually range from 20 to 90 Hz, Freeman prefers to call them gamma waves.

The fact that the bursts represent cooperative, interactive activity is not immediately clear in the EEG plots, because the burst segments differ in shape from tracing to tracing in a simultaneously recorded set. The average amplitude is not identical across the set – some versions of the carrier wave are shallow, and others are deep, but all of them curve up and down in near synchrony. The common behavior makes up between one quarter and three quarters of the total activity of the neurons giving rise to each trace.

Freeman states that it is not the shape of the carrier wave that reveals the identity of an odor. Indeed, the wave changes every time an animal inhales, even when the same odorant is repeatedly sniffed. The identity of an odorant is reliably discernible only in the bulb-wide spatial pattern of the carrier-wave amplitude (Fig. 5.7).

According to Freeman's proposal, the existence of a nerve cell assembly would help explain both the foreground-background problem and generalization-over-equivalent receptors. In the first instance, the assembly would confer "frontrunner" status on stimuli that past experience, stored in the Hebbian synapses, has made important to the individual. In the second instance, the assembly would ensure that information from any subset of receptors, regardless of where in the nose they were located, would spread immediately over the entire assembly and, from there, to the rest of the bulb.

Fig. 5.7 Simultaneous recordings from the olfactory bulb (**a**), front (**b**), and rear (**c**) parts of a cat's olfactory cortex. These show low-frequency waves interrupted by bursts – high-amplitude, high-frequency oscillations that are generated when odors are perceived. The average amplitude of a burst is some 100 μV. Each lasts a fraction of a second, for the interval between inhalation and exhalation (modified from Freeman (1991), by permission)

5.7 The Integration of the Overall Myogenic System in Brain-Body-Mind

Chapter 9 will describe important links and the role of the overall myogenic system with the brain and spinal cord by means of ultraslow oscillations. Further, in Chapter 24 a "string theory metaphor" will be presented for brain-body-mind, which also incorporates the organs of the overall myogenic system. This model strongly emphasizes the important role of the overall myogenic system in the machineries of mind.

Chapter 6
Dynamics of Sensory and Cognitive Processing

6.1 Introductory Issues

6.1.1 From Single Neuron to Neuron-Populations and the Brain

This chapter is one of the core chapters providing the application of the manifold system analysis methods that are included in the Brain Dynamics Research Program of Chapter 3.

The concept of brain oscillations has become one of the fastest developing branches of neuroscience. Accordingly, this chapter provides a model for the applications in Chapters 7–13.

The core of methodological ensembles includes the amplitude-frequency characteristics (AFCs), adaptive filters, and coherence function. The AFC and adaptive digital filters describe the resonant properties (enhanced frequency responses) of the brain, whereas the coherence function provide a measure for searching for links and/or coupling of oscillations between various parts of the brain.

In this chapter didactical sections will assist the reader in the understanding of the methods. Historical developments are also emphasized in the text.

Some basic knowledge is given concerning neural coding. The conventional neural coding is usually described in books on neuroscience and here the neat and effective description by Perkel and Bullock (1968) is used. Additionally, it is understood that oscillatory neural coding is a major brain code or building block of brain's functional organization. Accordingly, the information on neural coding is presented in two different sections. This description provides an important step, also for clinical and genetic studies, because Begleiter and Porjesz (2006) demonstrated the link between EEG-coding and genetically induced neural disorders (see Chapter 7).

The EEG-coding as a building block of neural systems merits important considerations in the scope of Descartes, who poses the question, "Are there few intelligible

E. Başar, *Brain-Body-Mind in the Nebulous Cartesian System: A Holistic Approach by Oscillations*, DOI 10.1007/978-1-4419-6136-5_6,

systems that govern the processes in the universe?" Fessard's question is also presented, which is basically similar to Descartes' view but tailored for brain function.

6.1.2 *From Ramon Ý Cajal to Vernon Mountcastle*

At the turn of the twentieth century, the morphological studies carried out by Ramon y Cajal (1911) and Charles Sherrington's physiological approach (1948) opened the way to the single-neuron doctrine by introducing the notion of one ultimate pontifical nerve-cell that integrates the CNS function. In this concept, integration was related to motor activity; the functional mapping was a type of movement mapping. Although this approach dominated twentieth century neuroscience, it had a major shortcoming in that cognitive functions and memory were not integrated into the neuron doctrine or in its derivatives; even the cerebellum was not included in Sherrington's model.

The single (or several) neuron hypothesis was proposed by Horace B. Barlow in 1972 and which he updated in 1995. Barlow's general principle is that single neurons detect peripheral events with features that are of behavioral significance to the organism. This is the feature-detector idea. However, in relation to this point Mountcastle states that, "This seems unlikely, however, that there are a sufficient number of feature detectors to account for the virtually infinite number of sensory stimuli we readily perceive, nor for solving the binding and relational problems in perceiving complex scenes (Mountcastle 1998).

The description of integration needs morphological descriptions, measurements in time-space, and the analysis of coherence functions. Therefore, to new concepts emerging from the study of brain oscillations, the role of time-space and coherence also gained considerable importance (Gray and Singer 1989).

The discovery of the EEG was followed by an explosion of publications related to brain function. The hope was renewed that a physical correlate of mental performances of the brain could now be tapped, a "psychic energy" in Hans Berger´s words (1929). However, from the 1960s the trend induced by Hans Berger and experimentally supported by Edgar Adrian (1934) remained in the shadow of neurophysiology research by using the single neuron approach and the methodology initiated by Adrian.

Suddenly a new paradigm change occurred; according to Mountcastle (1992) our perceptions are generated by the integration of the brain activity triggered by sensory stimuli with the activation of the neural images of past or current experience. The brain mechanisms involved in perception can now be studied directly, by means of measuring the changes in the electrical activity of the human brain with a large number of EEG recordings, or the use of multiple microelectrodes implanted in primates. The measurement of EEG oscillations is presently the most successful experimental paradigm used in perceptual neuroscience (Mountcastle 1998). Furthermore, Mountcastle (1992, 1998) states that the paradigm change introduced by using brain oscillations has become one of the most important conceptual and analytical tools for the understanding of cognitive processes. He proposes that a major task for neuroscience is to devise ways to study and analyze the activity of distributed systems in waking brains, in particular, human brains.

According to Luria (1966) mental functions too are similar to vegetative functions, a product of complex systems, a component part, which may be distributed through the structures of the brain. The task of neuroscience is, therefore, not to localize "centers," but rather, to identify the components of the various complex systems that interact to generate the mental functions. Luria called this task *dynamic localization*. A recent study testing the possible interplay between the working and long-term memory systems indicated the relevance of this dynamic localization (Sauseng et al. 2002). In a similar context, Lashley (1929) proposed that memories are in fact scattered across the entire brain rather than being concentrated in specific regions.

As a consequence of this chain of reasoning, the analytical and conceptual framework of the present chapter is based on the premise underlining Mountcastle's methodological advices and the conceptual statements of Luria and Lashley.

6.2 Neural Coding

The description of neural coding as outlined by Perkel and Bullock (1968) merits important consideration. The internal modes of communication of the nervous system are primarily electrical and chemical. Neural coding refers to the methods by which information is represented and transformed within the nervous system. As neural signals pass through successive synaptic junctions in the nervous system, the messages are dispersed and combined in new ways, and at each stage are transformed and recorded. Perkel and Bullock state that the formal properties of neural codes may be characterized by several independent aspects, as in the following:

1. The *referent* of a code is the signal or information being represented. It may be (a) an external physical quantity as in the case of primary receptors, or (b) a previously encoded neural signal, either chemical or electrical.
2. The *transformation* is the coding process itself, in which the afferent signal is transduced and transformed, usually combined with other signals. The transformation may be characterized in terms of the carrier of the signal, the representation scheme, the mechanism of this representation, and the reliability of the representation.
3. The *transmission* of the encoded signal includes the spatial and temporal aspects of conduction from its *source* to its *targets*.
4. The *interpretation of the signal* by its target neurons or effector cells is the last stage of the coding scheme.
5. Neural coding may be described on many levels: The individual *neuron*, the small circuit, and larger system codes involving firing rates of action potentials are highly common especially in the sensory and motor system.
6. In Perkel and Bullock's classification of neural coding, the descriptions of the higher level codes are those involving nerve impulses: labeled lines, time of occurrence, phase locking to a stimulus event, short or long-term firing frequency,

degree of variance of successive intervals between impulses, temporal patterns of impulses, and number of impulses or duration of a burst. In larger ensembles, candidate codes include the topographic distributions in a population of fibers, post stimulus, firing probabilities, and the larger population phenomena measured in EEG and event-related potentials (ERPs).

7. The most important comment made by Perkel and Bullock concerning neural coding is perhaps the following; the problems of neural coding are not separable from questions of neural functioning, at both cellular and higher levels. Coding underlies all neural functioning to the extent that the nervous system manipulates information. Form and function on the one hand, and representation and transfer of information on the other hand, are complementary aspects of nervous systems and must be investigated hand in hand.

6.3 EEG and Event-Related Oscillations as Information Codes in the Brain

6.3.1 Frequency Coding at Different Levels of Coding in the Brain

Perkel and Bullock (1968) created an extended list of the types of brain codes. Bullock further states that this list is not that of theoretically possible codes but modes of representation for which there is some physiological evidence. However, the EEG related integrative neuroscience approach is based on the description and interpretation of codes belonging to category #7 in Sect. 6.2. However, in this book common codings between EEG and evoked potentials will be used, namely the EEG frequencies as 2, 4, 10, 20, and 40 Hz.

Does multiple coding exist in such a way that evoked potentials or EEG can be considered to be compound potentials evoked internally or externally? In other words, do the frequency components of the ERP qualify as candidate codes? Experiments support an affirmative answer.

6.3.2 Do General Transfer Functions Exist in the Brain?

Alfred Fessard (1961) emphasized that the brain must not be considered simply as a juxtaposition of private lines, leading to a mosaic of independent cortical territories, one for each sense modality, with internal subdivisions corresponding to topical differentiations. The fundamental question of Fessard is the following one, What are the principles dominating the operations of heterosensory communications in the brain? To gain this knowledge requires an extensive use of multiple microelectrode recordings, together with a systematic computer processing of the data (see the works of

Eckhorn et al. 1988; Gray and Singer 1989). Further, Fessard (1961) indicated the necessity of discovering the principles that govern the most general or transfer functions of multi-unit homogeneous messages through neuronal networks.

The transfer function describes the ability of a network to increase or impede transmission of signals in given frequency channels. The *transfer function*, represented mathematically by frequency characteristics or wavelets (Başar 1980; Başar-Eroğlu et al. 1992) constitutes the main framework for signal processing and communication. The existence of general transfer functions would then be interpreted as the existence of networks distributed in the brain having similar frequency characteristics facilitating or optimizing the signal transmission in resonant frequency channels (Başar 1998, 1999). In an electrical system, an optimal transmission of signals is reached when subsystems are tuned to the same frequency range. Docs the brain have such subsystems tuned in to similar frequency ranges, or do common frequency modes exist in the brain?

The empirical results reviewed here imply a positive answer and provide a satisfactory framework in response to Fessard's question given in Chapter 1. Frequency selectivities in all brain tissues containing selectively distributed oscillatory networks (delta, theta, alpha, beta, and gamma) constitute and govern, mathematically, the general transfer functions of the brain. To fulfill Fessard's prediction, all brain tissues of both mammalian and invertebrates would have to react to sensory and cognitive inputs with oscillatory activity or with similar transfer functions. The degree of synchrony, amplitudes, locations, and durations or phase lags are continuously varying, but similar oscillations are most often present in the activated brain tissues (Başar 1999).

As to the process of coding explained in the previous section, the general transfer functions of the brain manifested in oscillations strongly indicates that frequency coding is one of the major candidates to analyze brain functioning.

6.3.3 Natural Frequencies of the Brain

The functional significance of oscillatory neural activity begins to emerge from the analysis of responses to well-defined events (event-related oscillations, phase- or time-locked to a sensory or cognitive event). Among other approaches, it is possible to investigate such oscillations by frequency domain analysis of ERP, based on the following hypothesis (Başar 1980, 1998).

The EEG consists of the activity of an ensemble of generators producing rhythmic activity in several frequency ranges. These oscillators are usually active in a random way. However, by application of sensory stimulation these generators are coupled and act together in a coherent way. This synchronization and enhancement of EEG activity gives rise to evoked or induced rhythms. Evoked potentials representing ensembles of neural population responses were considered to be a result of the transition from a disordered to an ordered state. The compound ERP manifests a superposition of evoked oscillations in the EEG frequencies ranging from delta to gamma (*natural frequencies of the brain* such as alpha, 8–13 Hz; theta, 3.5–7 Hz; delta, 0.5–3.5 Hz; and gamma, 30–70 Hz).

Chapter 4 describes how these natural frequencies have also been recorded in the vegetative system and spinal cord. Chapter 7 shows that these natural frequencies also belong to building blocks during the evolution of species from the Aplysia ganglia to the mammalian brain.

6.4 Emphasis of Multiple Oscillations in Brain Functioning

The functional importance of distributed multiple oscillations in the brain was published for the first time in a series of reports in the 1970s (Başar 1992; Başar et al. 1975a–c; Başar and Ungan 1973). Long distance coherences in the brain were also confirmed by Başar (1980) and Başar et al. (1979a, b).

The principle of superposition describes integration over the temporal axis as consisting of a relationship between the amplitude and phases of oscillations in various frequency bands. Furthermore, selectively distributed and selectively coherent oscillatory activities in neural populations illustrate integration over the spatial axis (Başar 1980). Consequently, integrative activity is a function of the coherences among spatial locations of the brain; these coherences vary according to the type of sensory and/or cognitive event and, possibly, the state of consciousness of the species (Başar 1999, 2004). Publications by Bressler and Kelso (2001), von Stein et al. (2000), and Varela et al. (2001) clearly describe the trend in which the concerted activity of alpha, theta, and delta and beta oscillations occur in distributed structures as reticular formation (RF), hippocampus (HI), thalamus, and sensory cortices.

As Mountcastle (1992) and Freeman (1999) state, neuroscience is ripe for a change. Accordingly, in the past decade an increasing number of reports and reviews on electrical oscillations in the brain were published in neuroscience literature. Now there is an increasing tendency to use the multiple oscillation approach. This chapter has the goal of embracing oscillations related to sensory and cognitive processing in all frequency channels, using recordings from human and animal subjects. Moreover, new results are included to verify the role of oscillatory dynamics in the coordination of function within the circulatory system, lymphatic system, and the overall myogenic system controlling peristalsis and all visceral organs in the ultra-slow frequency range (Başar and Weiss 1981).

6.4.1 The Views of Fuster and Klimesch on the Role of Oscillations in Memory Processing

From Fuster's (1997) point of view, memory reflects a distributed property of cortical systems. An important part of higher nervous function, such as perception, recognition, language, planning, problem solving, and decision-making, is interwoven with memory. Further, memory is a property of the neurobiological systems it serves and is inseparable from their other functions. By surveying the data presented in this section it can be hypothesized that the selectively distributed oscillatory systems (or networks) may provide a general communication framework and can be a useful concept for functional mapping of the brain (Mesulam 1990, 1994). The new experimental designs created by various research groups (Burgess and Gruzelier 2000; Egner and Gruzelier 2001; Haenschel et al. 2000; Klimesch 1999; Klimesch et al. 1994) are extremely important, because they add new steps to the general framework of function-related oscillations.

The experimental work of the Klimesch group showed the possibility of differentiating the role of alpha and theta oscillatory activity during memory tasks. The results from this group support the hypothesis that ERPs can be understood and described in terms of the superposition of several event-related oscillations recorded in various structures (Klimesch et al. 2000a; Doppelmayr et al. 2000). Moreover, their experiments include memory tasks differentiating oscillatory responses of human subjects with good and bad memory performers.

6.5 Selectively Distributed Oscillatory Systems in Brain Function: Distributed Multiple Oscillations in Brains

6.5.1 Concept, Definition and Methods

The functional significance of oscillatory neural activity begins to emerge from the analysis of responses to well-defined events (event-related oscillations, phase- or time-locked to a sensory or cognitive event). Among other approaches, it is possible to investigate such oscillations by frequency domain analysis of event-related potential (ERP), based on the hypothesis described in Sect. 6.3.3 (Başar 1980, 1998).

Time-locked responses of a specific frequency after stimulation can be identified by computing the AFCs of the averaged ERPs (Basar 1980; Röschke et al. 1995; Yordanova and Kolev 1997). The AFC describes the brain system's transfer properties, e.g., excitability and susceptibility, by revealing resonant as well as salient frequencies. It, therefore, does not simply represent the spectral power density characterizing the transient signal in the frequency domain but the predicted behavior of the system (brain) if sinusoidally modulated input signals of defined frequencies were applied as stimulation. Because it reflects the amplification in a given frequency channel, the AFC is expressed in relative units. Hence, the presence of a peak in the AFC reveals the resonant frequencies interpreted as the most preferred oscillations of the system during the response to a stimulus (see Glossary). To calculate the AFCs, the ERPs were first averaged and then transformed to the frequency domain by means of one-sided Fourier Transform (Laplace transform, see Başar 1980; Solodovnikov 1960) as shown in Fig. 6.1.

An AFC is illustrated in Fig. 6.2 demonstrating that auditory stimuli produce prominent resonant responses in the theta, alpha, and gamma (40 Hz) frequency bands in the cat hippocampus (from Başar 1980). The AFCs also serve to define filter limits for response-adaptive digital filtering of the averaged ERPs.

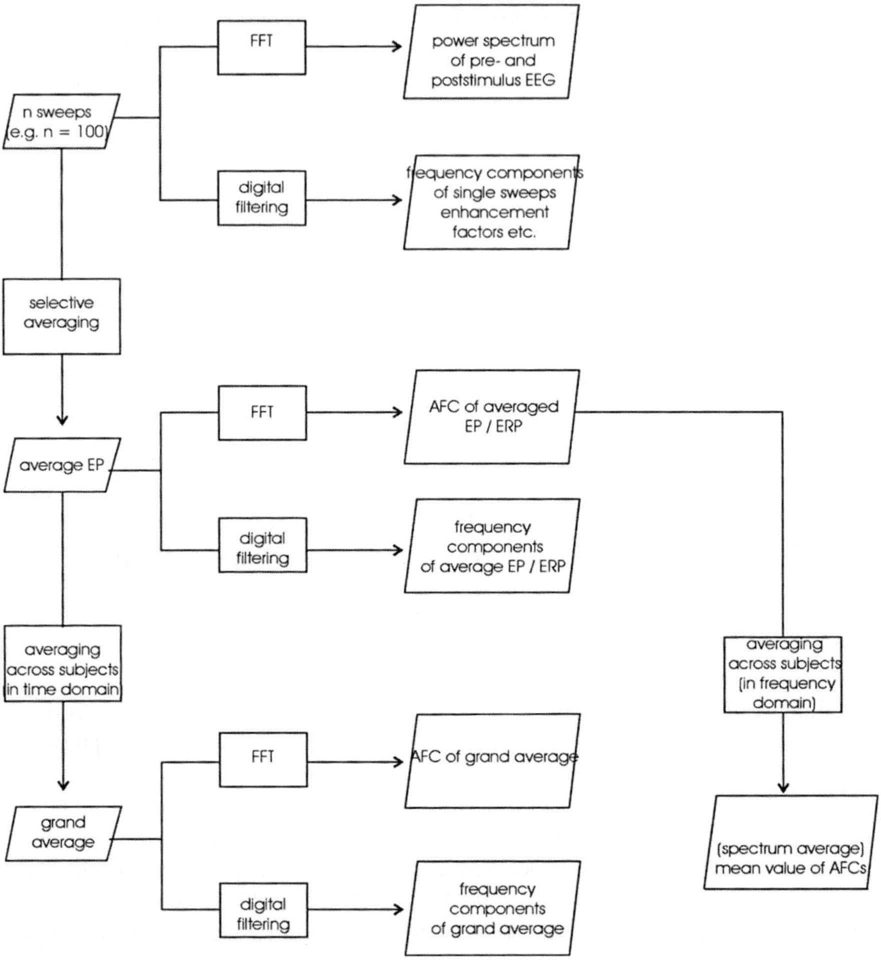

Fig. 6.1 Schematic illustration of the stepwise application of system-theoretical methods. For simplicity, the evaluation of wavelet decomposition and coherence functions are not included here (modified from Schürmann et al. 1997)

The filtered curves obtained in this way show the time course of the oscillatory activity in a certain frequency range (Fig. 6.3).

Gönder and Başar (1978) performed a comparative study of power spectral peaks both in spontaneous and single evoked activities in the cat brain. They also used a method based on the comparison of single EEG-EP epochs in a histogram distribution. The results showed a transition from disordered to ordered states in all EEG frequencies, even covering the gamma and the very highest frequency window up to 1,000 Hz (Fig. 6.4). These high frequency findings measured 30 years ago today find support by new findings in magnetoencephalography (MEG) (See Curio 2000 and Curio et al. 1997).

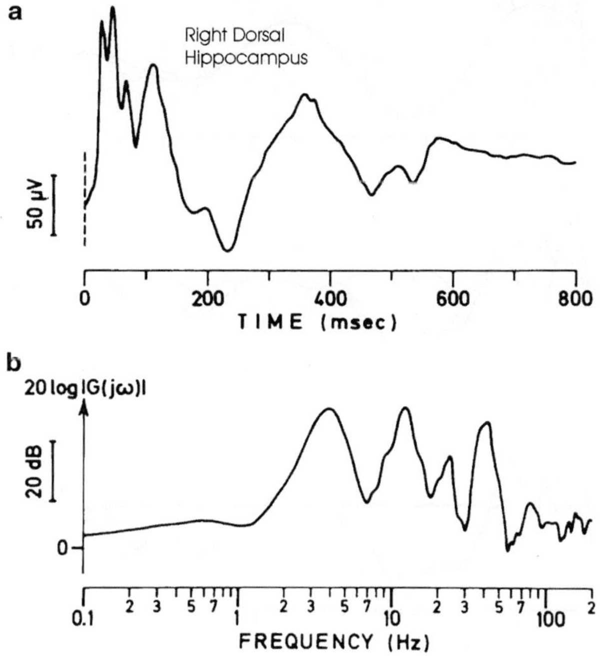

Fig. 6.2 Averaged EP (**a**) recorded from the right dorsal hippocampus of the cat after auditory stimulation in the form of a step function (3 s-long tone burst, with intensity of 2,000 Hz), negativity upward. (**b**) Amplitude frequency characteristics computed from the transient response (**a**). Double logarithmic presentation: along the abscissa – log (frequency), along the ordinate – relative amplitude in decibels (modified from Başar 1980)

Figure 6.5a shows the results of band pass filtering the auditory and visual evoked potentials recorded from the auditory cortex (Gyrus Ectosylvian anterior), Hippocampus (HI) and visual cortex, occipital cortex (OC) of a representative cat. Figure 6.5b shows the same results evaluated by the new wavelet analysis techniques applied to ERP analysis. Wavelet analysis confirms the results obtained using the AFCs and digital filtering. In addition, wavelet analysis can be used for signal retrieval and selection from a large number of sweeps recorded in a given physiological or psychological experiment (see Fig. 6.5b) (Başar 1999; Başar et al. 1999c; Demiralp et al. 1999). Note that the uppermost rows in Fig. 6.5a and b are identical, because there are wide-band filtered EPs from the same cat.

The basic observation on the oscillatory responses EPs in both Fig. 6.5a and b are generally in accordance. However, especially in higher frequency bands, the time localization capability of wavelet analysis was significantly better compared with the conventional band-pass filters. The ringing effects occurring in the conventional band-pass filtering techniques, which lead to oscillations before the stimulation time point, were smaller in alpha, beta, and gamma frequency ranges. The results support earlier studies (Başar 1999) and here two noteworthy results are described as follows:

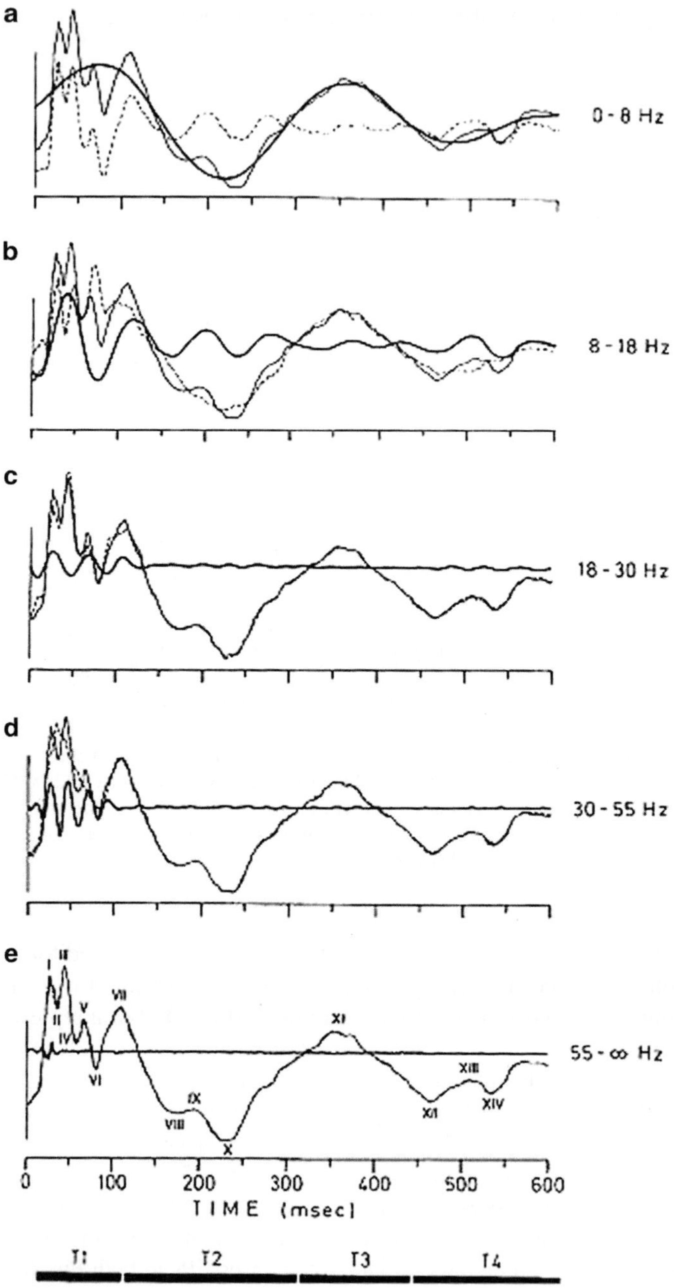

Fig. 6.3 Filtering of a hippocampal selectively averaged EP with different stop-band and pass-band. *Solid curves* are filtered, averaged EPs obtained with application of pass-band filters. *Dashed curves* are filtered, averaged EPs obtained with application of stop-band filters. The band limits (shown on the right of the averaged EPs) of the applied filters are chosen according to the TRFC method. The original selectively averaged EP is shown for comparison with all the filtered averaged EPs. Time sections T1–T4 are shown at the top of the illustration. Note the ample gamma response with comparable values to alpha response (modified from Başar 1980)

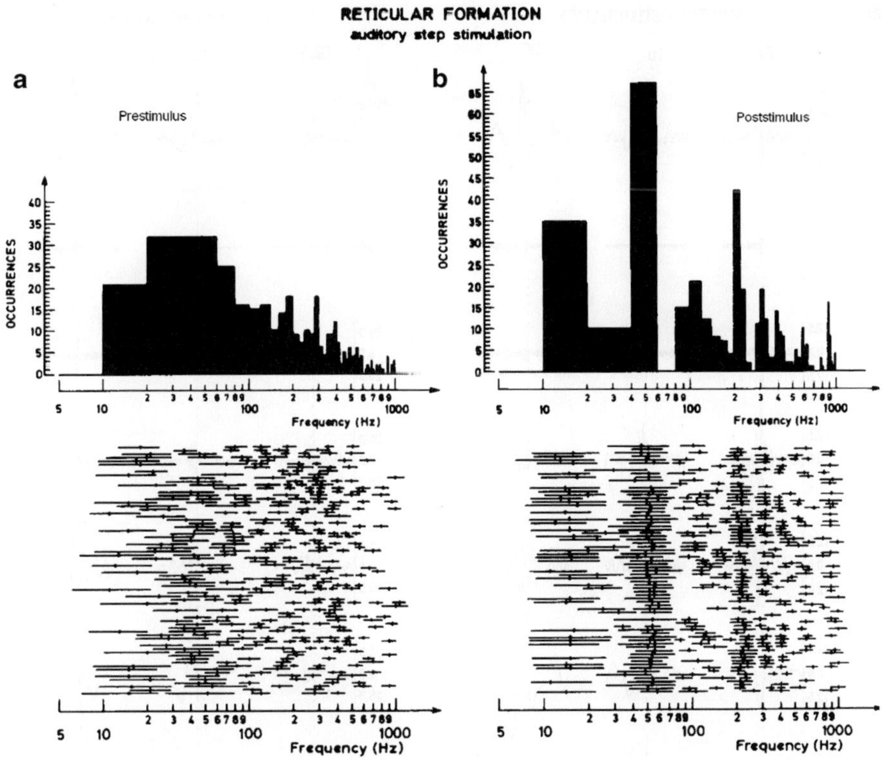

Fig. 6.4 Histograms showing frequency distribution. As revealed by (**a**) pre-stimulus power spectra, (**b**) post-stimulus instantaneous frequency characteristics of the RF. At the *bottom* of the figure, each spectral peak (or amplitude maximum) is represented by an approximate bandwidth (*horizontal line segment*) and an indicated center frequency. The histograms shown at the *top* of the figure are presented by plotting the number of center frequencies falling into each of a set of 20 Hz slots versus frequency. The number of power density-instantaneous frequency characteristic pairs, which are used to obtain the histograms, is 71 (modified from Gönder and Başar 1978)

1. The dominant frequency components of the visual EPs recorded in the OC area were in the delta and alpha ranges with comparable weights, whereas, the auditory EPs (recorded in GEA) had a dominant component in the alpha range.
2. Alpha response components were most pronounced in the responses of OC and GEA to adequate stimuli and in HI in visual modality, whereas their amplitudes radically decreased in the responses of both GEA and OC to inadequate stimuli.

The AFC method has an important advantage over wavelet analysis. It allows a global view of all frequency responses together, whereas in wavelet analysis the investigator rather arbitrarily defines windows. This search often leads to misinterpretations by orienting the search to special windows; for example, in most of the studies only the gamma band is selected.

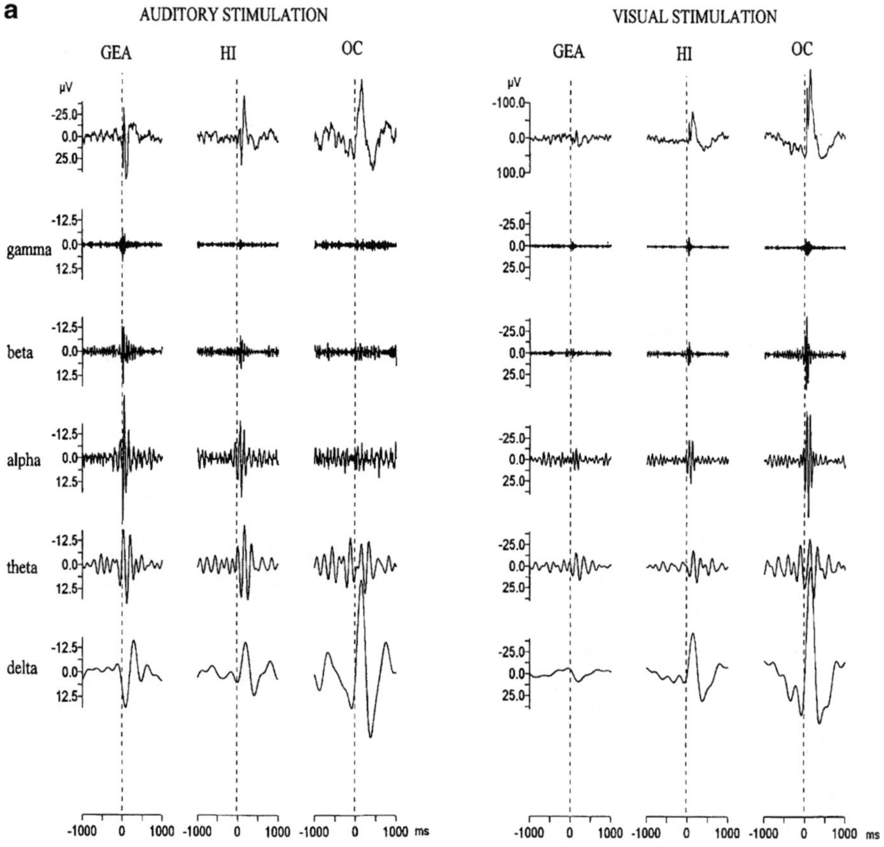

Fig. 6.5 (**a**) Results of band pass filtering in a typical animal. *Left*, auditory stimulation; *right*, visual stimulation. Each column refers to an electrode site (auditory cortex: GEA; visual cortex: OC; hippocampus: HI). The *uppermost row* shows the wide-band filtered curve. The remaining rows show the frequency components gamma (32–64 Hz), beta (16–32 Hz), alpha (8–16 Hz), theta (4–8 Hz), and delta (0.5–4 Hz) (modified from Başar et al. 1999a).

As will become clear, the combination of these methods yields results leading to the conclusion that, alpha, theta, delta, and gamma responses are brain responses related to psycho-physiological functions, in short, real signals (Başar 1998, 1999; Başar et al. 2001a, d). The intention here is to show that these oscillations have *multiple functions* and may act as *universal operators* or *codes* of brain activity. Besides frequency and site of oscillations, several other parameters are dependent on specific functions, namely enhancement; time locking, phase locking, delay and duration of oscillations. (For a review of the methods to assess these parameters, see for example, Kolev and Yordanova 1997).

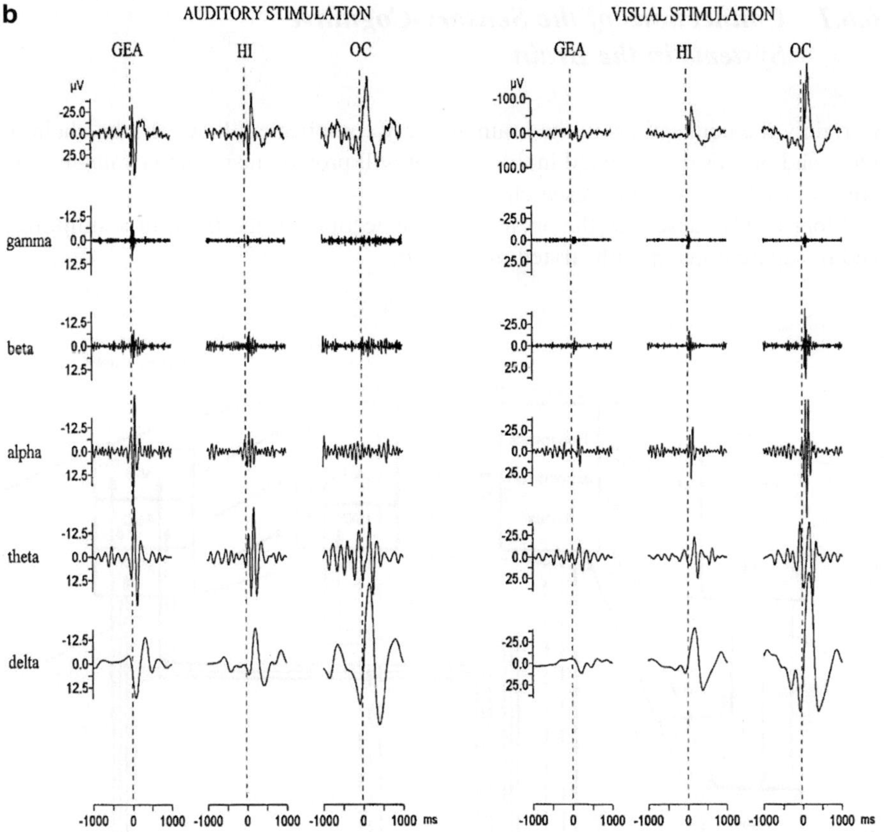

Fig. 6.5 (continued) (**b**) Results of wavelet decomposition in a typical animal. *Left*, auditory stimulation; *right*, visual stimulation. Each column refers to an electrode site (auditory cortex: GEA; visual cortex: OC; hippocampus: HI). The *uppermost row* shows the wide-band filtered curve. The remaining rows show the frequency components gamma (32–64 Hz), beta (16–32 Hz), alpha(8–16 Hz), theta (4–8 Hz), and delta(0.5–4 Hz) (modified from Başar et al. 1999a)

6.6 Remarks on Physiology of Selectively Distributed Oscillatory Processes

To proceed with the terminology of *physiologic mem*ory and *perceptual memory* it is necessary to have the knowledge related to EEG correlates of sensations, perception, learning, and remembering. Accordingly, at the beginning of this chapter the processes of activation of alpha, theta, and delta systems were described with schematic presentations.

6.6.1 *Connections of the Sensory-Cognitive Systems in the Brain*

Certainly the connections in the brain are very complicated; however, the chain of ideas and proposals presented in this chapter will provide insights to strategies that are also useful in memory research.

Flohr (1991) describes the anatomical connections in the brain in a simplified and transparent manner illustrated in Fig. 6.6.

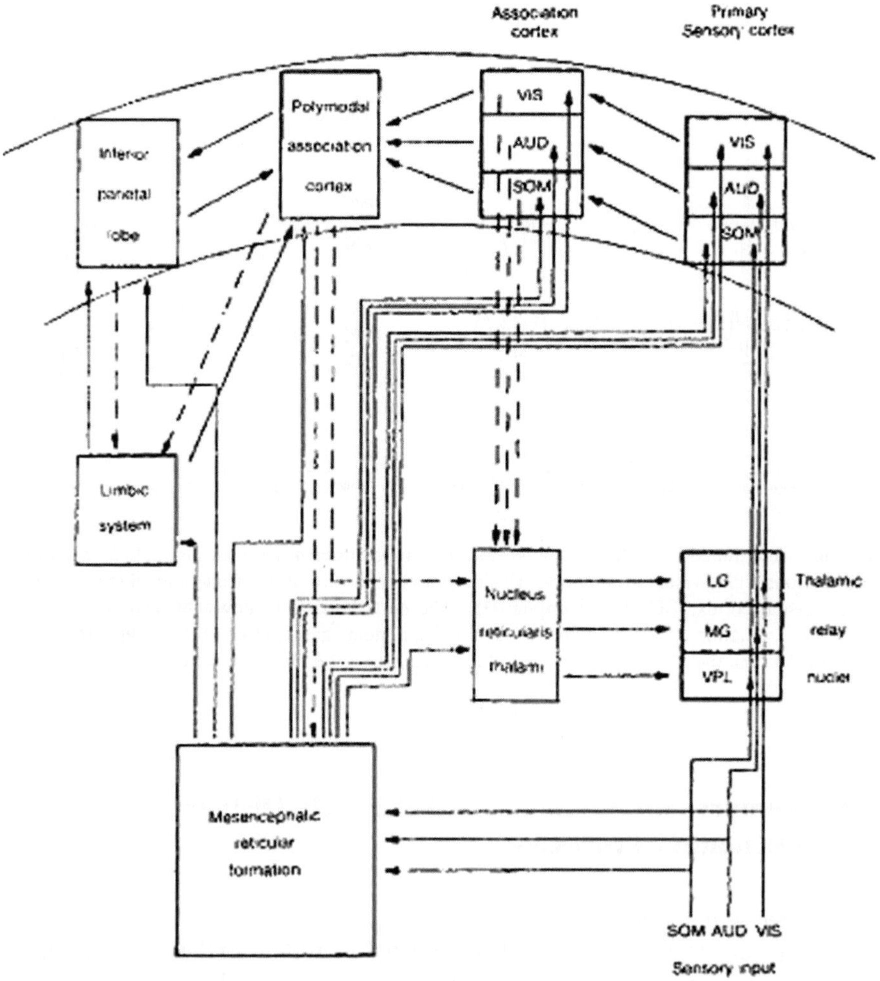

Fig. 6.6 Flow of information in the auditory, somatosensory and visual pathways, reticular formation, limbic system, and association areas of the cortex (sensory and cognitive neural pathways in this figure are modified from Flohr [1991])

1. *Specific afferents from sense organs reach specific thalamic nuclei before going to the primary cortical areas.* For instance, auditory information is transmitted through the medial geniculate nucleus to the primary auditory area; visual afferents are transmitted through the lateral geniculate nucleus to area 17 of the occipital cortex.
2. *Non-specific afferents reach the cortex from the mesencephalic formation.* It has now been established that reticular formation is connected to different nuclei with specific afferent connections. There is a second site where the reticular formation influences the processing of primary afferents; the thalamic relay nuclei. The nucleus reticularis thalami, a thin sheet of neurons, surrounds the dorsal thalamus and inhibits the thalamic relay nuclei. Its control function is, in turn, affected by collaterals of thalamocortical pathways, by collaterals from corticothalamic projections and by inhibitory afferents from the mesencephalic reticular formation. There are important connections within the cerebral cortex involving the association areas. Primary auditory, somatosensory and visual fields each project to adjacent unimodal association areas, which, in turn, project to secondary unimodal association fields. The unimodal association areas project to a number of polymodal sensory areas, lying in the cingulate gyrus, and parietal, temporal, and frontal lobes. The functions of these areas are vaguely described as a crossmodal association and synthesis. The polymodal association areas project to the inferior parietal lobe, which has been termed a *supramodal* area. Polymodal and supramodal regions have connections to the limbic system; these connections provide the anatomical substrate by which motivational states influence cortical processing of sensory stimuli.

Every sensation in the brain also induces cognitive loading, at least for matching processes. Furthermore, all the presented cognitive targets evoke sensations; the respective neural processes are interwoven and require, for final processing, at least three neural loops next to purely sensory connections in the simple sensory systems. These loops are:

1. *Secondary connections* to the cortex over reticular formation
2. *Secondary connections* over the limbic system
3. *Connections within the cortex* between association areas

6.6.1.1 Steps for a New Synthesis and the Problem of Binding

The new trend toward a treatise on brain oscillations implies the following immediate thoughts for functional analysis: (1) Not only single neurons, but neuron assemblies; (2) not only spikes of single neurons, but the oscillatory activity of *neurons and neural assemblies;* (3) not only movements, but cognitive processes and memory processes are interwoven in the integrative brain function. These concepts were not included in Sherrington's description of integrative brain activity, and neuroscience needs a new framework or theories and, possibly, new rules to analyze the integrative brain function. With these tenets new propositions or a new treatise will be outlined and this, in turn, may open new experimental avenues as was the case in the EEG-brain dynamics theories published 25 years ago (Başar 1980; Freeman 1975).

In the present chapter, only a few illustrative experiments are chosen to call attention to the large variety of frequency responses ranging from delta to gamma and from invertebrate ganglia to human cognitive responses. Publications considering the functional relevance of multiple oscillations and importance of selective distributions, delay, or prolongation of oscillations are rare (Bressler and Kelso 2001).

A fundamental unsolved problem in neuroscience concerns the manner in which the vast array of parallel processes occur in the brain at any given time; the diverse neural activities are bound together or integrated (Haig et al. 2000, and Chapter 7). For instance, a visual image of an object contains a collection of features that must be identified and segregated from those comprising other objects.

In the following three levels of related brain oscillations will be outlined.

1. A short chronological survey level related to brain oscillations.
2. The comparative level of various sensory cognitive-oscillatory neural populations both in human and animal recordings; functional correlates of oscillations and multiple oscillations.
3. The general level; superposition and distribution of selective oscillations and selectively distributed coherences.

6.7 A Survey of Work on EEG-Oscillations

6.7.1 Alpha Activity

1. Review of the studies of Andersen and Andersson (1968) and Başar (1999). A particular feature of the thalamic relay nuclei is their ability to convert a single afferent volley to a series of rhythmic discharges along the thalamocortical fibers. Adrian (1941) discovered that a single tactile stimulus elicited a series of waves in the thalamus, which he called thalamic after-discharges. Similar rhythmic activity was found by Bremer and Bonnet (1950) in the medial geniculate nucleus in response to a click. All these authors noted that the frequency of the evoked activity was around 10/s, i.e., similar to that of the spontaneous rhythmic cortical waves. Adrian (1941) maintained that the after-discharges consisted of bursts of spikes separated by slow waves. A peripheral stimulus elicited a series of 3–7 such cycles. By taking recordings from the white matter below the cortex, Adrian showed that the rhythmic discharge occurred in the thalamocortical fibers, indicating a thalamic origin of the after-discharges. Because of this rhythmic discharge in response to a single afferent volley, a series of waves are initiated in the cortex, appearing at a frequency of about 10/s (Bartley and Bishop 1933; Bishop 1933; Bishop et al. 1953; Jarcho 1949).

Adrian (1941) had already reported that rhythmic 10/s activity following a single afferent volley could be recorded within or at the dorsal surface of the thalamus, even if the appropriate cortical area was removed. In other words, the thalamic nuclei contain a mechanism for the transfer of a single volley to a rhythmic 10/s sequence without the presence of the cortical area to which the thalamocortical fibers project. Chang (1950) advanced the hypothesis that a corticothalamic reverberating circuit should be the basis for the evoked rhythmic activity. The arguments for this explanation were the presence of a similar rhythmic activity in the thalamus and cortex, and the difficulty of recording thalamic rhythmic activity after the removal of the appropriate cortical projection area. However, this theory is contradicted not only by the early reports from Adrian (1941) and Bremer and Bonnet (1950), but also by further observations by Adrian (1951) who critically tested the corticothalamic reverberating hypothesis, and by Galambos et al. (1952).

2. *Toward a Renaissance of Alphas.* The papers in the volume edited by Başar et al. (1997a, b) describe the physiological bases of the 10 Hz activities and their functional correlates, with emphasis on sensory, cognitive, and motor states that accompany the 10 Hz oscillations. This endeavor, which aimed to establish a new trend regarding the functional implications of the alpha activity, led to a "renaissance of alpha" 80 years after the first discovery by Hans Berger. Thus, a new nomenclature is proposed in the Başar et al. (1997b) special issue, the expressions *Alpha's* or *Brain's 10 Hz oscillations* were utilized to emphasize the multiplicity of phenomena in the alpha band (see also Klimesch 1999). Such usage helps to draw attention to the existence of multiple phenomena in the alpha band, which have hitherto been generally regarded as a single rhythm and therefore, a single phenomenon. In this approach, the mu and the tau rhythms are also regarded as components within the associated ensemble of phenomena that are classed as Alphas.

6.7.2 Earlier Experiments on Induced or Evoked Theta Oscillations

Ross Adey`s group (1960) started the early pioneering work on theta rhythms of the limbic system of the cat brain during conditioning. For the first time spectral and coherence functions were used to perform relevant experiments demonstrating that rhythmic field potentials of the cat brain are related to behavior (see also Elazar and Adey 1967; Miller 1991).

The use of the coherence function in comparing EEG activity in various nuclei of the brain was useful in refuting the view that "EEG is an epiphenomenon" (Adey 1989). The induced theta rhythm and the task-relevant coherence in the limbic system of the cat brain is a milestone in EEG research. During these experiments the cat hippocampal activity exhibits a transition from irregular activity to coherent induced rhythms. Such results have encouraged Başar and Özesmi (1972) and Başar and Ungan (1973) to choose the hippocampus as a model for a possible resonance theory of the brain.

The experiments performed by this group showed that such an explanatory model was feasible. (For resonance phenomena in the hippocampus, see also Miller 1991).

Miller (1991) summarized that hippocampal theta activity occurs in cats in association with locomotion and other body movements, in a manner similar to that seen in rats and other small mammals. However, there are a great many exceptions to these correlations, whereby theta activity can occur during immobility, or movement can occur without theta activity. Many of these exceptions occur in the course of learning. There are suggestions that theta activity occurs at the time when performance of a learned task is improving most rapidly, and declines as tasks become familiar. Apart from these results, there is a great deal of evidence that both spontaneous and conditioned orienting is accompanied by low-frequency theta activity in cats.

Petsche et al. (1962) followed up the early evidence from lesion experiments, which showed a septal involvement in the theta rhythm by recording from units in the septum of curarized rabbits. Theta activity was elicited by sensory stimulation or electrical stimulation of the reticular formation. A proportion of septal units showed regular rhythmic bursts of impulses, in phase with the hippocampal theta rhythms.

These findings have been confirmed many times. Vinogradova and Zolotukhina (1972), studying alert rabbits, found that only one third of neurons in both the medial and lateral septum showed theta bursts. Apostol and Creutzfeldt (1974), using curarized rabbits, simultaneously recorded the septal and hippocampal EEG and the spike activity of individual septal units during spontaneous rhythmic activity or rhythms evoked by sensory stimuli.

Brazier (1968) recorded activity in the hippocampus in 30 patients before surgery (28 cases of temporal lobe epilepsy, 2 non-epileptics). The hippocampal EEG showed peak power in the 2–4 Hz range, but the peak of coherence with EEG from other sites (e.g., parahippocampal gyrus) was in the 4–8 Hz range. Lieb et al. (1974) recorded from the hippocampus, amygdala, and parahippocampal gyrus. A small peak was seen at 8–10 Hz, variable between patients, with a broad peak at slower frequencies. Halgren et al. (1986) made some behavioral observations in a single case study. The hippocampus was synchronized at 5–6 Hz during quiet resting, when patients tensed all their muscles, or made simple alternating movements. Increased synchrony was also seen during testing of patients on a series of word meanings or a tapping sequence. Desynchronization occurred during speech or rapid breathing, tying a bow, imitating movements of the experimenter, or while the patient was giving a verbal description of the word.

6.7.3 Alpha Oscillations in Perception and Cognition: The Alphas

Alpha oscillations in functional EEG also gained importance in the last decade (Başar et al. 1997a, b). Observations at the cellular level are noteworthy. Evoked oscillations in the 8–10 Hz frequency range in visual cortex neurons upon visual stimulation suggest a relation to scalp-recordable alpha responses (Dinse et al. 1997; Silva et al. 1991) and thalamocortical networks oscillate in the alpha range (Steriade et al. 1992). Alpha responses from different structures in the cat brain are presented in Fig. 6.5. The sum of these observations permits a tentative interpretation of alpha as a functional and communicative signal with multiple functions. This interpretation of 10 Hz oscillations, at the cellular level, or in populations, might be comparable to the putative universal role of gamma responses in brain signaling.

6.7.3.1 Sensory Components

Taking for example the alpha response in cross-modality measurements in the cat brain, as mentioned in the previous paragraph, here the topographic differences of frequency components are considered. In particular, the results of measurements from auditory and visual areas are summarized. As auditory and visual stimuli were used, the conditions were either adequate stimulation (auditory cortex recording of auditory EP; visual cortex recording of visual EP) or inadequate stimulation (visual cortex recording of auditory EP and vice versa). Such experiments are referred to as *cross-modality* measurements (Başar 1998, 1999; Hartline 1987).

Panel A in Fig. 6.7 shows single-trial EPs filtered in the 8–15 Hz range. The left column refers to auditory stimulation with visual cortex recordings; the right column to visual stimulation with visual cortex recordings; i.e., inadequate versus adequate

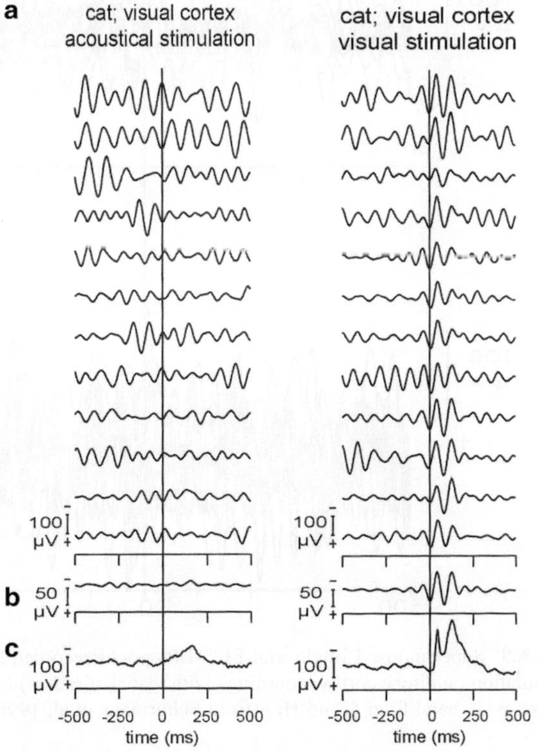

Fig. 6.7 EPs recorded from the cat brain by using intracranial electrodes (from Schürmann et al. 1996). (**a**) Single EEG-EP trials, filtered 8–15 Hz. (**b**) Averaged EP, filtered 8–15 Hz. (**c**) Averaged EP, wide-band filtered. *Left column*: inadequate stimulation (visual cortex recording with auditory stimulation). *Right column*: adequate stimulation (visual cortex recording with visual stimulation)

stimulation. Responses to visual – adequate – stimulation show amplitude increase and time- and phase-locking. A distinct response is also seen in the filtered averaged EP in panel B. The unfiltered averaged EP also shows an alpha-like waveform. In contrast, responses to auditory stimulation are inadequate and neither show amplitude increase or phase locking, nor can an alpha response be seen in the filtered average. There is a type of response in the unfiltered EP in panel C, but this is not an alpha response.

Thus, alpha responses were recorded with adequate stimuli in primary sensory areas. Adequate versus inadequate differences were larger for alpha responses than for theta responses, demonstrating the functional relevance of frequency components. As an aside, in cross-modality recordings from the auditory cortex (gyrus ectosylvianus anterior) of the cat brain, a complementary effect was observed in that large alpha enhancements were present in auditory EP recordings. In visual EP recordings from the auditory cortex, such alpha enhancements were not observed.

It is useful to compare the cat data with EEG and MEG recordings in humans. EEG measurements were performed in $n = 11$ subjects. Figure 6.8 shows filtered

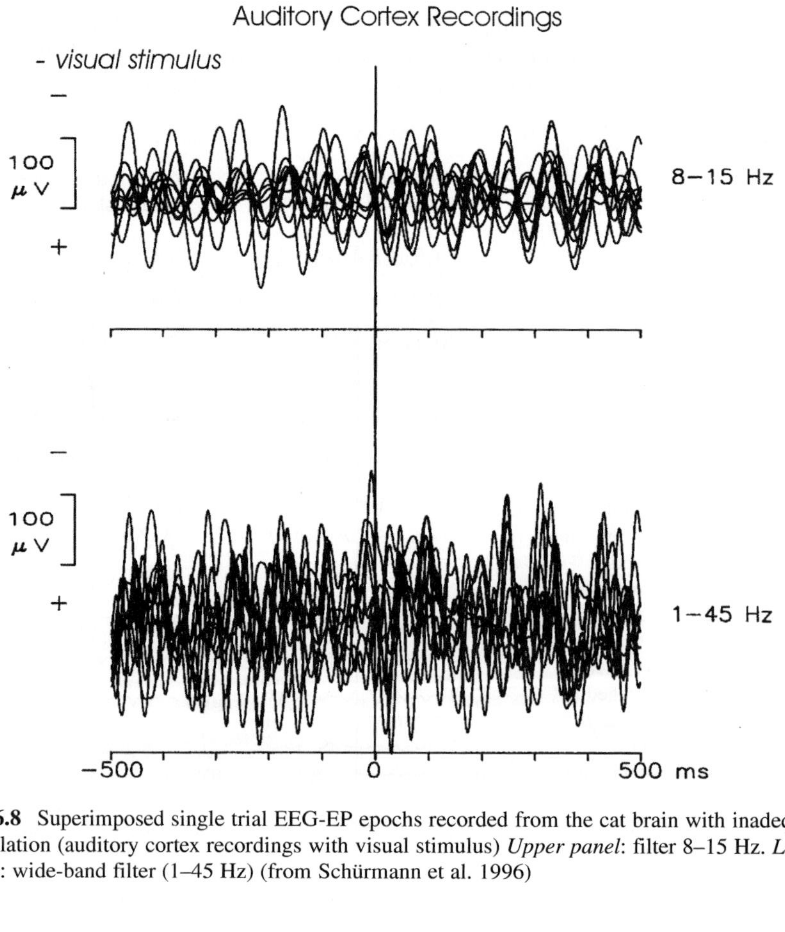

Fig. 6.8 Superimposed single trial EEG-EP epochs recorded from the cat brain with inadequate stimulation (auditory cortex recordings with visual stimulus) *Upper panel*: filter 8–15 Hz. *Lower panel*: wide-band filter (1–45 Hz) (from Schürmann et al. 1996)

curves computed from grand averages of occipital recordings (O_1). The upper half of the figure shows theta responses, whereas the lower half shows alpha responses. The alpha response to auditory stimulation (inadequate for the visual cortex, occipital located) is on the left, where the response is of low amplitude. The response to visual stimulation on the right, however, has a distinct alpha response. Note that the adequate-inadequate difference is less for the theta response. This supports the hypothesis given previously, as observed in cats: it is mainly the alpha response, which is dependent on whether or not a stimulus is adequate. A correlation between the alpha response and primary sensory processing is thus plausible both for human and for cat EEG-EP data.

Figure 6.9 shows oscillatory responses to auditory and visual stimulation at the occipital cortex.

MEG measurements were performed with both a BTI 7 channel MEG system (Saermark et al. 1992) and a PHILIPS 19-channel MEG system (Başar et al. 1992b; Schürmann et al. 1992). The methods used were similar to those for EEG recordings where possible. Auditory stimuli (2,000 Hz; 80 dB sound pressure level) were used and sensor positions were selected close to the auditory cortex and close to the visual cortex.

The data shown in Fig. 6.10 were obtained with the seven-channel system where the different positions required two experimental sessions. Panel A shows the temporal recordings, Panel B shows the occipital recordings, in both cases with

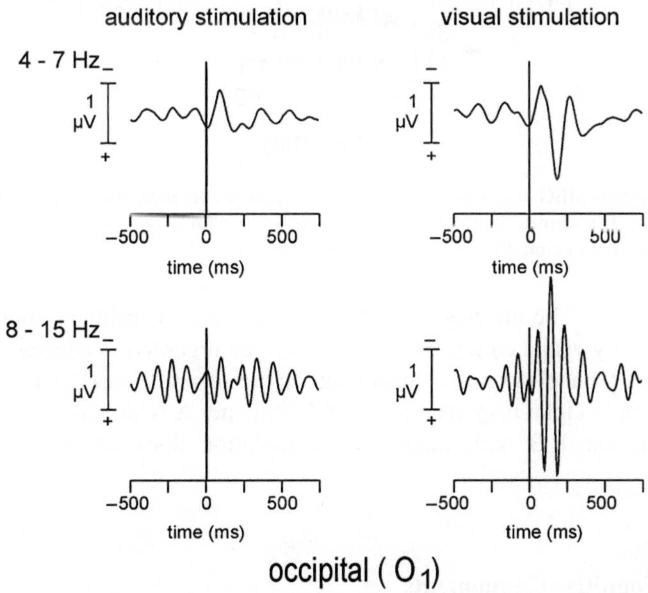

Fig. 6.9 Frequency components of grand average EPs ($n = 11$). *Top*: filter limits: 4–7 Hz, theta response. *Bottom*: filter limits: 8–15 Hz, alpha response. *Left*: acoustical stimulation. *Right*: visual stimulation (from Schürmann et al. 1996)

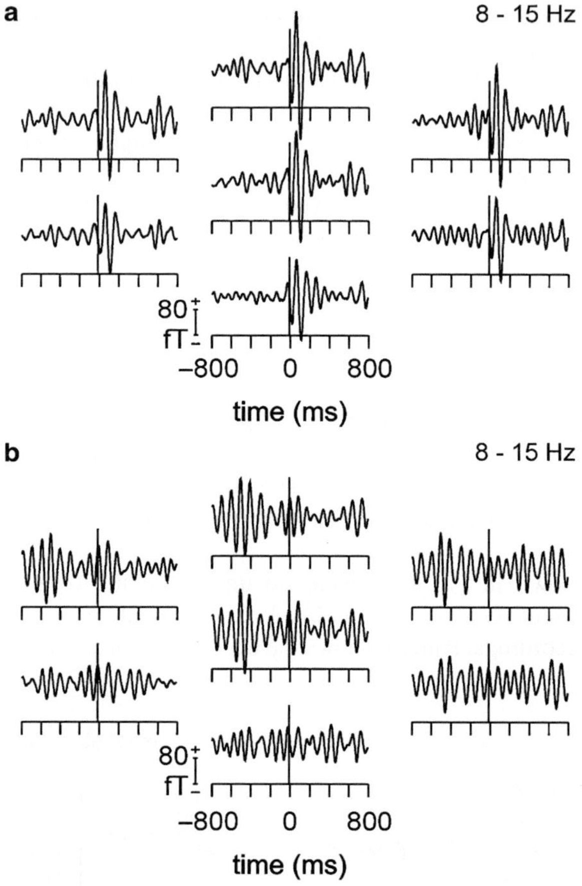

Fig. 6.10 Human MEG responses to auditory stimulation averaged evoked fields recorded in a typical subject (filter limits: 8–15 Hz). (**a**) Seven channels with pure temporal location. (**b**) Seven channels with pure occipital location (from Schürmann et al. 1996)

auditory stimuli. The underlying cortical areas being the primary auditory cortex and the primary visual cortex, auditory stimuli are regarded as adequate in the first case (Fig. 6.10 panel A) and as inadequate in the second case (Fig. 6.10, panel B). High amplitude alpha responses are visible in panel A with adequate stimulation. In contrast, panel B with inadequate stimulation does not show such alpha responses.

6.7.3.2 Cognitive Components

Cognitive targets significantly influence the alpha responses in P300: Using an oddball paradigm, prolonged event-related alpha oscillations up to 400 ms were observed as first published by Stampfer and Başar (1985) and Başar and Stampfer (1985) and later with a clear demonstration by means of single sweep analysis by Kolev et al. (1999) (see also Başar 1998, 1999). Memory related event-related alpha oscillations can be observed in well-trained subjects one second before an expected target (Maltseva et al. 2000). New results (Başar et al. 1997a, b; Klimesch et al. 1994) demonstrate that alpha activity is strongly correlated with working memory and probably with long term memory engrams.

The co-existence of evoked alpha oscillations with alpha blocking and event-related desynchronization (Pfurtscheller et al. 1997) hints at multiple processes being reflected in alpha oscillations. Examples of such co-existence are earlier measurements in which high amplitude spontaneous alpha activity coincided with alpha blocking while low amplitude alpha preceded EPs of high amplitude (Başar 1998; Klimesch et al. 2000a). Klimesch et al. (2000a) in new studies showed a plausible superposition of several types of alpha oscillations together in a schematic form. Furthermore, Krause et al. (2001) also report event-related synchronizations and desynchronization together. For more complete descriptions of function-related alpha the reader is referred to (Başar et al. 1997a, b) and some examples are given in the next paragraph.

6.7.3.3 Resonance in Brain Responses

Experiments have shown that damped alpha activity is not present in all parts of the brain or elicited by all types of stimuli. Only by the combination of EP frequency analysis, adequate stimuli, and appropriate electrode positions can such activities be demonstrated. The results underline the following properties of the neural tissues under study. In the 10 Hz frequency range (filter limits: 8–14 Hz) large enhancements of single visual EPs were recorded in the visual cortex (also reflected in the AFCs in the shape of a dominant 12 Hz peak). In the language of systems theory, significant (sharp) peaks in the amplitude characteristics of the transfer function characterize the resonant behavior of the system studied. This behavior can also be expressed as tuning the "device," or the resonant frequency channels being the "natural frequencies" of the system. For a general description of resonance phenomena in nature, see Appendix C.

6.7.3.4 Multiple Functions in the Alpha Frequency Window

Similar to the gamma band, the selectively distributed alpha system in the brain is interwoven with multiple functions and control functions:

1. The 10 Hz processes may facilitate association mechanisms in the brain: When a sensory or cognitive input elicits 10 Hz wave-trains in several brain structures, then it can be expected that this general activity can serve as resonating signals "par excellence" (Başar 1980).

2. Alpha activity controls EPs following experiments, several authors point out that the amplitude, time course, and frequency responses in EPs strongly depend on the amplitude of the pre-stimulus alpha activity (Başar et al. 1997a, b, 1998).

Makeig et al. (2002) published a relevant study to answer the question whether the averaged evoked potential (EP) is a tiny signal added to otherwise non-stimulus-related EEG oscillations, or is the EP a re-organization of ongoing EEG oscillations? Makeig's data, obtained in 15 subjects with approximately 3,000 trials per subject, substantially widen earlier experimental evidence of the role of phase re-ordering in EP generation.

However, phase re-ordering of spontaneous oscillations is only one of the phenomena indicating the dependence of EPs on spontaneous EEG oscillations. The second phenomenon is enhancement, i.e., an increase in amplitude of "spontaneous" pre-stimulus oscillations (see Figs. 6.7b and 6.10a). In particular, it was demonstrated that higher amplitude EPs were observed for trials with low-amplitude spontaneous EEG. Such findings may be explained in terms of *resonance in the EEG*, which is a basic property of brain tissue: responses to sensory-cognitive inputs that occur in the same frequencies as spontaneous oscillations.

6.7.4 Theta Oscillations in Perception and Cognition

Theta discharges are recorded from hippocampal neurons (Miller 1991), neurons, in n. accumbens, as well as cortical neurons. Hippocampal-cortical (frontal) networks operate in the theta frequency range. Theta cells in the hippocampus with multiple sensory behavioral correlates are described by Best and Ranck (1982). Theta EEG responses have been recorded in rats (Miller 1991), cats and humans (Başar 1999).

The following are some examples related to theta oscillations in perception and cognition:

1. Experimental data suggest that event-related theta oscillations are related to cognitive processing and cortico-hippocampal interaction (Başar 1999; Klimesch et al. 1994; Miller 1991).
2. Theta is the most stable component of the cat P300-like response (Başar 1998, 1999; Sakowitz et al. 2000).
3. Bimodal sensory stimulation induces large increases in the frontal theta response, thus demonstrating that complex events require frontal theta processing (Sakowitz et al. 2000; Başar 1998, 1999).
4. Event-related theta oscillations are prolonged and/or have a second time window approximately 300 ms after target stimuli in oddball paradigm experiments. The prolongation of theta is interpreted as being correlated with selective attention (Başar and Stampfer 1985; Stampfer and Başar 1985; Başar-Eroğlu et al. 1992).
5. Event-related theta oscillations are also observed after an inadequate stimulation, whereas event-related alpha oscillations do not exist if the stimulation is an inadequate one. Accordingly, the associative character for event-related theta

oscillations is more pronounced than for higher frequency event-related oscillations (Başar-Eroğlu et al. 1992).

6. *Orienting* is a coordinated response indicating alertness, arousal or readiness to process information; it is related to theta oscillations and manifested in cat experiments during exploration and searching and motor behavior (Başar 1998, 1999).
7. Time locked theta response reflects inter-individual differences in human memory performance (Doppelmayr et al. 2000).
8. Miller (1991) produced results on cortico-hippocampal signal processing support the functional role of theta transmission in all cognitive states related to *association.*
9. The review by Başar-Eroğlu and Demiralp (2001) indicates that theta response can be associated to several sensory and cognitive mechanisms, the distributed theta system of the brain being mostly assigned to associative processes.

Results on theta responses in recent studies are found in Chapter 7 and 13.

6.7.5 Delta Oscillations in Cognition

Thalamic neurons may discharge in the slow frequency range (Steriade et al. 1990). Slow potentials have been recorded in cortical neurons. Delta responses are recorded in cats and humans (Başar and Stampfer 1985; Schürmann et al. 1995). Experimental data hint at functional correlates roughly similar to those mentioned for theta oscillations, i.e., mainly in cognitive processing as in the following:

1. The responses to visual oddball targets have their highest response amplitude in parietal locations, whereas, for auditory target stimuli the highest delta response amplitudes are observed in central and frontal areas (Başar 1998, 1999; Schürmann et al. 1995).
2. *Cognitive functions:* The amplitude of the delta response is considerably increased during oddball experiments. Accordingly, it was concluded that the delta response is related to signal detection and decision making (Başar-Eroğlu et al. 1992).
3. *In response to stimuli at the hearing threshold*, delta oscillations are observed in human subjects consistent with the hypothetical relation to signal detection and decision-making (Başar 1998, 1999).
4. A waveform observed in response to deviant stimuli not attended by the subject, the mismatch negativity (Näätänen 1992) is shaped by a delayed delta response superimposed with a significant theta response (Karakaş et al. 2000b). Phase-locked delta responses are probably the major processing signals in the sleeping cat and human brains (Başar 1980, 1999; Röschke et al. 1995).

6.7.6 Activation of Alpha System with Light

When a subject (human or cat) has been visually stimulated, large alpha enhancements are recorded in the mesencephalic reticular formation, lateral geniculate nucleus and, in parallel, in the hippocampus. There are also large alpha enhancements in the visual and association cortices. Information flows to the polymodal association cortex and to the limbic system. The results have also shown that large theta enhancements were recorded in the thalamus, hippocampus, primary cortex, and association cortices including the frontal lobes (Başar 1999). The same Figs. 6.5a, b illustrate that alpha enhancements cannot be seen in the auditory areas of the cortex and thalamus on visual stimuli; however, theta enhancements are also present in these structures following light stimulation.

Large alpha and theta enhancements are observed in the limbic system, reticular formation, and lateral geniculate nucleus. No alpha enhancements are recorded in nuclei of the auditory pathways. Although the theta enhancements are one of the major response components also in thalamus and cortex, the alpha responses are not recorded in the auditory cortex and the medial geniculate nucleus.

Figure 6.11 describes schematically and globally the distribution of visual alpha responses in various structures of the cat brain according to empirical results.

6.7.7 Activation of the Alpha System with Auditory Stibmulation

Figure 6.12 explains a similar hypothetical flow of information on an auditory stimulation. In this case alpha enhancements dominate the structure in the auditory pathways, i.e., reticular formation and the hippocampus. The neurophysiologic processes in subcortical structures depict large alpha response components in reticular formation, medial geniculate nucleus, and auditory cortex. It is noteworthy that reticular formation and hippocampus show large alpha responses, whereas lateral geniculate nucleus and visual cortex show only that of theta responses. Theta enhancements are present in all structures, regardless of whether the stimulus is inadequate or adequate. The way of approaching the responses to auditory stimulation can also be applied to visual stimulation, but the largest alpha enhancements are marked in temporal, parietal and occipital areas.

Figure 6.12 describes schematically and globally the distribution of auditory alpha responses in various structures of the cat brain according to empirical results.

6.8 Superposition Principle and Theta and Delta Frequency Windows Shown with Examples in Cognitive Processes

According to the principle of superposition, the existence of different peaks in evoked oscillations does not necessitate the existence of different functional structures; similarly, the disappearance of peaks does not necessarily show that the functional groups have ceased their activity (Başar 1980; Başar and Ungan 1973). The basic application of the superposition principle is shown in the example of the auditory evoked potential of the cat hippocampus of Fig. 6.3. In the following, examples are given from recordings during cognitive processes.

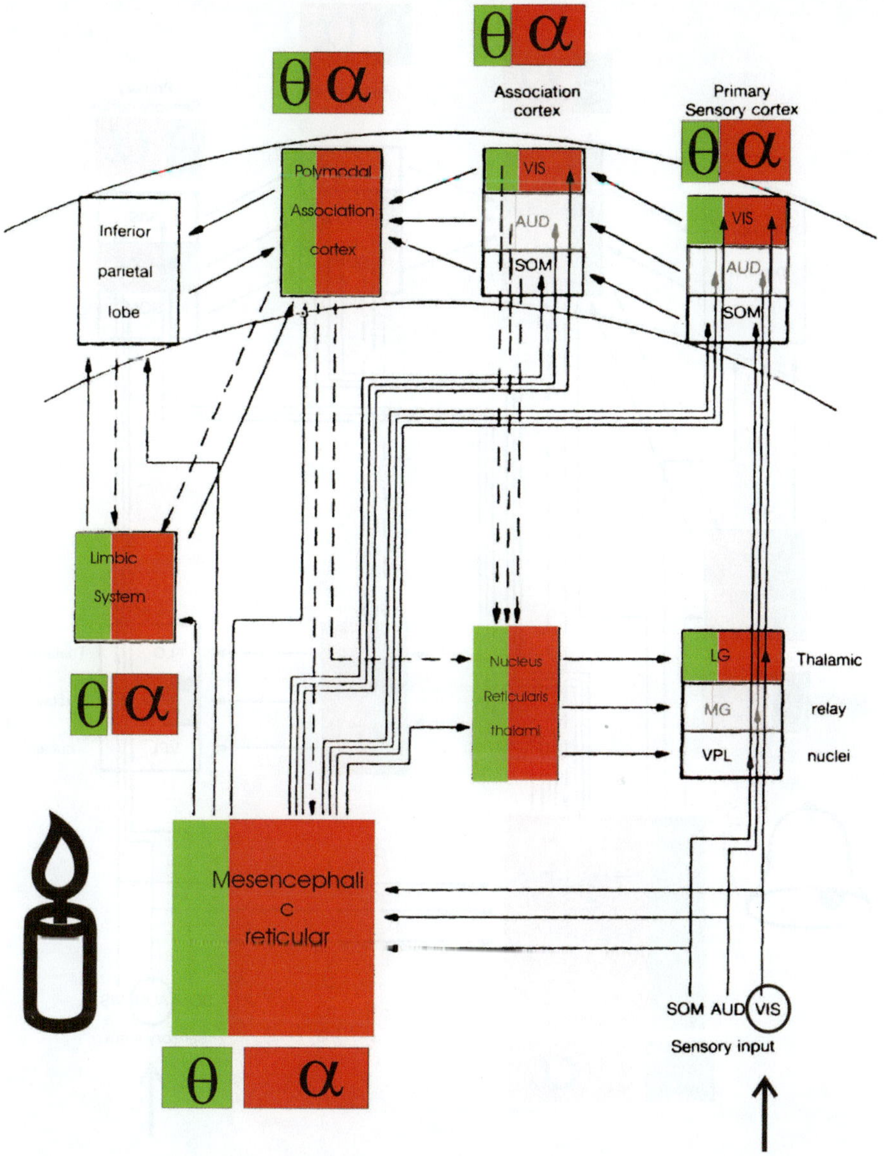

Fig. 6.11 Flow of oscillatory information in alpha and theta frequency channels in the visual pathway, reticular formation, limbic system, and association areas of the cortex. Letters *alpha* and *theta* indicate the existence of strong enhancements (sensory and cognitive neural pathways in this figure are modified from Flohr [1991])

Karakaş et al. (2000a, b) investigated the contribution of delta and theta responses to oddball P300 ERP components that were recorded from two topographical sites (Fz and Pz). The results showed that it is the interplay between

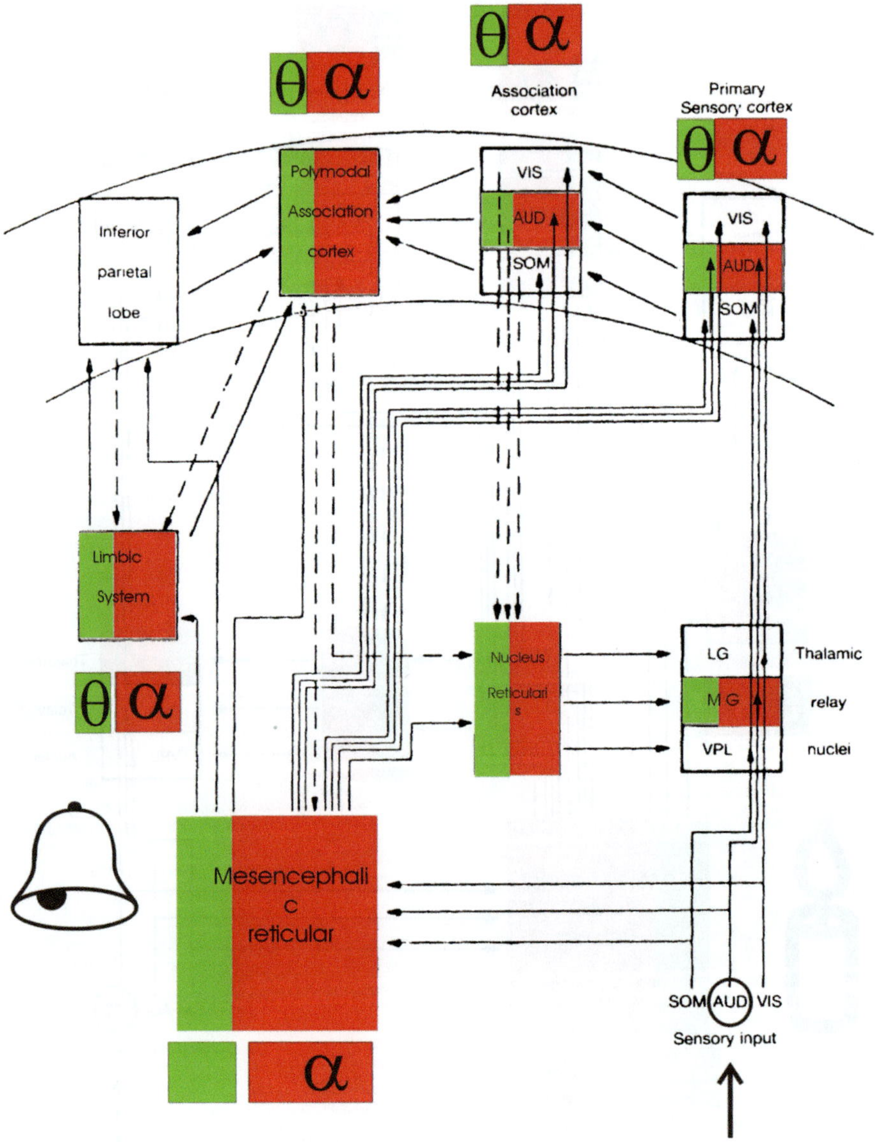

Fig. 6.12 Flow of oscillatory information in alpha and theta frequency channels in the auditory pathway, reticular formation, limbic system and association areas of the cortex. Letters *alpha* and *theta* indicate the existence of strong enhancements (sensory and cognitive neural pathways in this figure are modified from Flohr [1991])

the theta and the delta oscillations that produces the structure and the amplitude of the P300 component (Fig. 6.13).

Findings of the study by Karakaş et al. (2000b) further showed that the delta response contributes to the amplitude at the P300 latency and congruently, that it

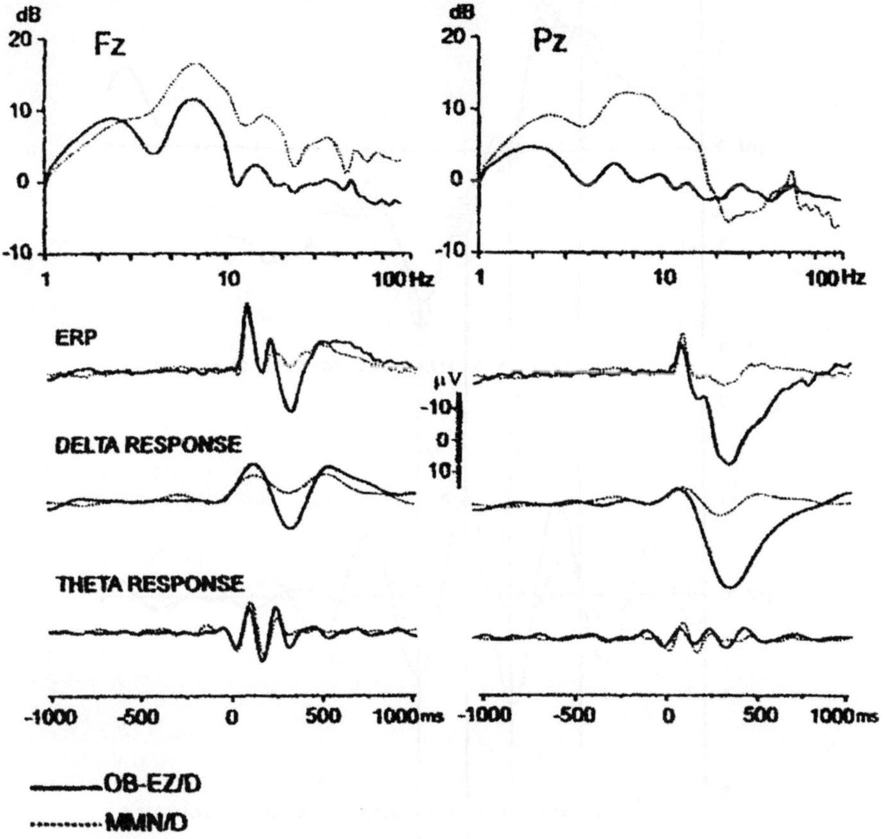

Fig. 6.13 From *top* to *bottom*: the amplitude frequency characteristics (abscissa: frequency in logarithmic scale; ordinate: potential amplitude, $|G(j\omega)|$, in decibels), grand average ($n = 42$) EEG-ERPs, filtered EEG-ERPs (abscissa: time in ms; ordinate: amplitude in μV). Stimulation applied at 0 ms time point. Curves for Fz (*left column*) and Pz (*right column*) are superimposed for MMN (*dotted line*) and OB (*continuous line*) paradigms (from Karakaş et al. 2000)

varies in amplitude with task-relevant responding that necessitates conscious stimulus evaluation and memory updating. Delta response thus represents the cognitive effort that involves stimulus-matching and decision with respect to the response to be made (Başar 1999; Başar-Eroğlu et al. 1992). Being recorded from various locations of the scalp, the delta response is, physiologically, a product of the distributed response systems of the brain (Başar 1999; Başar-Eroğlu et al. 1992).

Doppelmayr et al. (2000) measured theta responses in relation to memory performance. Their most relevant finding with respect to memory performance indicates that the increase in theta band power is significantly larger for successful as compared with unsuccessful encoding and retrieval attempts. Only for good memory performers the theta response oscillations appeared in preferred time windows after a target was presented in Fig. 6.14.

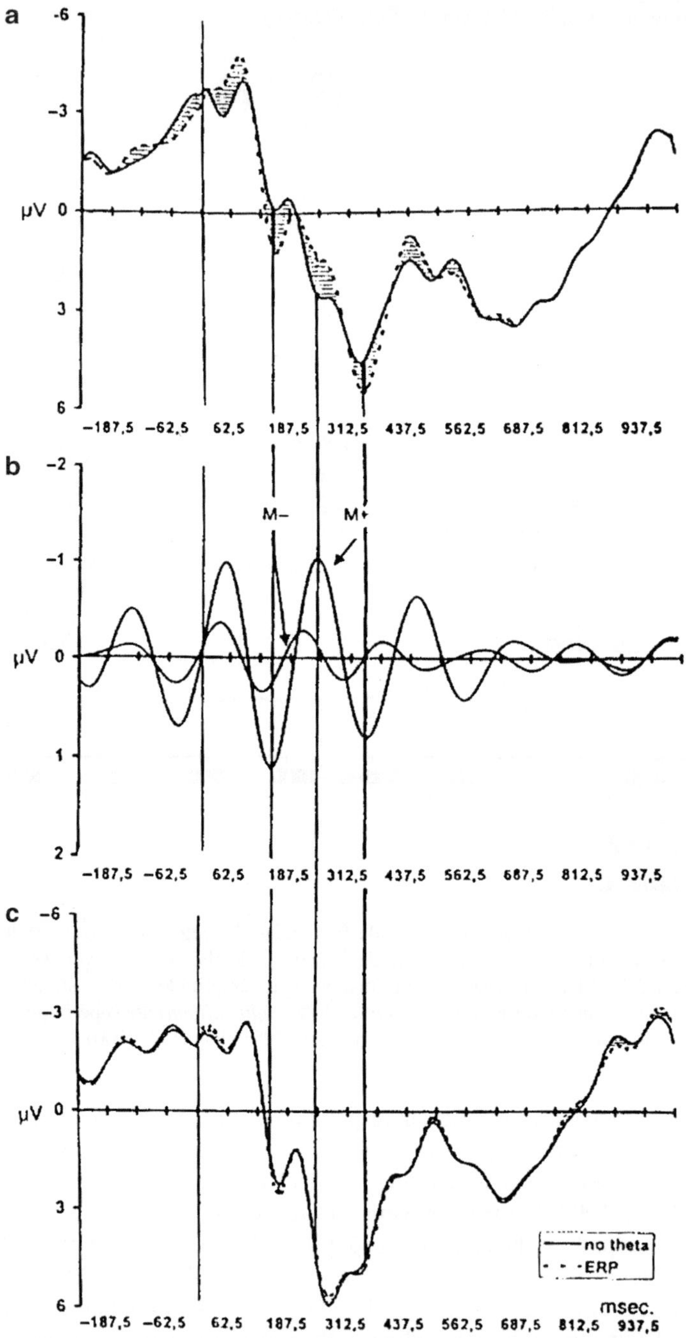

Fig. 6.14 The standard ERP (*dashed line*) and the ERP from which the theta ERP is subtracted (= ERP-θ; *bold line*) are shown for good and bad performers (**a, c**). The theta ERP is shown in (**b**). Note the large evoked theta for M+ in (**b**) and corresponding to this finding the large difference between the standard ERP and ERP-θ in (**a**) (from Doppelmayr 2000 by permission)

6.8.1 Activation of Theta and Delta Systems Following Cognitive Inputs

Again the same anatomical pictures of Fig. 6.6 are used to explain the electrical responses occurring in the region of 300 ms upon a cognitive target. For the sake of simplicity, we do not consider the delayed and prolonged 10 Hz responses and only take into consideration the dominance of delta, and theta responses. Experiments described by Başar-Eroğlu et al. (1992, 2001) confirmed large delta and theta enhancements around 300 ms following the stimuli. When the stimulation applied contains an event-related target, the hypothetical flow of information would show large theta and delta enhancements in all areas of the cortex, but probably also in the substructures. Large theta and delta responses and delayed alpha responses would be observed first at 300 ms after stimulation.

Figure 6.15 describes schematically and globally the distribution of auditory cognitive alpha theta and delta responses in various structures of the cat brain according to empirical results.

6.9 The Importance of Gamma Oscillations in Sensory, Cognitive and Motor Processes

6.9.1 Historical Survey

The empirical background of the gamma band dates back to Lord Adrian (1942) who reported that the application of odorous substances to the olfactory mucosa of the hedgehog induced a train of sinusoidal oscillations, within the 30–60 Hz range. Starting from 1942 to date, studies on the 40 Hz rhythmicity have passed through a total of four phases according to Başar-Eroğlu and colleagues (1996a, b). Initiated by Adrian's classical work, the induced character of the gamma band was studied in the first phase. The second phase took place between 1960 and 1980. The phase was characterized by the works of Freeman (1975), Başar and Özesmi (1972), Başar et al. (1980, 1979a, b) and Sheer (1976) in which a variety of functions were ascribed to gamma rhythmicity. The third phase started with the work of Galambos et al. (1981), which led to investigations concerning the sensory and cognitive correlates of gamma oscillation primarily in humans. The fourth phase started with the work of Gray and Singer (1987), which led to investigations of the 40 Hz at the cellular level. The present, fifth phase, is marked by the heterogeneity of the application of approaches and techniques focusing on solving the gamma puzzle (see also Karakaş and Başar 1998).

The empirical findings on the gamma band may be roughly classified into sensory (or obligatory) versus cognitive gamma responses. Some examples of sensory functions are:

1. *Sensory processes.* A phase-locked gamma oscillation is also a component of the human auditory and visual response as Fig. 6.16 shows. A new strategy involving the application of six cognitive paradigms showed that the 40 Hz response in the

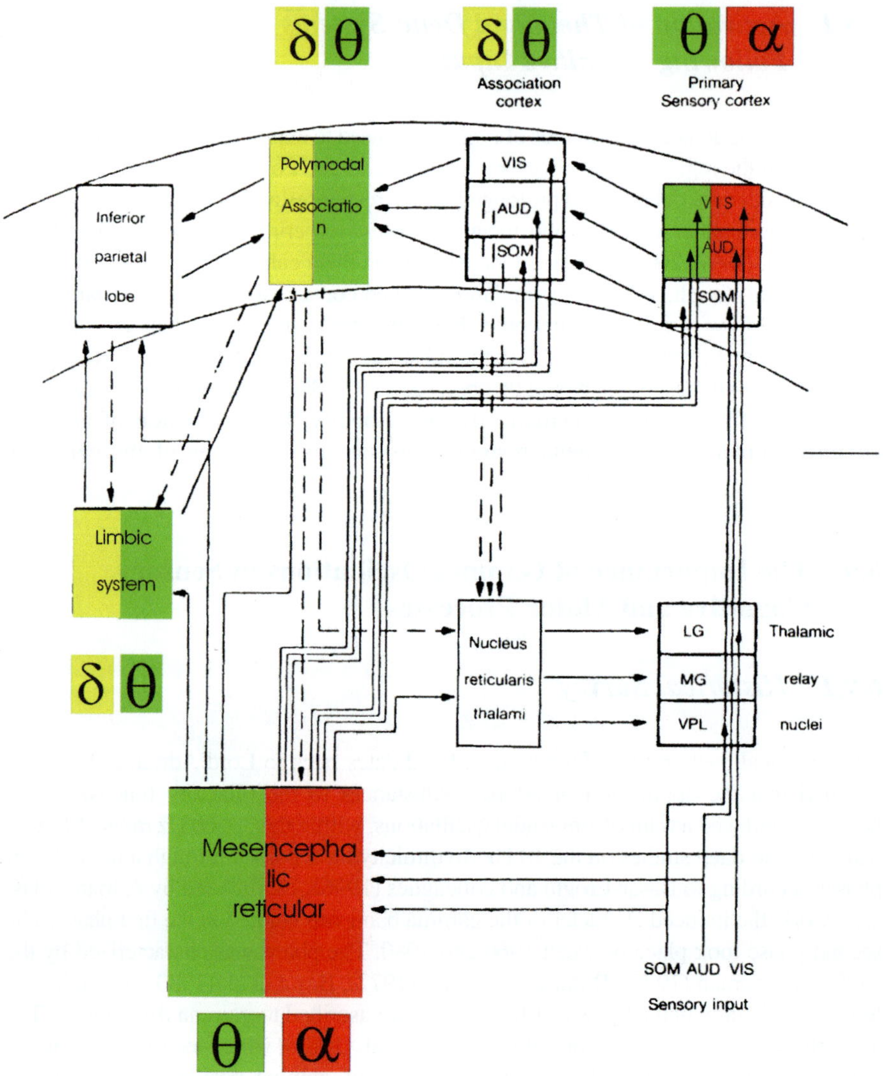

Fig. 6.15 Flow of oscillatory information in theta and delta frequency channels in the visual and auditory pathways (VIS, AUD) and limbic system. Letters *theta* and *delta* indicate the existence of strong enhancements (sensory and cognitive neural pathways in this figure modified from Flohr [1991])

100 ms after stimulations has a sensory origin, being independent of cognitive tasks (Karakaş and Başar 1998) (Fig. 6.16).

2. The auditory MEG gamma response is similar to human EEG responses with a close relationship to the middle latency auditory evoked response (Pantev et al. 1991).

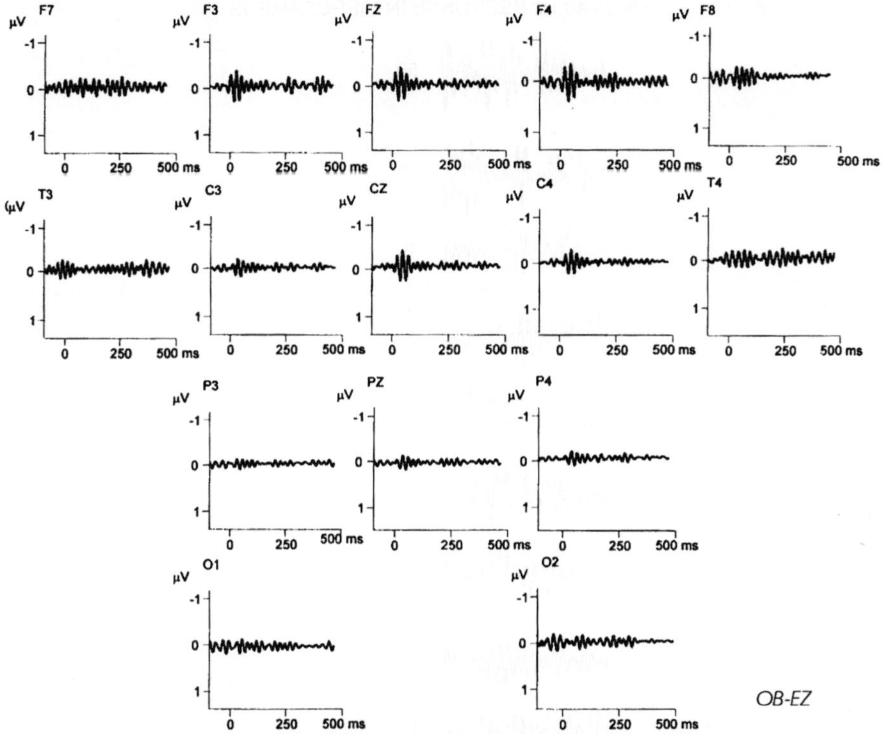

Fig. 6.16 Topography of the grand averages of filtered (28–46 Hz) EEG-ERPs from the target stimuli of the OB-EZ paradigm (from Karakaş and Başar 1998)

3. *Cognitive processes.* Several investigations dealt with cognitive processes related to gamma responses, some of them based on measuring the P300 wave. This positive deflection typically occurs in human ERPs in response to oddball stimuli or omitted stimuli interspersed as targets into a series of standard stimuli.

A P300-40 Hz component has been recorded in the cat hippocampus, reticular formation, and cortex (with omitted auditory stimuli as targets). This response occurs approximately 300 ms after stimulation, being superimposed with a slow wave of 4 Hz illustrated in Figs. 6.17 and 6.18 (Başar-Eroğlu and Başar 1991). Preliminary data indicate similar P300-40 Hz responses to oddball stimuli in humans (Başar-Eroğlu et al. 1992). However, a suppression of 40-Hz activity after target stimuli has also been reported (Fell et al. 1997). In a recent study the gamma band activity in an auditory oddball paradigm was analyzed with the wavelet transform. A late oscillatory peaking at 37 Hz with latency around 360 ms was observed, appearing only for target stimuli (Gurtubay et al. 2001). A study in the human limbic system also confirms the synchronization of gamma rhythms during cognitive tasks (Fell et al. 1997).

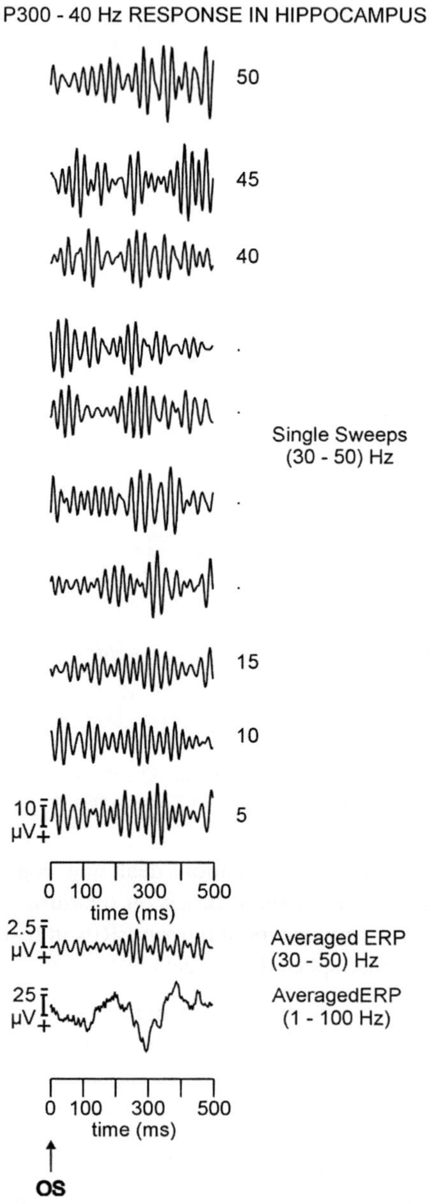

Fig. 6.17 Event-related potentials of the lower pyramidal layer (CA3) of hippocampus (one cat). *Top*: Single ERP sweeps (epochs) filtered at 30–50 Hz. *Middle*: Averaged ERP filtered at 30–50 Hz. *Bottom*: Unfiltered ERP, average of 50 artifact-free epochs (modified from Başar-Eroğlu et al. 1991)

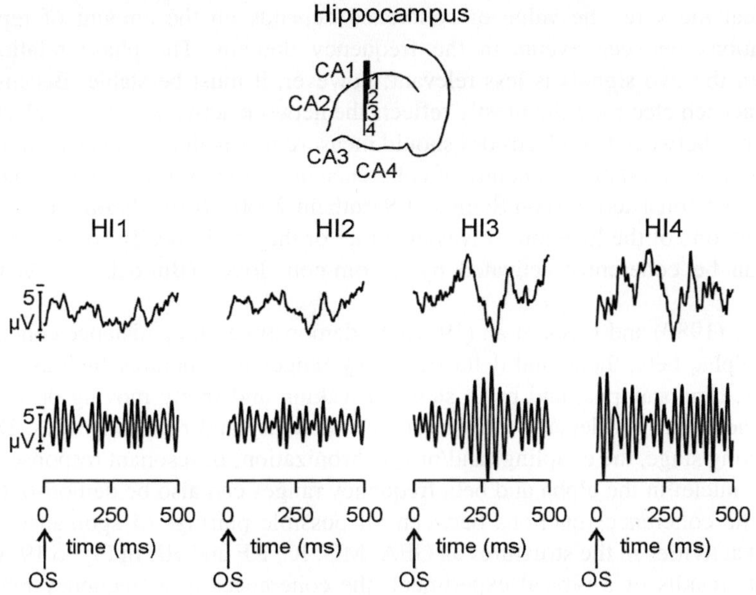

Fig. 6.18 Grand averages (mean values from 8 cats) of ERPs in various layers of hippocampus. *Top*: Location of multi-electrode; CA1 and CA3 corresponding to HI1 and HI3/HI4 respectively. *Middle*: Unfiltered ERPs. *Bottom*: Filtered ERPs (30–50 Hz; OS: omitted stimulation) (modified from Başar-Eroğlu et al. 1991)

6.9.1.1 Multiple Functions in the Gamma Band

The wide spectrum of experimental data being presented is, in accordance with a hypothetical selectively distributed parallel processing gamma system with multiple functions. Rather than being highly specific correlates of a single process, gamma oscillations might be important building blocks of electrical activity of the brain. Being related to multiple functions, they may (1) occur in different and distant structures, (2) act in parallel, and (3) show phase locking, time locking or weak time locking. Notably, simple electrical stimulation of isolated invertebrate ganglia evokes gamma oscillations (in the absence of perceptual binding or higher cognitive processes. see also Chapter 7). In conclusion, gamma oscillations possibly represent a universal code of CNS communication (Başar 1998, 1999).

6.10 Selectively Distributed and Selectively Coherent Oscillatory Networks

The description of integration needs morphological, functional interrelation in defined durations in the time-space; the degree of interactions between two signals can be measured by coherence (von Stein and Sarnthein 2000). Coherence is a statistical measure; the value of coherence depends on the amount of repeated correlations between events in the frequency domain. The phase relationship between the two signals is less relevant; however, it must be stable. Because the signal at each electrode site mostly reflects the network activity under the electrode, coherence between two electrodes should measure interactions between two neural populations. The statistical nature of coherence helps to unravel them from noise if they repeat consistently (von Stein and Sarnthein 2000). If two brain locations are coherent, one of the locations drives the other or they reciprocally cooperate. They also can be coherently activated by a common driver (Bullock and McClune 1989).

Başar (1980) and Başar et al. (1979a, b) demonstrated long distance coherences in the alpha, beta, theta, and delta frequency ranges in structures such as sensory cortices, hippocampus, and brain stem, in waking and freely moving or sleeping cats. Their strength depends on stimulation modality and recording sites. During the waking stage, the coupling (and/or synchronization) of resonant responses from various nuclei in the alpha and beta frequency ranges can also be demonstrated by using the coherence functions between all possible pairings of spontaneous and evoked activities in the structures as GEA, MG, IC, RF and HI. In Fig. 6.19, which presents results of a typical experiment, the coherences in a frequency range of 3–60 Hz for spontaneous activities and evoked potentials of all possible pairings of the studied brain structures are illustrated. Because of the reasonably large number of single sweeps included in the computation of averages and the spectral window used, the auto and cross-spectral amplitudes have been adequately smoothed and a significance value of 0.2 has been attained for all the curves. Therefore, the area under the coherence function is darkened only if the curve surpasses this value, to give emphasis to those parts of the curves above the significance level. It is immediately recognizable, that in the alpha (8–14 Hz) and beta (14–25 Hz) frequency ranges the coherence usually has high values between 0.5 and 0.9 for evoked responses. However, the coherence between spontaneous activities of the same pairings of nuclei definitely has lower values. In the alpha and beta frequency ranges, the coherence of the spontaneous activity barely reaches 0.3 in a few cases. In other words, there exists an important coherence increase upon stimulation. During slow wave sleep stages the evoked coherences between all brain structures were shifted to a slower delta frequency range (Başar 1980; Başar et al. 1979b). In Fig. 6.20 the coherences in the cat brain to visual stimuli are presented.

In search of experimental support for the hypothesized distributed alpha response system (Başar 1999), we measured EEG responses to visual stimuli were measured in the cat brain (with intracranial electrodes in cortical, thalamic and hippocampal sites). Alpha responses (10 Hz oscillations of 200–300 ms duration) were observed not only in occipital cortex and thalamus, but also in the hippocampus. Remarkably, hippocampal alpha amplitudes were higher in amplitude than thalamic ones, and the coherence of hippocampus-cortex was higher than the evoked coherence between thalamus and cortex (Schürmann et al. 2000).

(1) Coherences increased significantly on visual stimulation in delta, alpha, beta, and gamma frequency ranges; (2) hippocampal-cortical coherences were significantly

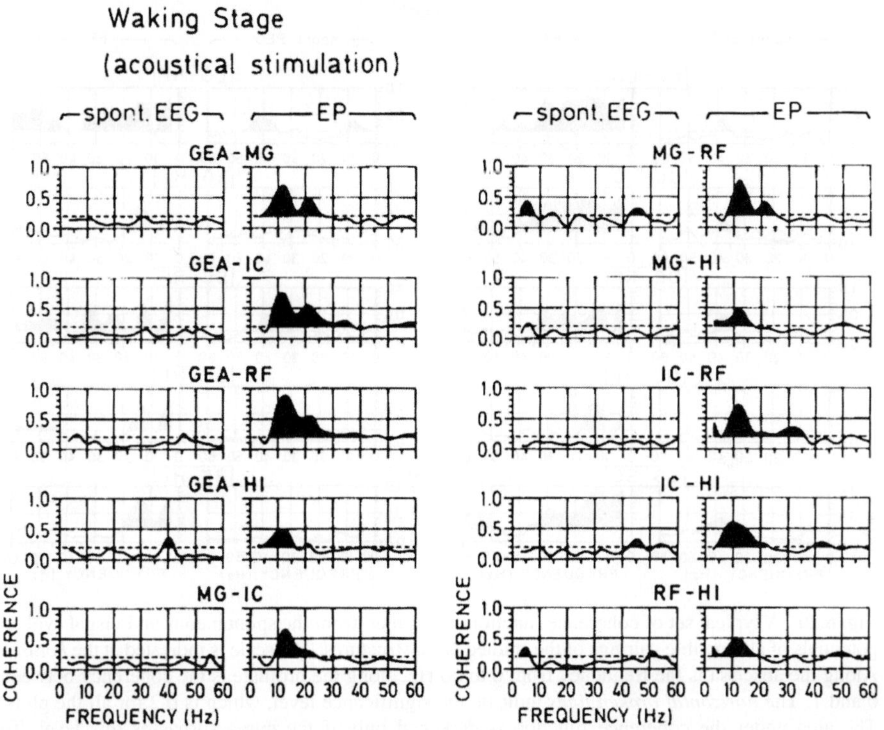

Fig. 6.19 A typical set of coherence functions computed from the spontaneous activities and EPs of all possible pairings of the studied brain structures during the waking stage. The scale is indicated at the *bottom*. Along the abscissa is the frequency from 0 to 60 Hz, along the ordinate is the coherence between 0 and 1. The *horizontal broken lines* indicate the significance level, which is 0.2 for all the plots. The area under the coherence function is darkened only if the curve surpasses this level. To facilitate a comparison between the coherence values computed from spontaneous and evoked parts of the EEG-EP EP-epochs the respective coherence functions are presented adjacently as couples for all the pairings of recording electrodes (from Başar 1980)

higher than thalamo-cortical ones for the alpha (HI-OC: 0.28 vs. LG-OC: 0.15), beta and gamma frequency ranges; (3) the effect of the interaction of channel and experiment factors, which refers to the differences of coherence changes upon stimulation between the two electrode pairs, was only significant in alpha and beta frequency ranges. Upon visual stimulation, the alpha coherence HI-OC increased to 0.41, whereas, the coherence LG-OC increased only to 0.18.

These findings also demonstrated that during sensory stimulation the recordings between various structures showed varied degrees of coherence, thus indicating the interaction between two structures (Kocsis et al. 2001; Schürmann et al. 2000). Varied degrees of coherences lead to the concept of selectively coherent oscillatory networks. The important role of functional integration and frequency response of long-range interactions in the alpha and theta ranges in processing of the mental context is also emphasized by confirming the author's long-standing view (Başar

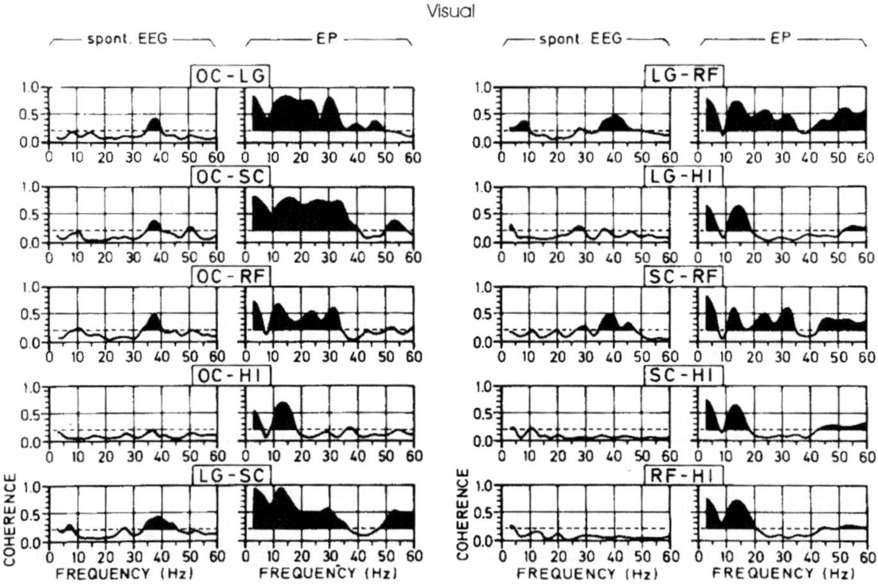

Fig. 6.20 A typical set of coherence functions computed from the spontaneous and visual evoked potentials of all possible pairings of the studied brain structures. The scale is indicated at the *bottom*. Along the abscissa is the frequency from 0 to 60 Hz, along the ordinate is the coherency between 0 and 1. The *horizontal broken lines* indicate the significance level, which is 0.2 for all the plots. The area under the coherence function is darkened only if the curve surpasses this level. To facilitate a comparison between the coherence values computed from spontaneous and evoked parts of the EEG-EP epochs, the respective coherence functions are presented adjacently as couples for all pairings of recording electrodes (from Başar 1980)

1980, 1999; von Stein and Sarnthein 2000). In the gamma frequency range, coherences between different spatial locations of the brain vary as these areas are activated with different classes of stimuli (haptic and visual) in an associative learning task (Miltner et al. 1999).

6.11 Interim Conclusions

The given examples of brain oscillatory activity converge into the following interim summary:

1. Event-related oscillations are real brain responses depending on stimulation modality, cognitive states, and showing clear differences in frequency, duration, and topography depending on brain state or function under study (Başar et al. 2001b).
2. A given function is manifested with multiple (not only one) oscillations, being selectively distributed in the brain (selectively distributed alpha, gamma, theta, and delta systems).

3. Vice versa, oscillatory responses are distributed and do not represent only one function but are involved in several functions.
4. The application of the principle of superposition will probably give more insight into the interpretation of changes in neuro-electricity during sensory cognitive processing.
5. Depending on the function, oscillatory responses show (1) Time-locking, (2) Phase-locking (3) Enhancement (4) Delay, and (5) prolongation (or several time windows).

6.11.1 A Brain-Body-Mind Interpretation Needs the Concept of Oscillatory Dynamics

At the beginning it was stated that the results of the present chapter will provide core information for the understanding of the entire book. The most general function of the brain is governed by alpha, beta, delta, theta, gamma oscillations that are selectively distributed in the brain. *Governing* means that in all functions, multiple oscillations are recorded in all tissues of the brain. However, oscillations are not manifested in every tissue with the same weight or linking properties. In basic processes (light stimuli, sound stimuli) responses are governed with, alpha, beta, delta, theta, and gamma responses. During cognitive processes, theta and delta components gain more weight in certain areas of the cortex. Once this is said, it is necessary to face the next important questions that are related to brain-body-mind understanding.

The basic processes in the cat and human brain have been presented; however, the minds of living beings in the animal world are certainly different. There are a number of emerging questions, to be answered in various chapters in this book. The questions are as follows:

1. The mind of a baby is certainly different from an adult, then what is the anatomy-electrophysiology constellation in babies. How are the basic processes modified during evolution? Has a snail ganglion or fish brain alpha responses. If so, are they different from the human alpha?
2. Are the alpha and theta responses of the brain different during complex processes as face recognition or emotional inputs?
3. What are dynamics in the brain of an older person, or in the pathologic brain?
4. What is the manifestation of transmitters in electrophysiology?
5. What is the role of transmitters in dysfunction of brain-mind-body integration?

These questions will be analyzed in Chapters 10–13.

Chapter 7
Dynamic Memory

7.1 Different Levels of Memory

7.1.1 Fuster's View on Memory Networks

According to Fuster (1995a), the cognitive functions of the frontal cortex, as with any part of the neocortex, consist of the activation and processing within and between networks of representation, or memory networks. Those networks are widely distributed and highly specific, defined by their synaptic structure and connectivity. Thus, the memory code is a relational code, and all memory is associative. In our opinion, one of the most important concepts proposed by Fuster (1995a) is that memory networks overlap and are diffusely interconnected with one another. Thus, a single neuron or group of neurons anywhere in the cortex can be part of many networks and thus many memories. This is why it is virtually impossible, by any method, to localize a memory.

7.1.2 A Tentative Model Related to EEG Activation

In all living beings, memory is incorporated within survival functions, ranging from the simplest reflex to the higher nervous activity including episodic and semantic memory processes. According to this judgment, within the hierarchy of memory functions, all the levels of living and survival processes are categorized. Three different levels of memory states are introduced in the scheme of Fig. 7.1.

Level I includes *inborn* (built-in memories), i.e., networks that are active during retrieval processes that are genetically coded and are usually not altered or less altered during life.

Level II includes *dynamic memory* states that are activated and interactive with integrative functions.

Level III includes *longer-term memory activation*. As we will see in the coming sections, no exact boundaries can be described between these levels.

E. Başar, *Brain-Body-Mind in the Nebulous Cartesian System: A Holistic Approach by Oscillations*, DOI 10.1007/978-1-4419-6136-5_7,

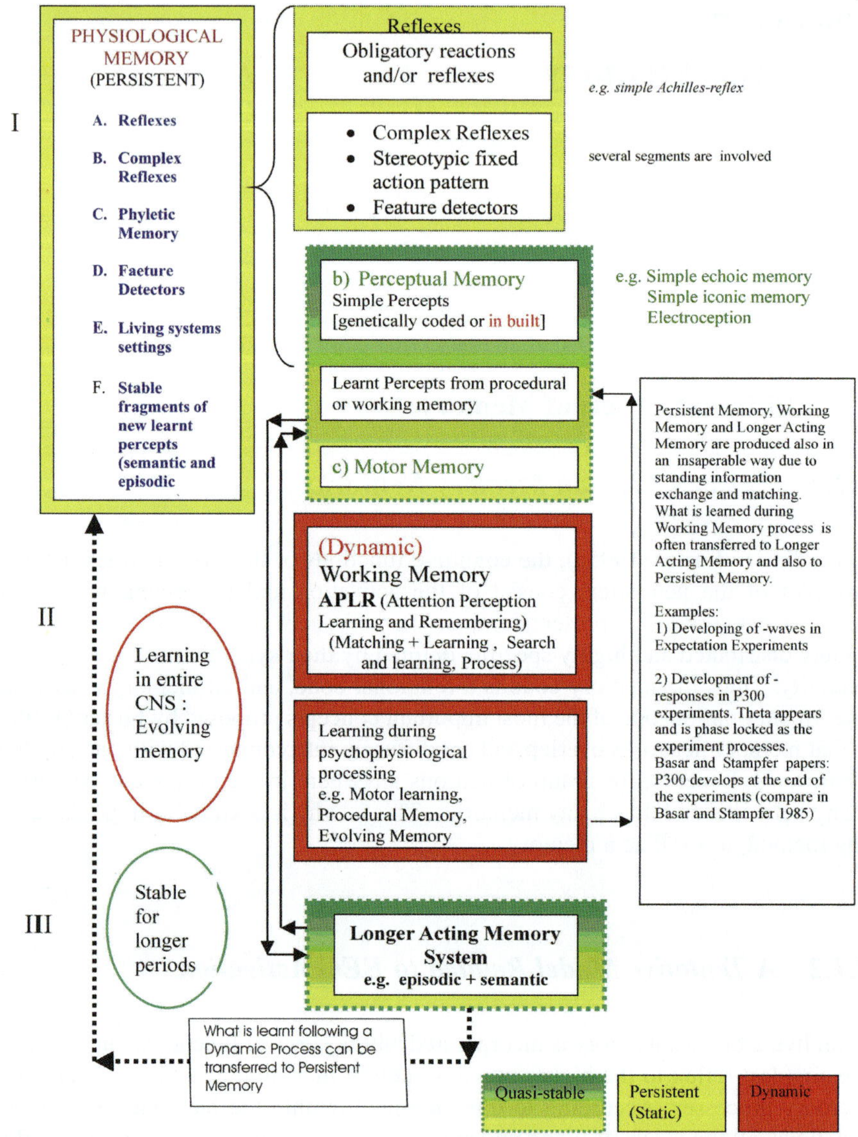

Fig. 7.1 Draft for a detailed scheme of memory levels, hierarchy, and transitions between memories. Level I: At the top of the illustrations is the persistent memory, of which the essential part is the physiological (or fundamental memory) composed of inborn (or built-in) memories. They are (1) Reflexes (a) monosynaptic, simple reflexes as Achilles reflex); (b) complex reflexes involving multiple segments in the spinal cord; (2) stereotypic fixed action patterns (e.g., flight reaction of *Aplysia*). Phyletic memory (echoic memory, iconic memory, ability of electroception) also belongs to physiological memory. Living systems setting as blood pressure of healthy living beings, smooth muscle reactions, heart beating also contribute to physiological memory. Motor memory also belongs to stable physiological memory. In this illustration, the *yellow background* is used to

7.1.3 Inborn (Built-in) Networks (Level I)

To establish a hierarchical classification of inborn memories, we propose six levels of inborn memory that can be classified as parts of the physiological (or fundamental-functional memory)

1. Simple reflexes
2. Complex reflexes
3. Stereotypic fixed action patterns
4. Phyletic memory
5. Feature detectors
6. Living system settings

7.1.3.1 Reflexes

The elements of reflexes are neurons and the reflex networks may be considered as reflex pathways linking inputs and outputs by way of specifiable transforms. There are many reflex-systems, some being as simple as those in the spinal cord described by Sherrington, others much more complex. Pavlov (1927) and Skinner (1938) gained insights into reflex processes supporting learning and memory. Sherrington (1948) provided physiological documentation of the neural pathways that connected stimulus with response. It was shown that a knee-jerk elicited by tapping the tendon below the patella is mediated by two types of neurons.

7.1.3.2 Stereotypical Fixed Action Patterns and Phyletic Memory

As a relevant example we mention the fixed action pattern of the escape actions of *Aplysia californica* in contact with a starfish, as illustrated in Fig. 7.2. Fuster (1995a) states the following: To understand the formation and topography of memory, it is useful to think of the primary and sensory motor areas of the cortex

Fig. 7.1 (continued) represent static components. Such memory types are persistent. In this illustration, the *green–yellow* background is associated with quasi-stable longer-term memory states. Newly learned percepts, acquired after the activation of procedural memory or working memory, are quasi-stable. There are new learned percepts during life; these are not inborn, but they can remain over very long periods as quasi-stable memory states. With time, they can be replaced or forgotten. Accordingly, in this illustration, the background is green–yellow. In other words, perceptual memory is categorized between Levels I and II. Level II indicates the level of dynamic processing and the working memory state. Dynamic changes in the APLR alliance are strongly associated with evolving memory. Following motor-learning or procedural memory, new engrams can be created. They are then transferred to perceptual or motor memory, as indicated with an *arrow*. Following learning during procedural or evolving memory states, the new memorized information is transferred to longer-term memory state III. Semantic and episodic memory is categorized at this level III, which is a quasi-stable level, marked with gray and yellow. The new learned material following a dynamic process is sometimes also transferred to persistent memory (from level III to level I), as the arrow indicates. Persistent memory is indicated as a separate block and colored *yellow*

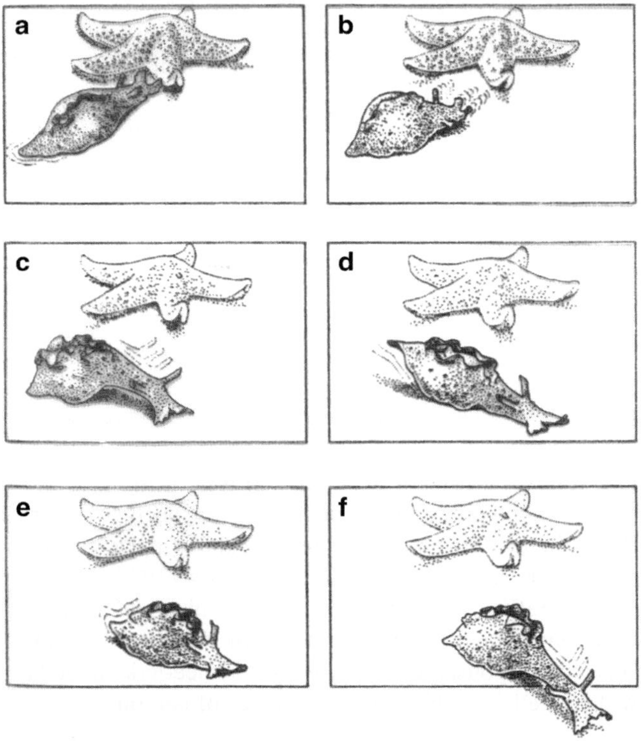

Fig. 7.2 Escape response of *Aplys, 6.13ia californica* to starfish *Astrometis sertulifera*. On contact with the starfish (**a**), the animal withdraws (**b**), turns away from the starfish (**c**), and escapes with rapid pedal waves (**d–f**). (Modified from Byrne et al., 1978)

that we may call *phyletic memory* or *memory of the species.* The structure of primary sensory and motor cortices may be considered a fund of memory that the species has acquired during evolution. This can be called memory because, like personal memory, it is information that has been acquired and stored, and can be retrieved (recalled) by sensory stimuli or the need to act. In the structure of primary systems, the phyletic memory contains the innate capacity to respond to, and to recall, the elementary features of sensation and movement that are common to the repertoires of all members of the species.

7.1.3.3 Feature Detectors

The primary features of stimuli, such as heat, force, light, sound, and chemical substances are selectively transduced at the peripheral ends of sets of sensory (afferent) nerve fibers. Different groups of those sensory fibers respond selectively at lower thresholds than do other groups to different forms of impinging energy.

This "tuning" is often called *feature detection* and is accomplished during evolution of species by the development of specific transducer mechanisms for different forms of energy, either in the nerve endings themselves or in complex sensory organs in which the afferent fibers terminate. Examples are the mammalian retina, cochlea and the pressure transducers of the primate hand skin (see the statements of Sokolov, 2001; Mountcastle, 1998).

7.1.4 What Is Physiological Memory? What Is Fundamental Memory?

Başar (2004) has proposed the use of the expression *physiological memory* to describe the ensemble of memories or living system settings that are necessary for all vital functions: Accordingly, physiological memory is a fundamental memory, because functioning of the CNS is impossible without this (see Fig. 7.1). Physiological memory is genetically coded and differentiated among species. It comprehends sensory memories as echoic and iconic memory; it also comprehends reflexes and stereotypic action patterns (see Fig. 7.2) and perceptual memory.

According to the results and flow charts in Figs. 6.3–6.5a, b; 6.6, 6.11 and 6.15, both sensory and perceptual memory mechanisms are manifested with multiple oscillations in the alpha,beta,gamma,theta,and delta frequency bands.

In all living systems, depending on the level of evolution or the type of species, there are several types of reflexes and also differentiated forms of sensory memory. We provide some examples to illustrate this differentiation.

Species of ray have poor vision, but have the ability of electroception. Lampreys do not see at all and have the ability of electroception. Mammals do not have the ability of electroception, but have well developed hearing and vision. Reflexes vary from the simplest Achilles reflex to stereotypical behavioral reflexes. All these types of fundamental reflexes belong to physiological functioning and are extremely important for survival. The physiology of electroception of ray species is explained by Bullock and Başar (1988) among others.

Recently, we concluded that recording of event related oscillations (EEG and ERPs) could provide an appropriate method to further the understanding of memory, both at the neurophysiological and psychological levels, because brain oscillations have similar frequency codes in the whole brain (Başar, 2004; Başar et al., 2000, 2001). Further, cognitive inputs evoke oscillatory responses within the brain that are similar to sensory oscillatory responses, both occurring in the same frequency channels of EEG. However, prolongation, amplitudes and topological distributions are different.

Accordingly, interactions in sensory-cognitive processing are rich, and thus separation of these processes is almost impossible. Remembering and memory are manifestations of various and multiple functional processes, depending on the complexity of the input to the CNS. The electrical response to a simple light flash is based on simple memory processes at the lowest hierarchy order.

7.1.5 *Living System Settings Incorporated in Physiological Memory*

7.1.5.1 Sympathetic System and EEG-Oscillations

Living system settings are ensembles of detectors and all types of mechanisms that serve living systems to maintain survival functions such as normative values of blood pressure, respiratory rhythms, cardiac pacemakers, and body temperature. Such mechanisms, which are important to maintain the body within the limits of healthy life qualities, should be categorized also into the level of persistent memory, because damage to these settings strongly affects higher levels of nervous activity and all levels of memory activation. Gebber et al. (1995a) reviewed a series of articles from their laboratory on the 10-Hz rhythmic sympathetic nerve discharges of cats and offered a hypothesis on its functional significance. In Chap. 9, the integration of oscillatory processes in the vegetative system and spinal cord and in the brain will be explained in detail.

7.1.6 *Genetic Factors Are Fundamental in Living System Settings and Physiological Memory*

The group of Begleiter and Porjesz introduced a fundamental approach to examining the genetic underpinnings of neural oscillations. This proposed that the genetic underpinnings of these oscillations are likely to stem from regulatory genes that control the neurochemical processes of the brain, and therefore influence neural function (Begleiter and Porjezs, 2006). The group's findings suggest that genetic analysis of human brain oscillations may identify genetic loci underlying the functional organization of human neuroelectric activity, and brain oscillations represent important correlates of human information processing and cognition. Further, these oscillations represent highly heritable traits that are less complex and more proximal to gene function than either diagnostic labels or traditional cognitive measures. Therefore, these oscillations may be utilized as phenotypes of cognition, as well as valuable tools for the understanding of some complex genetic disorders (Begleiter and Porjesz, 2006). These authors discuss recently identified genetic loci regarding both resting and evoked brain oscillations involving the GABAergic and cholinergic neurotransmitter systems of the brain (see also Kamarajan et al., 2004; Porjesz and Begleiter, 1996, 2003; Porjesz et al., 1998, 2002).

The advent of genomics and proteomics, combined with a fuller understanding of gene regulation, will open new horizons on the critical electrical events that are highly essential for human brain function.

Because the oscillatory system settings have a fundamental role in memory, the genetic settings also belong to Level I of the memory model described in Fig. 7.1.

Accordingly, future research may incorporate genetic factors within models of phyletic memory and physiological memory. In Chap.13 the importance of genetic studies will be explained in more detail.

7.1.6.1 Changes in Sensory Memory During Life

The visual cortex needs visual experience during a critical period of early life to achieve its full functional development. In the absence of critical experience, the cortex remains defective. As a consequence of this, either normal vision may not be attained or else the subject may need considerable training.

The literature contains a vast number of results indicating the change of oscillatory responses after acquiring new experiences. The examples in Chap. 11 demonstrate that, after learning and training in the early period of life, the alpha activity and alpha responsiveness do change. In the earlier period of life, there are no alpha responses to simple light stimulation; accordingly, the functioning of networks sensitive to light and sound (Başar, 1998) are susceptible to changes (see results in Chap. 11). The phyletic memory and the physiological memory are not perfectly stable throughout the lifetime.

During a memory process – either a short- or longer-term process – the perception of a sensory input is matched with information already stored in the neural tissue. When a simple light evokes alpha and gamma responses in the cortex, elementary oscillatory responses are also strongly associated with several memory processes at different hierarchical levels. The topology of the memories depending on the modality of the input must be different (see examples given previously: cross-modality experiments; measurements in cortical and subcortical structures). During a perception process, the brain performs a matching process with inborn sensory memories. Even complex percepts are interwoven with simple sensations. Because processing of simple sensations and complex percepts are inseparable, the physiological memory and perceptual memory must be also inseparable; they form one combined entity.

Başar et al. (2004) proposed the existence of an alliance of physiological and perceptual memories. This consideration has a crucial consequence: Although physiological and perceptual memories are presented separately in the schema of Fig. 7.1, they overlap in the functional processing.

Accordingly, physiological and perceptual memories do not depict a clear-cut functional hierarchy or absolute separation: The illustration should, in reality, show a continuum between physiological and perceptual memory. It should not present an absolute functional hierarchy or separation. This statement is in accordance with the described functioning of selectively distributed oscillations (see also Fuster's statement in Sect. 7.1)

By discussing the elements of the schematic presentation of Fig. 7.1, we will discover that it is quite impossible to define most memory types or levels as fully separated individual entities with rigid (frozen) boundaries in the time and topological spaces as in the case of perceptual memories. On the contrary, learned reflexes, the simplest sensory cognitive processes, and all types of memories have dynamic features and are steadily evolving as life continues. Following a learning process, our semantic memory is already altered or extended.

To date, few studies demonstrating these dynamic processes have been performed; experiments on the recognition of known and unknown faces belong to this type of experiments (Başar et al., 2004, 2006; Güntekin and Başar, 2009 and also

Chap. 12). Therefore, results and their interpretations have to be considered as preliminary and rough steps. Accordingly, multiple distributed memories cannot be treated in detail, and a perfect classification on all levels of distributed memories cannot be yet provided. It is probable that such a perfect system of classifications can never be achieved. Therefore, the model shown in Fig. 7.1 presents a proposal incorporating the dynamic memory and evolving memory.

7.1.7 Working Memory, Dynamic Memory (Level II)

During processing of many complex tasks, it is necessary to hold information in temporary storage to complete the task. The system used for this is referred to as *working memory* (Baddeley, 1996). Working memory is the temporary, ad hoc activation of an extensive network of short- or long-term perceptual components. That network would be, similar to perceptual memory, retrievable and expandable by a new stimulus or experience. Fuster (1995a) states that working memory has the same cortical substrate as the kind of short-term memory traditionally considered as the gateway to long-term memory (LTM).

The detection of a target signal during a P300 type of experiment also requires a type of working memory. The subject has to retain knowledge related to the target (such as nature of the target, frequency, shape, color etc.) during the experiment. The linkage between P300 amplitude and latency measures and working memory processes already has been established (Fabiani et al., 1990; Howard and Polich, 1985; Pratt et al., 1989; Sanquist et al., 1980; Scheffers and Johnson, 1994). The matching process after the detection of the target should rely, in this case, on working memory or the success of this type of memory P300 versus N100. It was shown that, during working memory and evolving memory, states of alpha, delta, and gamma oscillations are differentially activated (Başar, 2004).

7.1.8 Perceptual Memory

Several types of analysis categories are crucial in the functional interpretation of ERPs: The analysis of the stimulus itself: What can a stimulus evoke in the brain? It can evoke simple percepts, complex sensory percepts, bimodal percepts or memory related functions, etc.

According to Fuster (1995a), perceptual memory is memory acquired through the senses. It comprises all that is commonly understood as personal memory and knowledge, i.e., the representation of events, objects, persons, animals, facts, names, and concepts. Fuster further describes that, in the hierarchy of memories, at the bottom, memory originally acquired by sensory experience has become independent from it in cognitive operations. This means that perceptual memory partly belongs to built-in memory types, and that it evolves during life by adding newly stored percepts, thereby becoming richer in stored information. Therefore, perceptual memory comprises elements of phyletic memory of fundamental memory and also elements of semantic and episodic memories. Here, again, it is not possible to define exact boundaries in the hierarchies of perceptual, semantic, and episodic memories. To process percepts, the brain needs not only the built- in networks of elementary sensations but also new ones, formed from information obtained during life. We include perceptual memory also in category of quasi-stable memories, as indicated in the illustration in Fig. 7.1. This is because, according to the previous descriptions, the perceptual memory is partly inborn; however, complex percepts acquired during life are often not completely stabilized, but are quasi-stable. (For example, the grandmother percept is shaped by several events during life, and is not an inborn percept (Başar et al., 2004)). In Chap. 6 it was shown that alpha, gamma and theta responses to simple light or simple sounds are probably generated from built-in networks, because cross modality experiments demonstrate that alpha sensory response is evoked only by adequate stimuli in the cortex and thalamus (see Figs. 6.3–6.5, 6.11, 6.12, 6.15). The finding that alpha responses are recorded in the whole brain demonstrates that perceptual memory is distributed not only in primary areas of the cortex and thalamus, but also in the whole brain (see Figs. 6.15–6.18)

According to the considerations mentioned, we classify simple perceptual memory also as part of physiological memory. However, recognition of complex percepts that show individual differences are less strongly linked to the ensemble of physiological memory. The recognition of complex percepts is more overlapping or interwoven with the ensembles of subsets of evolving memory. The processes of recognition and of evolving memory are dynamic processes that are manifested by multiple oscillations.

One reason for mentioning working memory is that this memory function has been linked to the prefrontal region of the brain, i.e., part of the dorsolateral frontal lobe that comprises the anterior convexity of the cerebral hemispheres (Fuster, 1991; Goldman-Rakic and Friedman, 1991). However, most studies of working memory have used non-human primates, making it difficult to draw direct neuroanatomical comparisons with humans. Nonetheless, there is strong evidence from non-human primates that the prefrontal region plays a crucial role in working memory, and it is likely that this relationship will, at least to some extent, also apply to humans (see Baddeley, 1986; Başar, 2004).

7.1.9 Incorporation of Oscillatory Codes in Physiological Memory Consisting of Phyletic, Sensory, and Perceptual Memory

In Fig. 7.1 we tentatively presented indications that the physiological memory incorporates phyletic, sensory and perceptual memories. Are these memory types or activated memory states interwoven with similar frequency codes?

Several publications show that the mammalian brain responds to simple visual and auditory stimuli with selectively distributed alpha, theta, beta, gamma, and delta responses (see also results in Chap. 6).

As we explained in earlier publications (reviewed by Başar, 1999), alpha, theta, and gamma responses are also manifestations of phyletic memory, because they are inborn and possibly wired responses. This is clearly indicated in the fourth column of Table 7.1, which describes the hierarchy of activated memories.

According to the results of the present book, perceptual memory is manifested with multiple oscillations in the alpha, beta, gamma, theta, and delta frequency bands.

Our empirical evaluation suggests that all these memory types are interwoven and/or tuned with the frequency codes of the EEG-oscillations. This can facilitate the transition between memory states and communication in the brain. Besides this, strong links or alliance with all integrative functions in the brain could be rapidly progressed.

The neurons-brain theory (Başar, 2004) describes a systematic account of measurements of electrical activity of neural assemblies, and also incorporates proposals derived directly from measurements. Accordingly, in Table 7.1 we underline some principles on electrical activity of neural assemblies that are derived from measurements.

1. The brain has natural frequencies or oscillations, measurable at the cellular and population levels.
2. Brain oscillatory responses are real responses related to function (Başar et al., 2001).
3. Oscillations are selectively distributed and selectively coherent.
4. Oscillations are related to multiple functions, and a given function is often manifested by means of multiple oscillations (Klimesch, 1999; Klimesch et al., 2000a). The principle of superposition is confirmed by several publications (Karakaş et al., 2000 a, b).

A natural consequence of the neurons- brain theory (or brain assemblies theory) is the super synergy in oscillatory dynamics. The proposition or a model for the electrical manifestation of the grandmother percept, is based on strong evidence: Research shows that the increasing complexity of percepts is accompanied by an increased number of multiple oscillations in parallel with the increased number of activated neural populations, although the proposal related to the perception of the grandmother is anchored in tenable arguments (see Chap. 12)

According to results described in throughout chapters of the present book, selectively distributed and selectively coherent oscillatory networks in the delta, theta, alpha, beta, and gamma bands play a major role in brain functioning. Sensory and cognitive events evoke superimposed multiple oscillations that are transferred to spatially distributed tissues almost in parallel with various degrees of amplitude, latency, duration, synchronization, and coherence. Sakowitz et al. (2001) reported significant increases in gamma response amplitude within distributed areas of the brain and a 100% frontal theta enhancement by bimodal stimulation compared with

Table 7.1 Hierarchy of activated memories

Principals Of Neuro: Electricity (measured)	Functional level	Examples: (measured)	Activated memory (tentative hierarchy)	Some references
1. Brain shows oscillations neural activity Oscillations are almost invatriant in evolution Oscillatory responses are real brain responses	Very Simple Functions or Fuctional Building Blocks ↓	α-response γ-β-response (components of auditory and visual responses) δ-response at the visual and hearing threshold	Phletic memory Simple sensory memory Decision making	Başar (1992) Başar et al. (2000) Başar and schürmann (1994) Demiralp et al. (1999) Gruzelier et al. (1996) Karakaş and Başar (1998)
2. Each Oscillatory Activity respresent multiple functions vice versa Each function is represented by multiple oscillations Superposition principle	Integrative Neurophysiology and cognition ↓	Frontal θ, occipital α Oddball P300: superposition of δθαγ responses α-expectation	{ 1. Encoding, association { 2. Working memory 3. Decision making 4. Upgrading memory { 5. Dynamic memory	Başar (1999) Başar et al. (1987) Başar and Stampfer (1985) Başar-Eroğlu et al. (1992) Karakaş et al. (2000) Klimesch (1999) Maltseva et al. (2000)
3. Multiple oscillations are selectively distributed in the brain as parallel processing	Very complex function		Memory is selectively distributed	Başar et al. (1975, 2000) Country et al. (1997) Fuster (1995, 1997) Goldman-Rakic (1997) Gruzelier (1996)
4. Selectively distributed parallel oscillatory systems are intergrated in function Various topology dependent enhancements, phase locking delays, prolongations in paralel processing		Proposal e.g. perception of grandmother Gestalt	Episodic ↓↑ Semantic memory	Başar et al. (2000) Burgess (2002) Doppelmayer (2000) Klimesch (1999, 2000)
Increasing number of neural populations		Increasing complexity at the functiona level	Increasing memory complexity	

The term *theory* may be used to signify any hypothesis, whether confirmed or not, or may be restricted to hypotheses that have been so strongly confirmed as to become part of the accepted doctrine of a particular science. In its most appropriate use, it signifies a systematic account of some field of study, derived from a set of general propositions. These propositions may be taken as postulates, as in pure mathematics, or they may be principles more or less strongly confirmed by experiences, as in natural science (from Encyclopedia Britannica) unimodal. It was reported (Başar et al., 1993) that retrieval of visual experience from short-term memory is associated with 40 Hz activity. The coherences between different spatial locations of the brain vary as these areas are activated with different classes of stimuli (haptic and visual) in an associative learning task (Başar, 1988). Such findings experimentally substantiate the submechanisms that have been outlined. Accordingly, we suggest that complex percepts (such as the visual image of one's grandmother) are formed and/or manifested by means of the ensemble of oscillatory superbinding dynamics.

The column adjacent to the proposal of Grandmother or to Perception of Gestalt (see Table 7.1) presents the complex memory function (episodic and semantic memory). This arrangement of the table is in accordance with references indicating an increasing number of involved populations and frequency windows with increased complexity at the functional level.

7.1.10 What Is Motor Memory and Procedural Memory Developed During Life?

According to Fuster (1995a) motor memory consists of representations of motor action in all its forms, from skeletal movement to spoken language. Motor memory, like perceptual memory, is evoked and acquired through the senses. However, once acquired, it is largely presented in the neocortex and frontal lobe. The most automatic and firmly established aspects of motor memory are represented outside of the neocortex. Basal ganglia and the cerebellum are the most fundamental structures related to motor memory. Perception and motor action are interrelated, and both are part of many representational networks.

According to Baddeley (1986), procedural memory refers to the acquisition of skills, whether perceptual-motor skills, such as those of riding a bicycle or driving a car, or cognitive skills, such as in advanced reading or problem solving. They clearly comprise an important area of learning and do, of course, represent the archetypal example of procedural learning. Baddeley describes that skills can be divided into two types: *continuous*, in which each component of the skill serves as a cue to the next, as in cycling or steering a car; and *discontinuous*, in which a series of discrete stimulus-response links are involved, as in typing. In general, continuous skills seem to show little or no forgetting, whereas forgetting clearly occurs with discontinuous tasks.

Specifically, the basal ganglia have been linked to various forms of non-declarative memory, particularly those types of memory that are dependent upon a motor act for their realization, or what has been termed *procedural memory* (e.g., riding a bicycle, skating). The cerebellum, another motor-related structure situated behind the brainstem at the base of the brain, participates along with the basal ganglia in many types of procedural learning and memory (e.g., Başar, 2004; Glickstein, 1993; Thompson, 1986, 1990). In the illustration shown in Fig. 7.1, perceptual memory and procedural memory are in the category of quasi-stable memories. There are also changes of oscillatory response activity during aging, as will be described in Chapter 11.

7.2 Dynamic Memory in Whole Brain: Memory States Instead of Memories

7.2.1 Alpha, Theta, and Delta Oscillatory Processes During APLR

Processes of attention, perception, learning and remembering (APLR) are simultaneous and interwoven. This is described in detail at the level of EEG-oscillations in Başar (2004), Baddeley (1996), and Desimone (1996) at the single cellular level. Figure 7.3 shows that the evolving memory is accompanied by the dynamic shaping of theta oscillations in a P300 type experiment. The interplay between delta and theta activities is completely changed at the end of the experiments. We also mention the dynamic shaping of the alpha-oscillations during learning-and dynamic memory experiments, which was clearly shown by Başar et al. (1997). Hayek (1952)

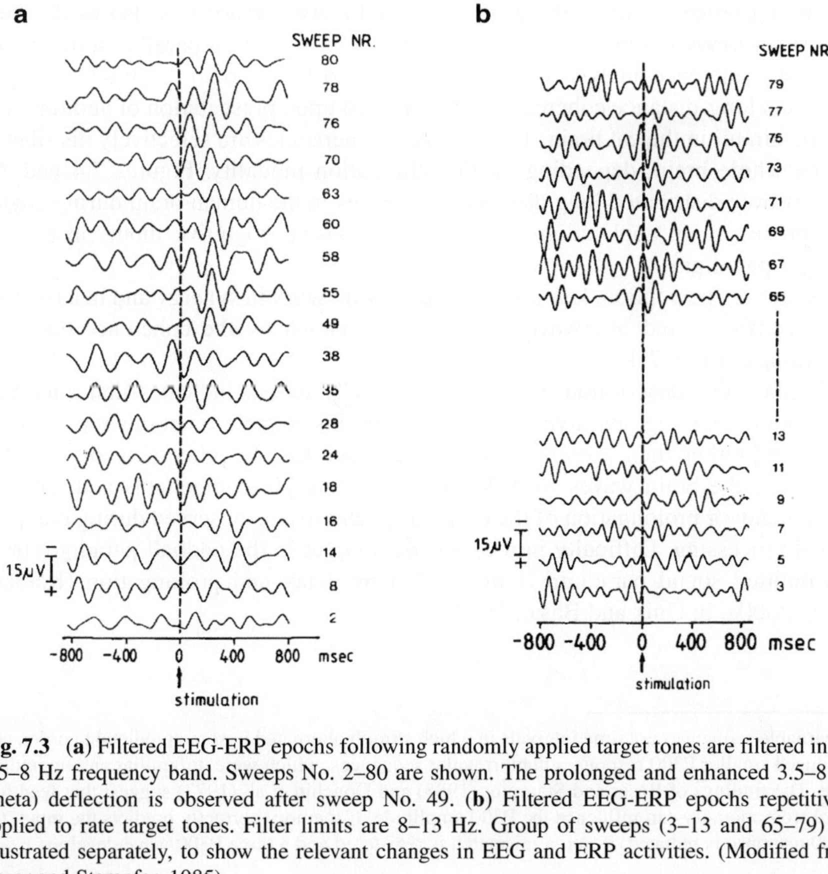

Fig. 7.3 (**a**) Filtered EEG-ERP epochs following randomly applied target tones are filtered in the 3.5–8 Hz frequency band. Sweeps No. 2–80 are shown. The prolonged and enhanced 3.5–8 Hz (theta) deflection is observed after sweep No. 49. (**b**) Filtered EEG-ERP epochs repetitively applied to rate target tones. Filter limits are 8–13 Hz. Group of sweeps (3–13 and 65–79) are illustrated separately, to show the relevant changes in EEG and ERP activities. (Modified from Başar and Stampfer, 1985)

presented the concept of cortical memory network in the context of the main topic, which is not memory itself but, significantly, perception; perception as the *source* of memory and as the *product* of memory. We introduced APLR-alliance as a consequence of the measurements: It is not a theoretical construct, but is derived from the empirical evidence.

7.2.2 Are Dynamic EEG-Templates Created During Processing of the Alliance of APLR? Do Such Templates Build a (Virtual) Short-Term Storage of the New Learned Material?

During learning tasks, oscillatory activity does change when subjects acquire new knowledge.[1] These changes are manifested by newly created EEG templates (Fig. 7.6). Further, such changes are probably consequences of matching processes, and are at least temporarily stored *in* the brain. As a metaphor, it can be stated that (1) during processes of working memory oscillatory components also work, often in a superimposed way, and that (2) the oscillations work in parallel in the whole brain, as coherence analysis has indicated. Figs. 6.19 and 6.20 illustrate that alpha and theta long-distance coherences are increased upon presentation of auditory and visual stimuli in the cat brain. These evoked coherencies are selectively distributed in the whole brain, depending on the stimulation modality. Figures 7.4 and 7.5 demonstrate distributed long distance coherences in the human brain during cognitive processes on application of auditory and visual cognitive inputs in a P300 oddball paradigm.

Attention, perception, learning and memory do work in synergy and in a reciprocally activating and interwoven way, as also shown by the coherence data and recordings in Fig. 7.4.

Further, this description, which is empirically based, indicates that attention, perception, learning, and dynamic memory are inseparable functions representing an entity, rather then presenting separable processes. In processing of complex functions, the brain waves work longer, have multiple components and, accordingly, delay or prolongation of the oscillatory activities is observed during complex signal processing. Difficulty in recognizing a target in the oddball paradigm or in the omitted sound paradigm is manifested by delay and prolongation (Karakaş et al., 2000a, b; Öniz and Başar, 2009).

[1] Learnable sequences are time intervals in which stimuli alternated in some predictable order, and produced smaller P300 responses than irregular sequences, which were unfamiliar and unpredictable. The findings of Başar and Stampfer (1985) and Donchin et al. (1973) suggest that feed forward from memory can influence the P300 amplitude. If memory correctly predicts the input, the P300 response is reduced; if not, a mismatch is registered and a large P300 wave develops.

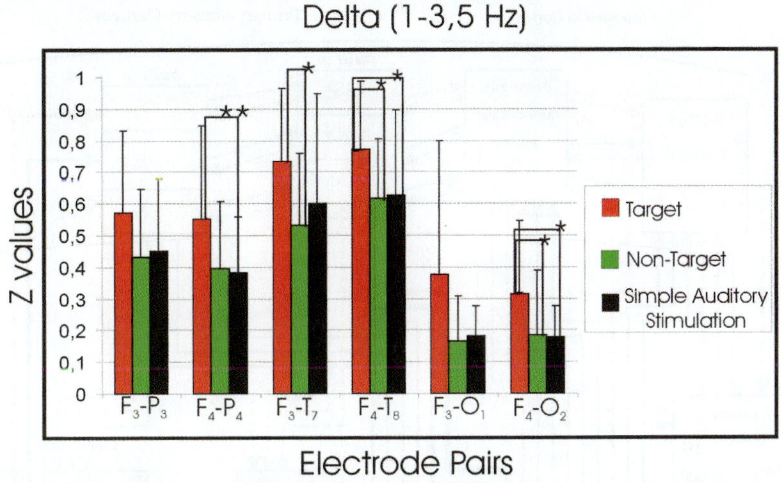

Fig. 7.4 Mean Z values of target, non-target and simple auditory stimulation responses of delta (1–3.5 Hz) frequency range, where *double asterisk* represents $p < 0.01$, *asterisk* represents $p < 0.05$

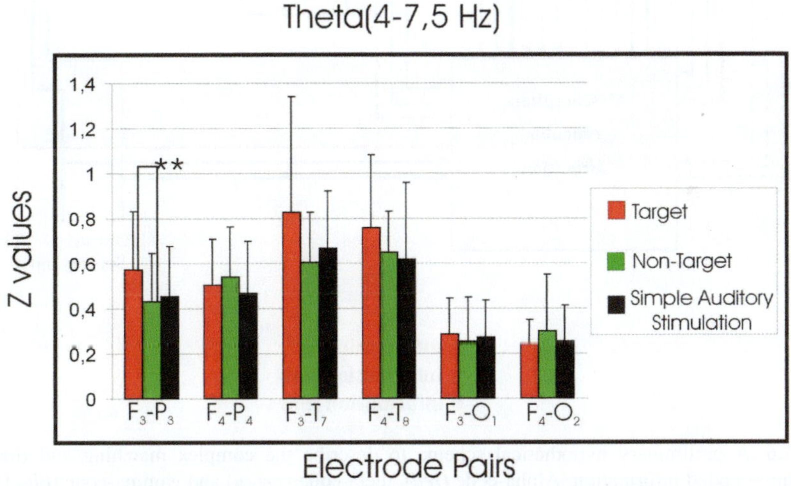

Fig. 7.5 Mean *Z* values of target, non-target and simple auditory stimulation responses of theta frequency range (4–7.5 Hz), where *double asterisk* represents $p < 0.01$

7.2.3 Recent Examples of Brain Oscillations in the Cognitive Processes of Healthy Subjects

A large number of studies have been published concerning the cognitive processes of healthy subjects. The most important studies are related to oddball paradigms, presenting results on target and standard stimuli, working paradigm, and

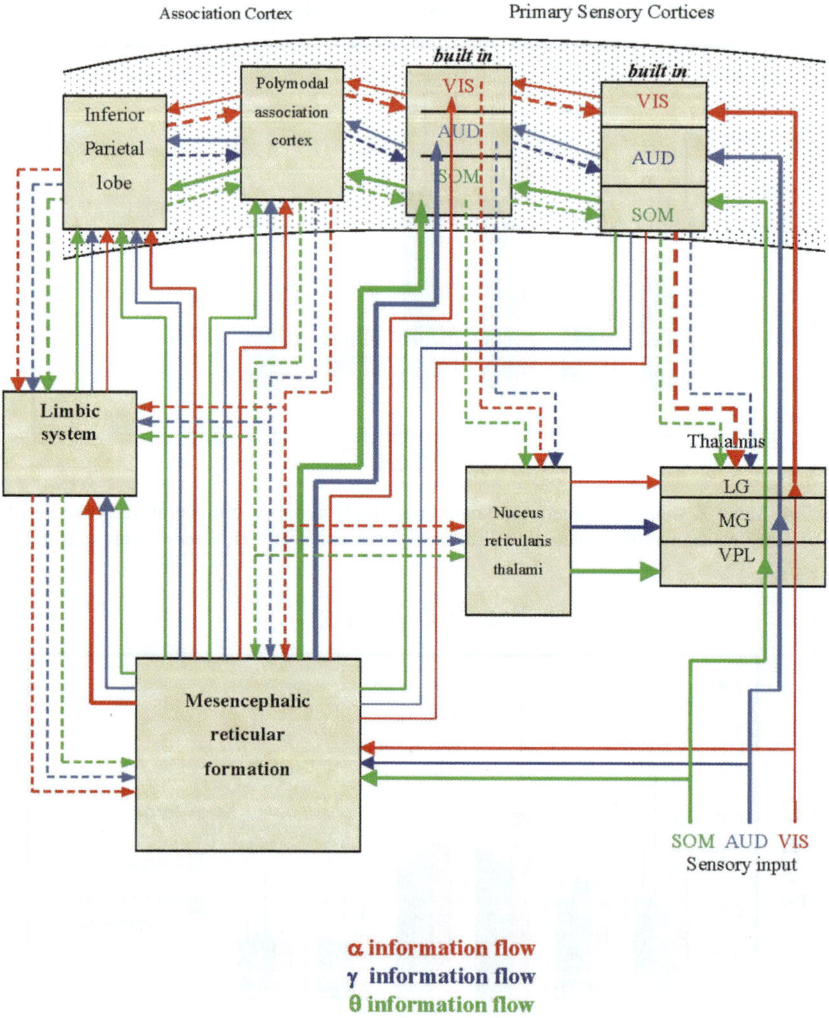

Fig. 7.6 A preliminary hypothetical scheme to describe the complex matching and flow of oscillation-coded information. Alpha-code (*red*), theta-code (*green*) and gamma-code (*blue*)

simple auditory and visual evoked potentials. The enhancement of delta and theta responses, prolongation of alpha oscillations, and the appearance of a second theta response window are some of the relevant features of the target response (Başar, 1998, 1999; Başar et al., 1997a–c; Başar-Eroğlu et al., 1992; Doppelmayr et al., 2005; Klimesch, 1996, 1999; Klimesch et al., 1994, 1997, 1998; Sauseng et al.,

2005a, b; Yordanova and Kolev, 1998a, b; Yordanova et al., 2003). During the oddball paradigm, gamma responses have also been recorded by several authors. A number of studies centered on working memory, memory in general, and episodic memory. The differentiation of brain oscillations and human memory in the alpha and theta bands were reported by Klimesch et al. (1997). Furthermore, evidence of theta activity during episodic memory retrieval was noted by Klimesch et al. (1994, 1997). Stam (2000) published the dynamics of potentials in theta and alpha frequency bands during memory performance in humans. Weiss and Rappelsberger (2000) carried out a number of studies on the correlates of memory processes and long range EEG synchronization. Several studies also analyzed the coherence function (Petsche and Etlinger, 1998). (For further information, refer to reviews by Başar, 2007; Başar et al. 2001 ; Gevins, 1998.) The recognition of faces and facial expressions are amongst the most complex functions in the cognitive processes (Başar et al., 2006, 2007; Güntekin and Başar, 2007), as is the recognition of ambiguous figures (Başar-Eroğlu et al., 1996; Isoglu-Alkaç et al., 2000; Mathes et al., 2006; Strüber et al., 2000). Reviewing a number of publications in this area, it was concluded that cognitive processes are manifested by multiple oscillations (Başar et al., 2001, 2006; Klimesch et al., 2004, 2007)

Sauseng et al. (2005a, b) analyzed the evoked theta and upper alpha desynchronization during a special memory task, which was designed to study the transfer of information between both memory systems. The results showed that, during attempts to retrieve information from long-term memory, evoked theta oscillations spread from anterior to posterior recording sites. When information is retrieved, the direction reverses and theta spreads to frontal sites. This time point, when direction reverses, varies between subjects to a large extent but is significantly correlated with memory performance and the onset of upper alpha desynchronization. The conclusion is that this phenomenon reflects the transfer of information between the working memory and long-term memory.

Demiralp et al. (2007a) analyzed human event-related EEG oscillations recorded in a memory-related paradigm in which 13 subjects perceived known and unknown visual stimuli. The paradigm revealed event-related oscillations in the gamma range, which depended significantly on the phase of simultaneous theta activity.

Jensen and Tesche (2002) recorded neuromagnetic responses from 10 subjects performing the Sternberg task. Subjects were required to retain a list of 1, 3, 5, or 7 visually presented digits during a 3 s retention period, in which the authors observed ongoing frontal theta activity in the 7–8.5 Hz band recorded by SQUID-sensors over the frontal brain areas. The activity in the theta band increased parametrically with the number of items retained in the working memory. A time-frequency analysis revealed that the task-dependent theta was present during the retention period and during memory scanning. Following the memory task, the theta activity was reduced. These results suggest that theta oscillations generated in the frontal brain region play an active role in memory maintenance.

7.2.4 *Are All Functions of the Brain Linked with Memory?*

According to Goldman-Rakic (1996),[2] working memory is the ability to hold an item of information transiently in mind in the service of comprehension, thinking and planning. Working memory encompasses both storage and processing functions. It serves as a workspace for holding items of information in mind as they are recalled, manipulated, and/or associated to other ideas and incoming information.

In earlier publications we have shown that, during experiments with cognitive loadings, the oscillatory pre-stimulus and post-stimulus activities reflect constant work of the brain in progress. Alpha activity before expected cognitive targets goes to a state of order – the time course of the alpha activity is ordered and aligned; amplitudes of oscillations are increased during cognitive demands and memory loads. Further, the alpha oscillations are phase locked to the expected target signal (Başar, 1989; Başar et al., 1997; Barry et al., 2003). If we bring these results together with the statement by Goldman-Rakic (1996), we come to our main theme by considering all brain functions and memory as an entity: Possibly we should more appropriately talk about memory and brain functions – alliance as a more general categorization of APLR-alliance. We should better say that all brain functions are memory-linked or all memory is function.

This means: the qualitative and quantitative properties of the alpha activity are changing and evolving during experimental sessions, thus giving the message that alpha activity is involved with constant *wo*rk. This work reflects parallel and reverberating multiple-components: attention, perception, learning and remembering are interrelated and interwoven.

Memory types such as procedural memory, working memory, motor learning are categorized within Level II of Fig. 7.1. Motor memory is more difficult to categorize solely on one level, and so is categorized between Levels I and II.

7.3 Complex or Multiple-Matching Evolving Memory, and the APLR-Alliance

Pribram (1963) defined the memory as any set of events that makes available to an organism something of a situation after that situation no longer obtains. By the term *thought*, Pribram refers to the active uncertainty produced when an ordered set of memories mismatches the current novelties of the situation. According to Fuster (1995a), neural memory is special in several ways, among which is its capacity not only to retain information but also to utilize it for adaptive purposes. In this sense neural memory becomes connatural with learning, from which it is operationally difficult to distinguish, although the term learning usually refers to the process of

[2]Regional and cellular fractionation of working memory, Proc. Natl. Acad. Sci, Vol 93. pp.13473–13480, Nov 96, Colloquium Paper).

acquiring memory. The schematic illustration of Fig. 7.1 also takes into account the processes explained in the statements by Fuster (1995a) and Pribram (1963). According to our experimental work, we extended the concepts of Hayek (1952) and Fuster (1995a), with the inseparability of perception and memory; and by proposing that processes of attention, perception, learning, and memory are difficult to distinguish, and that they are interwoven and inseparable.

Helmholtz (1866, 1962) introduced the notion of mental constructs that are thought to be generated by past experience and to be stored and recalled from memory. Accordingly, percepts are thought to be produced by the comparison of recalled and evoked neural images. What is going on in the brain when such images are evoked? Do these images also evoke electrical potentials in parallel to recalling?

Mountcastle (1998) goes a step further, "Perhaps what we perceive are patterns of neural activity recalled from the memory for the matching *operation*, rather than the activity evoked directly by sensory stimuli themselves?" It will come as no surprise that it has proved difficult to demonstrate direct neurophysiologic evidence for the processes of unconscious inference. These sets of propositions are included as assumptions to several of the major unsolved problems of neuroscience: How are experiences stored in and recalled from memory? How are neural populations matched and compared? How does the chosen match flow through the conscious experience? Nevertheless, the idea of unconscious experience and matching identification remain themes of some later psychological theories of perception (Mackay, 1970).

What is happening in everyday cognitive process such as recognizing a familiar object? The basic idea here is that, after a sensory code is established, the information in *longer-term memory* or persistent memory that is required to identify the perceived object is accessed. If the matching process yields a positive result, the object is recognized.

It is also evident that, during a process of evolving memory, the APLR-alliance undergoes a complex multiple matching process, which develops with the following steps:

1. What will be learned has to be matched (compared) with the stored percepts (or events) of earlier experience.
2. In a new sequence, attention will be paid to new learned percepts.
3. In the ensuing matching process, these percepts will be (re-)matched with new sensory-cognitive inputs.

We also have to say this: If the brain is able to match a coming input with the just newly learned experience, at least temporarily created EEG traces are needed. Without temporary traces, recurrent inputs during an oddball experiment cannot be compared with previously learned material.

Fuster (1997) describes that components of perceptual memory are retrievable and expandable by a new stimulus or experience. Working memory is presumably the same cortical substrate as the kind of short-term memory traditionally considered the gateway to long-term-memory. The following remark within Fuster's description is crucial, "Hierarchical organization, however, does not imply that the various individual memories are rigidly stocked and stored in separate cortical domains. Rather, different types of memories – for example episodic, semantic or procedural – are probably interlinked in mixed networks that span different levels of perceptual and motor hierarchies."

7.3.1 Multiple and Complex Matching Processes: Reciprocal Activation of Alpha, Delta, Theta, and Gamma Circuits in the Whole Brain. Reentry

According to Damasio (1997), memory depends on several brain systems working in concert across many levels of neural organization; "Memory is a constant work in progress." If this is so, what type of processing should occur at the EEG-level during this constant work in progress? To elucidate the role of EEG-related memory, we propose the existence of complex, recurrent (or reverberating) mechanisms. These proposed mechanisms are founded on experimental results of perpetual changes of alpha, theta, delta, gamma, and beta oscillations and their dynamical superposition.

The temporarily created templates seemingly provide a dynamic (storage) copy to be compared with new input signals. Further, new sensory input to the brain is matched with oscillatory networks including memory traces of "phyletic memory" and of the "memory acquired during life." We should also emphasize that response oscillations strongly depend on the state of the pre-stimulus oscillations during working memory processes (see Barry et al., 2003a; Rahn et al., 1993).

According to Tononi et al. (1992), the notion of reentry extends concepts such as simple feedback, feed forward, or recurrent circuits. Reentry is inherently parallel, has a statistical nature, and has a distributed nature, meaning that it can occur simultaneously within and across several different areas via multiple parallel and reciprocal connections (as seen in corticocortical, corticothalamic, and thalamocortical radiations); it can also occur via more complex arrangements linking the cortex with the hippocampus or with the basal ganglia.

Başar and Stampfer (1985) and Başar (1988) mentioned the possibility of such recurrent and reentrant neural populations by means of recording of EEG-oscillations without using the expression "reentry" (Edelman, 1978, 1987). The gross electrical activity of a neural population selected to fire (to be activated) shows a high degree of variability and plasticity. During experiments with cognitive load and memory activation the EEG was altered before stimulation, this giving rise in turn to new types of responses, because EEG activity preceded the stimulation controls ERPs (Barry et al., 2003; Başar et al., 1998; Rahn and Başar, 1993a, b). According to Barry et al. (2003), stimulation creates states of preference. It should be emphasized that the control of the pre-stimulus activity on responsiveness or excitability has complex behavior. The alpha, delta, theta activities that create preference states depend on the modality of sensory-cognitive input and also on the site of the cortex. (More references to this topic can be found in Barry et al., 2003 and Başar et al., 1998.) The selectively distributed oscillations in the cortex, thalamus, hippocampus, and various areas of the brain stem were demonstrated by Başar et al. (1975a–c) in the cat brain. These authors used a different terminology at that time, "Synchronized selectivity in the whole brain." Accordingly, it can be hypothesized that, during sensory-cognitive processing the oscillatory networks at different levels of the brain are activated as recurrent or reentrant circuits. This behavior can be measured during various types of exogenous or endogenous stimuli as the plasticity and evolving behavior of oscillatory pre-stimulus or post-stimulus activity in the experiments described by Başar and Stampfer (1985). It is possible that reentrant signals play an essential role in creating learnable sequences, as enhancement, alignment of oscillations, and their transition to preferred phase angles prior expected stimuli show.

In Fig. 7.1 the loops also indicate this recurrent behavior. Note that measurements of EEG-oscillations provide the only possibility of checking the existence of reentrant circuits in evolving memory and related types of behavior. The following issue is also important: Başar (1980, 1998) and Başar et al. (1997c) demonstrated that if a brain structure shows spontaneous oscillatory activity, then this structure is susceptible to react, or is responsive in the same frequency channel of spontaneous oscillations (principle of Brain response susceptibility,[3] first indicated by Sato (1963); Sato et al. (1971, 1977). This concept will be explained in Chap. 11 with the fourth example of missing alpha response in 3-year-old children.

7.3.2 Prolonged Oscillations, Delays, and Coherent States During Complex Matching in the Whole Brain

The results of experiments on selectively distributed oscillatory alpha and theta networks in the whole brain were explained by means of schematic illustrations. Sensory cognitive inputs activate not only in the cortex, but also in the whole brain alpha.

We denoted this process as *selectively distributed increases of coherence.* Table 7.1, includes the differentiated activation of memory in several hierarchical processes. Fuster underlies that (1) working memory, which has also been called operant memory, is an operant concept of active memory, and also (2) that active memory is a state rather

[3]The response susceptibility of a brain structure depends mostly on its own intrinsic rhythmic activity (Başar, 1980, 1983a, b, 1992; Narici et al., 1990). A brain system could react to external or internal stimuli, producing those rhythms or frequency components that have already been present in its intrinsic (natural) or spontaneous activity, i.e., if in a given frequency range the spontaneous brain rhythms are missing, they will be absent in the evoked rhythmicities and vice versa.

The concept of response susceptibility is strongly connected with the rule of excitement states of neuronal populations, as suggested by Başar (1980, 1983a, b, 1992). According to this rule, if a neuronal population is able to produce spontaneous activity in a given frequency range, then this structural group can be brought to a state of excitement in the same frequency range by sensory stimuli. This means: excitability is related to spontaneity. As an immediate consequence, common features of response susceptibility of brain structures are related to general common tuning and generalized resonances in various structures of the brain in alpha, theta, delta, beta, and gamma frequency bands.

than a system of memory. The dynamic changes in the APLR-alliance are manifested by the activity of selectively distributed networks in the brain. Networks are innumerable; they are selectively activated on input to the CNS. The constant dynamic interaction between attention, perception, learning, and remembering are represented by selective multiple oscillations and long distance coherence in various frequency channels that take place on sensory cognitive stimulation and /or during experiments related to the APLR-alliance (For long distance coherences, see also Chaps. 6 and 13).

By novelty or incertitude of the sensory-cognitive input, the number of mismatches is certainly increased, because new percepts are not reflected in earlier oscillatory templates, but in the newly stored oscillatory templates. Therefore, for all brain structures, it takes time to match it and the process of remembering has longer time duration. Because every morning I see my secretary, it is no problem to remember her name and who she is; conversely, when I have to remember the name of a person whom I have not seen for years, the process of remembering requires more time. In cases of novelty or incertitude, our brains work with delays or prolonged behavior and need more time (or they work longer). Parallel to these delays and prolongations in the brain's work, we observe delays or prolongation of oscillations (see also Özgören et al., 2005; Öniz and Başar, 2009).

P 300 oscillatory responses (superposition of oscillatory responses) in alpha, theta, and gamma have longer duration, whereas the evoked responses on simple sensory input have shorter and damped oscillatory responses (second theta window, alpha prolongation by difficult tasks, delta delay in P300). The response susceptibility of a brain structure depends mostly on its own intrinsic rhythmic activity (Başar, 1980, 1983a, b, 1992; Narici et al., 1990). A brain system could react to external or internal stimuli producing those rhythms or frequency components, which have already been present in its intrinsic (natural) or spontaneous activity; i.e., if in a given frequency range the spontaneous brain rhythms are missing, they will be absent in the event related oscillations, and vice versa. In a similar way, when incertitude is reduced and the matching does not require longer duration, the oscillations are shorter, as shown by Başar-Eroğlu et al. (2002).

When the whole brain is involved in the remembering process, or in dynamic-reciprocal acting[4] of APLR-alliance, reverberation of signals between brain structures on sensory stimulation is most probable. Therefore, the process of matching takes longer. There are probably cross-talks or reverberations between different structures.

Relevant increases of coherence of theta and alpha responses in the whole brain, between cortex, brain stem, hippocampus, and thalamus, were recorded on sensory-cognitive inputs to the brain (Başar, 1999; Başar et al., 1998). Because the signal at each electrode location mostly reflects the network activity under the electrode, the coherence between two electrodes should measure interactions between two neural populations. In cases of superbinding of all frequencies, we have to observe a relevant increase of overall coherence, and this is the case, as experimental results demonstrate (Başar, 1990; Başar et al., 1979).

---

[4] One could also describe this as mutual effect.

A hypothetical scheme to describe the complex matching is explained in the following section.

7.4 Matching of Multiple Oscillations in the Whole Brain

In the foregoing sections we explained the notion of matching, mentioning the first steps by Helmholtz, followed by the crucial statements by Mountcastle (1998) and Klimesch (1999). In the following, we hypothesize that the matching processes take place by means of superposition (parallel processing) and of serial processing between different brain areas. Further, we assume that the information flow uses multiple frequency codes that are those of EEG-oscillations. Resonance phenomena show that various brain structures – including long distance structures – are tuned to the same frequency codes and that, upon endogenous or exogenous stimulation, links or coherences are enhanced within multiple frequency windows. This means that almost all brain areas are tuned to be activated or resonate with the EEG frequency codes (Başar, 1999; Başar et al., 1999). Therefore, it is reasonable to hypothesize multiple-frequency codes for matching processes. This fact once established, we do not include the possibility of other matching codes, for example the process of activation of feature detectors by simple impulse stimulation.

A preliminary hypothetical scheme to describe the complex matching and flow of oscillation-coded information is presented in Fig. 7.6. For the sake of simplicity only alpha-code (red), theta-code (green) and gamma-code (blue) are represented here. For a more complete description, delta-code, beta code, (and additionally alpha1 and alpha2 codes) also should be inserted into the illustration.

This illustration has been derived again from the schema of Fig. 7.1, which roughly describes the information flow in the CNS.

It is important to note that the frequency-coded information flow occurs as (1) serial-processing and (2) parallel-processing.

The matching processing means the following: The inborn alpha (phyletic memory), theta, and gamma networks facilitate the adequate information flow in these given frequency channels. The received information is matched in all networks to evaluate whether the signals coming from peripheral organs are adjusted or tuned with the alpha, theta, gamma, and delta and beta frequency codes. Here a simple-matching is the comparison with a unique frequency code, for example, the alpha matching. In the proposed complex-matching procedure, comparisons or matching with all frequency codes (alpha, beta, gamma, and theta) are fully or partly activated as parallel or serial processing, depending on the nature of performed integrative functions. Because the oscillatory networks are selectively distributed, the participation of individual frequency codes has varying degrees of amplitudes. Once an ample alpha signal has been brought to "ignition," all structures can be excited by giving rise to a huge reciprocal alpha resonance. Through a high load of exogenous or endogenous sensory-cognitive input to the brain, the occurrence of a huge resonance in the brain can take place. (For resonance phenomena in the brain, see Başar, 1980, 1999 and especially Appendix C.)

In case of serial processing, delays can occur, usually in the range of 500 ms, which is the maximal time-interval needed for information to flow from one structure to another (see Başar, 1999; Libet, 1991; Miller, 1991). The alpha, theta, and gamma signals can also circulate between diverse structures when the inputs – endogenous or exogenous ones – contain information that is difficult or complicated to process, and accordingly, multiple-matching of circulating of all frequency-coded signals with a great number of brain structures in the whole brain is needed. This may be the cause of prolonged oscillations and of the late alpha, theta, or gamma windows around 300–500 ms. In this case, evidently, the whole brain works longer. We should also emphasize that serial and parallel processing may occur in all levels together, in a reciprocal activating manner.

Figure 7.6 provides a hypothetical illustration of the multiple matching processes during oddball tasks or tasks requiring working memory or other cognitive tasks. As the experiments by Başar-Eroğlu et al. (1992), Karakaş et al. (2000a, b) and Güntekin and Başar (2010) have shown, the ERPs and pre-stimulus EEG segments are dominated by delta and theta oscillations in P300 responses.

The process of matching during cognitive load in the theta and delta frequency range is justified with the recent findings of Güntekin and Başar (2010). Fig. 7.7 illustrates the grand averages of coherences values for target, non-target and simple auditory stimulation responses for left hemisphere electrode pairs (left hemisphere: F_3-T_7, F_3-P_3, F_3-O_1) in the 1–15 Hz frequency range. Figure 7.8 illustrates the grand averages of coherences values for target, non-target, and simple auditory stimulation responses for right hemisphere electrode pairs (right hemisphere: F_4-T_8, F_4-P_4, F_4-O_2) in the 1–15 Hz frequency range. Blue lines represent the grand average of coherence values for frontotemporal electrode pairs; orange lines represent the grand average of coherence values for frontoparietal electrode pairs; pink lines represent the grand average of coherence values for frontooccipital electrode pairs.

The grand averages of coherence values on target stimulation are illustrated in the upper part of the diagrams; the grand averages of coherence values on non-target stimulation are illustrated in the middle part of the diagrams; the grand averages of coherence values on simple auditory stimulation are illustrated in the lower part of the diagrams. As seen in Figs. 7.7 and 7.8, frontotemporal coherences were higher than the frontoparietal and frontooccipital coherence values for all modalities (target, non-target, simple auditory stimulation). Furthermore, the frontoparietal coherence values are higher than frontooccipital coherence values for all modalities (target, non-target, simple auditory stimulation). The peaks of delta, theta, and alpha frequency ranges of grand averages of coherence values for target responses can easily be detected, and they are >0.55 for frontotemporal electrode pairs (F_3-T_7, F_4-T_8) (Figs. 7.7 and 7.8). However, the grand averages of coherence values for non-target and simple auditory stimulation responses do not exceed 0.50 at any location and they are particularly low at the left hemisphere (Fig. 7.7).

These results imply the following interpretation: After application of simple or visual stimulation, separated and distant structures in the brain depict mutual resonant responses. The mutual resonances are manifested by increased coherences.

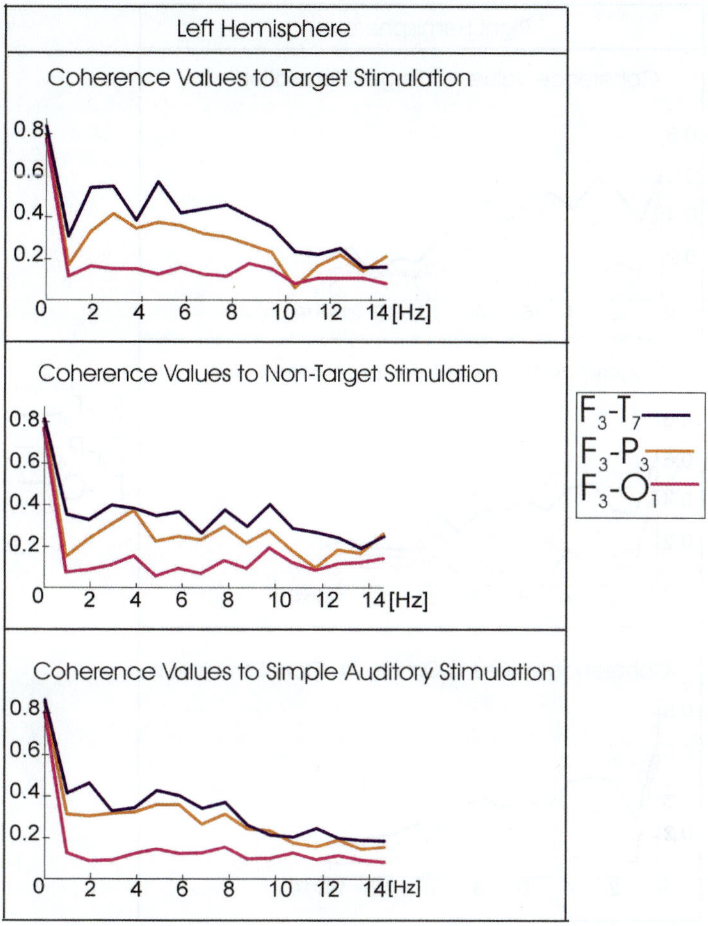

Fig. 7.7 Grand averages of coherences for target, non-target, and simple auditory stimulation responses for left hemisphere electrode pairs (left hemisphere: F3-T7, F3-P3, F3-O1; right hemisphere: F4-T8, F4-P4, F4-O2)

Whereas alpha and theta coherences govern responses to simple sensory stimulation, delta coherences reach the maximal values on target stimulation in the oddball paradigm. In long distant electrodes the target delta response is considerably tuned, thus showing the efficiency of a cross-talk and accordingly of delta matching

Accordingly, in Fig. 7.9, thick lines of theta (green) and very thick lines of delta (yellow) information processing have been added to show the dominance of theta and delta resonance.

We also mention a recent framework proposed by Herrmann et al. (2004) that relates gamma oscillations to two underlying processes: the comparison of memory contents with stimulus-related information, and the utilization of signals derived from this comparison. This model attempts to explain early gamma-band responses in terms of the match between bottom-up and top-down information.

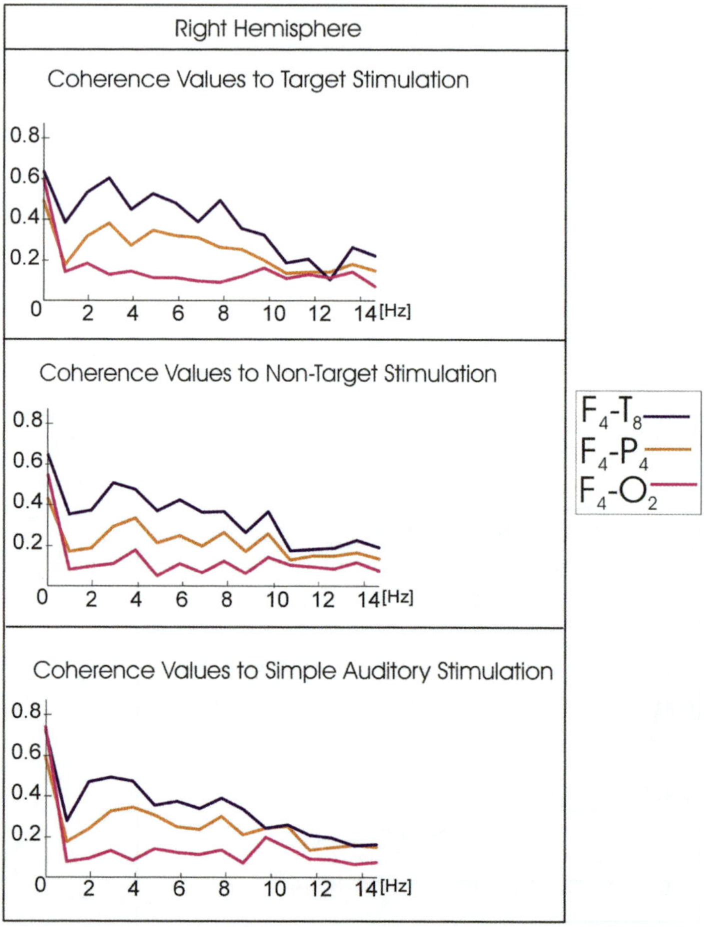

Fig. 7.8 Grand averages of coherences for target, non-target, and simple auditory stimulation responses for right hemisphere electrode pairs (right hemisphere: F4-T8, F4-P4, F4-O2)

7.5 Longer-Acting-Memory and Transition to "Persistent Memory" in the Whole Brain

7.5.1 Evolving Memory Is Identified as Multiple-Level Functioning in CNS

The expression *evolving memory* or *memory building* was introduced as the activation of memory states or processes, augmented with new learned (memorized) episodes created by constant and reciprocally-active APLR-alliance. The evolving memory also includes the interplay in APLR operations that are manifested as

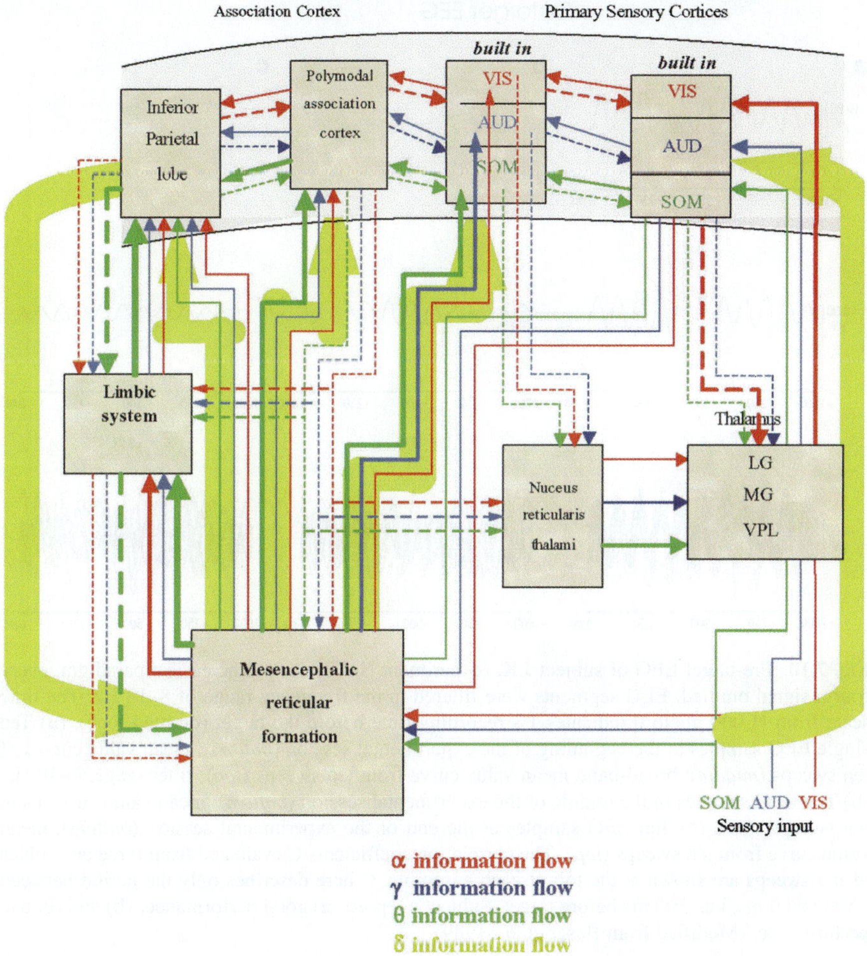

Fig. 7.9 A preliminary hypothetical scheme to describe the complex matching, and flow of oscillation-coded information. Alpha-code (*red*), theta-code (*green*) and gamma-code (*blue*). Thick lines theta (*green*) and very thick lines delta (*yellow*) representing information processing have been added to show the dominance of theta and delta resonance

selectively distributed oscillations in the whole brain. We have seen that the oscillatory theta or alpha response related to cognitive inputs does evolve during the APLR-Process. An example of the augmentation of knowledge, or the learned material in CNS, is the occurrence of regular and increased alpha activities or types of alpha-templates, as demonstrated in Fig. 7.7 and reported in earlier studies (see, e.g., Başar, 2004).

Figures 7.10–7.13 explain results of measurements of a subject (J.K). The subject was listening to repetitive tones (targets); every fifth target was omitted during

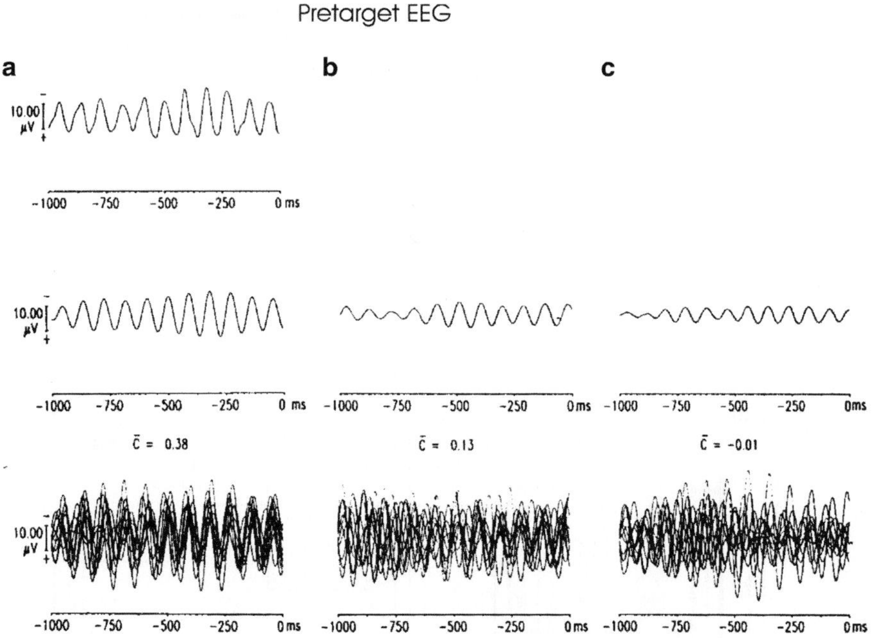

Fig. 7.10 Pre-target EEG of subject J.K. (experiment No. 3) during the easiest paradigm, every fourth signal omitted. EEG segments were filtered in the frequency range of 8–13 Hz. The time scale from -1,000 ms to 0 indicates 1 s recording time before the target (omitted tone). (**a**) Ten single EEG samples at the beginning of the experimental session (*bottom*); mean value curves of ten sweeps (*middle*); broad-band mean value curve from ten sweeps (*top*); filter range, 1–30 Hz. (**b**) Ten EEG samples in the middle of the experimental session (*bottom*); mean value curve from ten sweeps (*top*). (**c**) Ten EEG samples at the end of the experimental session (*bottom*); mean value curve from ten sweeps (*top*). The correlation coefficients C evaluated from three ensembles of ten sweeps are shown at the top of each ensemble. C here describes only the period between –500 and 0 ms, i.e., 500 ms before target. Subject's report: (**a**) good performance; (**b**) and (**c**) bad performance. (Modified from Başar et al., 1989)

the easiest paradigm. In the most difficult paradigm every fourth to seventh target was omitted. After the subjects had learned the paradigm, regular pre-stimulus alpha oscillations, which are phase-locked to the target, were measured (Figs. 7.10 and 7.13). On the contrary, during the most difficult paradigm, regular alpha oscillations were not observed (Fig. 7.12). These measurements show that during learning and working, memory process alpha activity behaves as a type of memory-related and internally triggered reverse evoked potential, thus demonstrating the active role exerted by EEG oscillations during a process of attention, perception, learning, and remembering (for more details see Başar, 2004).

According to the description in Fig. 7.14, event related changes in oscillations provide relevant changes and extensions in the electrical manifestations of evolving memory. In this case, it is clear that the new learned material is transformed to

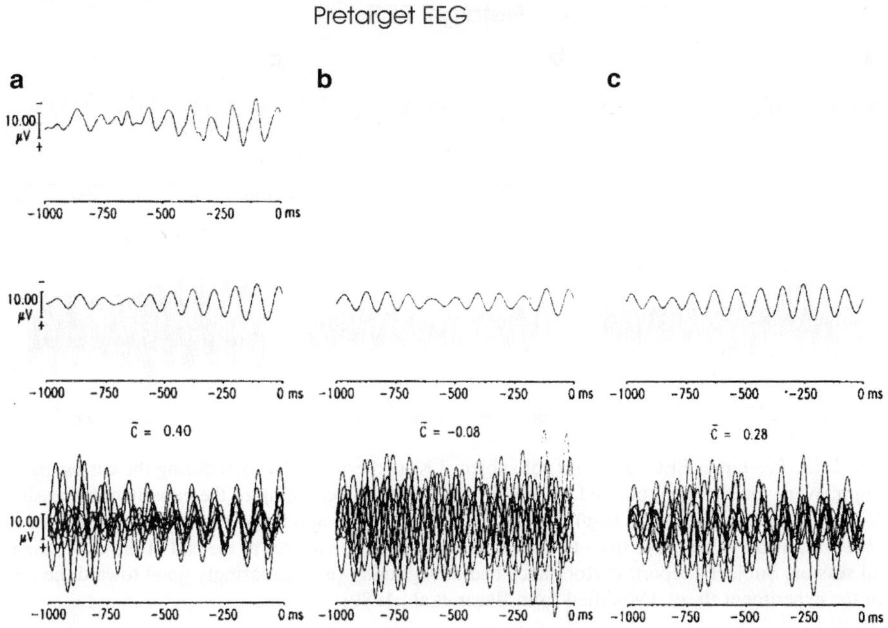

Fig. 7.11 Pre-target EEG of subject J.K. (experiment No. 15) during the same (easiest) paradigm. Explanation as for Fig.3.8; this shows a repetition after several months. Subject's report: (**a**) and (**c**) good performance; (**b**) bad performance. (Modified from Başar et al., 1989)

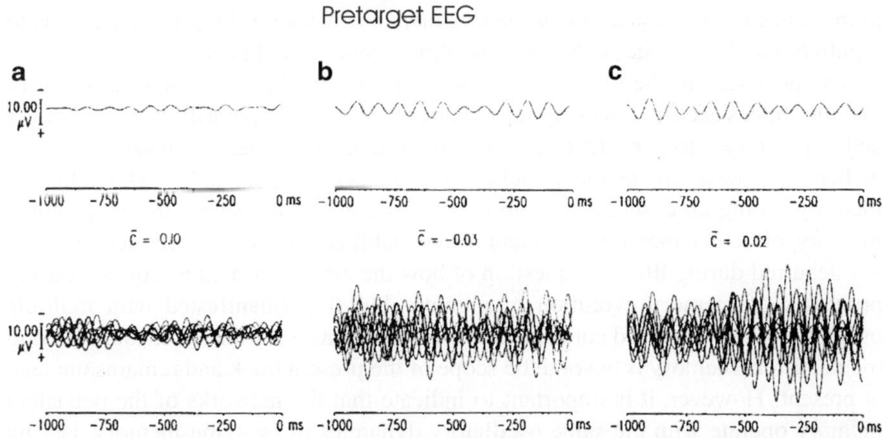

Fig. 7.12 Pre-target EEG of subject J.K. (experiment No. 16) during the most difficult paradigm, every fourth to seventh signal omitted. EEG segments were filtered in the frequency range of 8–13 Hz. The time scale from –1,000 ms to 0 indicates 1 s recording time before the target (omitted light). (**a**) Ten single EEG samples at the beginning of the experimental session (*bottom*); mean value curve (*top*). (**b**) Ten EEG samples in the middle of the experimental session (*bottom*); mean value curve from ten sweeps (*top*). (**c**) Ten EEG samples at the end of the experimental session. The correlation coefficients C here describe only the period from –500 to 0 ms. Subject's report: Tried to do well. (Modified from Başar et al., 1989)

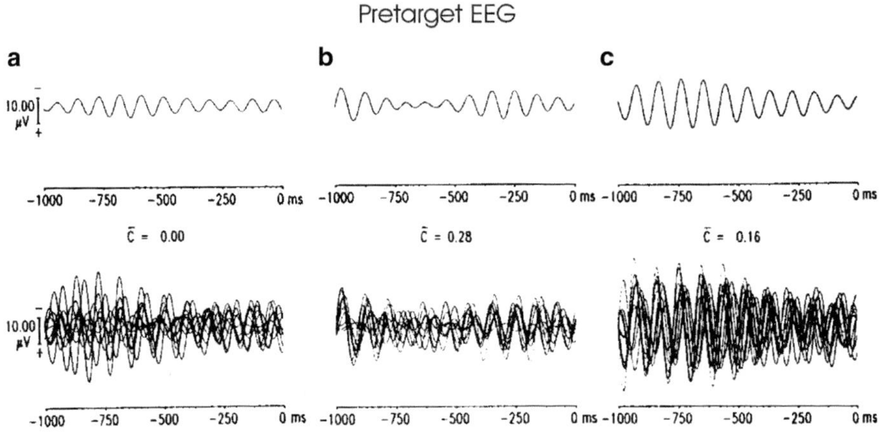

Fig. 7.13 Pre-target EEG (8–13 Hz) of subject J.K. (experiment No. 19) during the easiest paradigm, every fourth signal omitted. (**a**) EEG samples at the beginning of the experimental session (*bottom*); mean value curve (*top*). (**b**) EEG samples in the middle of the experimental session (*bottom*); mean value curve from ten sweeps (*top*). (**c**) EEG samples at the end of the experimental session. Subject's report: Performance bad at beginning (**a**), increasingly good toward the end of the experiment (**b**, **c**). (Modified from Başar et al., 1989)

LTM, at least for longer time intervals in comparison with working memory. Surprisingly, time periods of working memory and the duration of storage in LTM are not clearly defined in the literature. In our opinion, it would be more appropriate to introduce the expression *longer acting memory* instead of long-term memory, to highlight the differentiation between working memory and persistent memory.

The next step in the hierarchy of the memories in Fig. 7.1 is the transition of memory traces acquired in everyday experiences – and temporarily stored in longer acting memory – to persistent memory. According to the memory levels described in Fig. 7.1, persistent memory includes the inborn memory as the physiological memory (being an ensemble of sub-memories comprising echoic memory, iconic memory, olfactory memory, etc.) and of the stabilized parts of longer acting memory acquired during life. The question of how the new information acquired during processes of memory evolution or memory building (manifested with multiple oscillations and enhanced coherence in the whole brain) is transferred and stored in the persistent memory is beyond the scope of the present book and remains unclear, at present. However, it is important to indicate that the networks of the persistent memory operate with the same oscillatory dynamics of evolving memory, i.e., by using the same basic oscillatory codes alpha, beta, theta, etc. This possibly indicates that frequency codes may also be transferred to persistent memory, or the frequency codes play an essential role during the transition. They are probably invariant building blocks also partly contributing to the development of memory traces (or engrams), in which the principle of susceptibility plays a major facilitation of signal transfer. According to Tranel and Damasio (1995), there is good reason to believe

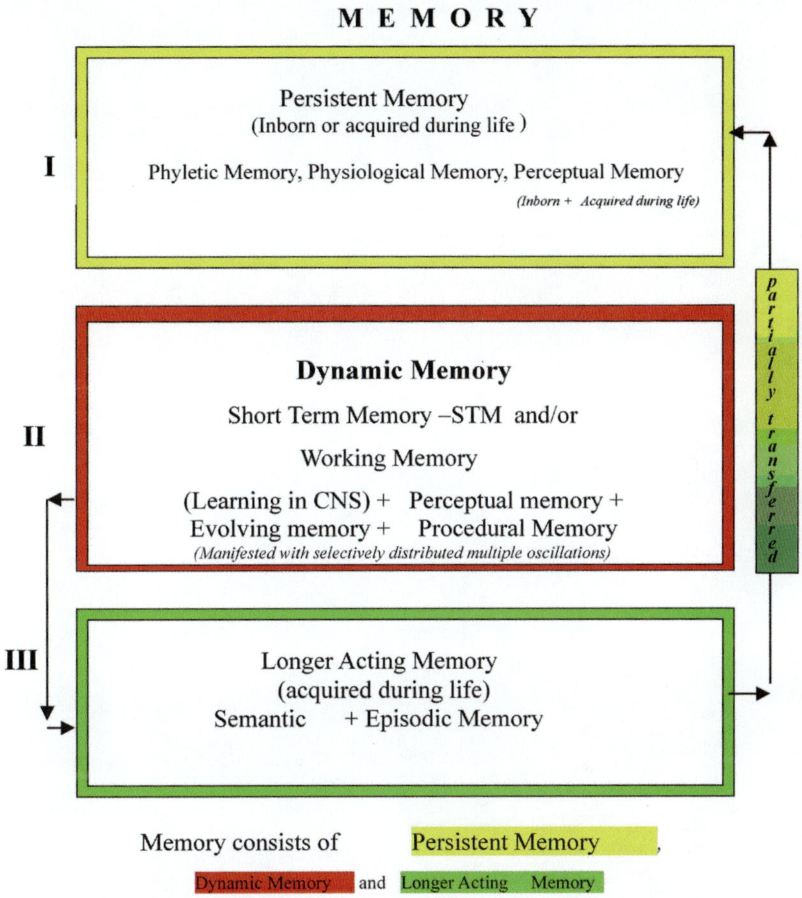

Fig. 7.14 A global scheme indicating memory levels and transitions between them. The persistent memory is indicated as a separate block and colored in yellow as an additional description to the scheme in Fig. 7.1 This scheme is a simpler version of Fig. 7.1 Again, *red background* indicates dynamic processes dominated by working memory system, and states of APLR-alliance. The working memory system is involved with the whole-brainwork. Procedural memory also belongs to this category. Learned memory traces are the transferred to more stable or quasi-stable longer activated memory. The persistent memory is explicitly described in this scheme as an additional point to Fig. 7.1 It is composed of physiological memory, stabilized traces of dynamic memory, and longer acting memory, which is quasi stable and can be partly transferred to the persistent memory (compare also with Fig. 7.1, Level III in activities is portrayed in Fig. 7.1)

that, depending on the type of memory under consideration, the storage may be selectional or instructional, i.e.,it may depend more or less on selection from a pre-existing repertoire of neuron circuit states (cf. Edelman, 1987; Shenoy et al., 1993).

The last steps related to the transition between memory states is schematically described in Fig. 7.14, which also should be compared with Fig. 7.1.

Fig. 7.14 A global scheme indicating memory levels and transitions between them. The persistent memory is indicated as a separate block and colored in yellow as an additional description to the scheme in Fig. 7.1. This scheme is a simpler version of Fig. 7.1. Again, we [illegible] memory systems formatted by working memory system and short [illegible]. The working memory system is involved with the whole brain etc. Procedural memory also belongs to this category. [illegible] equivalent to Fig. 7.1 is composed of [illegible] memory, [illegible] and long-lasting memory, which [illegible] in Fig. 7.13

that, depending on the type of memory under consideration, the storage may be selectional or instructional, i.e., it may depend more or less on selection from a pre-existing repertoire of neuron circuit states (e.g. Edelman, 1987; Shenoy et al., 2013).

The first steps related to the transition between memory states is schematically described in Fig. 7.14, which also should be compared with Fig. [illegible]

Chapter 8
Whole-Brain Work

8.1 The Theory of the Whole-Brain Work: An Approach to Brain Function by Means of Brain Dynamics

Chronological evolution of our conceptual framework evolved in the last 20–25 years and is based on empirical foundations from several laboratories.

The theory of whole-brain work proposes that integrative brain function is based on the coexistence and cooperative action of many interwoven and interacting sub-mechanisms. In its extension, the theory includes mechanisms that consist of super-synergy, superbinding and reciprocal interaction of attention, perception, learning, and remembering (APLR-alliance). This theory is based on empirical evidence and can be described in a number of sub-mechanisms.

In the following discussion, these mechanisms were grouped under four structural and/or functional levels.

8.1.1 Level A: From Single Neurons to Oscillatory Dynamics of Neural Populations

1. The neuron is the basic signaling element of the brain.
2. Since morphologically different neurons or neural networks are excitable upon sensory-cognitive stimulation, the type of the neuronal assembly does not play a major role in the frequency tuning of oscillatory networks. Research has shown that neural populations in the cerebral *cortex, hippocampus*, and *cerebellum* are all tuned to the very same frequency ranges, although these structures have completely different neural organizations (Başar 1998, 1999; Eckhorn et al. 1988; Llinas 1988; Singer 1989; Steriade et al. 1992). It is therefore suggested that all brain networks communicate by means of the same set of frequency codes of EEG-oscillations.
3. Intrinsic oscillatory activity of single neurons forms the basis of the natural frequencies of neural assemblies. Oscillatory activity of the neural assemblies of the brain consists of the alpha, beta, gamma, theta, and delta frequencies.

E. Başar, *Brain-Body-Mind in the Nebulous Cartesian System: A Holistic Approach by Oscillations*, DOI 10.1007/978-1-4419-6136-5_8,

These frequencies are the natural frequencies and thus the real responses of the brain (Başar et al. 2001a–c).

4. *Feature detectors* (Sokolov 2001), *place cells*, and *memory cells* (Fuster 1995a) are empirically established neural elements. However, a crucial turning point occurred with so-called "grandmother" experiments, which showed that large groups of neural populations were selectively activated on complex semantic and episodic inputs to the brain, and that complex percepts cannot be processed only by means of cardinal cells (Başar 2004; Bullock 1992; Edelman 1978, and experiments described in Sect.8.1.3). In attempts to describe the integrative functions of the brain, these experiments, and other similar studies, replaced the functional role of the single neurons with neural assemblies (Başar et al. 2001a). The emphasis on neural assemblies is the major point that differentiates our theory from Sherrington's neuron doctrine and Barlow's new perception doctrine (Barlow 1995).
5. Sokolov (2001) has excellently described and also constructively criticized the role of feature detectors. However, integrative functioning of the brain requires the selectively distributed and coherent neural populations in concert with the feature detectors.
6. The brain has *response susceptibilities*. These susceptibilities mostly originate from its intrinsic rhythmic activity, i.e., its spontaneous activity (Başar 1980, 1983a, b; Başar et al. 1992; Narici et al. 1990). A brain system responds to external or internal stimuli with those rhythms or frequency components that are among its intrinsic (natural) rhythms. Accordingly, if a given frequency range does not exist in its spontaneous activity, it will also be absent in the evoked activity. Conversely, if activity in a given frequency range does not exist in the evoked activity, it will also be absent in the spontaneous activity.

The rule of response susceptibility is demonstrated and explained in detail in Chapter 11 as well as included in the discussion of Chapter 21.

7. There is an inverse relationship between EEG and event-related potentials. The amplitude of the EEG thus serves as a control parameter for responsiveness of the brain, which can be obtained in the form of evoked potentials or event-related potentials (Barry et al. 2003; Başar 1998; Başar et al. 2003; Rahn and Başar 1993a).
8. The EEG is a *quasi-deterministic or a chaotic signal* and should not be considered as simple background noise (see also Chapter 14). This characteristic, and the concept of response susceptibility, lead to the conclusion that the oscillatory activity that forms the EEG governs the most general transfer functions in the brain (Başar 1990).
9. Oscillatory neural tissues that are selectively distributed in the whole brain are activated on sensory-cognitive input. The oscillatory activity of neural tissues may be described through a number of response parameters. Different tasks and the functions that they elicit are represented by different configuration of parameters. Because of this characteristic, the same frequency range is used in the brain to perform not just one but multiple functions. The response parameters of the oscillatory activity is as follows: enhancement (amplitude), delay (latency), blocking or

desynchronization, prolongation (duration), degree of coherence between different oscillations and degree of entropy (Başar 2004; Başar et al. 1999a, b; Kocsis et al. 2001; Miltner et al. 1999; Neuper and Pfurtscheller 1998a, b; Pfurtscheller 1997, 2001; Pfurtscheller et al. 1999, 2006; Rosso et al. 2001, 2002; Schürmann et al. 2000).

10. The number of oscillations and the ensemble of parameters that are obtained under a given condition increase as the complexity of the stimulus increases, or as the recognition of the stimulus becomes difficult (Başar 1980, 1999; Başar et al. 2000, 2001a). For comparison of stimulation with complex signals, the reader is referred to Chapter 12.

8.1.2 Level B: Super-Synergy of Neural Assemblies

According to the theory of whole-brain work, super-synergy consists of the following mechanisms (numbers 11–18):

11. In simple binding, there is temporal coherence between cells in cortical columns. This has been demonstrated by several authors (Eckhorn et al. 1988; Gray and Singer 1989).
12. Each function is represented in the brain by the superposition of the oscillations in various frequency ranges. The values of the oscillations vary across a number of response parameters (Principle 9). The comparative polarity and phase angle of different oscillations are decisive in producing function-specific configurations. Neuron assemblies do not obey the *all-or-none rule* that the single neurons obey (Chen and Herrmann 2001; Karakaş et al. 2000a, b; Klimesch et al. 2000a, b).
13. The *superposition* principle indicates synergy between the alpha, beta, gamma, theta, and delta oscillations during performance of sensory-cognitive tasks. Thus, according to the superposition principle, integrative brain function operates through the combined action of multiple oscillations (see also Sects.8.1.3 and 8.1.4).
14. The *response susceptibility* of the brain activates resonant communications in the brain by facilitating electrical processing between networks (Başar 2004; Başar et al. 1997a, b). This could also be interpreted as a general tuning process between neural populations and feature detectors (Sokolov 2001).
15. *Parallel processing in the brain shows selectivity*. The selectivity in parallel processing is produced by variations in the *degree of spatial coherences* that occur over long distances between brain structures/neural assemblies (Başar 1980, 1983a, b; Başar et al. 1999a; Kocsis et al. 2001; Miltner et al. 1999; Schürmann et al. 2000).
16. Temporal and spatial changes of entropy in the brain demonstrate that the oscillatory activity is a controlling factor in the functions of the brain (Beim-Graben 2001; Beim-Graben et al. 2000; Quiroga et al. 2001; Yordanova et al. 2002).
17. The *superbinding* mechanism can be denoted, according to the previous explanations, as an ensemble of mechanisms consisting of "superposition,

activation of selectively distributed oscillatory systems, and the existence of selectively distributed long distance coherences." The concept of *super-synergy* includes superbinding and, additionally, entropy, and the role of EEG-oscillations as control parameter in the brain's responsiveness.

18. *Complex matching in the whole brain* is proposed as a mechanism contributing to the machineries of memory and remembering (see especially Figs.7.5 and 7.6).

8.1.3 Level C: Integration of Attention, Perception, Learning, and Remembering

Extension of the theory of whole-brain work to cognitive processing is governed by the following principles:

19. *All brain functions are inseparable from memory function* (Fuster 1995, 1997; Hayek 1952). As in all integrative brain functions, memory is manifested as multiple and superimposed oscillations. A specific superposition of oscillations, each of which is characterized by the response parameters in Item 9, represents the configuration that is specific to the given type of memory.
20. *Attention, perception, learning and remembering* (APLR-alliance) are interrelated. As the grandmother experiments demonstrated (Başar 2004; Başar et al. 2003), memory-related oscillations are selectively distributed within the brain. They have dynamic properties and evolve on exogenous and endogenous input to the brain. Memory states have no exact boundaries along the time space. There is a hierarchical order that takes place on a continuum, but the boundaries of memory states merge into each other. Memory functions, from the simplest sensory memories to the most complex semantic and episodic memories, are manifested in distributed multiple oscillations in the whole brain.
21. In our theoretical framework, we introduced the expression evolving memory or memory building. The critical factor in memory building is the APLR-alliance. This concept represents a constant reciprocal activation within its sub-processes. Evolving memory has a controlling role in integrative brain functions (Barry et al. 2003a; Edelman 1978; Tononi et al. 1992). The hierarchy of memories is not manifested with separable states, since the memory manifests rapid transitions. Therefore, we suggest using the term memory states rather than memory stores,a concept in which memory is considered to take place in successive stages. These explanations do not apply, however, to persistent memory, which can be inborn or obtained through over-learned engrams or habits (see also Chapter 7).

8.1.4 Level D: Causality in Brain Responsiveness

To discover the cause of an event is to discover something among its temporal antecedents such that, if it had not been present, the event would not have occurred. In the introduction causality was described as conceptualized by Newton, Galileo, and Einstein. The present section considers causality as it pertains specifically to the responsiveness of the brain. The theory of whole-brain work presently considers three groups of factors as causes of the brain responses.

8.1.4.1 Genetically Fixed Causal Factors

The brain and the CNS-ganglia contain genetically coded networks. The phyletic memory networks that are inborn play essential roles in the responsiveness of neural populations. Accordingly, (1) occipital networks in the mammalian brain respond to light stimulation with enhanced 12 Hz oscillations (Başar 2004). On the other hand, temporal auditory areas that do not react to light stimulation respond to auditory stimuli with 10 Hz enhanced oscillations. (2) The ray brain reacts with 10 Hz oscillations to electrical stimuli (electroception); the human brain, on the other hand, does not have this ability (Başar 2004). (3) Like alpha networks, there are selectively distributed gamma networks in the brain. These networks show obligatory responses to sensory stimuli (Karakaş and Başar 1998). (4) Reflexes are genetically coded. The so-called "pre-potent responses" (Miller 2000) in reflexive actions also partially represent this type of causality. (5) The findings of Sokolov (1975) on the orienting response and the genetically fixed causal factors have to be emphasized: There are expectation cells that fire on expected input; sensory-reporting cells, which fire in response to actual stimulus; and comparator cells, which fire whenever there is a discrepancy between stimuli (Başar 2004).

The group of Begleiter and Porjesz (2006) recently launched a fundamental approach to examine the genetic underpinnings of neural oscillations. It is proposed that the genetic underpinnings of these oscillations are likely to stem from regulatory genes, which control the neurochemical processes of the brain, and therefore influence neural function. According to the work of this group, genetic analysis of human brain oscillations may identify genetic loci underlying the functional organization of human neuroelectric activity, and brain oscillations represent important correlates of human information processing and cognition.

Present behavior influences immediate future behavior. The plasticity in this adaptive behavior is demonstrated in the oscillations, showing that oscillatory plasticity is an additional causal factor in brain responsiveness. In auditory and visual memory task experiments, the EEG oscillations manifest a high degree of plasticity: The reciprocal activation of the APLR alliance (Başar 2004) also affects the future responsiveness of the brain, attesting to the presence of *oscillatory plasticity* in the higher cognitive processes.

The brain theory explained in the present chapter stems from the accumulation of findings over many years, and is also a consequence of Chapter 6 and 7. In the coming chapter, this theory will be linked with the coordination in the vegetative system.

To understand what is meant by the "mind," Part III describes several types of brains, thus opening the way to the differentiation of mind levels.

The framework established in this chapter is re-evaluated in the concluding chapters of Part VI.

Chapter 9
Does the Brain-Body-Mind Work as a Dynamic Syncytium?

9.1 General Principles: Descartes' Suggestion and Fessard's Extended Question

In the rationale of this book, philosophical background and philosophical questions dominate. Descartes suggested that few intelligible systems could dominate all processes in nature. In the chapter related to the description of the central nervous system (CNS), significant reference is made to the question presented by A. Fessard (1961), who asked, "Are there some general transfer functions dominating processes in the brain?" Chapter 10 extended Fessard's question to cover neurophysiologic events during the evolution of species. Because the present chapter describes a holistic approach to brain-body-mind, we extend Fessard's question again, to combine physiological processes of the CNS with the vegetative system. We now search for common rules and common transfer functions for the link that may demonstrate brain-body integration. Term *body* may be defined as the vegetative system together with all organs of the body that are linked to the brain electrically, by means of cranial nerves; and with transmitters by means of biochemical pathways.

9.1.1 How to Approach General Transfer Functions?

The present chapter provides material, discussion, and conclusions related to brain-body integration. Accordingly, it is one of the key chapters that constitutes the essence of the hypothesis presented in this book. Chapter 1 stated that this book aims to surpass the conventional descriptions of mind: The aim is to forge a synthesis, bridging cognitive processes and biological settings within the whole body. Chapter 5 explained the overall myogenic system in detail. This system is distributed throughout the body and manifests concerted oscillatory behavior in all visceral organs, the arterial system, and all related peristaltic functions in the body. Further, Chapter 7 and 8 mentioned the role of physiological settings, especially the overall myogenic system, as one of the most important causal factors influencing memory.

E. Başar, *Brain-Body-Mind in the Nebulous Cartesian System: A Holistic Approach by Oscillations*, DOI 10.1007/978-1-4419-6136-5_9,

The present chapter aims to complete the picture with a tentative synthesis, bringing together several physiological mechanisms that integrate the vegetative and CNS. Moreover, cognitive processes is also discussed as one of the factors influencing the link between the vegetative system and the CNS. In the didactical Box 1.1 the cranial nerves that are bridging elements of the vegetative system and the CNS are briefly explained. For further information, the reader is referred to works on physiology (Ganong 2001; Guyton 1971). The dynamics of brain-body integration are explained in five essential steps:

1. Oscillatory behavior of the overall myogenic system and oscillatory behavior in the brain stem.
2. Coherence of oscillations in the heart, sympathetic nerves, and spinal cord.
3. Distributed oscillations in basal ganglia and the influence of transmitters in oscillatory behavior.
4. Coordinated oscillation in brain stem and respiration, and coordination of respiration with oscillatory activity in the olfactory bulb.
5. The role of transmitters in modulation of the oscillatory activity.

After the description of these five steps, information related to changes of physiological settings and their influence on the brain is also provided. Within the scope of the present book, it is not possible to provide extensive information on each of the mentioned steps.

9.1.2 Cranial Nerves

Table 9.1 globally illustrates major functions of the cranial nerves. Readers who may be interested in more detailed information are referred to works on physiology. This short section aims to explain anatomical and neural links between brain and body. Figure 4.1 illustrates the links between the spinal cord and the autonomic nervous system.

9.2 Overall Myogenic System Revisited

Başar and Weiss (1981) measured and reviewed mechanisms of control, auto-oscillations of blood flow, contractility of the vasculature, forced oscillations in the peripheral circulatory system, spectral activity of peristaltic organs, and dynamics of the lymph nodes and the lymphatic system and found that all of these subsystems showed spectral properties (oscillatory activity) in the ultraslow frequency range of 0.01, 0.04–0.06, and 0.1 Hz. These authors introduced the concept of the overall myogenic system after quantifying and treating the performance of smooth muscle dynamics, by considering this as an overall and coordinating and driver system. The overall myogenic system incorporates.

Table 9.1 Cranial nerves and their major functions

Cranial nerve	Major functions
I Olfactory	Smell
II Optic	Vision
III Oculomotor	Eyelid and eyeball movement
IV Trochlear	Innervates superior oblique
	Turns eye downward and laterally
V Trigeminal	Chewing
	Face and mouth touch and pain
VI Abducens	Turns eye laterally
VII Facial	Controls most facial expressions
	Secretion of tears and saliva
	Taste
VIII Vestibulocochlear (auditory)	Hearing
	Equilibrium sensation
IX Glossopharyngeal	Taste senses
	Carotid blood pressure
X Vagus	Senses aortic blood pressure
	Slows heart rate
	Stimulates digestive organs
	Taste
XI Spinal Accessory	Controls trapezius and sternocleidomastoid
	Controls swallowing movements
XII Hypoglossal	Controls tongue movements

1. The vascular system, with all the arteries, arterioles etc. in the systemic circulation.
2. The lymphatic system, with lymphatic vessels and nodes.
3. The visceral system, which performs the visceral functions of the vegetative system and peristalsis in vegetative function.
4. A schematic presentation of the overall myogenic system is described in Chapter 5.
5. The smooth muscle cells are building blocks and basic effectors of the overall myogenic system (see Fig. 9.2a, showing the organization of the overall myogenic system, and Fig. 9.2b, schematically describing the effects of overall myogenic control). According to the findings of Başar and Weiss (1981), the overall myogenic coordination occurs in the common frequency ranges of 0.01, 0.04–0.05, and 0.1 Hz, thus corresponding or overlapping with the frequency range of new measurements reported in the literature (Allers et al. 2002).
6. Aladjalova (1957, 1964) demonstrated ultraslow periodicities in the brain as early as 1957. Subsequently, a number of publications indicated the existence of these ultraslow oscillations and possible links with the electroencephalography (EEG) (Ruskin et al. 2002). Multi-second oscillations in firing rate, with periods in the range of 2–60 s (mean 20–35 s) are present in 50–90% of spike trains from basal ganglia neurons recorded from locally anesthetized, immobilized rats. To determine whether these periodic oscillations are associated with similar periodicities in cortical activity, transcortical EEG activity was recorded in conjunction

with single- or dual-unit neuronal activity in the subthalamic nucleus (STN) or the globus pallidus (GP), and the data were analyzed with spectral and wavelet analyses (Allers et al. 2002). Ultraslow oscillations in firing rates of 31% of the STN neurons and 46% of the GP neurons with periodicities significantly correlated with bursts of theta (4–7 Hz) activity in transcortical EEG.

These examples demonstrate conclusively the possibility of ultraslow wave oscillatory coordination between myogenic organs and the brain.

9.3 Sympathetic Nerves of the Heart and the Spinal Cord

In the last decade the group led by Gebber in Michigan studied the frequency links of the spinal cord to the heart and kidneys. Such studies are rare; however, they provide core information for the oscillatory integration between the CNS and the vegetative system.

Gebber et al. (1999) used time and frequency domain analyses to examine the changes in the relationships between the discharges of the inferior cardiac (CN) and vertebral (VN) postganglionic sympathetic nerves produced by electrical activation of the midbrain periaqueductal gray (PAG) in urethane-anesthetized, baroreceptor-denervated cats. Further, CN-VN *coherence* and *phase angle* in the 10 Hz band served as measures of the coupling of the central oscillators controlling these nerves. The 10 Hz rhythm in CN and VN discharges was entrained 1:1 to electrical stimuli applied to the PAG at frequencies between 7 and 12 Hz. CN 10 Hz discharges were increased, and VN 10 Hz discharges were decreased when the frequency of PAG stimulation was equal to or above that of the free-running rhythm. In contrast, stimulation of the same PAG sites at lower frequencies increased, albeit disproportionately, the 10 Hz discharges of both nerves. In either case, PAG stimulation significantly increased the phase angle between the two signals (VN 10 Hz activity lagged CN activity); coherence values relating their discharges were little affected. However, the increase in phase angle was significantly more pronounced when the 10 Hz discharges of the two nerves were reciprocally affected. Importantly, partialization of the phase spectrum using the PAG stimuli did not reverse the change in CN-VN phase angle. Gebber et al. (1999) discussed the possibility that the increase in the CN-VN phase angle reflected changes in the phase relations between coupled oscillators in the brain stem.

Barman and Gebber (1993) recorded a variable mixture of 10 Hz and 2–6-Hz discharges from sympathetic nerves in decerebrate cats. Although medullary lateral tegmental field (LTF) neurons are considered to be a source of the 2–6-Hz oscillation in sympathetic nerve discharge (SND), their role in the control of the 10 Hz rhythm has not been critically evaluated. This issue served as the focus of the current study. In the first series of experiments, spike-triggered averaging of inferior cardiac SND was used in an attempt to identify LTF neurons with activity correlated to the 10 Hz rhythm in SND. The discharges of only one of the 120 LTF neurons studied were correlated to this component of SND. In contrast, 17 of 79 neurons had activity correlated to the 2–6-Hz oscillation in experiments in which this component of SND was prominent. These data indicate that LTF neurons neither receive input from, nor are components of, the 10 Hz rhythm generator. In a second series of experiments, muscimol was microinjected into the LTF bilaterally. Chemical inactivation of the LTF either eliminated the 10 Hz rhythm or reduced the power and peak frequency in this band of SND. These data support the view that LTF neurons have a permissive role in governing the 10 Hz rhythm in SND, probably by acting on elements of the rhythm generator located elsewhere. As expected, muscimol microinjections reduced the power in the 2–6-Hz band in SND in some experiments.

In the early studies of Michigan group (Barman et al. 1995) coherence analysis revealed that the 10 Hz rhythm in SND was not correlated to that in either inferior olivary activity of decerebrate cats or in neocortical spindles of urethane-anesthetized cats. Also, the discharges of some ventrolateral medullary and raphe neurons contained a 10 Hz rhythm that was not correlated to that in SND. These data support the hypothesis that a 10 Hz rhythm reflects the organization of a brainstem network that specifically governs sympathetic outflow.

To evaluate the role of pontine neurons, Barman et al. (1997) hypothesized whether pontine neurons are elements of the network responsible for the 10 Hz rhythm in SND. The first series of experiments tested whether chemical inactivation of neurons in the rostral dorsolateral pons (RDLP) or caudal ventrolateral pons (CVLP) affected inferior cardiac postganglionic SND of urethane-anesthetized cats. Muscimol microinjections into either region eliminated the 10 Hz rhythm in SND. According to this author support the view that pontine neurons are involved in the expression of this rhythm. Additional experiments were designed to determine whether pontine neurons are activity correlated to the 10 Hz rhythm in SND or whether they merely provide a tonic (nonrhythmic) driving input to the rhythm generator. Coherence analysis revealed that local field potentials recorded from the RDLP or CVLP had a 10 Hz component that was significantly correlated to SND. Additionally, spike-triggered averaging and coherence analysis showed that the naturally occurring discharges of individual RDLP or CVLP neurons were correlated to the 10 Hz rhythm in SND. Barman et al. (1997) concluded that these data support the hypothesis that RDLP and CVLP neurons are essential for the expression of the 10 Hz rhythm in SND and that they are elements of, or receive input from, the rhythm generator. The question here is, "Where is the rhythm generator?"

Ultraslow oscillations in firing rate, with periods in the range of 2–60 s, and averaging 20–35 s, are present in 50–90% of spike trains from neurons in basal ganglia nuclei (STN, GP, entopeduncular nucleus, and substantia nigra pars reticulata) recorded from locally anesthetized, immobilized rats (Allers et al. 2000; Ruskin et al. 1999b). Similar oscillations have also been observed in spike trains recorded from the basal ganglia of wake monkeys (Wichmann et al. 2000). Oscillations in rate with periods within this time range (2–60 s appear to be too slow to be involved in the fine aspects of motor control, but pharmacological studies have hinted that they may play a role in processes such as those involved in synaptic plasticity or attentional state (discussed in Allers et al. 2002). Stimulants such as cocaine, methylphenidate, and amphetamine increase the frequency of these oscillations within GP spike trains (Ruskin et al. 2001a, b), whereas general anesthesia virtually eliminates them (Allers et al. 2000; Ruskin et al. 1999b, 2001b). The fact that dopamine receptor stimulation affects the frequencies of these oscillations in a similar manner in all the basal ganglia nuclei studied (Allers et al. 2000; Ruskin et al. 1999b) suggests that there may be a considerable degree of coherence in oscillations in this time range in the basal ganglia. This idea is supported by additional studies that have shown that approximately 30% of simultaneously recorded pairs of basal ganglia neurons, with individual neurons located in either the same or different basal ganglia nuclei, show firing rate multisecond oscillations with matching periods (Allers et al. 1999).

These observations have led to the hypothesis that multisecond oscillations could act to organize the propagation and synchronization of faster oscillatory activity in distributed circuits. This study combines single- and dual-unit recording studies in the basal ganglia with transcortical EEG and depth electrode recordings in the frontal and parietal cortices and hippocampus, to examine the relationship between multisecond oscillations in the basal ganglia and synchronized activity at higher frequencies in the cortex and hippocampus. The results indicate that multisecond oscillations in STN and GP firing rates can be correlated with bursts of theta rhythm activity (4–7 Hz) in transcortical EEG recordings, which appear to reflect synchronized activity in the theta range in the hippocampus.

To explore the source of the synchronized activity producing the transcortically recorded theta signal, localized field potentials in the ipsilateral hippocampus and frontal or parietal cortices were recorded simultaneously with GP unit activity. Dorsal hippocampal field potentials (6 of 7) recordings exhibited bursts of theta that correlated with multisecond oscillations in GP firing rates (Fig. 9.1). Frontal (1 of 7) and parietal (0 of 3) cortical field potentials rarely showed this correlation.

Previous studies have shown that these ultraslow oscillations are not present in the basal ganglia of rats systemically anesthetized with chloral hydrate, ketamine, or urethane (Allers et al. 2000; Ruskin et al. 1999b, 2001b). In contrast, oscillations in the 1 Hz range are prominent in recordings from many brain regions in ketamine- or urethane-anesthetized rats. Correlated oscillations in the 1 Hz frequency range were originally observed in thalamocortical circuits (Steriade 2001), but have also been found in the GP, STN, and cortex of ketamine- or urethane-anesthetized rats (Magill et al. 2000), as well as in the nucleus accumbency and hippocampus (Goto and O'Donnell 2001). According to Allers et al. (2002), these observations indicate that mechanisms exist for the widespread propagation of both slow (1 Hz) and ultraslow oscillatory activity throughout basal ganglia, thalamocortical, and limbic areas, although it is not clear how similar the mechanisms involved in generation and propagation of these different oscillations may be.

A number of studies have focused on the time-limited emergence of correlated multisecond (ultraslow) oscillations in neural activity in early postnatal development and their potential role in the establishment of activity-based synaptic connectivity (Feller 1999). The mature organism, however, also clearly maintains mechanisms for generating correlated multisecond periodicities, as evidenced by the present observations, which combine data showing that oscillations in this frequency range modulate a number of processes including heart rate (Cooley et al. 1998), EEG activity patterns in sleep (cyclic alternating pattern) (Terzano et al. 2000), and

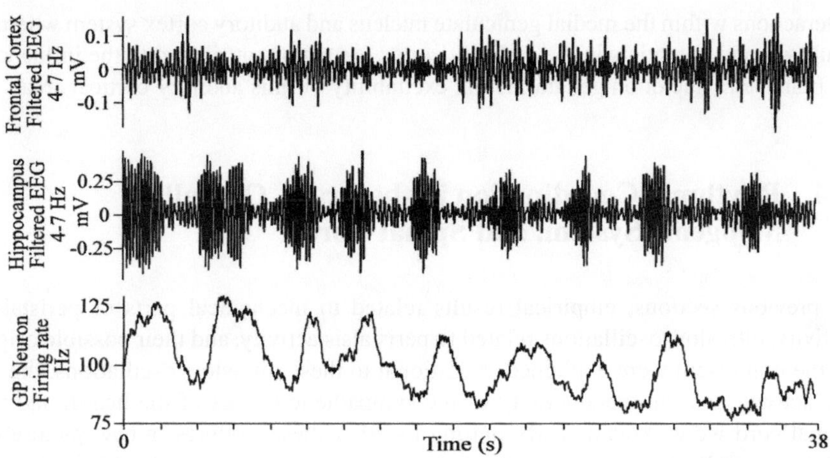

Fig. 9.1 Relationship between GP firing rate and hippocampal and cortical theta activity. *Top* and *middle traces* are digitally filtered (4–7 Hz) depth recordings from frontal cortex and hippocampus. *Bottom trace* is a mean frequency plot of a GP spike train. Note the multisecond periodicity of bursts of hippocampal theta rhythm power that correlate with a similar periodicity in firing of the GP neuron. Theta power in frontal cortex is variable, without apparent correlation to GP neuronal spiking rate (modified from Allers et al)

coherence in cortical fMRI signals (Biswal et al. 1995). According to Allers et al. (2002), the observed variability of phase relationships in the slow oscillations in EEG theta power and GP/STN firing rates and the absence of direct anatomical connections between the GP and the hippocampus, suggest that mechanisms underlying the correlated oscillations in basal ganglia/hippocampal activity involve a complex pattern of connectivity. It also important to note that Buzsáki and co-workers (Penttonen et al. 1999) have demonstrated multisecond (ultraslow) periodicities in hippocampal excitability in around 0.025 Hz, evident in recordings from the behavior of drug-free Sprague-Dawley rats, and point out that the network properties of the hippocampus may be capable of generating slow oscillations in excitability.

9.3.1 Dynamics of Ultra-Slow Potentials in the Auditory Cortex

Several new publications indicated the presence of ultra-slow activity(>0.5 Hz in cortical slides of the auditory system of the brain. Flippov et al. (2007) studied the extra-cellular infra-slow brain activity in the medial geniculate nucleus and the primary auditory cortex. Auditory evoked changes in ultra-slow activity in the range of seconds reflected specific mechanisms of acoustic information processing in the auditory system. Additional experiments have implanted chronic electrodes in the medial geniculate nucleous and primary cortex (Flippov et al., 2007). The results of these authors suggested that infra-slow activity occurs in specific mechanisms of interactions within the medial geniculate nucleus and auditory cortex system whereas multi-second potentials in the auditory cortex are mainly attributed to the influences of brain stem nuclei on general neural excitability of this auditory cortical area.

9.4 Rhythmic Coordination in the Brain, Overall Myogenic System, and Spinal Cord

In previous sections, empirical results related to mechanical parts of peristalsis activity, ultraslow oscillations related to peristalsis activity, and their possible origin in the brain stem were explained. Additional to these ultraslow oscillations, the 10 Hz activity and the theta activity of the sympathetic nerves of the hearth and the spinal cord were explained. By putting together these findings, a new picture on relative coordination among peristalsis, overall myogenic control, and the brain's electrical activity is emerging. We have given few examples from the circulation, respiration, lymphatic system, and the heart. The analysis of oscillatory brain dynamics can be performed experimentally in an easier and more concrete way. To observe the brain oscillatory responses to sensory and cognitive inputs, the subjects are stimulated or cognitively loaded with given and defined causal stimulation (Chapter 6). This type of measurement can be performed with healthy subjects under clear experimental conditions. The situation is quite different in the analysis of circulatory respiratory systems. Almost all measurements in vegetative systems are performed with the organs in vitro or in vivo under anesthesia. In some experiments, the stimulations used are electrical signals; in other measurements, mechanical stimulations are used, for example pressure stimulation. In most cases the stimulation signals are coming from hidden sources. Because of these experimental limitations, a systematic classification of outputs in the measured vegetative parameters has not yet been achieved. Nevertheless, in all areas of the brain and body frequencies of the overall myogenic system and frequencies of EEG oscillations are observed. There is strong experimental evidence that EEG oscillations and ultraslow oscillations are natural frequencies of the CNS and the vegetative system. There are several difficulties related to experimental conditions. It is usually not possible to measure mechanical contractions together with the electrical activity. By taking all these facts together, a statistical interpretation of the dynamic behavior is imposed: Several interacting effects and mutual excitation between the CNS and the vegetative system open the way to communication and control in the integration of brain and body. A sudden blood pressure increase in the circulatory system most probably gives rise to increased signals in appropriate relevant receptors in the arterial system (carotid artery) and also in the appropriate parts of the brain stem.

One thing is clear: In brain-body integration, there exists a general response susceptibility or resonance susceptibility. Interestingly, the structures of the CNS (including brain and cerebellum) and structures of the vegetative system (including all: vasculature, peristaltic organs, digestion organs, genital organs, and lymphatic system) have common resonance susceptibilities. Accordingly, we assume the existence of a common excitability/resonant frequencies in brain-body integration (see also Fig. 15.2). In Chapter 23 and 24 we will again talk about mutual excitability and resonances in unifying models.

9.5 Globally Coupled Oscillators in Brain-Body-Mind Integration

In Sect. 9.3 it was pointed out that the brain itself and the autonomous system manifest three categories of oscillations: the higher frequency band (>40 Hz), slow oscillations in the range of EEG frequencies, and ultraslow oscillations in the frequency window between 0.001 Hz and 1 Hz, mostly known as the rhythms of peristaltic organs and vasculature. The findings of ultraslow activity in the brain and the EEG frequencies in the spinal cord and the heart (Barman and Gebber 1993) strongly suggest the existence of oscillatory links in the brain, spinal cord, and all organs of the body. However, it is very difficult to conceive a concrete model for the one-to-one connections and synchronization of oscillations among all parts of the body. According to this reasoning, a good Gedanken-Model is provided with the concept of globally coupled oscillators. The concept of globally coupled oscillators, as illustrated in Fig. 9.2 (Pikovsky et al. 2001), also seems to be an appropriate Gedanken-Model (model of thought) for oscillatory dynamics of the brain/body.

If phenomena in large ensembles of oscillators, where each element interacts with all others are assessed, the mechanism that is usually denoted as global or all-to-all coupling needs to be described.

Suppose that there are a number of similar self-sustained oscillators influenced by a common external force. These oscillators can be different, but they should have close frequencies (in the example, 9, 11, 12 Hz). Then, another external force can synchronize all or almost all oscillators in the ensemble and, as a result, the oscillators move coherently with the same frequency (but probably with a phase shift). Sensory organs are known to consist of a large number of neurons, firing at different rates. In the case of external stimulation, the group of neurons with frequencies close to that of the stimulus can be entrained to impose the brain to produce a genuine signal. In the scheme presented in Fig. 9.2, all oscillators are driven by a common force, and this activity can entrain many oscillators in the field. The external force, which is globally coupling, is not predetermined, but arises from interaction within the ensemble. In describing a technical system, it is easier to precisely predict and indicate resonance phenomena. When we translate this viewpoint to the brain-body, we first mention that the sensory organs, the heart, and the visceral organs have common frequency channels, which are able to produce mutual resonances.

Accordingly, innate oscillators of the body that produce driving signals most probably elicit mutual interactions and mutual resonances between a number of organs such as the brain and the spinal cord. These processes can be denoted as hidden mechanisms. Accordingly, the understanding of globally coupled oscillators in the brain stem, the cerebral cortex, the spinal cord, and all types of peristalsis organs cannot be precisely described. There are too many driving forces, receptive fields, and different functions leading to forced and synchronized oscillators that

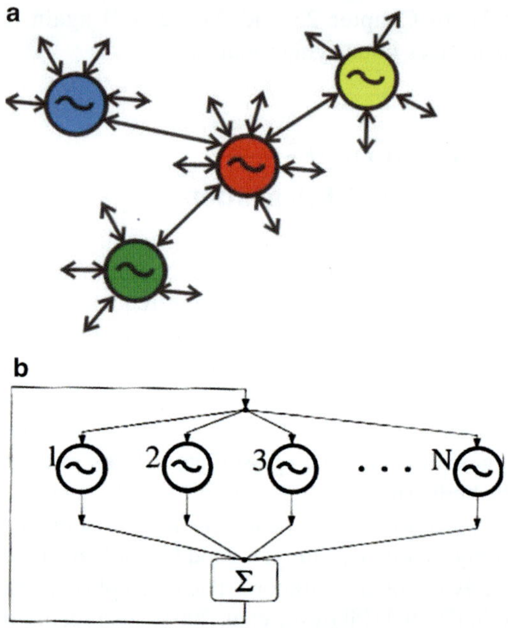

Fig. 9.2 (**a**) Each oscillator in a large population interacts with all others. Such an interaction is denoted as all-to-all, or global coupling. (**b**) An equivalent representation of globally coupled oscillators. Each element of the ensemble is driven by the mean field that is formed by all elements (modified from Pikovsky et al. 2001)

also give rise to mutual interactions and mutual resonances. Despite this influence on oscillatory components from neural and vegetative effectors, it is generally thought that general transfer functions of the link in the processes of brain-body interaction consist of mutual resonance phenomena in the ultraslow frequency range and in the frequency bands of EEG oscillations. Accordingly, the holistic description of brain-body-mind requires knowledge of oscillatory *dynamics* and the influence of the vegetative system, as affected by neurotransmitters. The globally coupled oscillators described in Fig. 4 can be expanded by taking into account the links between the brain, the spinal cord, and the autonomous system. In this case, three different assemblies of globally coupled oscillators are involved in the processing of mutual links and communication. Dynamics of cognitive processes is also governed by oscillatory dynamics (shown in Chapter 6 and 7); therefore, it can be concluded that oscillatory components play a key role in brain-body-mind integration. The role of transmitters in pathological situations of the mind is described by Bowden (2008),Chapter 3 and 13, and Başar (2008). Transmitters that greatly influence oscillatory dynamics should, therefore, be included within the main parameters of brain-body-mind integration.

The concluding Chapter 23 discusses the model of globally coupled oscillators in the integration of the brain, spinal cord, and vegetative organs. Furthermore, this model is brought to a more tentative general statistical configuration inspired by string theory (Chapter 24).

Part III
The Brain in Different States: Evolution, Maturing, Emotion, and Pathology

Prelude to Part III

In Part II, methods and the concept of oscillatory brain dynamics were applied so as to partly describe brain function in sensory-cognitive-memory processes, and in turn to lay the psychophysiology groundwork to approach the brain-body-mind. In Part III, attention is switched to other types of measurements.

Does an Aplysia have a "typical" mind? Does it share common electrophysiological components with all the studied brains? Does it share common features despite a variety of these brains? To see changes (or a mutation of electrophysiological activity) during the evolution of the species, Bullock (Fig. 1) and Başar published a series of papers that described studies in which they measured the electrical activity of invertebrates such as Aplysia and Helix pomatia as well as low vertebrates such as goldfish and rays.[1]

The child brain is different from the adult brain, especially in the morphology of the frontal cortex and the frequency range (10 Hz). Do 10-Hz oscillations play a major role in the shaping of mind? How similar are the minds of Alzheimer's, bipolar, and schizophrenic patients? Part III discusses these questions.

During the evolution of the species the central nervous systems of various living beings went through great changes in anatomical structure. Thus, evolution showed a type of dynamic behavior. A second type of dynamic change takes place during maturation of the human brain. The brain of a fetus, a newborn baby, a 3-year-old child, and an adult exhibit morphological differences, especially in the frontal lobes (Solms and Turnbull 2002). Important changes during the evolution of the species and the maturation of the human brain are accompanied by intelligent behavior, and the final stages of both developments mark the appearance of creative processes in relation to intuition and the ability to perform synthesis. Neither babies nor low-level living beings possess creativity, decision-making ability, and intuition.

---

[1] Ted Bullock is known as the father of neuroethology. Following the publication of Başar's *EEG–Brain Dynamics* (1980) in 1982 T.H. Bullock invited E. Başar to La Jolla for a series of joint experiments that also continued in Lübeck during a period of almost 25 years. According to Başar, professor Bullock was a living handbook in the field of anatomy and physiology of invertebrates and low vertebrates. One of the most important publications of this joint research was E. Başar and T.H. Bullock, *Induced Rhythms in the Brain* (1992).

THEODORE HOLMES BULLOCK

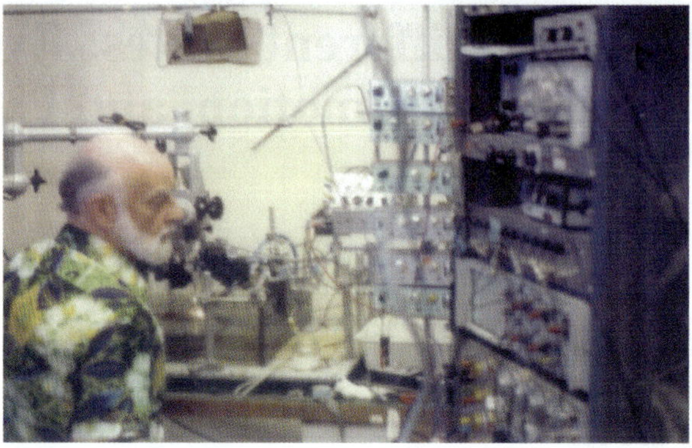

Fig. 1 Photo of Theodor Holmes Bullock (1915–2005) in his laboratory in the summer of 1984

In the next two chapters a comparison is made between anatomical and electrical properties of brains that are less capable of thinking or have intuitive behavior and the human brain. Intelligence, intuition, and all higher intellectual functions are also important properties of brains. Without analyzing anatomical, electrophysiological, and biochemical differences among diverse types of brains during evolution, it is not possible to fully understand brain function. How does a change in the neocortex create variations in brain functioning?

One of the most important revolutionary developments in biological sciences was Charles Darwins (Fig. 2) Introduction to *The Evolution of the Species*. Darwin worked within a "transformist" framework of the living world initiated earlier by Jean-Baptiste de Lamarck (see also Changeux 2004). Darwin's theory rests on two fundamental ideas. The first is the concept of *heritable variation*, which appears spontaneously and at random in individual members of a population and is immediately transmitted through descendants. The second is the idea of *natural selection*, which results from a struggle for life.

The brilliant French philosopher Henri Bergson, who studied the work of Charles Darwin, came to the conclusion that the *superiority of the human brain* in comparison with lower species lies in its ability for *intuitive and creative thinking*. Bergson described three types of mental abilities that developed during the evolution of the species: *instinct, intelligence, and intuition* (Bergson 1907; see Chap. 17). Although instincts are observed in low living beings such as invertebrates, intelligent behavior also belongs to the functional properties of lower vertebrates and mammals. However, only human beings have intuition. This is also, according to René Descartes (1840) and John Locke (1690), what makes human beings different from other species.

At the beginning of the twentieth century Bergson's proposal could not be verified by means of electric recordings. However, Bergson's view has gained in importance through recent empirical studies and metaphysical essays in which the results were extended to the role of the brain's electrical alpha activity during the evolution of the species (Başar and Güntekin 2009).

Fig. 2 Charles Darwin (February 12, 1809–April 19, 1882)

The emotional brain is analyzed in Chap. 12, and the pathological brain in Chap. 13. Emotional inputs considerably change the dynamics of electrical activity; therefore, these changes should be considered important keys toward a theory of mind.

Chapter 13 studies the brains of Alzheimer's, bipolar, and schizophrenic patients. How do changes in oscillatory processes and neurotransmitters affect breaks or complete changes of mind? The results presented in Chap. 13 are considered again in Chap. 22, in which quasi-invariants in body-brain-mind functioning are re-evaluated.

The abundant results in Part III add great value to the reasoning about brain-body-mind outlined in the unifying Chaps. 22 and 25.

The knowledge offered to the reader in the present chapter, along with a comparison of the results, is intended to elucidate new components for syntheses in the concluding chapters 22, 23, 24 and 25 of Part VI.

Chapter 10
The Brain in Evolution of Species and Darwin's Theory

10.1 A Unifying Step in Brain Function: The General Transfer Functions in the Brain According to Fessard

Fessard (1961) stressed that the brain must not be considered simply as a juxtaposition of private lines leading to a mosaic of independent cortical territories, one for each sense modality, with internal subdivisions corresponding to topical differentiations. What are principles dominating the operations of hetero-sensory communications in the brain? Fessard (1961) indicated the necessity of discovering principles that govern *the most general – or transfer – functions* of multi-unit homogeneous messages through neuronal networks. The transfer function describes the ability of a network to increase or impede transmission of signals in given frequency channels. The *transfer function*, represented mathematically by frequency characteristics or wavelets (Başar 1980; Başar-Eroğlu et al. 1992) constitutes the *main* framework for *signal processing and communication.* The existence of general transfer functions would then be interpreted as the existence of networks distributed in the brain having similar frequency characteristics facilitating or optimizing the signal transmission in resonant frequency channels (Başar 1998; see also Chap. 6).

According to Fessard's proposal, all brain tissue, both mammalian and invertebrate, would have to react to sensitive and cognitive inputs with similar oscillatory activity or transfer functions. The degree of synchrony, amplitudes, locations, and durations or phase lags varies continuously, but similar oscillations are most often present in the activated brain tissues (Başar 1999). As to the process of coding explained in Chap. 6, the general transfer functions of the brain manifested in oscillations strongly indicates that frequency coding is one of the major candidates to govern brain functioning.

In the scope of the new Cartesian system we now try to extend our question of unifying to the mutation of electrical signals from the brain to the evolution of the species.

E. Başar, *Brain-Body-Mind in the Nebulous Cartesian System: A Holistic Approach by Oscillations*, DOI 10.1007/978-1-4419-6136-5_10,

10.1.1 Do Some General Transfer Functions in Electrical Activity of Nervous System Exist During the Evolution of the Species?

An analysis of the electrophysiology of species consisting of recordings of spontaneous electrical activity and evoked potentials has been carried out by the author jointly with the laboratory of Bullock in San Diego (Başar et al. 1999b; Bullock and Başar 1988). In addition to the conventional electrophysiological recordings, the efficient method of oscillatory brain dynamics was used. The experiments used the isolated ganglia of Aplysia, *Helix pomatia*, and the brains of low vertebrates such as goldfish and ray. Cortical and subcortical structures of the cat brain and the scalp recordings from the human brain were also analyzed. This examination of brain oscillations included delta, theta, alpha, beta, and gamma oscillations.

During the evolution of the species the central nervous systems (CNSs) of various living beings, from invertebrates to low vertebrates, vertebrates, mammalians, and humans, went through great changes in anatomical structures and biochemical pathways, which were accompanied by changes in electrical activity. Thus, evolution also showed a type of dynamic behavior.

Is there commonality at all stages of evolution or does the functional mechanism change itself during the development from simple to complicated neuronal networks? Does the development to complexity involve new possibilities of information transfer? Based on these questions, comparative analyses of evoked potentials in different invertebrates and vertebrates are described. Except for Aplysia and Helix, the sensory stimulus modalities applied are physiological (acoustical or, in the case of the ray, a low-voltage electrical field). In Aplysia and Helix, non-physiological electrical stimulation was given to a connective or nerve.

10.2 Dynamics of Potentials from the Brain of Invertebrates

10.2.1 Introduction

Since Adrian (1931, 1937) and, Adrian and Matthews (1934), among others, drew attention to some evidence of similarity, and differences between invertebrates (insects) and mammals, little progress has been made. Bullock (1945, 1974, 1983, 1984a, b) and Bullock and Horridge (1965) pointed out that there was evidence both for widespread similarity among vertebrates and general differences between them and invertebrates.

Many invertebrate ganglia are simple-structured nerve cell populations consisting of a relatively small number of cells. Lower level ganglia, analogous to spinal cord, plus a few brains of the less advanced invertebrates, consist of up to a few thousand neurons. The brains of most insects, lobsters, and cephalopods have fairly to quite complex structures and tens to hundreds of thousands of cells. The *H. pomatia* is intermediate, around $>10^4$. The cat or human brain, a complex, three-dimensional network, is comprised of an astronomical number of neurons (on the order of up to 10^{12}). The significance of studying an invertebrate CNS as a brain model may be supported by Bullock and Başar's (1988) observation:

> When ongoing compound field potentials in the brains of invertebrates and vertebrates are compared, the recorded differences between vertebrates and most invertebrates are not due to brain size, cell size, number or density, lamination, or single cell power spectrum, but are primarily due to assembly properties, i.e., to cooperativity.

Therefore, the following questions have been addressed: (1) Are there some common components among the frequency characteristics of evoked potentials of vertebrate and invertebrates representing different stages of evolution? (2) Are there changes of coherence in the neural tissue during the evolution of the species?

10.2.2 Anatomy and Physiology of the Invertebrate (Gastropod) Nervous System

The ganglia of the CNS are organized such that, together with the commissures and connectives, they form a circumesophageal ring around the intestine and esophagus. Figure 10.1 illustrates the anatomy of the Helix as a prototype. The complement of the ganglia comprises the following series of node-like clusters of cells and neuropiles with connectives and commissures: (1) a pair of cerebral ganglia; (2) a pair of buccal ganglia; (3) a pair of pleural ganglia; (4) a pair of pedal ganglia; (5) a pair of parietal ganglia; and (6) the visceral ganglion.

The most important cellular building blocks of the gastropod CNS are the unipolar ganglion cells. The ganglia of gastropods, such as Aplysia and Helix, are built from only a small number of neurons (several thousand according to Bullock and Horridge 1965) (Fig. 10.2).

In the CNS of the Helix are nerve cells of different sizes varying from the smallest cells (6–7 μm) (Nabias 1894) in the globuli cell mass of the procerebrum to the largest (so-called giant cells) ranging from 260 μm (Kunze 1921) to 400 μm (Böhmig 1883) in the parietal. Giant cells are found in all the central ganglia in small numbers: fewer in the buccal (one), pleural (one to two) and cerebral ganglia; parietal (nine to ten) and pedal (ten) most in the visceral (22), (Bullock and Horridge 1965; Kunze 1917). Since the records of spontaneous field activity in the Helix ganglia, as in Aplysia, are dominated by either small or modest-sized spikes, a question is raised: Are these spikes triggered by field potentials, or are the field potentials the result of the spikes (after-potentials)? According to the observation of Bullock and Başar (1988) in Aplysia the low amplitude slow waves are not attributable to envelopes of spikes, but may come from the after-potentials of the larger units.

Fig. 10.1 The central nervous system and the major nerves of a young *Helix aspersa* (Kerkut et al. 1975). See Fig. 10.2 for the neural composition of ganglia

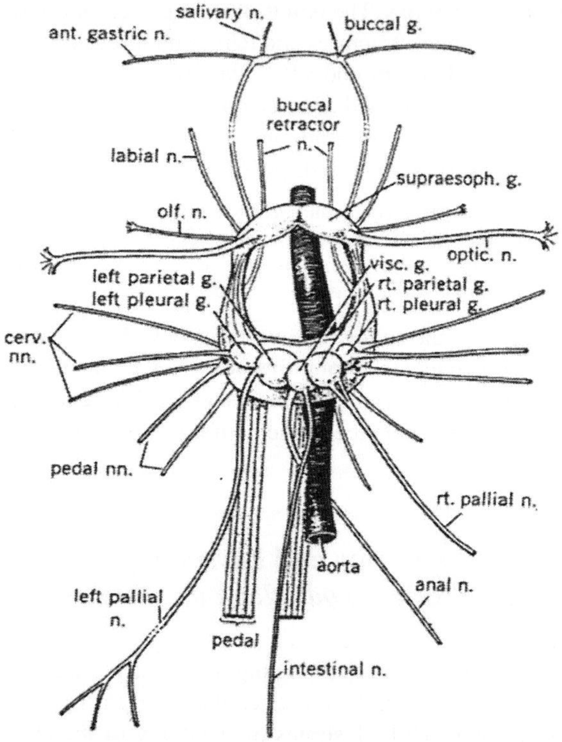

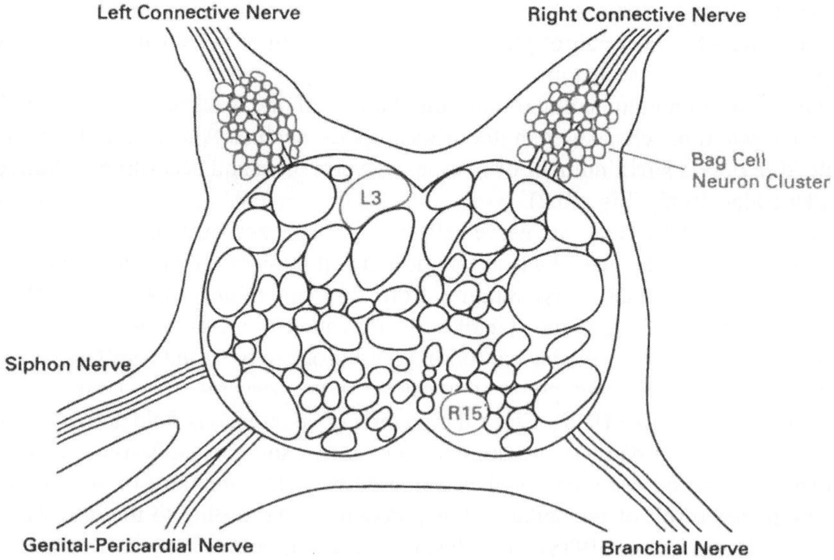

Fig. 10.2 Identified neurons and groups of neurons in the Aplysia abdominal ganglion. The drawing is from one by Eric Kandel and colleagues (Kandel 1976) of the dorsal surface of the Aplysia abdominal ganglion (Levitan and Kaczmarek 2002)

10.2.3 The Relationship Between the EEG of Vertebrates and Field Potential Fluctuations of Invertebrates

When comparing the slow rhythmical potential fluctuations of the Helix and Aplysia ganglia with those of the higher vertebrates, it must be kept in mind that the CNS is totally disconnected from the peripheral organs. This means that the sensory association fields no longer have exogenous inputs in natural forms. Yet, distinct fluctuations are revealed in the low amplitude field potentials of the invertebrate ganglia when the signals are arbitrarily pass-band filtered: For *H. pomatia* in the 1–4, 4–8, 8–15, 15–30, 30–48, and 52–125 Hz ranges (Röschke and Başar 1988; Schütt et al. 1992) and for Aplysia in the 2–5, 5–l0, 10–20, and 40–80 Hz range (Bullock and Başar 1988). (See Fig. 7.3a for all frequency windows and Fig. 7.3b for 10 Hz windows.)

The different types of pass-band filters chosen for Aplysia and Helix are simply arbitrary and not caused by any consistent, reproducible species differences in fluctuations. These fluctuations are solely of the intrinsic cellular activities of small cell populations independent of sensory modality (Fig. 10.3a).

The relative weakness of slow waves in most invertebrates might result from relatively little synchrony among neurons. In contrast to Aplysia and presumably most other invertebrates, vertebrates have a significant degree of synchronization of low-frequency components (<25 Hz), judging by the mean coherence between electrodes <1 mm apart. Another basic difference is that the power spectra of higher centers in arthropods and gastropods are much stronger ranging from 50 to 500 Hz than in the vertebrate higher centers, resembling the cerebellum and spinal cord (Bullock and Başar 1988).

Evidence provided by these findings strongly imply that the 8–15 Hz activity (alpha band), which is known as one of the most important frequency components of mammalian brain activity, is also generated in the much smaller and less developed CNS, such as the isolated invertebrate ganglia.

These recordings provide the following evidence.

Irregular and low-amplitude alpha activity is recorded in the invertebrate ganglia. The 10-Hz activity is also a component of the electrical activity. However, the regularly shaped high-amplitude alpha of higher living beings is not encountered here (see Fig. 10.3b).

Experiments with some neurotransmitters (Schütt et al. 1992) have provided additional evidence that the transmitter of the cholinergic system, acetylcholine, which is known to be involved in sensory as well as cognitive processes of the mammalian brain, induces frequency responses in the 2–20 Hz range (Fig. 10.4).

10.2.3.1 Potentials Evoked by Means of Electrical Stimulation

Aplysia

The potentials evoked in the Aplysia cerebral ganglion by electrical stimulation to the connective, show characteristic responses that can be roughly grouped into two

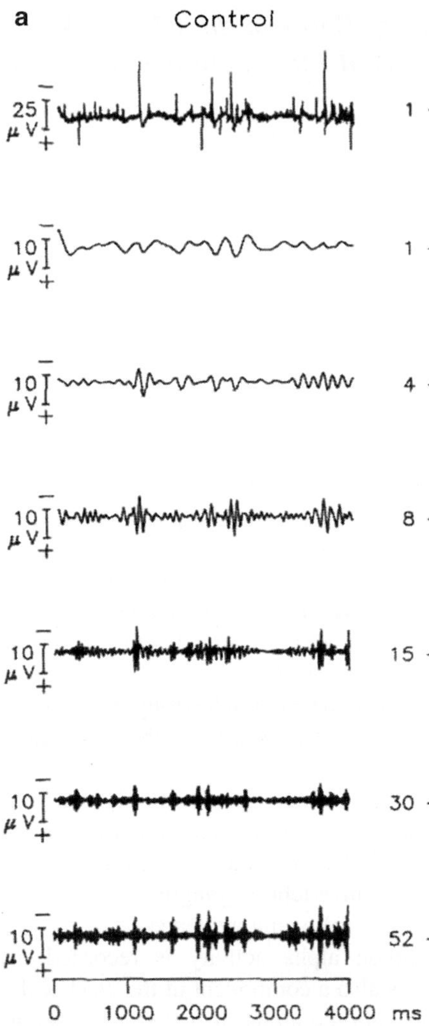

Fig. 10.3 (**a**) *Helix pomatia.* A typical time signal (approx. 4 s) of the ongoing activity with a modest number of spikes in the neutrophil of the isolated visceral ganglion; wideband (1–125 Hz) and pass band components (1–4, 4–8, 8–15, 15–30, 30–48, and 52–80 Hz) (Schütt et al. 1992). (**b**) *H. pomatia.* The field potential fluctuations in different ganglia; a sample of 2 s each. *Top:* the cerebral ganglion. *Bottom:* the visceral ganglion. *Thin line:* wide band component (1–50 Hz). *Thick line:* narrow-band component (8–15 Hz). Note that occasionally regular 10 Hz oscillations of maximal 10 μ amplitude can be recorded (Shütt et al. 1992)

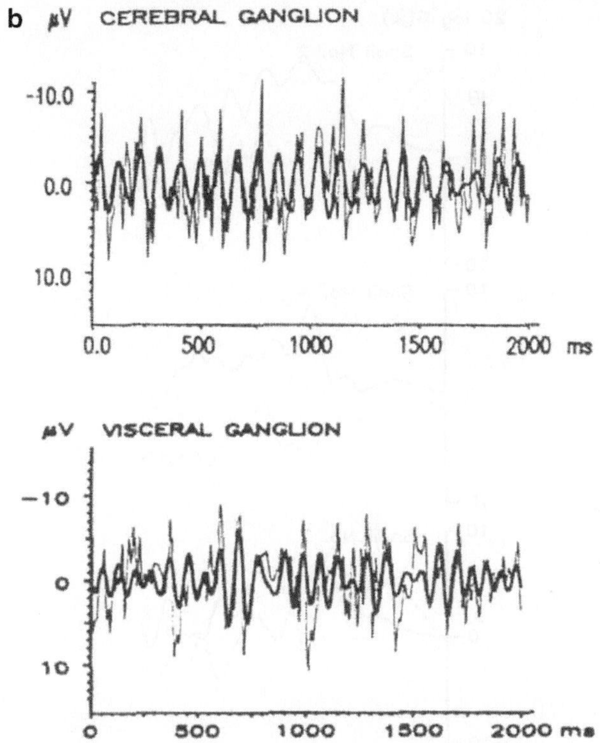

Fig. 10.3 (continued)

types. One group ($n = 8$) exhibits clearly stimulus evoked amplitude enhancement in the range of 2–20 Hz and additional resonance in the ranges of 40 Hz as well as 80 and 100 Hz. The other group ($n = 5$) all elicit, in contrast, clearly separated resonances in the ranges of 5, 10, and 25 Hz and additional local maxima between 60 and 100 Hz. The amplitude characteristics of two typical responses from both groups are presented in Fig. 6.7(Sturbeck 1988). Sturbeck draws particular attention to the 40-Hz resonance, which also exists in this simplest neuronal network.

Helix pomatia

Sturbeck (1988) recorded potentials evoked by electrical stimuli (maximal intensity, 30 mA, of 1 ms duration) from five different ganglia (right and left pleural, right and left parietal and visceral) stimulating three different nerves (*N. pallialis dexter, N. analis*, and *N. pallialis sinister*). The right pleural ganglion responds, accompanied by amplitude enhancement, with a power increase in the frequency bands of 2–10 and 20–30 Hz. The right parietal ganglion, on the other hand, elicits a power increase in the pre-stimulus spontaneous activity which is evenly distributed

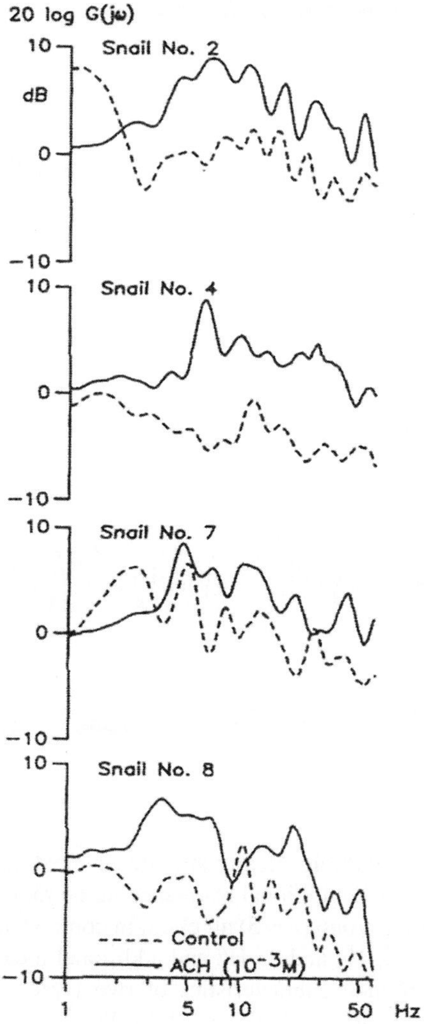

Fig. 10.4 Power spectra of the ACh-induced activity in four preparations. Each curve represents a power spectrum averaged from 20 epochs (81.92 s) (Schütt et al. 1992)

over the entire range of 1–100 Hz. The frequency amplitude characteristics of the post-stimulus evoked activity show maximizations in the approximate ranges 2, 13–23, and 40–65 Hz (Fig. 10.5).

After long years of study in a joint endeavor in San Diego and Lübeck, results show that electrical stimulation in Helix seems to evoke diverse neuronal population characteristic field potential fluctuations similar to those in mammals, having peaks at the same frequencies in the bands 1–8, 8–15, 15–30, 30–48, and 52–100 Hz.

The gamma component particularly, which is more or less pronounced in all the examined structures, should be mentioned here because this band is repeatedly

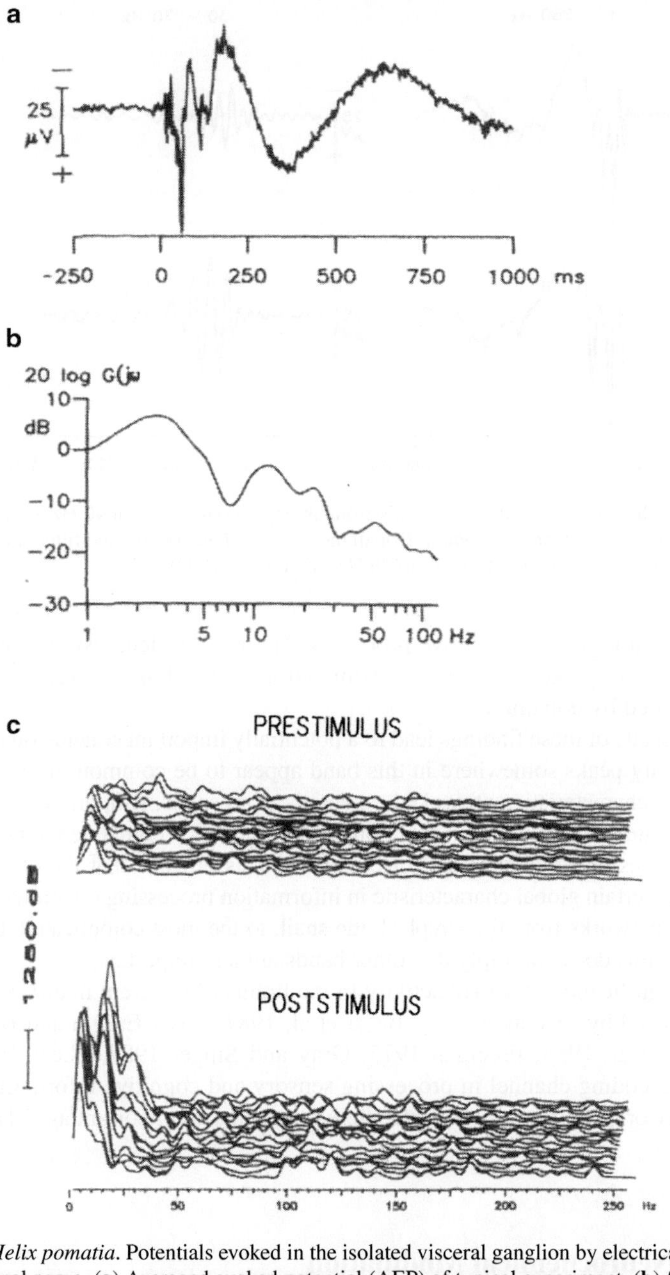

Fig. 10.5 *Helix pomatia*. Potentials evoked in the isolated visceral ganglion by electrical stimulation of the anal nerve. (**a**) Averaged evoked potential (AEP) of typical experiment, (**b**) frequency-amplitude characteristic of the AEP, (**c**) typical single power spectra of a preparation with an especially high number of bursting spikes. Pre- and post-stimulus activities are separately presented. Note that evoked activity is clearly visible up to 250 Hz (Shütt et al. 1992)

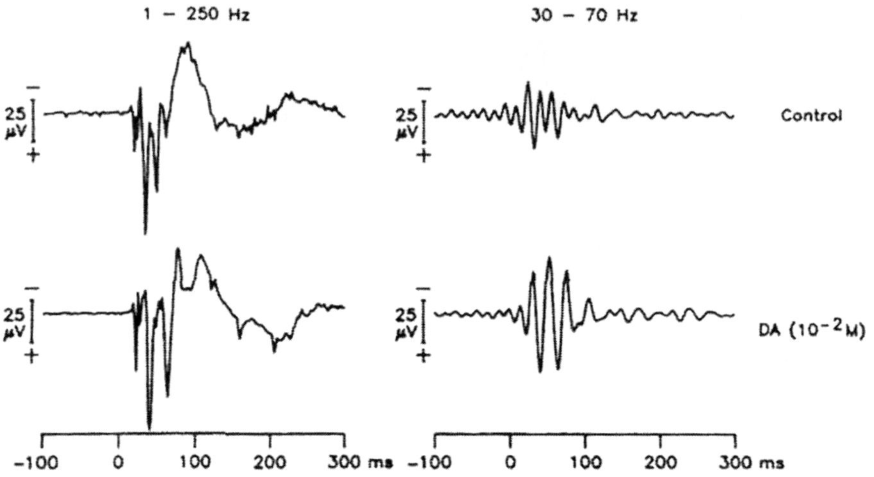

Fig. 10.6 The averaged evoked potentials from an experiment in which 40 Hz activity increases after dopamine. *Top:* Control. *Bottom:* Dopamine (10^{-2} M). *Left:* Wide-band filtered from 1 to 250 Hz. *Right:* Pass-band filtered from 30 to 70 Hz (Shütt et al. 1992)

discussed in regard to cognitive processes. Figure 10.6 demonstrates 40-Hz burst activity that is evoked by electrical stimulation in the Helix visceral ganglion and is modulated by dopamine.

The results of these findings lead to a potentially important conclusion that prominent activity peaks somewhere in this band appear to be common in the frequency-amplitude characteristics of evoked potentials from snails, fish, and mammals. This may indicate a widely ranging function of this frequency component among the different classes of vertebrates and invertebrates. It is obvious that the 40-Hz activity manifests certain global characteristic in information processing over a wide range of neuronal networks from the simplest, the snail, to the most complicated, the human. However, this does not imply that other bands are less important.

The significance of 40-Hz activity in the brains of different mammals has been hypothesized by several authors (Başar et al. 1987; Başar-Eroğlu and Başar 1991; Eckhorn et al. 1988; Freeman 1975; Gray and Singer 1987; Sheer 1984) as an important coding channel in processing sensory and cognitive information in neuronal networks; and what is reported here represents an important agenda for future research.

10.3 Neurochemical Modulation

In the regulatory mechanisms of rhythmic phenomena, a significant role is attributed to the chemical substances of the nervous system. These are produced in the neural elements and, after being released from time to time, they can take part in the regulation of rhythms both centrally and at the periphery. For instance, the excitability of the nerve cells is modulated at the synaptic levels by the release of neurotransmitters, such as acetylcholine (ACh), dopamine (DA), noradrenaline (NA), and serotonin (5-HT). These neuromodulators and others have long been known to be involved in motor and cognitive processes in mammals (see also Chaps. 3 and 12).

Concerning the invertebrate, evidence of the chemical control of spontaneous activity of isolated ganglia in insects was provided by Prosser as early as 1938. In insects, for instance, ACh is involved as a transmitter in the synaptic process of the information processing system (Gerschenfeld et al. 1967). A number of accounts in the literature report the involvement of these neurohormones in endogenous signal processing in the brains of invertebrate such as gastropods and insects (including, Gelperin 1989; Hiripi and Salanki 1973; Kerkut et al. 1975; Twarog and Roeder 1957).

Figure 10.7 describes the global influence of four important transmitters by means of amplitude frequencies. More efficient changes can be seen by using the filtering techniques shown in Fig. 10.8.

The selective increase of the 4–8, 8–15, and 30–70 Hz responses observed in the Helix under the influence of ACh may be a possible reason for what may happen in the neuronal assemblies of the Helix brain by exogenous stimulation. ACh may have a similar function in the mammalian brain. Especially, the increase in the 4–8 and 30–70 Hz components may add an explanation to the cat hippocampal event-related 3.5–8 Hz (theta) and 40 Hz (gamma) activities (Başar-Eroğlu et al. 1992), which could be assumed to be of a cholinergic origin.

These basic transmitters tend to excite or inhibit activity in general. Slow waves (< approx. 50 Hz) are the automatic consequence of activity that is known mostly as firing. The only observation not expected would be differential effects, e.g., enhancing one band and reducing another.

From Schütt's (unpublished) observations it can be tentatively concluded that all the important invariant components evidenced in the brains of the different species studied here, from mammals to snails, may be modulated by each of the neurotransmitters in a manner similar to what occurs in the Helix ganglion. All four transmitters (ACh, DA, NA, and 5-HT) modulate the response by selectively enhancing the 40 Hz activity by 20–50% at the concentrations of 10^{-5}–10^{-3} M. ACh, the transmitter of the cholinergic system, not only increases this invariant component (40 Hz activity), but also intensifies the 4–8 Hz (theta) and 8–15 Hz (alpha) activities by 30%. NA has an additional capability of intensifying the 15–30 Hz activity (beta activity) at 10^{-5} M (Table 10.1).

As far as ongoing induced activities are concerned, all but NA increases the activity, DA mainly the low-frequency range, and Ach as well as 5-HT over the wide-band range up to 80 Hz. Only NA depresses the ongoing field potential. Interestingly, all transmitters actively take part in regulating the 4–8 Hz and 8–15 Hz components: Ach, DA, and 5-HT enhance and NA depresses the activity. Three transmitters (Ach, NA, and 5-HT) seem to be involved in influencing the 15–30 Hz component, whereas NA only negatively affects this range. ACH and 5-HT appear

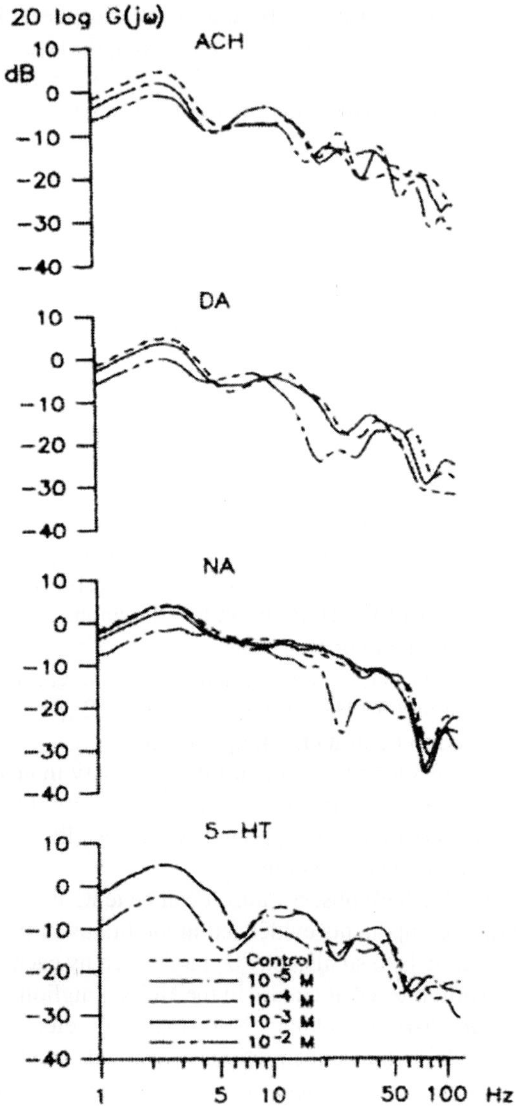

Fig. 10.7 *Helix pomatia*. Amplitude-frequency characteristics of potentials evoked electrically in the isolated visceral ganglion under the influence of neurotransmitters. Grand averages of the experiments shown in Fig. 7.11(5-HT: serotonin; ACH: acetylcholine; DA: dopamine; NA: noradrenalin) (Shütt et al. 1992)

to be very similar in the way they induce responses, particularly in their ability to increase the 15–30 and 30–48 Hz components, but the low frequency component of 1–4 Hz can only be increased by 5-HT.

With regard to stimulus evoked potentials, all four transmitters selectively enhance the 40 Hz activity (NA only at lower concentrations). This is an essential

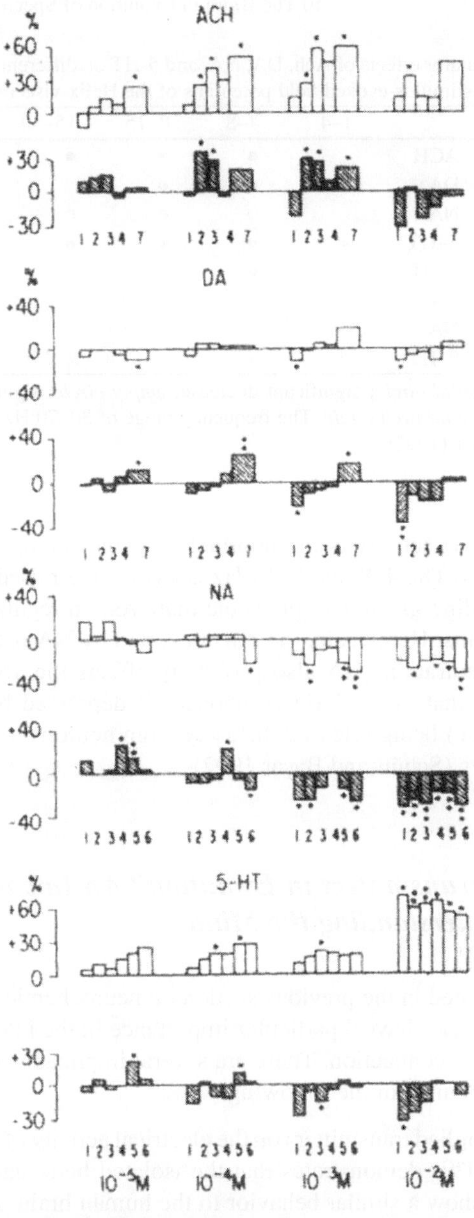

Fig. 10.8 *Helix pomatia*. The isolated visceral ganglion. The effects of Ach, DA, NA, and 5-HT on electrically evoked potentials. Pre-stimulus ongoing activity (RMS-voltage within 500 ms before stimulus) and post-stimulus maximum evoked potentials (within 500 ms after stimulus) were analyzed separately. Each value indicates a mean percent change from control of pre-stimulus activity, empty column, and that of post-stimulus maximum amplitude, the shadowed column, averaged from several preparations. Pass-band filters applied, are defined as: (1) 1–4 Hz; (2) 4–8 Hz; (3) 8–15 Hz; (4) 15–30 Hz; (5) 30–48 Hz; (6) 52–90 Hz; (7) 30–70 Hz. Differences are based on: ACH, 120 epochs (five snails); DA, 200 epochs (five snails); NA, 160 epochs (four snails); 5-HT, 160 epochs (five snails). Significance levels: $*p < 0.05$; $**p < 0.01$ (Shütt et al. 1992)

Table 10.1 The modulator effects of Ach, DA, NA, and 5-HT on different frequency components of ongoing as well as stimulus-evoked field potentials of the Helix visceral ganglion

		1–4	4–8	8–15	15–30	30–48	52–90
On-going	ACH		●	●	●	●	●
	DA	●	●	●			
	NA		○	○	○		
	5–HT	●	●	●	●	●	●
Stimulus-evolved	ACH		●	●		●[a]	
	DA	○				●[a]	
	NA	○	○				○
	5–HT	○		○		●	

Significant increase: *solid circle*; significant decrease: *empty circle*; significant increase only at lower concentrations: *shadowed circle*. The frequency range of 30–70 Hz was selected
[a]From Schütt and Başar (1992)

fining, because 40 Hz activity is an important frequency component in sensory and cognitive responses. The 4–8 and 8–15 Hz activities are raised only by Ach. This means that the cholinergic system plays the main role in regulating these invariant response components. Here again, the effect exerted by NA is largely suppression, but at lower concentrations NA also positively affects the 15–30 and 30–48 Hz components. Note that the 1–4 Hz component is depressed by such transmitters (DA, NA, and 5-HT) being released diffusely from neurons to modulate the functioning of pathways (Schütt and Başar 1992).

10.3.1 Neurotransmitters in Evolution? An Important Invariant for Understanding the Mind

The findings presented in the previous section on neurochemical modulation in the evolution of the species have a particular importance in the integrative approach to the brain-body-mind connection. There are several implications related to the findings; a few are explained in the following:

1. The effects of applied transmitters on the electrical activity of invertebrate ganglia are enormous. This demonstrates that the isolated helix ganglia with maximal 2,000 neurons show a similar behavior to the human brain. Accordingly, it may be stated that transmitters can be considered as *invariant control and signaling elements* of the nervous systems during the evolution of the species. This is similar to the fact that the electrical oscillation also shows "invariant" behavior. There is certainly a mutation of degree of synchrony, intensity of amplitudes, and quantification of release of transmitters. However, they belong to core *common mechanisms* during evolution.
2. The same transmitters also play a key functional control in the vegetative system in humans. This point is important because the transmitters should be considered the most essential structural elements in brain-body integration.

3. In attempting to approach "the mind" and mind changes, it is necessary also to analyze pathologic brains. Are the minds of Alzheimer's patients and patients with bipolar disorders not different from the minds of healthy subjects?

In Chap. 13, it can be seen that change in transmitter release plays a crucial role. For example, the acetylcholine application causes important increase of theta response in Alzheimer's similar to the snail ganglion or the cat brain.

It can be tentatively assumed that to understand the mind and brain-body-mind integration, it is necessary to strongly consider the structural importance of effects of transmitters and their influence to oscillations during the evolution of the species and pathological states.

10.4 Dynamics of Potentials from the Brain of Anamniotes (Low Vertebrates): Goldfish and Ray

Bullock and Başar (1988) reported that

> when the extra-cellular ongoing, spontaneous potentials recorded from the surface or the depth of the cerebrum are compared in terms of gross wave form and power spectral density, there are no systematic differences within the vertebrate classes (mammals, reptiles, birds, amphibians, bony and cartilaginous fishes).

They also commented

> that the differences, if any, in brain electrical activity between fish and mammals are smaller, in spite of the major histological differences, than those in different states and loci within the same individual.

For instance, there are no differences in the parameters of ongoing activity among various bony fishes and elasmobranches. On the other hand, cerebellar cortex, tegmentum, medulla, and cord show quite different forms of ongoing activity (Bullock and Başar 1988).

Turning to potentials elicited by stimuli or cognitive events, sensory or evoked potentials, the use of descriptors of systems theory analysis may help us gain further insight into the questions of differences and similarities in brain function between anamniotes (fish) and mammals. A study was made of sensory evoked potentials in two representative fish, the ray (*Platyrhinoidis triseriata* from the class *Elasmobranchia*) and the goldfish (*Carassius auratus* from the class *Teleostei*) with the purpose of comparing them with those of mammals (cat and human).

A number of authors have recorded evoked potentials in different structures of the ray brain (Bleckmann et al. 1989; Bullock 1979; Bullock and Corwin 1979; Platt et al. 1974; Schweitzer 1986). In our research, the animal was stimulated with adequate pulses having a gradient field between 20 and 45 μV/cm. The experimental set-up to produce this electrical field is described by Schweizer (1986). The preparation and electrode implantation is outlined by Schweizer (1986) and Bullock and Başar (1988). The fish were acoustically stimulated by clicks at 70 dB through a loudspeaker in the air. In some experiments this stimulus was applied in random intervals of 2–3 s and in the others in a constant rate mode.

10.4.1 The Oscillatory Processes in the Ray and Goldfish

Four experiments carried out with different fish are analyzed in the following. A weak electrical field of 20–45 μV/cm was applied as stimulus, and the evoked potentials were recorded simultaneously from medulla and mesencephalon. The frequency characteristics averaged from five experiments are shown in Figs. 10.9 and 10.10.

1. Midbrain (Fig. 10.9a). The power maximum lays in the range of approx. 28 Hz and a second, less dominant maximum in the theta range of 5 Hz.
2. Medulla (Fig. 10.9b). Three distinct maxima of similar amplitude at 9, 16, and 24 Hz are visible, indicating resonance at these frequencies. Additional maxima at 40 and 50 Hz, lower in amplitude, are also visible.

From these results it may be stated that evoked potentials with similar resonance characteristics can also be recorded in the ray brain by stimulation of the electroreception. Stable resonance phenomena found in the midbrain were in the theta and beta ranges, and those in the medulla in the alpha and beta ranges with an additional indication in the gamma (40 Hz) range.

The experiments were performed with five different animals. After acoustical stimulation (70-dB clicks) evoked potentials were recorded from three different brain structures: the medulla, mesencephalon, and telencephalon. Figure 10.10a–c

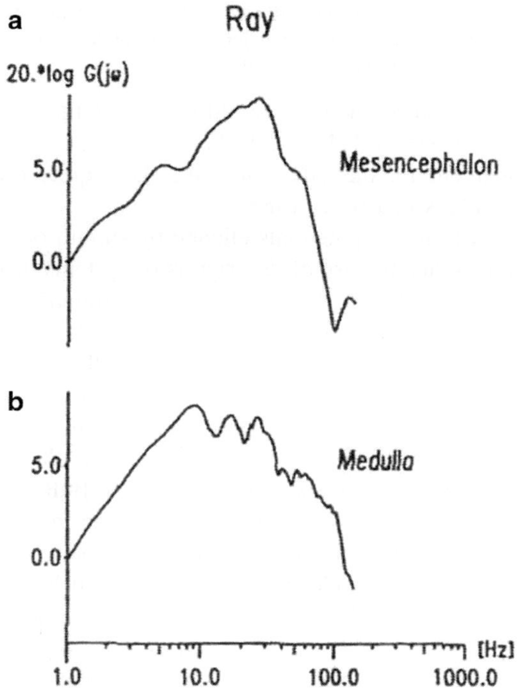

Fig. 10.9 Ray. Amplitude-frequency characteristics of evoked potentials averaged from four experiments (470 epochs): (**a**) midbrain, (**b**) medulla (Schütt et al. 1992)

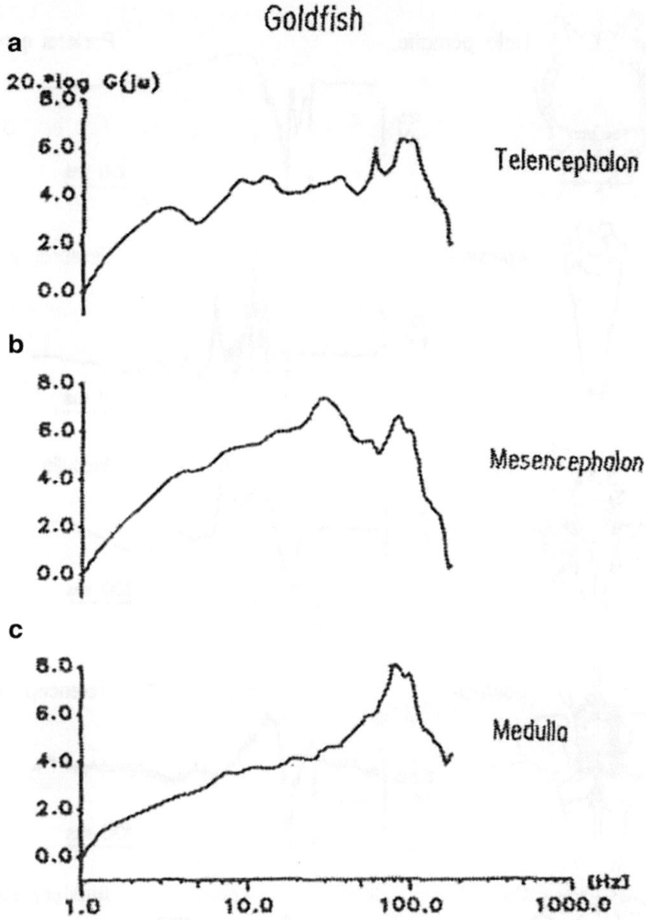

Fig. 10.10 Goldfish. Amplitude-frequency characteristics of evoked potentials averaged from five experiments (500 epochs): (**a**) telencephalon, (**b**) mesencephalon, (**c**) medulla (Shutt et al. 1992)

illustrates the frequency characteristics averaged from the five experiments. In each experiment 100 single evoked potentials were recorded (a typical averaged evoked potential of the ray is shown in Fig. 10.11).

In all three brain centers investigated, a double-peaked Fourier component, dominant and stable, appears in the higher frequency range around 80–100 Hz. However, the patterns in the other frequency ranges vary depending on loci. In this particular loci of the telencephalon (a), prominent peaks exist at 3, 9, and 12, and 35 Hz. (Echteler and Saidel 1980, described AEPs with different dynamics and form in a second region of the telencephalon.) In the present description the loci of the midbrain (b), another stable component is clearly distinguishable as a pronounced peak at 30 Hz. The other components around 4, 7, and 13 Hz are also

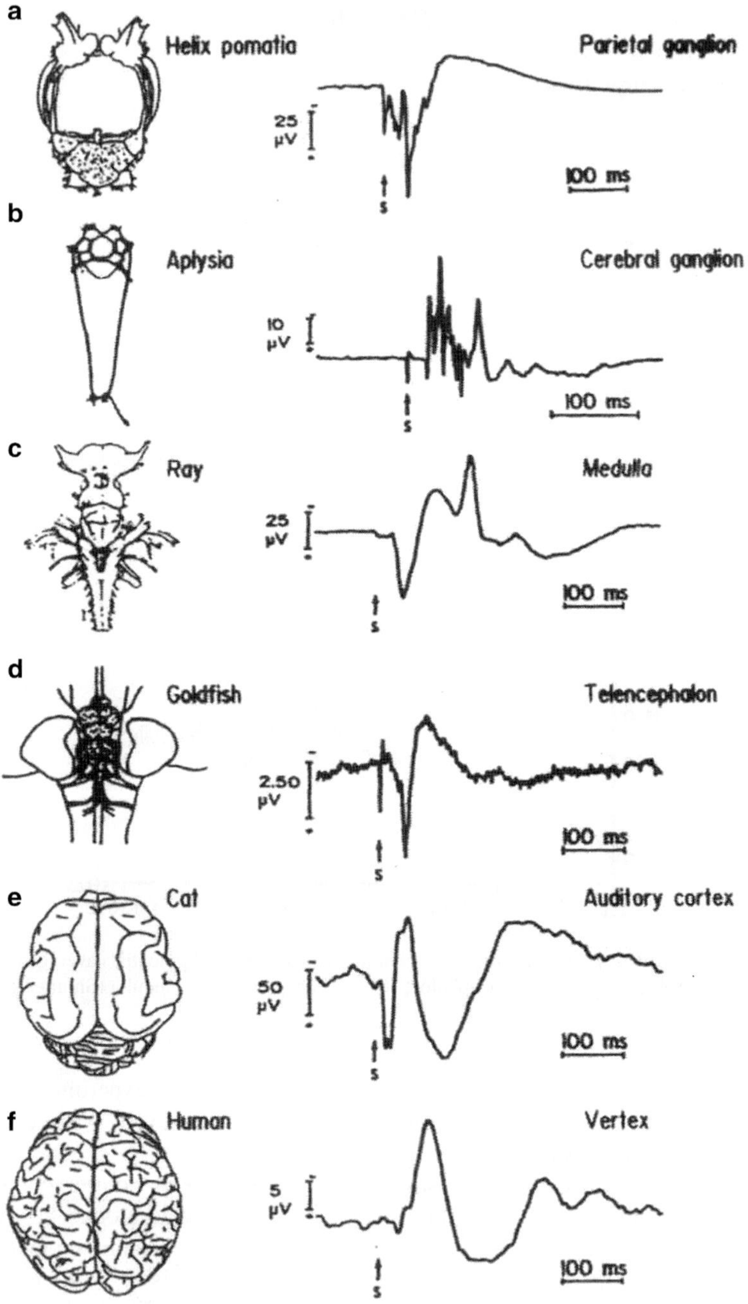

Fig. 10.11 The brains investigated are schematically presented together with the characteristic averaged evoked potentials from different loci, s: stimulus. (**a, b**) Direct shock to afferent nerve, (**c**) physiological stimulus of electroreceptors, (**d–f**) acoustical stimulation (Schütt et al. 1992)

indicated, but only as shoulders in the slope of the curve. In the loci sampled in the medulla, the theta component disappears totally, but alpha as well as beta maxima are recognizable as minor peaks.

From these results, a conclusion may be drawn that characteristic frequencies can occur in the ranges of 4, 10, 30–35, and 80–100 Hz in acoustically evoked potentials in different brain structures of the bony fish. These resonances, however, elicit different amplitude levels in the transfer function depending on the structures. The "characteristic" frequency components observed in the samples of different brain structures of the ray and the goldfish are presented again in Table 10.2 together with those of the invertebrate samples and mammals (cat and human).

What is noticeable is that the predominant frequency components of the mammalian cortex are usually found within the range not higher than that of 40 Hz. However, the reticular formation (cat) shows a dominant frequency component around 85 Hz. It is of interest that a similar high frequency apparent invariant (80–100 Hz) exists in goldfish medulla and, to a lesser degree in ray medulla, although

Table 10.2 Comparative presentation of maxima in frequency characteristics in different species

Frequency band (Hz)	1–8	8–15	15–30	30–48	52–100
species	Maxima in frequency characteristics (Hz)				
Helix					
Pleural r.	5	–	20	–	–
Parietal r.	2	–	16	40	65
Visceral r.	2	–	15	35	60
Parietal l.	2		to	50	–
Pleural l.	2	9	18	40	75
Aplysia					
Group I	2	to	20	40	100
Group II	3		25	–	60
Ray					
Mesencephalon	5	–	28	–	–
Medulla	–	9	16, 24	40	–
Goldfish					
Telencephalon	3	10	–	35	90, 100
Mesencephalon	–	–	–	30	80, 100
Medulla	–	–	–	–	80, 90
Cat					
GEA	5	–	18	40	80
RF	5	11	25	40	85
HI	5	12	–	45	–
Human					
Cz	4	8	25	40	–
O	3, 5	10	25	45	75
P	6	10	20	40	70
F	4	8	20	35, 45	–

Modified from Başar et al. (1999)

Table 10.3 Grand averages of coherence (C) for nearest neighbor pairs of electrodes 9–10 mm apart

Filter band (Hz)	2–5	5–8	8–13	13–20	20–35	35–50
Subdural C10 mm	0.47	0.47	0.43	0.40	0.34	0.38
Depth probe C9 mm	0.48	0.61	0.55	0.49	0.49	0.51

Data pooled from samples in the awake, alert state, for six standard frequency bands. C10 mm from subdural recordings is based on $n = 2{,}876$ values from 14 to 22 pairs and at least three 20-s epochs from each of the seven patients using awake data (modified from Bullock et al. 1995a)

only seen as an indication. The 15–30 Hz invariant also exists in all structures examined.

In the human brain alpha activity indicates high level wavelet entropy. Earlier studies by Bullock also showed that the alpha coherency in invertebrate ganglia is totally absent and reaches higher values from low vertebrates to the human brain (Bullock and Başar 1988; Table 10.3).

10.4.2 *Similarities and Differences During the Evolution of Species*

Is there commonality at all stages of evolution, or does the functional mechanism change during development from simple to complicated neuronal networks? Does the development of complexity involve new possibilities of information transfer? Comparative analyses of evoked potentials in different invertebrates and vertebrates have been carried out based on these questions. The species studied are two invertebrates, sea slug (*A. californica*) and land snail (*H. pomatia*); two lower vertebrates, ray and goldfish; and two higher vertebrates, cat and human. Except for Aplysia and Helix,sensory stimulus modalities applied are "physiological" (acoustical or, in the case of the ray, low voltage electrical field). In Aplysia and Helix "nonphysiological" electrical stimulation was given to a connective or a nerve. Examples of the averaged evoked potentials of the mentioned species are shown schematically in Fig. 10.11. It must be pointed out, however, that shapes of averaged evoked potentials may vary depending on different conditions. The spectral peaks from different structures of the species are listed in Table 10.2.

As the first approximation, these components should be considered invariants for the particular recording loci and afferents stimulated because they are found not only in individual frequency-amplitude characteristics, but also in the average of a number of frequency characteristic records (860 in Aplysia). Concerning highly developed vertebrate brains, not only human cortical field potentials measured by scalp surface electrodes, but also cat cortical signals recorded by intracortical semi-microelectrodes show dominant frequency components up to the 40 Hz range. The response signals from the samples of the subcortical structure, reticular formation, in the cat exhibit a frequency component around 85 Hz. Interestingly, a similar high frequency invariant (80–100 Hz) is particularly pronounced in certain loci of goldfish and ray medulla.

According to Sturbeck's observation in Helix (1988), practically all frequency bands seemed to occur except the 10 Hz range (Table 10.2). This under-representation of the 10 Hz response in Helix that he observed may, however, result from the rather strong stimulus intensity (four times the threshold). In the later study with the Helix visceral ganglion (Schütt and Başar 1992) it could be shown that, when the stimulus is about two times the threshold (15 mA), this structure always responds to electrical stimulus, under the research conditions, with roughly a 10 Hz frequency component accompanied by the other important invariant components such as 3, 20, 40, and 60 Hz. This additional evidence in Helix can perhaps fortify the notion that the 10 Hz component may exist in the brains of a wide range of species from human and cat to snail. In the Aplysia,cerebral ganglion could be found in the potential evoked when a certain nerve was electrically stimulated under study conditions and all frequency components (e.g., 5, 10, 25, 40, 60, and 100 Hz), implying the corresponding valleys between them.

To summarize, principally all frequency components exist throughout the species. This observation may be supported by Bullock and Başar's statement (1988) that invertebrates (insects, crustaceans, gastropods, and cephalopods) also show both fast and slow evoked activity. As our research observations have shown, not only the most frequently discussed 8–13 Hz (alpha) and 40 Hz (gamma) components, but other frequency components as well have equally important weight for the frequency-oriented interpretation of neuronal network characteristics. Not only the 60–100 Hz apparent invariant, but also the low frequency components and the 15–30 Hz (beta activity) seem to have been maintained in the course of evolution.

This chapter has described in some detail the electrophysiological manifestation of the CNS during the evolution of the species. Invertebrate ganglia type of electrical activity has already been recorded that is similar to mammalian or human brain. It is important to note that it is possible to find invariant codes during evolution. Although the major emphasis in this book is not focused on the evolution of the brain and its physiology, the description of brain-body-mind gains considerable importance if ways can be found to differentiate highly developed brains with cognitive and intuitive behavior from lower brains with less intelligence and/or pure cognition. The relative volume of brains or anatomically complicated structure building in various species is also an important landmark in the understanding of the parallels between the mechanism of thought and the brain's structure. This important problem is revisited in Chap. 19 on creative evolution. The diversity of kinds of neurons must have some roles and consequences in the sum of all activity of organized assemblies. Uniquely among all the organ systems of the body, the nervous system has units with widely different or only slightly but importantly different receptive fields and projection fields as well as sub-threshold behavior and spiking properties, from never spiking to two kinds of spikes (Bullock 1980).

According to Bullock (2002), synchrony among some proportion of the active cells must be a major variable. Bullock further mentions that several distinct kinds of synchronization co-exist, largely by the phase locking of slow fluctuations. The same author supposes that the major part of the communication between cells is to non-synaptic field effects besides the classical spikes and synapses. It is further supposed that several or many codes that include non-spike codes operate in parallel in the brain.

Synergy among all processes in nature is an important mechanism, as studied in detail by Herman Haken (1977) (see Chap. 2), who used a laser dynamics approach. This is a concept that embraces many approaches and several measures. It is also called *cooperativity*,as it considers any aspect of the assembly to be an interactive group. *Coherence* is a first-order, linear form of cooperativity for pairs of places at each frequency in the Fourier space. As mentioned later in the book, cooperativity in brain structures and the entropy of functional processes in the brain possibly play a crucial role in the evolution of the species; therefore, the results of Bullock's research team are highly pertinent to describe the evolution of the species.

In the measurements made by Bullock and McClune (1989) and Bullock et al. (1995a,b) on average, the coherence is high between electrodes of a few cell diameters apart. In the rabbit cortex it is commonly low and various, widely second by second and pair by pair a few millimeters apart. Average coherence falls rapidly with the distance in recordings with small electrodes to a level within approximately 10 mm. This is true for all frequencies except for theta during recordings of high theta or for alpha during recordings of high alpha. Somehow the average coherence of local field potentials may attain quite significant results on the scalp, even at separation as great as 14 mm between electrodes.

10.4.3 Importance of Synchrony: The Work of T.H. Bullock

According to Bullock, synchrony is a prominent candidate for a general difference. This has been estimated in a preliminary way in the sampling of coherence as a function of the distance between pairs of electrodes referenced to a common and demonstrably inactive electrode. Averaging over many sample periods of 5–10 s and many pairs of electrodes of each separation, the mean in mammals falls from approx. 0.8–1 (perfect coherence) at 1–2 mm to a level indistinguishable from a stochastic control at a separation of approx. 10–20 mm in rabbits and human beings (macrocroelectrodes on the pia or dura mater). This is true for all frequencies between approx. 2 and 50 Hz, the more sensitive measure is the distance for mean coherence equal to 0.5, but it demands large-scale averaging of many pairs with different separation, and is impractical in most situations. Bullock found that it lies roughly in the range of 5–10 mm in rabbit cortex; approx. 3–4 mm in the lizard *(Gecko)* cortex, only 1–3 mm in a ray optic tectum (*Platyrhinoides*), and <1 mm in the gastropod, Aplysia (Bullock 1989b). It will require many more samples to establish whether this is a valid trend. For the present it can only be suggested that there may be a significant *evolution in the degree and distribution of slow wave synchrony.*

Oscillatory spontaneous and evoked responses of invertebrates and low vertebrates have been given in detail. The reason why this detailed information has been given is twofold. First, it can be seen that when the strength of amplitude, distribution, and duration of the electrical activities are discarded, it is easy to grasp that basically EEG-like oscillations can be found in all the brain studies in this book. This information has crucial value in the search for codes for brain functioning. It seems that during evolution the almost invariant oscillatory properties are among the fingerprints in evolution. More applications of several types of stimuli were administered: strong electrical stimuli to isolated ganglia, weak electrical stimuli to ray, and acoustical stimuli to goldfish. In addition, these visual and auditory cognitive stimuli are used for P300 experiments and also facial stimuli to understand face recognition. It is a marvel to find face locking, coherence, delays, and prolongations in the response manifestations of oscillations in different EEG frequency windows with varied degrees of amplitude. Further, the response oscillations are highly topology dependent (see Chapters 6, 7, 12, and 13). In future studies the questions can be asked as to whether these oscillatory activities are also relevant codes for information transfer between species. Are engrams or vital information transferred from lower to higher species so as to maintain the fundamental properties of living beings? The second reason is that detailed information related to oscillatory dynamics in the mammalian and human brain can be found easily in neuroscience literature. The detailed presentations here can be important for leitmotifs throughout the book and theories that have been developed in the search for the evolution of the nervous systems and functions of the nervous and vegetative systems.

It is necessary to mention a basic finding related to the electrical activity in the brain stem and reticular formation of mammalians. The mammalian cerebral cortex operates mostly with slower frequencies then 40 Hz. On the contrary, structures such as reticular formation, inferior and superior colliculi, and cerebellum also operate by means of high frequencies similar to those of invertebrate ganglia. The immediate point that arises is that brain stem structures are not associated with higher mental processes, as is the case with the cerebral cortex. Invertebrates do not show higher cognitive performance similar to human beings, as can be seen in the last chapters of this book. Human alpha activity is possibly associated with higher mental activity, which does not exist in lower living beings.

10.5 Concluding Remarks on the Evolution of the Brain

Figure 10.12 shows the evolution of the relative surface area of the frontal cortex in higher mammals. In the cat brain, the frontal cortex occupies 3.5% of the relative surface area. In the dog it is 7%, in the chimpanzee the prefrontal cortex area is 17%, and finally in the human brain the prefrontal cortex occupies 29% of the relative surface. In other words, the development of cognitive processes shows an important parallel with the increase in size of the prefrontal cortex. Chapter 11 reveals that babies have relatively smaller prefrontal areas, whereas in the adult brain the morphology and the size of the frontal cortex is immensely changed. Also there are crucial changes in alpha activity during maturation of the human brain. Although it is not possible to find exact correlations between parallels in macro- and micro-evolution, these parallels are interesting.

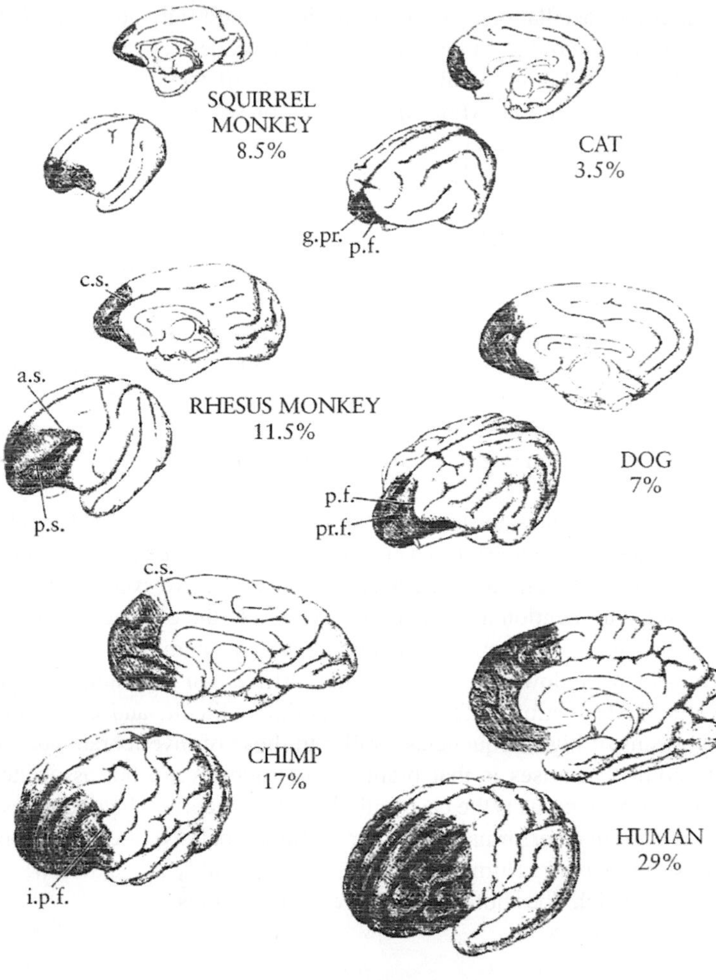

Fig. 10.12 Evolution of the relative surface area of the prefrontal cortex in the higher mammals. Note the dramatic increase from cat and dog to monkey, and then to human, in a territory of the brain that is crucially involved in cognitive functions, the frontal lobe (*shaded in black*). The percentages indicate the surface area of the prefrontal cortex as a proportion of the cortical surface area (modified from 2004)

10.5.1 A Global Scheme of Alpha Response in the Evolution of the Species and Maturation of the Brain

Figure 10.13 illustrates global waveform shapes of alpha activity during the evolution of the species and maturation. An exact scaling of time and amplitude is not given; only approximate measures are described that globally reflect the differentiation in the amplitude of alpha spontaneous activity and responses. Only low amplitude and irregular spontaneous activity is observed in the ganglia of invertebrates (left side of the illustration). Because of the irregular behavior of the oscillatory activity, the entropy is high. In low vertebrates (fish brain), the alpha activity is higher and more regular in comparison with the recordings of invertebrate ganglia. The cat brain, which shows higher and more regular alpha activity, is not included in this illustration.

In the human brain the amplitudes figure of the alpha activity depict drastic changes during maturation (right panel of the figure). Children do not have alpha activity until the age of 3 years. There are also no frontal or occipital alpha responses during this period of life. According to the results of Başar-Eroğlu et al. (1994), in scalp recordings of some babies a type of weak alpha activity is already observed in the occipital locations at the beginning of the third year of life. An important development is observed in the adult, in whom large occipital alpha responses are seen. However, frontal alpha responses of the adults are low and sometimes are not observed at all in the adult brain. A reversal between occipital and frontal recordings is observed in subjects older than 55 years – the occipital alpha responses are decreased, and frontal alpha responses are highly increased (Başar et al. 1997c).

Two major types of changes in the alpha activity are observed in Fig. 10.13:

1. An increase of amplitudes and a decrease of alpha entropy (i.e., decrease in the shape of alpha) during the evolution of the species.
2. Higher alpha activity during maturation of the brain and a decrease of occipital alpha activity in older subjects with a parallel increase of frontal alpha activity.

In both cases, morphological changes in the brains are observed.

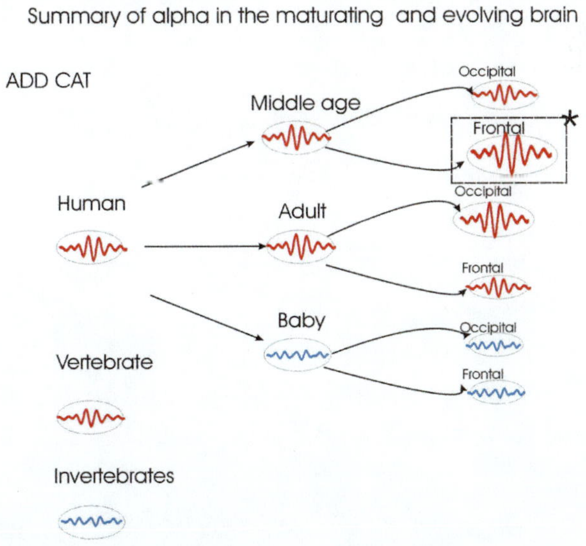

Fig. 10.13 Globally illustrated waveform shapes of alpha activity during the evolution of the species and maturation of the human brain

10.5.2 Remarks on Darwin's Theory

Heritable variation is one of the most important concepts in Darwin's theory, i.e., some properties of living beings are transmitted through descent. During this transmission, anatomical and physiological properties that are inherited undergo mutation. Is it possible to also say this about the electrical activity of ganglia and brains? By comparing the illustrations and tables in this chapter it can be said that delta, theta, alpha, beta, and gamma activities show a heritable variation. Amplitude, shape, and synchrony of the electrical activity show qualitative similarities; however, there are also major quantitative dissimilarities. According to these experimental facts, it is possible to assume that the electrophysiological properties of the evolution of the species are in global accordance with Darwin's theory. Alpha is observed at the beginning (invertebrate level) and at the end (human brain). According to the results presented, there are quantitative changes as the alpha activity and its synchrony increase.

The importance of the evolution of oscillatory dynamics is discussed in Chap. 17, which covers the role of alpha activity in the evolving brain (also see Chap. 10).

Chapter 11
The Maturing Brain

11.1 Introduction

Chapter 6 provided an analysis of the essentials of dynamics in the sensory and cognitive processes of the brain. This includes basic studies in the both the human and animal brain, demonstrating that oscillatory activities in the delta, theta, alpha, beta, and gamma bands govern control and communication processes in the brain. Accordingly, these oscillations constitute the "most general transfer functions" of brain function. The Preface proposed one of the guiding themes of the book, which is the introduction of a new Cartesian system with multiple causalities. To achieve this, it is necessary to understand the changes that occur during the maturation processes of the brain.

Chapter 10 on the evolution of the species added an important dimension for changes in brain dynamics by answering the following questions: What are essential features of the anatomy of invertebrate ganglia and electrical activity of the ganglia? What are essential features of electrical activity in the cat and human brain? Despite important anatomical changes and changes in the amplitude and degree of synchrony of oscillatory responses, an ensemble of common features provided a transfer of hereditary information from Aplysia to the human brain: There are common frequency codes (or quasi-invariants as denoted in Part VI, Chapter 22) that organize functions in diverse brains.

There are also important changes in the structure and electrical activity in the human brain from the fetus to the brain of elder subjects. How do these changes affect the transfer functions, i.e., communication processes in the maturing brain? A 2-year-old child does not have a full command of his or her native language and cannot solve mathematical problems. Also, a 2-year-old child does not show alpha activity. Furthermore, the frontal lobes are not completely developed; synaptic organization is in the process of steady development. This chapter includes measurements of three groups; 3-year-old children, adults, and middle-age people. To provide the basis of the ontogenesis of the maturing brain, the next section gives an anatomical description.

E. Başar, *Brain-Body-Mind in the Nebulous Cartesian System: A Holistic Approach by Oscillations*, DOI 10.1007/978-1-4419-6136-5_11,

11.2 Changes in Structure and Synaptic Organization of the Human Brain

In ontogeny, similar to the evolution of the species (Fig. 11.1), the neocortex develops much more in size and volume than any other neuronal structure of the brain. Also in the human, from embryo to adult, the relative growth of the white matter underlying the cortex with its connective structure by far surpasses that of white matter elsewhere in the central nervous system (CNS). Neuron generation seems to have been completed in the human neocortex by the end of the second trimester of gestation. Afterward, however, cortical neurons continue to grow in size even at the time

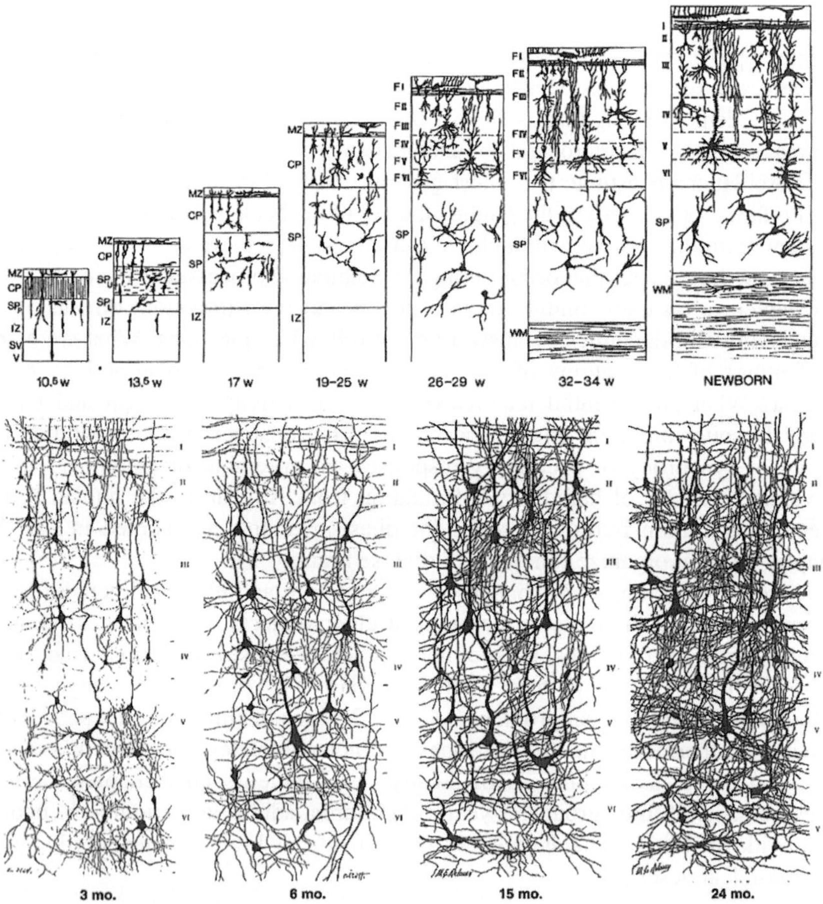

Fig. 11.1 Development of neurons in the human cortex. *Top:* Prenatal period from 10.5 weeks to birth. From Mrzljak et al. (1990). *Bottom:* Postnatal period at 3, 6, 15, and 24 months (modified from Conel 1963)

the infant is born; some neurons are still developing and migrating to their final location. Growing neurons develop their axons, which branch out to develop collaterals (see Fig. 11.1). Synaptogenesis begins in the third trimester of the pregnancy and continues until the age of 2 years. According to Changeux (2004), the main lines of the cellular architecture of the cerebral cortex are established before birth. This architecture is largely determined by the developmental genes and those responsible for the formation of nerve connections and the propagation of signals. According to Changeux (2004), about half of all adult humans synapses are formed after birth, and their number continues to change, rising and then falling until death. In humans the postnatal development of the brain lasts considerably longer than in other mammals. Cranial capacity increases 4.3 times after birth in humans. Moreover, cranial capacity reaches 70% of the adult human volume 3 years after birth. It is important to note that alpha activity in the child starts 3 years after birth, almost parallel with the development of speech. In human beings the rapid phase of synaptic growth is shorter in a sensory area such as the visual cortex, where it continues to grow until 2 or 3 years after birth, then in association areas such as the prefrontal cortex, where it grows up to 10 years after birth. Changeux (2004) further states that this observation has great importance from a functional point of view. The prefrontal cortex, which is very rich in neurons in layers two and three, plays a central role in cognitive functions.

In their book, *The Brain and the Inner World*, Solms and Turnbull (2002) indicate that the frontal cortex is crucial for the retrieval of memory. In this context, it is notable that the frontal cortex, no less than the hippocampus, is poorly developed in the first 2 years of life. There is a substantial growth spurt in the frontal cortex at around 2 years of age, and then a second spurt at about 5 years. Further, the frontal cortical volume continues to expand throughout adolescence. In the first few years the level of the organization of the frontal system may be considered so poor that the organized retrieval process is not available to the young child. The growth of volume and synaptic organization is seemingly accompanied with important changes of alpha activity in primary visual areas and crucial changes in frontal areas (association).

These relevant findings on anatomical organization of the neocortex are revisited later in this chapter.

11.2.1 The Aim of This Chapter

The age of the human subject is one of the most important factors influencing the amplitude and frequency of the electroencephalogram (EEG) (Dustman et al. 1993; Katada et al. 1981; Niedermeyer 1993; Obrist 1976). Within the brain response susceptibility concept, the evoked oscillations also can be expected to undergo important changes with increasing age in children and adults. To demonstrate these changes, a comparative analysis is made for oscillatory evoked potentials (EPs) of 3-year-old children who do not have developed occipital 10 Hz rhythm, adult subjects with expressed alpha (18–32 years old), and elder subjects with reduced occipital alpha (55 years old). The major results provide a model to show that the spontaneous and evoked alpha is interrelated.

The concept of *brain response susceptibility* is effectively supported by the evaluation of the comparative data (for the hypothesis the reader is referred to Chaps. 7, 8, 23, and 24).

11.2.2 Spontaneous and Evoked Alpha Activity at Occipital Sites in Three Age Groups

Figure 11.2a illustrates the instantaneous power spectra at occipital recordings of three subjects from each of the three age groups: 3-year-old child, young adult, and middle-aged adult. As known from earlier studies (Eeg-Olofsson 1971; Niedermeyer 1993; Petersén and Eeg-Olofsson 1971), the EEG in 3-year-old children does not have spontaneous activity in the 10 Hz frequency range. In fact, no 10 Hz activity was recorded in the EEG of a 3-year-old child (Fig. 11.2a). In contrast to the results from children, young adults had distinct and ample 10–12 Hz activity in the occipital recording (Fig. 11.2a, middle panel).

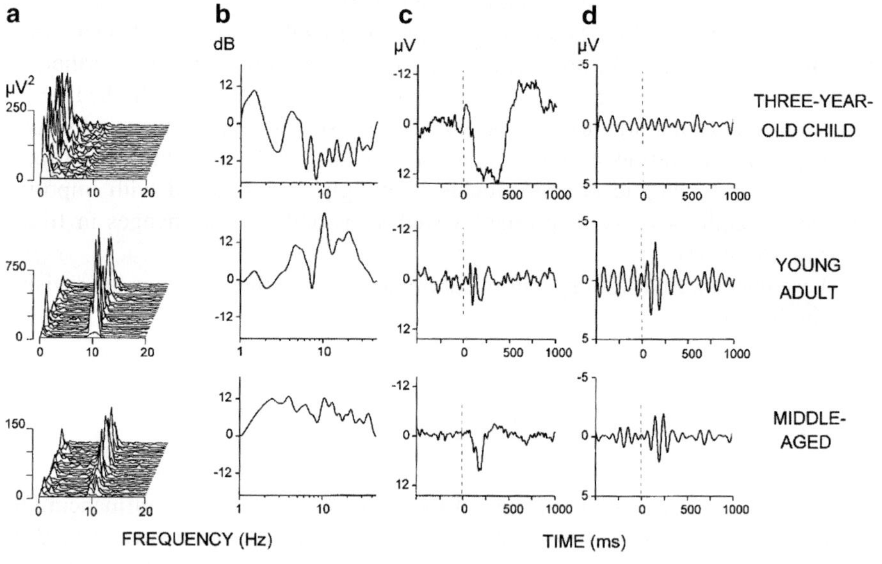

Fig. 11.2 Averaged VEPs of three representative subjects: 3-year-old child, young adult, and middle-aged adult. (**a**) Instantaneous power spectra of consecutive 2 s long EEG epochs, (**b**) amplitude-frequency characteristics of visual evoked potentials, (**c**) averaged VEPs, (**d**) filtered in the range of 8–15 Hz. All recordings are from the left occipital site O1. Stimulus onset occurs at 0 ms (from Başar et al. 1997c)

Results from experiments on middle-aged subjects showed a reduction in 10 Hz activity in the occipital areas. Figure 11.2a (bottom) illustrates the power spectra of a 55-year-old subject. The 10 Hz activity of this subject is apparent in comparison with the 3-year-old child, but drastically reduced in comparison with the young adult. Figure 11.2b shows the amplitude-frequency characteristics (AFCs), and Fig. 11.2c displays the averaged EPs on visual stimulation for the three subjects. Figure 11.2d illustrates the filtered average visual EP responses with band limits of 8–15 Hz. No alpha responses (defined as the oscillatory brain activity in the 8–15 Hz frequency range within approximately 200–300 ms following external stimulation) are recorded in the visual EPs of children. In young adults, the group mean amplitude of the peak-to-peak alpha responses in the averaged visual EPs was 4.5 μV. In the visual EPs of middle-aged adults, the peak-to-peak alpha response was 3.1 μV.

11.2.3 A Comparative Analysis of Frontal Vs. Occipital 10 Hz Activity in Young and Middle-Aged Adults

Figure 11.3a presents stack plots of power spectra of the spontaneous EEG of young and middle-aged adults to enable comparison between their frontal and occipital alpha activity. Although young adults had low 10 Hz activity at the frontal site (or sometimes no activity) as a rule, their posterior alpha had relatively high amplitude. (In this example the young adult manifests alpha power at approximately 100 μV^2 for the frontal, and more than three times higher for the occipital recording.) In the middle-aged adult, a most important phenomenon was observed; in the occipital recording (O1), alpha power was maximally 20 μV^2, whereas at the frontal (F3) recording site alpha power was approximately 80 μV^2.

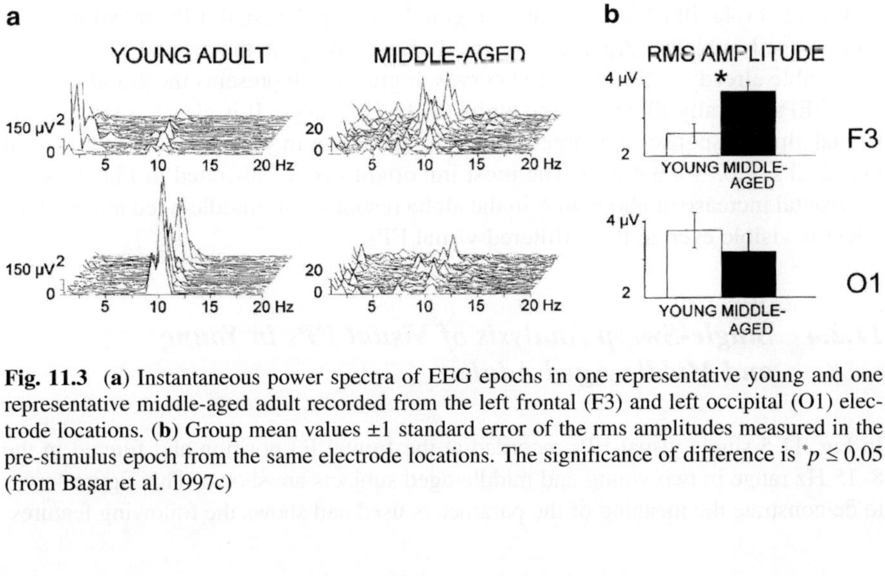

Fig. 11.3 (**a**) Instantaneous power spectra of EEG epochs in one representative young and one representative middle-aged adult recorded from the left frontal (F3) and left occipital (O1) electrode locations. (**b**) Group mean values ±1 standard error of the rms amplitudes measured in the pre-stimulus epoch from the same electrode locations. The significance of difference is $^*p \leq 0.05$ (from Başar et al. 1997c)

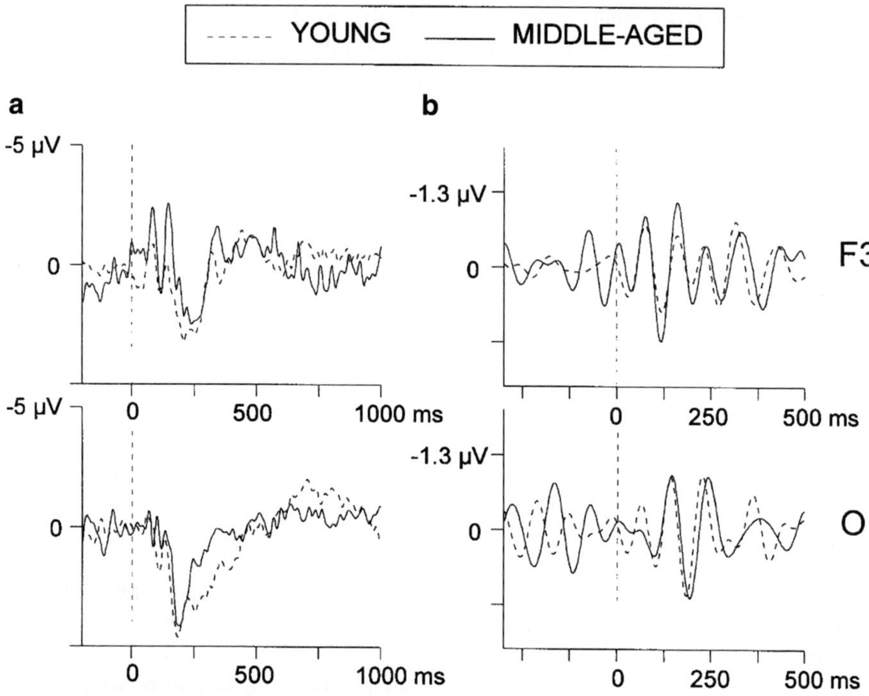

Fig. 11.4 Grand average VEPs in young and middle-aged adults from left frontal (F3) and left occipital (O1) locations. (**a**) Unfiltered, and (**b**) filtered in the range of 8–15 Hz. Stimulus onset occurs at 0 ms (from Başar et al. 1997c)

A significant increase of frontal alpha amplitude was obtained for middle-aged subjects as also revealed from the mean group rms values in Fig. 11.3b. In contrast, lower rms values were produced by the older subject for the occipital recordings.

Figure 11.4a illustrates unfiltered grand averaged visual EPs in young and middle-aged subjects. An increase in 10 Hz activity at the frontal (F3) site is observable already in the unfiltered curves. Figure 11.4b presents the grand average visual EPs digitally filtered in the alpha (8–15 Hz) range. It is clearly seen that the frontal alpha responses are larger in middle-aged than in young adults (in averaged visual EPs, 4.63 vs. 3.4 μV). The most important result illustrated in Fig. 11.4b is the frontal increase of about 40% in the alpha responses of middle-aged adults. This effect is visible even in the unfiltered visual EPs.

11.2.4 Single-Sweep Analysis of Visual EPs in Young and Middle-Aged Adults

In Fig. 11.5 single visual EPs recorded at the frontal F3 position and filtered in the 8–15 Hz range in two young and middle-aged subjects are shown. This figure serves to demonstrate the meaning of the parameters used and shows the following features:

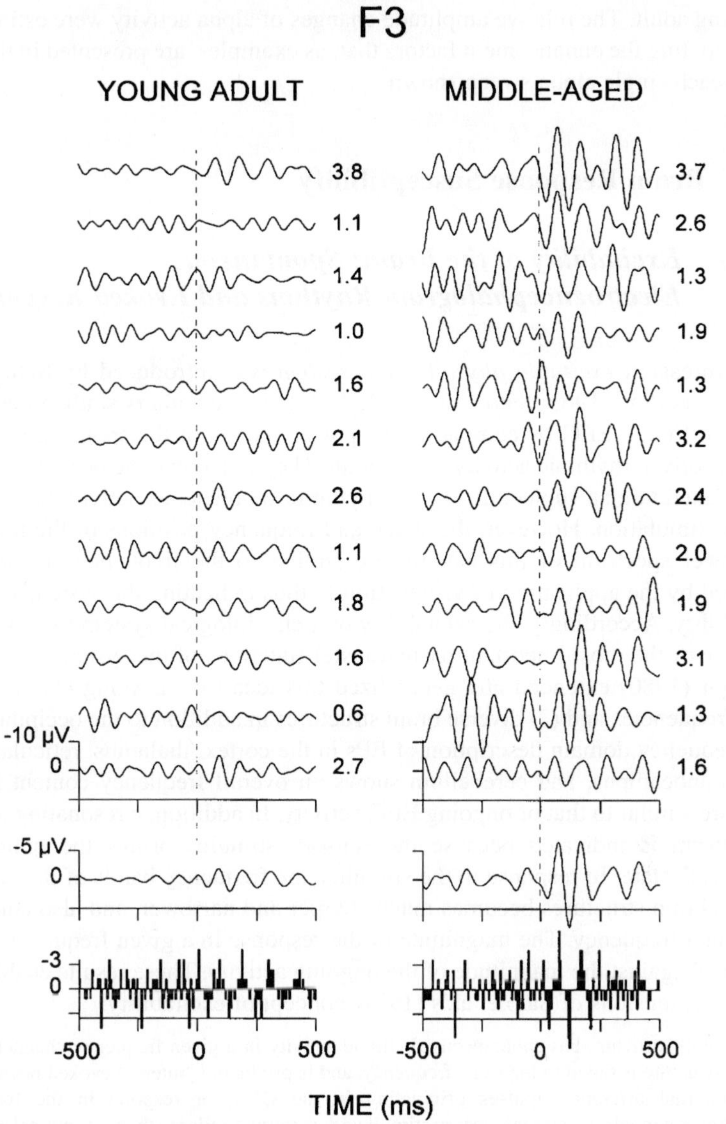

Fig. 11.5 Single sweeps from left frontal lead (F3) in one representative young and one middle-aged adult, filtered in the 8–15 Hz frequency range. The numbers on the right side of each single sweep are the corresponding enhancement factors (from Başar et al. 1997)

- Maximal alpha amplitudes within 0–300 ms are higher in the middle-aged than in young adult person.
- Consecutive alpha responses of the elder adult are better synchronized or much more congruent than those of the young adult.
- Although the pre-stimulus alpha activity is higher in the elder subject, the relative amplitude enhancement of the response is still somewhat higher than in the

young adult. The relative amplitude changes of alpha activity were estimated by computing the enhancement factors that, as examples, are presented in the figure for each single alpha sweep shown.

11.3 Brain Response Susceptibility

11.3.1 Excitability of the Brain: Spontaneous Electroencephalogram Rhythms and Evoked Responses

The expression *excitable physiological system* was introduced by Sato and his co-workers (Sato 1963; Sato et al. 1971, 1977). These authors studied the relation of visual EPs to EEG by comparing the power spectra of the spontaneous activity with the driven rhythmic activity of the brain. They found that the occipital recording of a relaxed human subject displays a rhythmic spontaneous activity before photic flicker stimulation. However, the shape and frequency positions of the maxima in the power spectrum of this activity are similar to the frequency characteristics obtained by the application of visual stimuli, thus indicating the system's selective excitability. Accordingly, the excitability of a physiological system may be considered one of the most important basic transfer functions of the system.

Başar (1980) extended and generalized this idea by analyzing (1) data for all EEG frequencies and (2) diverse brain structures in addition to the occipital cortex. The frequency domain description of EPs in the cortex, thalamus, reticular formation, hippocampus, and cerebellum shows an overall frequency content for each structure similar to that of ongoing EEG activity. In addition, a resonating universal mechanism is indicated because the sensory stimulus brings the brain into a "coherent" state. In response to the stimulus, the frequency bands of the activity in various brain structures becomes much sharper and narrower, and also coherent in phase and frequency. The magnitude of the response in a given frequency range is enhanced against the magnitude of the ongoing activity. These results enabled us to develop extensions of Sato et al.'s (1971) concept of excitability:

> If a brain structure has spontaneous rhythmic activity in a given frequency channel, then this structure is tuned to the same frequency, and is producing "internal evoked potentials" to internal afferent impulses originating in the CNS, or respond in the form of evoked potentials to external sensory stimuli with patterns similar to those of internal evoked potentials. Thus, knowledge of the spontaneous or ongoing activity preceding stimulation must be considered as an important prerequisite for evaluation of evoked potentials (Başar 1980).

The response susceptibility of a brain structure depends mostly on its own intrinsic rhythmic activity (Başar 1980, 1983a, b, 1992; Narici et al. 1990). A brain system could react to external or internal stimuli producing those rhythms or frequency components that are already present in its intrinsic (natural) or spontaneous activity, i.e., if the spontaneous brain rhythms are missing in a given frequency range, they will be absent in the evoked rhythms and vice versa.

11.3.2 Electroencephalogram in Children Might Provide a Useful Natural Model for Testing the Hypothesis of Brain Response Susceptibility

The intrinsic (spontaneous) oscillatory brain activity depends on several factors:

1. Age
2. Topology
3. Vigilance and/or cognitive states
4. Pathology

The effects of some of these factors are used in the following to demonstrate brain response susceptibility.

By measuring EPs in 3-year-old children, the developing brain was used as a "natural model" to investigate the following question: How do brain systems respond to external stimulation if their intrinsic (spontaneous) rhythms are different (or not yet developed) in comparison with EEG rhythms in adults? Spontaneous EEG activity in children and adults was hypothesized to represent different types of intrinsic background rhythmic activity with respect to both alpha and slow frequency ranges (Fig. 11.6). According to the concept of brain systems response susceptibility, it was expected that:

1. Upon sensory stimulation, the evoked rhythms (post-stimulus enhancement, time, and frequency-locking during the post-stimulus period) as well as the corresponding evoked frequency EP components, will differ between children and adults if their spontaneous EEG rhythms are different; ***and
2. The evoked rhythms in children and adults should reflect their spontaneous EEG frequency patterns, respectively.

11.3.3 Aging and Topology-Related Changes in Alpha Activity and Brain Response Susceptibility

Young adults manifested somewhat higher alpha amplitudes during ongoing EEG over occipital brain areas. Accordingly, only at occipital locations were they able to produce larger and better synchronized alpha responses than middle-aged adults. In contrast, over the frontal brain regions, middle-aged adults had significantly higher and much more strongly phase-locked alpha responses than young adults, which were accompanied by higher pre-stimulus alpha power in the group of elder subjects.

Several reports indicate that a shift occurs in the alpha activity toward the more anterior sites of the brain with increasing age in adults (Fisch 1991). The results from analysis of spontaneous and pre-stimulus alpha activity in young and middle-aged adults are in line with these previous reports. The most important step in this

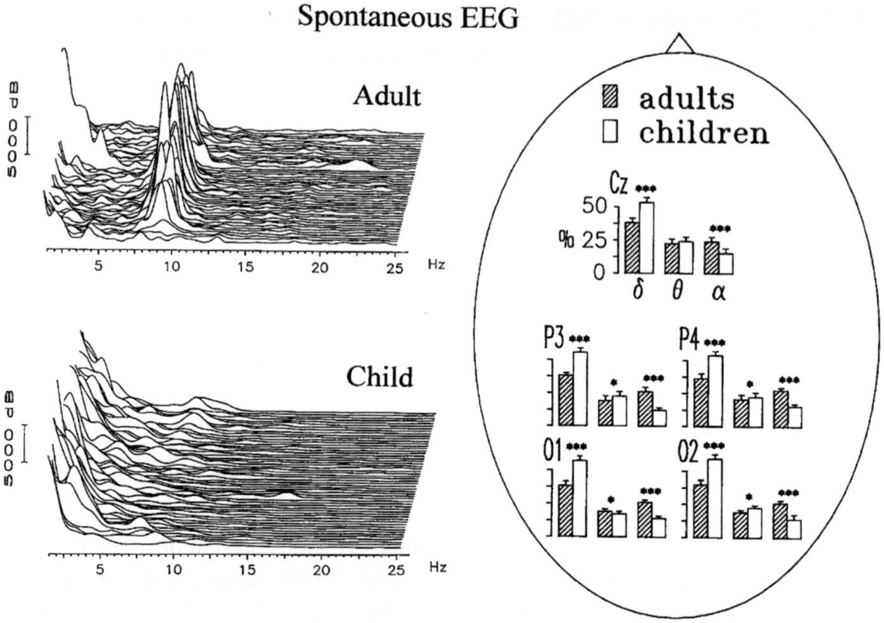

Fig. 11.6 Spontaneous EEG in 3-year-old children and adults. (*Left panel*) Instantaneous power spectra of EEG epochs recorded in one representative adult and one representative 3-year-old child (left occipital lead O1). Calculations are performed according to the method of compressed spectral arrays. Each curve presents a record of 1 s duration. Note the differences in the spectral characteristics in adults and children, as well as the different dynamics during the recording. (*Right panel*) Mean group values (+1 SE) of the root mean square amplitudes in the spontaneous EEG for delta (0.5–3.5 Hz), theta (4–7 Hz), and alpha (8–15 Hz) ranges, measured in percent from the sum of the amplitudes in all frequency ranges. The significance of difference is designated as: $^{*}p < 0.05$, $^{**}p < 0.01$, and $^{***}p < 0.001$ (with modifications from Başar-Eroğlu et al. 1994)

analysis is the finding that the age-related changes in the ongoing EEG are parallel with corresponding alterations in the visual and auditory evoked alpha responses. These results, as well as the results from children (Başar-Eroğlu et al. 1994; Kolev et al. 1994; Yordanova and Kolev 1996), show that EPs are controlled by spontaneous 10 Hz activity.

11.4 Conclusion: Importance of Maturation in Brain-Mind

This chapter gives a global description of the ontogenesis and synaptic organization of the human cortex as well as oscillatory spontaneous activity and responses in three important age groups. The results and interpretation of results have two important consequences.

1. It was stated at the beginning of the chapter that one of the major aims of this book is the development of a new Cartesian system to approach brain-body-mind integration. To do this, several chapters describe various properties of the brain related to brain-mind integration.

Does a child have a different type of mind than an adult? Does an elder subject have a different type of experience and "different type of mind" than a young adult? The findings emphasize the need for a "hyperspace" and presentations at different levels for the comparison of cognitive processes in babies, young adults, and elder subjects. Further, it can be seen that pathologies considerably change anatomical, biochemical, and electrical brain properties (see Chap. 13). This has the following implication: To be able to perform a real comparison of cognitive processes in children, adults, and older subjects, it is necessary to take various standardizations into account. This is discussed in Chaps. 14–16 as well.

2. It is also intriguing to observe that alpha activity grows in amplitude during the evolution of the species and maturation of the brain (Fig.17.2). Further, as humans age, the high amplitude alpha activity moves from the posterior to the frontal brain. Section 11.2 shows that synaptic organization of frontal areas does occur later in the maturating process. In comparison with the brains of animals, Bullock et al. (1995b) has shown that coherence is higher in the human cortex in comparison with lower animals. Because alpha activity is not found in the frontal areas in the brain of a young child, it is not possible to find high coherence. Altogether it seems that existence of mature alpha activity is of a major importance for the development of associative behavior of brain structures. Therefore, Chap. 10 and the present chapter introduce relevant thoughts on the connectivity and differentiation of mind in the brain.

11.4.1 Important Comment to the Parallelism of Alpha Activity During Maturation of the Human Brain and the Evolution of the Species

At this point it is suggested that the reader take into consideration an important parallel between the maturing of the brain (microevolution) with the increase of alpha activity during evolution of species (macroevolution). An increase in alpha activity is observed during the transition of the child brain to the adult brain. Moreover, alpha activity shifts from posterior areas to the frontal cortex in the brains of elder subjects. Also, the largest alpha activity is observed in the frontal areas in these subjects. In terms of evolution theory, this can be tentatively considered a *mutation* of alpha activity in the human brain. Frontal alpha activity becomes richer with the development of cognitive processes and the semantic experience of the maturing brain. The same situation is observed during the evolution of the species: 10 Hz oscillations in invertebrates have small amplitudes, and no coherent activity in the isolated ganglia is observed. In the human brain alpha activity shows high amplitudes and regular shapes. Are there parallels in the evolution of the species and the maturing brain? This is a very important question related to a type of maturation of alpha activity during evolution of the species. Is the increase of alpha activity a sign of augmentation of cognitive processes? Chap. 17 returns to this very important question.

Chapter 12
Oscillatory Dynamics of the Emotional Brain: Links of Emotion to Episodic Memory

12.1 Emotions: Introduction

Previous chapters have shown how one can approach functional correlates of brain oscillations; that the oscillations in alpha, beta, theta, etc., are dependent on the modality of sensory–cognitive events; that, depending both on topology and input modality, the brain response oscillations react with various ensembles of multiple oscillations. Chaps. 6 and 7 explained that a simple light or auditory signal evoke multiple oscillations selectively distributed in various structures of the brain. Further, even a simple light signal requires phyletic memory. Measurements of short-term memory by means of the oddball paradigm showed that the brain can differentiate a diverse range of tasks and electrophysiological recordings provide an efficient tool to detect differences during various states of the working brain.

It is clear that the brain can perform a number of more difficult differentiations than those related to simple light or auditory signals. We can differentiate a sea from a mountain landscape; classical music from jazz; a table from a tree. Such percepts are easy to distinguish in comparison with the differentiation of facial expressions. No doubt, the differentiation of a tree from a table is less difficult than that of a smiling face from an angry face. Certainly differentiation of facial expressions is one of the higher cognitive abilities of the brain. Recognition of known and unknown faces is the basic step. However, the task of face recognition also includes the recognition of facial expressions. The next step is gender differences. In the analysis of facial expressions, we also confront another task, or as Mark Solms and Oliver Turnbull (2002) discuss, we include in our analysis the sixth sense, *emotions*.

What, then, could be the highest level of nervous activity of the brain? We assume that *intuition*, which is not existent in other species, can be considered the highest level of brain functioning. Chap.17 discusses intuition, related to the evolution of species, in which it is assumed that the most developed electrical signal of the brain, namely, alpha activity, can be assigned to intuitive behavior.

E. Başar, *Brain-Body-Mind in the Nebulous Cartesian System: A Holistic Approach by Oscillations*, DOI 10.1007/978-1-4419-6136-5_12,

12.1.1 What Is Emotion? Definition and General Philosophy

There are several definitions of emotion within the neuroscience literature, yet none of them is generally established in the general view. According to James (1890), emotions are cognitive responses to information from the periphery. According to Solms and Turnbull (2002), emotion is akin to a sensory modality that provides information about the current state of body self, as opposed to the state of the object world. Emotion is the aspect of consciousness that is left if all externally derived contents are removed. These authors stated: "If you were deprived of all sensory images (drawn from present and past perception), you will still be conscious." The emotion that one perceives is also subjective: What one perceives when feeling an emotion is one's own subjective response to an event itself. "Emotion is a perception of the state of the subject, not the object world" (Solms and Turnbull 2002).

Le Doux (1999) proposed that emotions or feelings are *consciousness* of *unconscious processes*. However, he states that despite millennia of preoccupation with every facet of human emotion, we are still far from explaining this part of our mental experience in a rigorous physiological sense.

According to Damasio (1994, p. 139):

> Emotion is the combination of a mental evaluative process, simple or complex, with dispositional responses to that process, mostly toward the body proper, resulting in an emotional body state, but also toward the body proper, resulting in an emotional body state, but also toward brain itself, resulting in additional mental changes.

Damasio proposed two types of emotions: (1) primary, and (2) secondary emotions. He stated that primary emotions depend on limbic system circuitry, the amygdala and anterior cingulate being the prime layers. Primary emotions are the most basic emotions; they are the emotions that William James described. James proposed that when we see a bear we do not experience fear until after we run away from the bear; he stated that the conscious experience of emotion occurs after the cortex receives signals from the bodily changes.

12.2 Oscillatory Dynamics of Emotion

As stated by Solms and Turnbull (2002), emotion can be considered as a sixth sense, because it controls various brain functions and, accordingly, it is not possible to describe the electrophysiology of memory by means of limited experimental description in a short chapter. However, we introduce here three examples related to emotional behavior to demonstrate that the oscillatory dynamics of the brain clearly reflect its ability to differentiate complex emotional behavior. First, the dynamics of recognition of a picture of one's grandmother is presented; second, pictures of facial expressions; and third, recognition of the face of a "known and loved" person.

The results are described in detail in the coming sections. Because such studies remain rare, it is useful to demonstrate that significant changes in brain oscillation are recorded for these highly complex cognitive–emotional processes.

12.2.1 Grandmother Experiments, Experimental Strategy, and Procedure for Recognition of Known and Unknown Faces

We used a strategy consisting of the application of three different types of stimulations:

1. With simple light stimulation as the control signal, its luminance was approximately at the same level as for pictures two and three described in the following (app. 30 cd/m2).
2. The picture of an "unknown face" – an anonymous elderly lady
3. The picture of a "known face" – the subject's own grandmother

A total of 26 subjects in the age range 15–36 years (17 females and 9 males) participated in the study. They had normal or corrected to normal binocular visual acuity and were right-handed. The pictures were presented in black and white (17 × 17 cm) and displayed on a screen at a distance of 120 cm from the subjects. The stimulus duration was set to 1,000 ms, with intervals varying between 3.5 and 7.5 s. The subjects were instructed to minimize blinking and eye movements, and they sat in a soundproof and dimly illuminated echo-free room. Fig. 12.1 illustrates the photographs of the own grand mother and of an anonymous face approximately in the same age.

Here we indicate only the most important and relevant consequence: The whole brain and all oscillations are activated during recognition or remembering one's own grandmother and an anonymous face. The ensemble of responses behaves like a three-dimensional construct, consisting of *temporal, spatial, and frequency spaces*. The responses to an anonymous face or the face of a subject's own grandmother are not represented solely by one location and a unique frequency, or at the same position along the temporal axis, as differentiated delay and prolongation of multiple oscillations that are selectively distributed on the whole cortex show.

In the whole cortex, the pictures of the grandmother and the anonymous face elicit various degrees of enhanced responses in several frequency windows and recording sites. Accordingly, the distributed delta, theta, alpha, beta, and gamma systems are selectively activated. Globally seen, selectively distributed enhancements of amplitudes in response to the grandmother picture originate from a greater number of populations of activated neurons. This greater involvement of neural populations in all frequencies probably requires ample and more widely distributed memory activation.

Figure 12.2 shows the histograms of peak amplitudes in three frequency ranges and four locations. Table 12.1 shows statistical evaluation of responses in detail. We only emphasize the significant difference in responses to a grandmother picture and to an anonymous face in the theta frequency range (Figs. 12.1–12.3).

Table 12.1 The first statistical evaluation related to the experiments

Main and interaction effects	Results of repeated measures		
	Main effects		Interaction effect
Maximum amplitude	Condition	Location	
Delta-band	***	***	***
Theta-band	**	–	–
Alpha-band	–	**	***

*$p < 0.05$, **$p < 0.01$, ***$p < 0.001$

Results detailed analysis: oneway-anova (+Posthoc-tests Bonferroni and Scheffe)

Alpha	
Location (Scheffe test): condition *light* significant difference between maximum amplitudes of	f3 and o1 ($p < 0.005$)
	f3 and o2 ($p < 0.0001$)
	f4 and o1 ($p < 0.0001$)
	f4 and o2 ($p < 0.0001$)
Condition (Bonferroni-correction): position *f3* significant difference between maximum amplitude of	Face and light ($p < 0.05$)
f4 significant difference between maximum amplitude of	Face and light ($p < 0.01$)
	Grandmother and light ($p < 0.01$)
Delta	
Location (Scheffe): condition *grandma* significant difference	f3 and o2 ($p < 0.005$)
Face	f4 and o2 ($p < 0.005$)
	f3 and o2 ($p < 0.005$)
	f4 and o2 ($p < 0.005$)
Condition (Bonferroni): position *o1* significant difference between max. Amplitude of	Grandmother and light ($p < 0.01$)
o2	Grandmother and face ($p < 0.01$)
	Grandmother and light ($p < 0.0001$)
	Face and light ($p < 0.001$)
Theta	
No location effect	
Condition (Bonferroni): at position *f3* significant difference between max. Amplitude of	Grandmother and face ($p < 0.05$)

12.2.2 *The Efficiency of the Grandmother Paradigm for Differentiation of Memory Components or States*

This paradigm shows that it can serve as an appropriate strategy to check mental disorders resulting from memory reduction or complete memory loss. The refinement of the evaluation of results and possible interpretations cannot be achieved by simple application of ERPs. Combination and comparison of results using the whole strategy, different recording areas in all frequency windows, consideration of enhancement, delays, and prolongations are most useful parameters, as stated in Chap. 6. Analysis of all these factors is required for a deeper understanding of memory activation or reactivated memory components.

Fig. 12.1 Photographs of grandmother and unknown person presented to one of the subjects

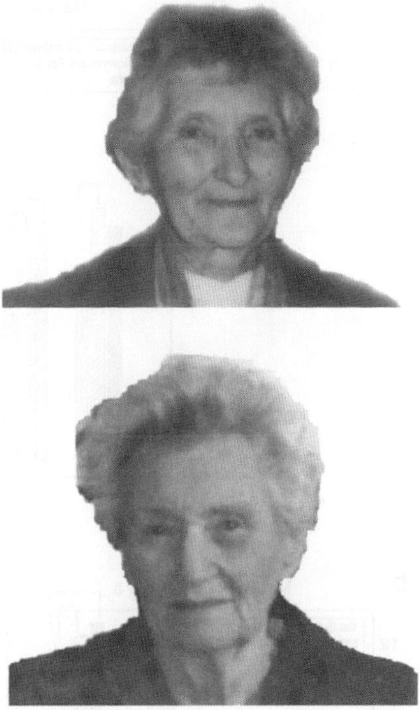

The components of memories are working in a given template employing all parts of the brain. Frontal theta is the major operating rhythm; frontal theta response plays a major role in differentiating semantic and episodic memories. Alpha activity is the major operating occipital rhythm; it plays a major role in differentiating the memory related to grandmother-episodic memory and the semantic memory activated by anonymous faces.

12.2.3 Does Activation of Larger Neural Populations Indicate the Reactivation of Episodic Memory Components?

When viewed globally, selectively distributed enhancements of amplitudes in response to a grandmother picture originate from a greater number of activated neural populations. The involvement of neural populations at all frequencies probably requires a more ample and high-level selectivity in distribution of memory activation. Fuster (1997) presented the idea that the cortical dynamics of evoking episodic memory are identical to that of evoking a familiar stimulus, such as the cue in a delay task. Although the cue is represented in the *posterior cortex*, the prefrontal cortex is essential for its retention toward prospective action. Accordingly, the prefrontal cortex is very important for the sequencing of behavior, thinking, and speech. All three require working memory.

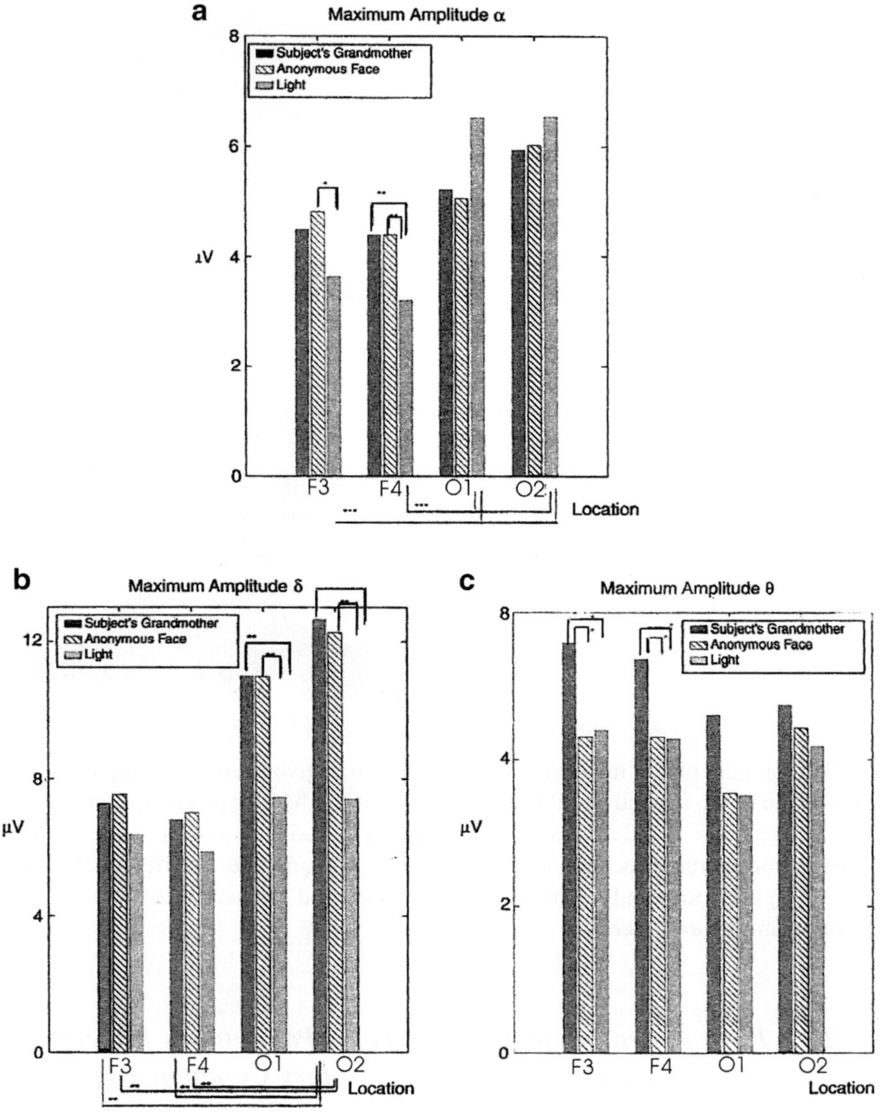

Fig. 12.2 First statistical evaluation. (**a**) Alpha responses. (**b**) Delta responses. (**c**) Theta responses

The grandmother picture (a well-known face) activates delta and alpha oscillations in posterior areas and, parallel to this, theta and alpha oscillations in frontal areas. In a way, this phenomenon shows joint activation of posterior and frontal areas by processing episodes. However, our data show that other cortical areas are also activated.

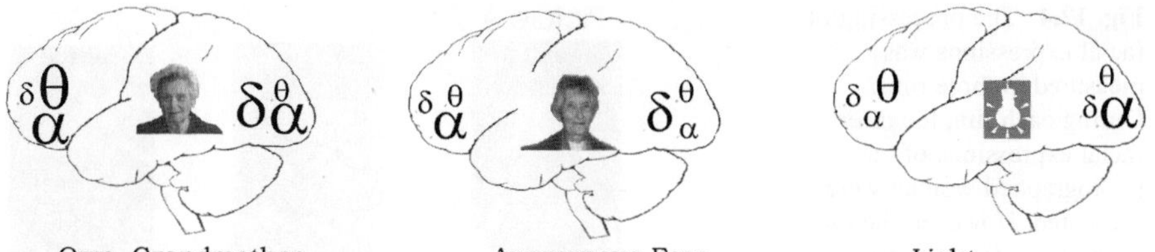

Fig. 12.3 This figure presents a qualitative overview of the different global responses to three types of stimuli, showing the large occipital alpha and large frontal theta activity

12.3 Oscillatory Dynamics of Facial Expressions

The present section describes selectively distributed oscillations on presentation of different facial expressions. Face processing involves different interwoven aspects, such as emotions, memory, emotional memory, and affections. It follows that the neuronal correlates associated with processing of faces are potentially modulated by differences in emotional expression, valence as well as familiarity, and affectionate involvement with the person behind the face (e.g., Başar et al. 2007). Accordingly, the differentiation of facial expressions is one of the most complex processes in integrative neurophysiology.

The number of reports is increasing in this field. However, most of the methodological approaches are based on conventional evoked potential analyses and applications of fMRI data analysis.

In the analysis of electrophysiology of facial percepts, the researcher is confronted with face processing, which comprises: (1) perceptual and memory processes required for the recognition of the complex stimulation as representing a face; (2) the identification of the particular face in view, and (3) the analysis of its facial expression (McCarthy 2000). In addition to the processes pointed out, the valence and arousal dimensions that the subjects express are the prominent features of facial expression analyses.

Güntekin and Başar (2007a,b,c) studied electrophysiology of facial expression by using an ensemble of photographs presented by Ekman and Friesen (1976) with the expressions "angry," "happy," and neutral" (Fig. 12.4).

The most important differences in participants' reactions on stimulation of angry and happy face expressions are manifested in the alpha and beta frequency bands. It was observed that, at the posterior sites (T_5, P_3, P_4, O_1, and O_2), the amplitude of the alpha responses on angry face stimulation was higher than on stimulation with the happy face. At F_3, C_z, and C_3, significantly higher amplitude beta responses were found during angry face stimulation compared with the happy face stimulation (Güntekin and Başar 2007c). There was no difference between responses to face expressions within the delta and theta frequency windows (Güntekin and Başar 2007c).

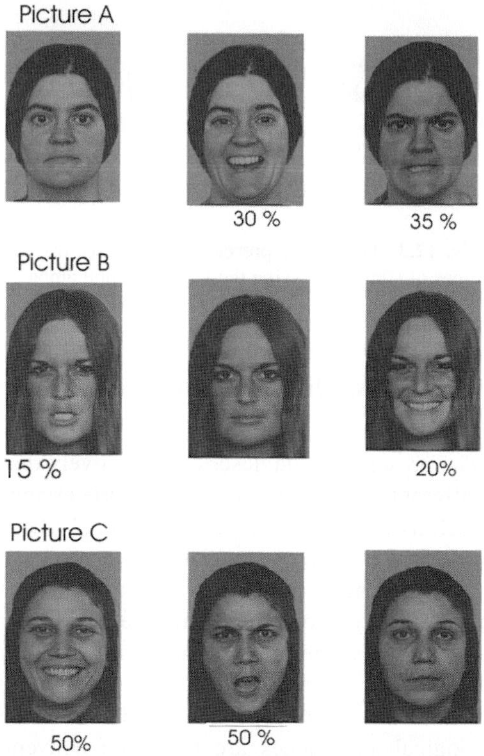

Fig. 12.4 The processing of facial expressions was measured in three runs. During each run, the three facial expressions of one photographed woman were presented. However, the order of the three facial expressions (angry, happy, and neutral) differed between runs, to reduce interactions between facial expression and learning, habituation, or fatigue (modified from Güntekin and Başar 2007b)

These authors showed that the delta amplitude was significantly higher at the temporal-parietal-occipital areas in comparison with frontal-central areas. Başar et al. (2006, 2007) also presented results showing higher occipital delta responses to both known and unknown faces. Balconi and Lucchiari (2006) observed maximal delta synchronization in the posterior regions for all types of facial expression stimuli. According to all these results it can be stated that temporal-parietal-occipital delta responses play an important role in both the recognition of faces and facial expressions.

The data presented in this chapter show that frequency-space distributions of theta responses are completely different for facial expressions in comparison of differentiation of known and unknown faces. During recognition of facial expressions, occipital theta response reaches a high degree of synchronization (Güntekin and Başar 2007c) and higher amplitudes in comparison with the paradigm of recognition of known and unknown faces (Başar et al. 2006). Further, Güntekin and Başar (2007c) showed that temporal–occipital theta responses are higher in comparison with central theta responses. Therefore, it can be concluded that the brain imposes different strategies during various face-perception tasks. There are few studies in the literature that reported higher occipital theta responses in general. Gladwin and De Jong (2005) found increased occipital theta response for alternation versus repetition trials for a visual task. Başar et al. (2007) indicated

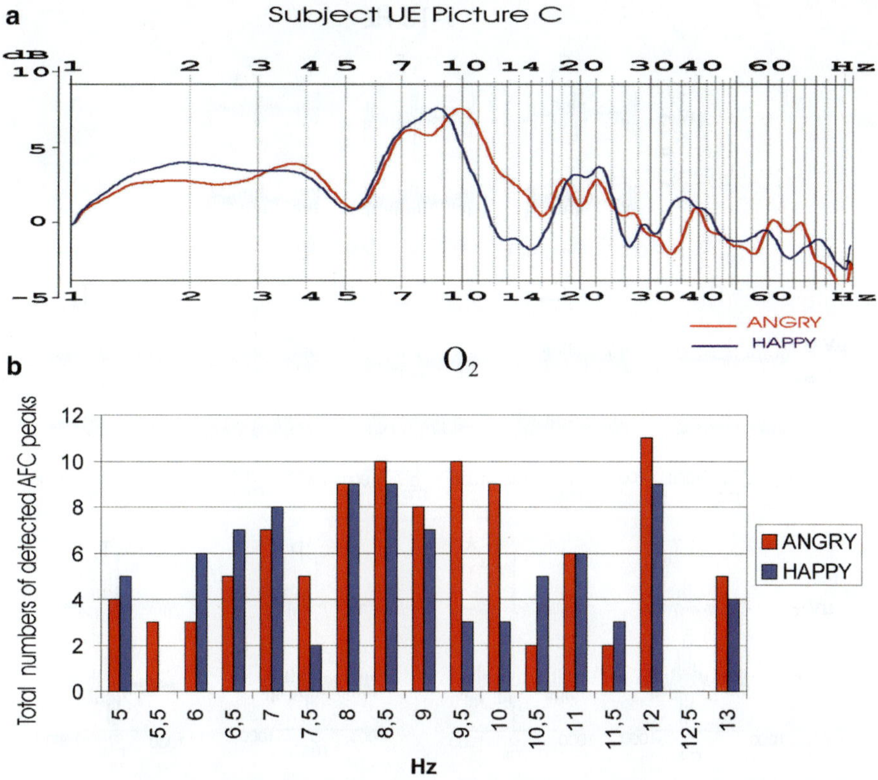

Fig. 12.5 (**a**) The graph shows the AFCs of a representative subject at the location O_2 on angry face and happy face presentations. The existence of various maxima can be observed (modified from Güntekin and Başar 2007). (**b**) Histogram illustrating the number of subjects, showing a peak at frequencies between 5 and 13 Hz in the occipital region for 60 measurements. Three sets of measurements were taken for each of the 20 subjects (modified from Güntekin and Başar 2007)

that occipital delta and frontal theta responses play an important role in face recognition processes.

Alpha responses showed a relevant ambiguity on stimulus using different facial expression, as Fig. 12.5a shows (see also Güntekin and Başar 2007c). The alpha response was determined, according to the AFC analysis, as the peak-to-peak maximum oscillatory component of an evoked potential in the 9–13 Hz frequency range.

A single example was chosen, specifically the location O_2, because the differentiation of angry and happy faces was manifested by several topologically selected frequencies. In this short report, we mention only limited results; a detailed description of the alpha response will be treated in future publications.

Differences can be seen between both types of stimulations, especially in the alpha and beta-gamma frequency range. For this subject, the angry face evoked 10 Hz responses, whereas the happy face evoked 8.5 Hz responses. In the gamma frequency range, a peak at 40 Hz in response to the angry face and 35 Hz in

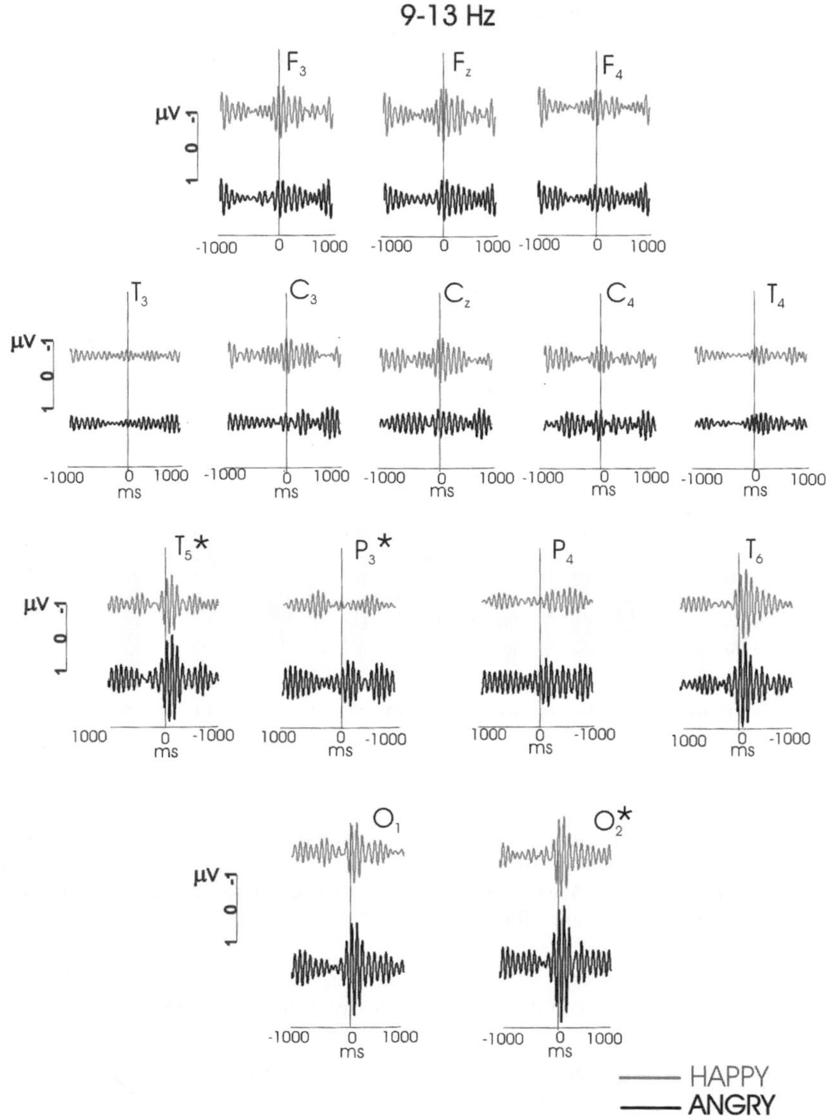

Fig. 12.6 The peak-to-peak amplitudes of the alpha responses were significantly larger following angry face stimulations compared with happy face stimulations at T_5 ($p = 0.005$), P_3 ($p = 0.023$), and O_2 ($p = 0.021$) locations (modified from Güntekin and Başar 2007)

response to the happy face was observed. Results from experiments using 20 subjects – with three sessions each – constituted the basis of the histogram in Fig. 12.5b (modified from Güntekin and Başar 2007c).

One of the most marked differences between angry versus happy face stimulation can be seen in the alpha range (9–13 Hz). The number of peak amplitudes on

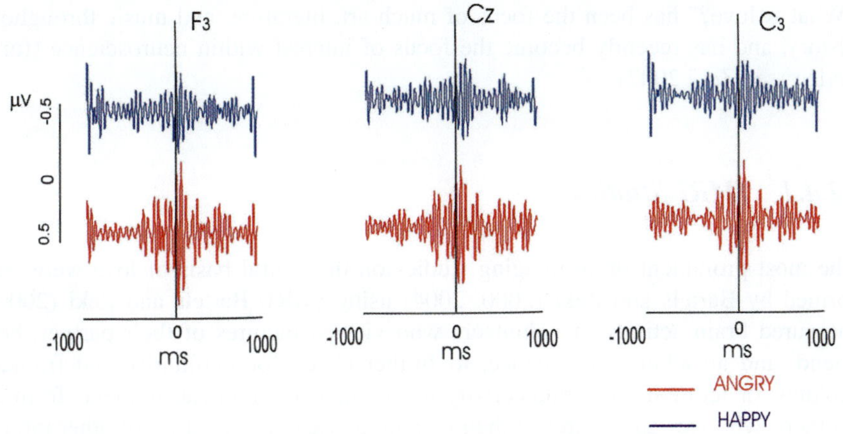

Fig. 12.7 At F_3 ($p = 0.008$), C_z ($p = 0.044$), and C_3 ($p = 0.014$), significantly higher amplitude beta responses were found during angry face stimulation compared with happy face stimulation (modified from Güntekin and Başar 2007)

angry face stimulation was significantly higher than on happy face stimulation in the alpha (9–13 Hz) frequency range ($p = 0.04$, $\chi^2 = 4.20$) (modified from Güntekin and Başar 2007c).

The topology of the grand averaged alpha responses is presented in Fig. 12.6. It is seen that the amplitude of the alpha responses on "angry face" stimulation is higher than on stimulation with the "happy face" at the posterior sites (T5, P_3, P_4, O_1, and O_2). It should be emphasized that significant differences in alpha amplitudes between angry and happy expressions were only found if the stimuli with highest individual valence ratings were selected for the ERP data analysis (Güntekin and Başar 2007c).

According to the results of amplitude frequency characteristics, the beta response was defined as the peak to peak maximum oscillatory component of an evoked potential in the 15–24 Hz frequency range. In the 15–24 Hz frequency range, we found significant differences in amplitude at the frontal and central electrode sites. At F_3, C_z, and C_3, significantly higher amplitude beta responses were found during angry face stimulation compared with happy face stimulation (see Fig. 12.7).

12.4 Oscillatory Dynamics of a "Loved Person" Versus Unknown Faces and Simple Light Stimuli

For the survival of a population of any species, one of the most important factors is the increase of the population (Darwin 1859). It is also assumed, here, that "love" is a prerequisite for conjunction of females and males. Furthermore, the question,

"What is love?" has been the focus of much art, literature, and music throughout history, and has recently become the focus of interest within neuroscience (for a review, see Zeki 2007).

12.4.1 *fMRI Studies*

The most prominent brain imaging studies on the neural basis of love were performed by Bartels and Zeki (2000, 2004) using fMRI. Bartels and Zeki (2000), measured brain activity in volunteers who viewed pictures of their partner, best friend, and an adult acquaintance, to further check for familiarity and friendly feelings. In addition, this brain activity was compared with measurements from 22 mothers, who viewed pictures of their own infants and also pictures of other infants with whom they had been acquainted for the same period (Bartels and Zeki 2004). This determined the brain activation related to maternal and romantic love, while checking for the effects of familiarity and only friendly feelings. An intriguing finding was the overlap between the brain areas activated during feelings of romantic love for a partner and those involved in maternal love for one's own children. Although, the observed correlates of romantic love share brain areas with other closely linked emotional states (Bartels and Zeki 2004; Zeki 2007) the findings emphasize that romantic love, with all its complexity, still appears to be very distinctive in its neural activity pattern. These studies found that viewing a picture of the beloved engages a complex network involving areas in the cortex (including the medial insula, anterior cingulate, and hippocampus) and the subcortex (including parts of the striatum and probably also the nucleus accumbens), which together constitute core regions of the reward system, and have been related to dopamine release and a "feel-good" state (Zeki 2007). Furthermore, these authors stress that feelings of love are not localized, but involve areas with rich connectivity to other sites in the brain, e.g., including connections with the frontal, parietal, and middle temporal cortex. Thus, Bartels and Zeki (2004) conclude that strong emotional connections between persons engage a strong sense of reward and, at the same time, inhibit negative emotions and affect the brain circuits involved in making social judgments about that person.

12.4.2 *Electrophysiology Studies*

Although fMRI studies merit important consideration, a word of caution should be given because of the very low temporal resolution (Grill-Spector et al. 1999). Accordingly, Başar et al. (2008) used strategies of oscillatory brain dynamics. For this purpose, pictures were presented of a loved person to 26 female subjects, and their elicited responses were compared with their responses to pictures showing faces of a known and appreciated person and pictures showing an unknown person. Başar et al. (2008) argued that a blocked presentation of each stimulus type (i.e., a picture of the boyfriend, a close friend, an unknown male, and light stimulation) would be used in one experiment. This should allow the investigation of brain activity related to different, relatively persistent states of affect. In the second experiment, a randomized presentation of the stimuli was used, so as to consider the possible effects of strong adaptation or habituation caused by the presentation of the stimuli in blocks. Accordingly, in the second experiment, stimuli were presented randomly. In the second experiment, the persistence of emotional states was also taken into account by using relatively long ISIs. For details on the experimental procedure, the reader is referred to Başar et al. (2008).

12.4.2.1 Event-Related Delta Responses

Figure 12.8 shows the event-related potentials (ERPs) among the various conditions so as to give a conventional description of the results.

The main significant differences between the various conditions and picture sets were found in the delta frequency band. Figure 12.8 shows the grand averaged delta responses for the block-design (left column) and the random presentation set (right column).

When given in the form of the block-design, the different face pictures and light stimuli elicited responses similar to classic visual evoked responses, with maximum amplitude over occipital locations peaking between 250 and 280 ms after onset. In contrast, the random presentations led to a broader delta response comparable to the ERP component P3, with the largest amplitudes over parietal–occipital locations.

In the block-design, significant smaller amplitudes posterior for light stimuli were found compared with all other conditions. In addition, the unknown male face elicited significantly reduced amplitudes over the anterior regions (see Fig. 12.9, left column).

For the random set, as in the block-design the smallest amplitudes were elicited by light stimuli over the posterior areas.

12.4.2.2 Event-Related Theta Responses

In both random and block-design sets, the largest peak-to-peak amplitudes in the theta band were observed over the anterior locations between 100 and 280 ms after stimulus onset. Condition differences were only found in the block-design set (Fig. 12.7, upper panel), indicating larger amplitudes in the anterior locations and smaller amplitudes in the posterior locations during light stimulation compared with all other conditions.

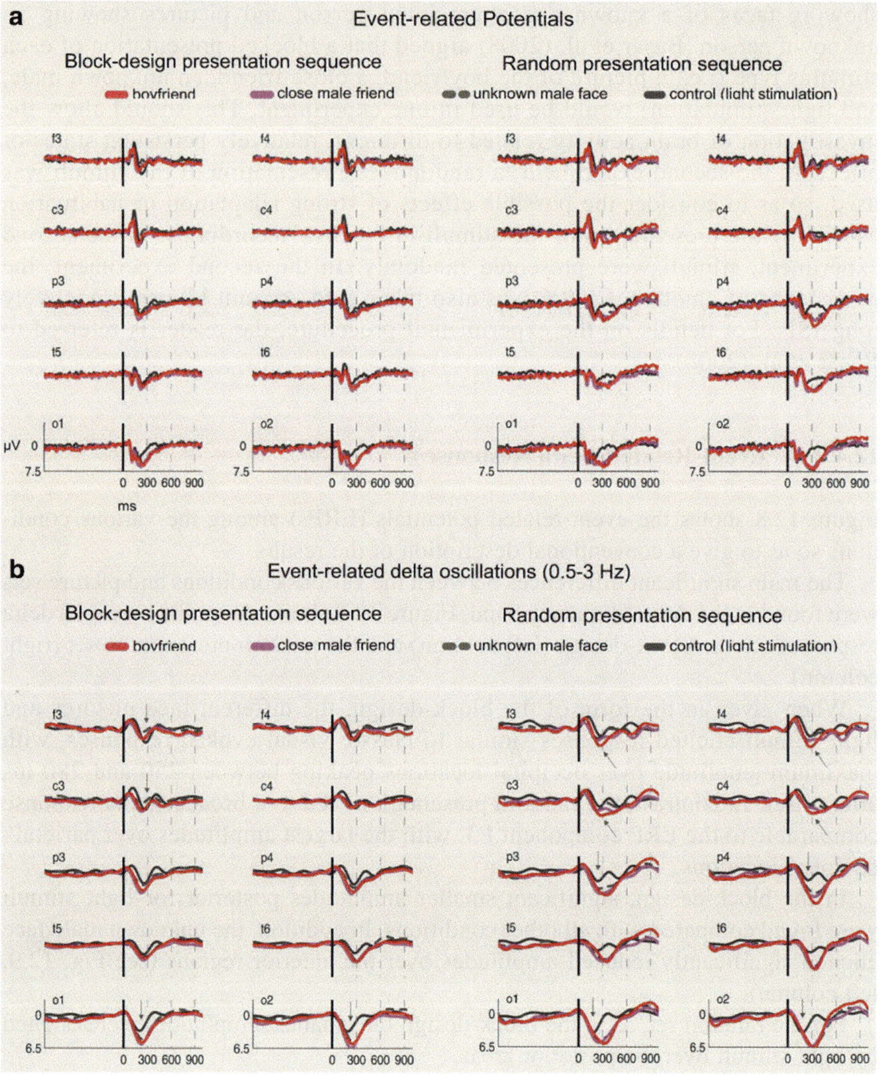

Fig. 12.8 (**a**) Event-related potentials. (**b**) Grand averaged event-related delta oscillations (*lower part*) in the four conditions superimposed within the block-design (*left column*) and the random (*right column*) presentation experiments

12.4.3 Interim Discussion

Significant differences between angry and happy face stimulations in the alpha (9–13 Hz) and beta (15–24 Hz) frequency ranges were found in the "face expression" experiment. The amplitude of alpha responses was significantly higher on angry face stimulation than during happy face stimulation at posterior locations

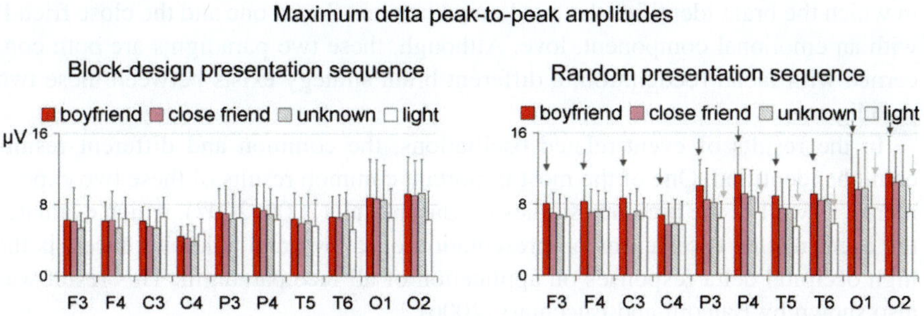

Fig. 12.9 Mean values and standard deviations of the maximum post-stimulus peak-to-peak delta amplitude values in the four conditions within the block-design (*left column*) and the random (*right column*) presentation experiments

(specifically T_5, P_3, and O_2). Furthermore, the amplitude of beta responses at F_3, C_z, and C_3 were also increased on angry face stimulation when compared with responses to happy face stimulation. The theta response was significantly larger for right temporal and occipital electrodes than for central electrodes for all facial expressions. The delta response was significantly larger for temporal (T_6), parietal, and occipital electrodes than for frontal and central electrodes for all facial expressions.

In the experiment using pictures of a loved person, the main differences between the various face types (loved one, close friend, unknown male, and light stimuli) as well as the random or blocked picture presentation sequences were found in the delta frequency band. In addition, differences between the control condition (light stimuli) and the face stimuli were found in the delta and theta frequency bands. The specific findings were as follows:

1. Small delta amplitudes were found over posterior regions for light stimuli compared with all other conditions, irrespective of the selected way of presentation, random or block-design picture sequence.
2. In the theta band, larger anterior but smaller posterior amplitudes during light stimulation were also observed when compared with all other conditions; however, only when employing blocked picture sequences (Fig. 12.7).
3. The unknown male face elicited significantly reduced delta amplitudes over the anterior regions, when each face type was shown in blocks of 30 trials each.
4. Random presentation of the different faces led to significantly larger overall amplitudes while stimulation with the boyfriend's face compared with all other faces and the control light stimuli. This effect was largest over anterior locations.

These two experiments analyzed different types of emotional states. In the analysis of event-related oscillations on presentation of "face expression" the paradigm evaluates the identification of facial expression. However, there is also an emotional component, which could be identified as feeling positive or negative. In the analysis of event-related oscillations on presentation of a loved person, the paradigm is the identification of a known person, but there is an also emotional component, in which the brain identifies the two known persons (loved one and the close friend) with an emotional component, love. Although, these two paradigms are both concerned with facial recognition, a different brain strategy exists between these two paradigms.

In the results of event-related oscillations, the common and different results could be identified. One of the most important common results of these two experiments, as well as the previous studies by Başar et al. (2006, 2007), which evaluates the event-related oscillations on presentation of known and unknown faces, is the high occipital delta responses on application of all face paradigm. This result was also shown by Balconi and Lucchiari (2006).

The difference in event-related oscillations between these two studies (facial expression vs. loved person) is the following: Significant differences between angry and happy face stimulations in the alpha frequency range (9–13 Hz) at the temporal-parietal-occipital regions and beta range (15–24 Hz) at the frontal–central regions were found in the "face expression" experiment. The theta response was significantly larger for right temporal and occipital electrodes than for central electrodes for all facial expressions; this was not the case for the loved person experiments or the previous experiments by Başar, which analyzed the known and unknown face.

12.4.4 Dynamic Localization

When we consider the main difference of the loved person experiment, a major difference is higher response of delta amplitudes over the anterior regions on the presentation of the loved person synthesis in the foregoing chapter, it was consistently indicated that it is impossible to correlate a unique oscillatory response to a given function. Furthermore, it is almost impossible to simply localize a function. Our results supported Luria's view (1966), which stated that mental functions are similar to vegetative functions. This means that mental functions are a product of complex systems; a component part, which may be distributed through the structures of the brain. The task of neuroscience, therefore, is not to "localize centers," but rather to identify the components of the various complex systems that interact to generate the mental functions. Luria named this task "dynamic localization."

The results of emotion-based experiments support and extend the concept that integrative brain functions are based on multiple oscillations. It is important to emphasize that the analysis of conventional ERPs and single frequencies may lead to restricted interpretations (Başar 1980, 1999). This view is supported by several recent publications (Başar 1999; Gruzelier 1996; Klimesch et al. 2000a; Makeig et al. 2002). Different functions are often correlated with different oscillations at distinct locations (Başar et al. 2001; Leiberg et al. 2006; Sakowitz et al. 2001, 2005).

In a study using single cell recordings Quiroga et al. (2005) reported subsets of neurons that are selectively activated in the human medial temporal lobe; these authors recorded activity in restrictive operative conditions and did not have the chance to record activity in the fusiform area and occipital or frontal cortices.

12.5 Integration of Episodic Memory and Emotion

Several studies have combined emotion and memory. It is clear that emotion has effects on memory and memory has effects on emotion. It can be assumed that without memory we would only have primary/basic emotions of the kind displayed by even basic animals. According to a memory model by Başar (2004) that is explained in Chap. 7, memory has three fundamental features. (Memory has more features, but the three fundamental functions are discussed here, for the purpose of simplicity.) The first one is persistent memory (phylogenetic), the second is dynamic memory, and the third is longer-acting memory.

12.5.1 Links Between Emotion and Persistent Memory

Solms stated that "basic emotion command systems" evolved over time. The basic emotions exist because they have established survival value. In situations of biological significance, these emotions provide ways of reacting that increase the likelihood that the organism will survive and reproduce and therefore propagate its genes. Further, Solms proposed that we share with all other mammals the basic emotion command system. Therefore, the basic emotions define a set of common biological "values" that unite us all in our struggle with the task of life (Solms and Turnbull 2002).

Damasio (1994) proposed two kinds of emotions, as described: primary and secondary. For primary emotions he stated:

> To what degree are emotional reactions wired in at birth? I would say that neither animals nor humans are, of necessity, innately wired for "bear fear", or "eagle fear". One possibility I have no problem with is that we are wired to respond with an emotion, in pre-organized fashion when certain features of stimuli in the world or in our bodies are perceived, alone or in combination. Examples of such features include size (as in large animals), large span (as in flying eagles); type of motion (as in reptiles); certain sounds (such as growling); certain configurations of body state (as in the pain felt during a heart attack).

As Damasio (1994) described, we do not fear a bear from birth because we do not know that it is something that represents a danger, but we are afraid of huge objects and sudden or high voices.

The experiments by Cannon and Bard (1925) and Bard (1928) also suggest that there are basic/primary or phylogenetic emotions. In their experiment, they used cats from which the whole cortex was removed. The cats show coordinated rage with their hypothalamus intact, but when their hypothalamus was also removed they did not show any coordinated rage.

The hypothesis in the literature supports the idea that we have basic/primary/ phylogenetic emotions. This type of emotion is coded genetically and is universal for human as well as animal species (although there are some differences). We have the ability or susceptibility of having emotions. Accordingly, our persistent memories keep the information of these emotions. Our emotions are also linked with our dynamic memories, as explained in the next section.

12.5.2 Links Between Emotion and Dynamic Memory

The other type of emotion that Damasio (1994) described is the "secondary emotions." In this type of emotion, the process begins with the conscious, deliberate considerations that are entrained about a person or situation. These considerations are expressed as mental images organized in a thought process. The first process is the sensory process that is distributed topologically. At the second process, networks in the prefrontal cortex automatically and involuntarily respond to signals arising from the processing of the images. At the third process, the response of the prefrontal cortex is signaled to the amygdala and the anterior cingulate and is followed by changes in the bodily state, which is called the "emotional bodily state." This state is then signaled back to the limbic and somatosensory systems.

Solms and Turnbull (2002) described four emotional responses – seeking, rage, fear, and panic. Further, they suggested that it is not enough to have only four emotional responses. They stated that these emotions, coupled with a handful of automatic, stereotyped behaviors, are required to cope with the vast complexities of everyday mammalian life.

According to the literature, the second income is that the emotions also have a dynamic character, just as dynamic memory does. This means that the emotions have a very strong relationship with the dynamic memory. Attention, perception, learning, and remembering, which are the key features of dynamic memory, are also very important for secondary/dynamic emotion (according to our point of view). Humans have a wide range of emotions because of their capability of dynamic memory. Damasio (1994) stated that secondary emotions/dynamic emotions can change from person to person and according to his or her individual experiences. The characteristics of dynamic memory vary among individuals. If there is a strong link between the dynamic memory and the secondary/dynamic emotions, it is appropriate to examine this strong link to explain the varying degrees of emotions.

12.5.3 Links Between Emotion and Earlier Episodes and Longer-Acting Memory

In addition to these types of emotions described in the literature, the primary/basic (persistent emotion in our view), and the secondary (dynamic emotion in our view), we propose a third type of emotion, longer-acting emotions, which have strong connections with the longer-acting memory. These types of emotions are dependent on longer-acting memories and influence an individual's social life. An example of this kind of emotion is the love felt for one's mother, father, husband, wife, close friend, or so on. Longer-acting emotions are the emotions that an individual cannot feel toward another person without a memory of him or her. This type of emotion is also very closely related to dynamic emotion. The dynamic emotions construct the longer-acting emotions. For example, you meet someone new; at first glance you may feel neutral, positive, or negative toward her. This meeting could be kept in your memory as a neutral, positive, or

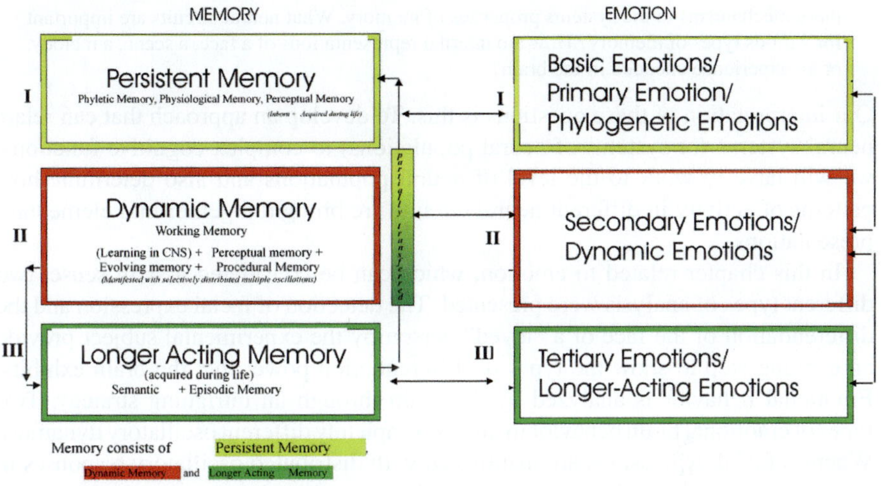

Fig. 12.10 Link between memory and emotion

negative memory. As your relationship develops, the memories that you associate with her will increase each day – some of them neutral, some positive, and some negative. After a while, you will have developed a positive or a negative emotion toward her. The character of this emotion will depend on the sum (which is not possible to calculate with our mathematics) of positive or negative memories that you have experienced with her.

We propose the following. To understand emotions, we have to subdivide them into categories. Love and fear, and anger and hate are not the same types of emotions. The question arises about the extent to which animals have emotions. According to this viewpoint, the more memory they have more emotions they have. Accordingly, humans have the largest memory, so they have the largest range of and capacity for emotions. Simple animals have little emotion, because they have little memory.

Another question presented in the literature is, "Are emotions universal?" According to Ekman (1992b) and Darwin (1872), emotions are universal. We propose that the primary/phylogenetic emotions are universal, but the secondary/dynamic as well as longer-acting emotions depend on the individual's memory. These individual differences also cause the differentiations observed among cultures (Fig. 12.10).

12.6 Future of Emotion Experiments, Link with the Intuitive Brain, and Metaphysical Thoughts

In his important book on memory and the mind, Eric Kandel (2006) asks the very important questions:

> Where is the new science of mind heading in the years ahead? In the study of memory storage, we are now at the foothills of a great mountain range. We have some understanding of the cellular and molecular mechanisms of memory storage, but we need to move from

> these mechanisms to the systems properties of memory. What neural circuits are important for various types of memory? How are internal representations of a face, a scene, a melody, or an experience encoded in the brain?

Our interpretation of these questions is this. To develop an approach that can relate neural systems (or systems of neural populations) to complex cognitive functions, we will have to work to the level of neural populations and also determine how patterns of activity in different neural circuits are brought together into elementary presentations.

In this chapter related to emotion, which can be considered a sixth sense, two different types of analysis were presented. The detection of facial expression and the differentiation of the face of a "loved" person by the experimental subject provide an exciting step to show the types of differentiation power that the brain exhibits. Emotional behavior is analyzed by the brain through an intriguing strategy. Two types of emotional brain behavior manifest completely different oscillatory dynamics. Whereas facial expressions are manifested with distributed oscillatory responses in the whole cortex, including frontal-parietal-temporal-occipital areas, the love for a partner is manifested only in frontal areas and solely in one frequency channel, the delta channel. Such findings are new, and a large number of paradigms are needed to express the electrophysiology of emotion more exactly. With the present results, we have an interesting lens for looking through this window of the sixth sense. The fMRI measurements also show the possibility to differentiate faces or objects. However, the time resolution of fMRI are presently very limited and emotional reactions within the first second of brain responses is extremely important. It can be also stated that other types of paradigms, by using emotion-inducing patterns, will give other types of oscillatory responses. It is also important to state that emotions are measurable, and emotional inputs cause the brain to perform different types of strategies.

By taking these results into account, the expression *emotional brain* gains importance and must be strongly considered in researching brain-body-mind integration. Therefore, it is included in the schematic illustrations within Chaps.16 and 17 for presentation of the Cartesian system in hyperspace.

Le Doux (1999) stated that memory is generally understood to be the process by which we bring back to mind some earlier conscious experience. The original learning and the remembering in this case are both conscious events. Further, this author is of the opinion that emotional and declarative memories are stored and retrieved in parallel, and their activities are integrated seamlessly into our conscious experience. "Emotion is not just unconscious memory: It exerts a powerful influence on declarative memory and other thought processes." Thus, emotions or feelings are conscious products of unconscious processes.

Chapter 20 discusses the importance of episodic memory, explained by means of grandmother experiments. According to the findings of electrophysiological measurements, there is the possibility of explaining phenomena that are at the boundary of conscious and unconscious processes.

Chapter 20 also discusses ways of analyzing "travelling back to the past," according to a question presented by Eric Kandel. How might the important scope of Eric Kandel's work be approached experimentally? In the present chapter we provided basic electrophysiological evidence that it is possible to measure manifestations of emotions that are linked to episodic memory and emotional episodes. These results clearly manifest the electrical correlates of traveling backward to the past, as in the paradigm of Marcel Proust with the episode of Madeleine, described in Chap. 20. In the present chapter, we have grasped several significant concepts of how the brain's alpha, theta, and delta responses are interwoven with emotional/episodic memory. Could these empirical results serve as a key to our unconscious-conscious processes or provide crucial knowledge for intuitive processes? To be able to remember the face of one's grandmother or the taste of a biscuit, we need to travel back in time. This travel takes place in a fraction of a second. Is this the heterogeneous time, "the duration" described by Henri Bergson? We return to these crucial questions in Chaps.18 and 20.

Chapter 13
Pathologic Brain: Impairment of Mind Based on Break of Oscillations and Modulation of Neurotransmitter Release

13.1 The Importance of Pathology in Understanding Brain Function and Mind

In physics, two perfect clocks at two distant points should show exactly the same time. However, there are, according to Einstein, *good clocks* and *bad clocks*. "Good" clocks are synchronized and always show exactly the same time even when they are spatially separated. In the brain, various clocks are oscillating with several discrete frequencies. They are synchronized or partially synchronized by executing diverse types of brain functions. In this chapter, the insufficient synchronization between oscillators or "clocks" of different neural populations in neuropsychiatric patients is described. Metaphorically speaking, this means that clocks in the pathological brain are often "bad clocks," as they are not synchronized. In the theory of relativity "asynchrony" of clocks plays an important role in processes observed between places separated by a long distance.

In 1861, the French physician Pierre Paul Broca conducted a post-mortem examination of the brains of two patients who died as consequence of hemorrhages in the left hemisphere. Because these patients had severe language impairment, this examination was a primary step for the later discovery of the so-called "Broca speech area" and the results opened an important research area related to language-function. Additionally, this anatomical research provided one of the most important models for studying brain function with the help of observed pathologies. Some pathologies reflect dysfunctions as breaks of a physiological signal, others as reduced release of transmitters, or histological degeneration in neural populations.

The discovery of so-called "Vagustoff" (acetylcholine) by Otto Loewi is one of the most important developments in basic and clinic medical science, because this discovery opened the way to understanding the importance of transmitters (see also Chapter 2 and Chapter 3, also the dream of Loewi in chapter 19).

This chapter, which is the key chapter related to *impairment of mind*, makes use of both cited examples. What types of cognitive impairment are correlated with brain oscillations? What might a break of oscillations or bad clocks and coherences

E. Başar, *Brain-Body-Mind in the Nebulous Cartesian System: A Holistic Approach by Oscillations*, DOI 10.1007/978-1-4419-6136-5_13,

indicate? These questions are of basic importance and in the future may provide an opportunity to use oscillatory dynamics as biomarkers in clinical studies. More important is the possibility to see how the break of oscillations or coherences may cause cognitive damage or impairment of the mind.

Studies on the interplay of brain oscillations and transmitters are possibly one of the most important future candidates in the analysis of brain-body-mind. chapter 22 addresses this important point.

13.2 Importance of Brain Oscillations and Neurotransmitters in Impairment of Mind

According to Lenz et al. (2008), even in the new era of functional neuroimaging, including MRI and fMRI, the electroencephalogram (EEG) is still an important tool for brain neurology and psychiatry. The EEG reflects the electrical activity of large populations of synchronized neurons; therefore, some diseases can be identified more easily with an EEG than with functional imaging. The total literature on neuropathology and the effects of transmitters is broad. However, reports on brain oscillations and pathologies are limited; thus, this chapter is only a limited review embracing the keywords neuropathology, oscillations, and transmitters.

This chapter focuses on the changes in oscillatory dynamics, mostly in patients with Alzheimer's disease (AD), schizophrenia, bipolar disorders, and mild cognitive impairment. Begleiter and Porjesz (2006) proposed that the genetic principles of these oscillations are likely to stem from the regulatory genes that control the neurochemical processes of the brain and therefore influence neural function. Genetic analyses of human brain oscillations may identify genetic loci underlying the functional organization of human neuroelectric activity. "The new trends in cognitive neuroscience make it possible to study neural network dynamics in the human brain in health and disease; they have, therefore, strongly contributed to the study of predisposition and brain dysfunction in psychiatric populations" (Banaschewski and Brandeis 2007; Ford et al. 2007; Herrmann and Demiralp 2005; Porjesz et al. 2005; Van der Stelt and Belger 2007).

Genetic loci that have been recently identified regarding both resting and evoked brain oscillations involving the GABAergic and cholinergic neurotransmitter systems of the brain were analyzed in the pioneering studies carried out by Porjesz et al. (2005). According to their results, oscillations also represent highly heritable traits that are less complex and more proximal to gene function than either diagnostic labels or traditional cognitive measures. Therefore, these oscillations may be utilized as "phenotypes of cognition" and valuable tools for understanding some complex genetic disorders. It is concluded that the advent of genomics and proteomics, and a fuller understanding of gene regulation, will open new horizons on these critical electrical events so essential for human brain function.

13.3 The Basic Properties of Neurotransmitters

It was long thought that a given neuron released only one kind of neurotransmitter, but today many experiments have shown that a single neuron can produce several different neurotransmitters. Chapter 3 discusses the best-known transmitters that are involved in functions, such as causing blood vessels to contract and the heart rate to increase. Norepinephrine plays a role in mood disorders such as manic depression. Additional to the results presented in this chapter, the reader is also directed to a series of experiments by Bullock's research team in California and Başar's group in Lübeck covering comparative research on invertebrates and low vertebrates.

Schütt and Başar (1992), Schütt et al. (1992), Başar et al. (1999b), and Bullock and Başar (1988) also examined the effect of transmitters such as acetylcholine, dopamine, noradrenalin, and serotonin on the isolated ganglia of Helix pomatia (snail) and showed changes in the oscillatory dynamics of these ganglia. These results are briefly presented in Chap. 10. We emphasize two results pertinent to the present chapter: The application of acetylcholine (ACh) induced a large increase in the theta response in the isolated visceral ganglion. Dopamine induced a crucial change in the oscillatory response that was recorded in the gamma frequency band following the electrical stimulation in the helix visceral ganglion, as shown in Fig. 13.1 (modified from Schütt and Başar 1992), which shows the averaged evoked potentials from an experiment in which 40 Hz activity increases after the administration of dopamine (control, top; with dopamine [10^{-2} M], bottom; wide-band filtered 1–250 Hz, left; pass-band filtered 30–70 Hz, right).

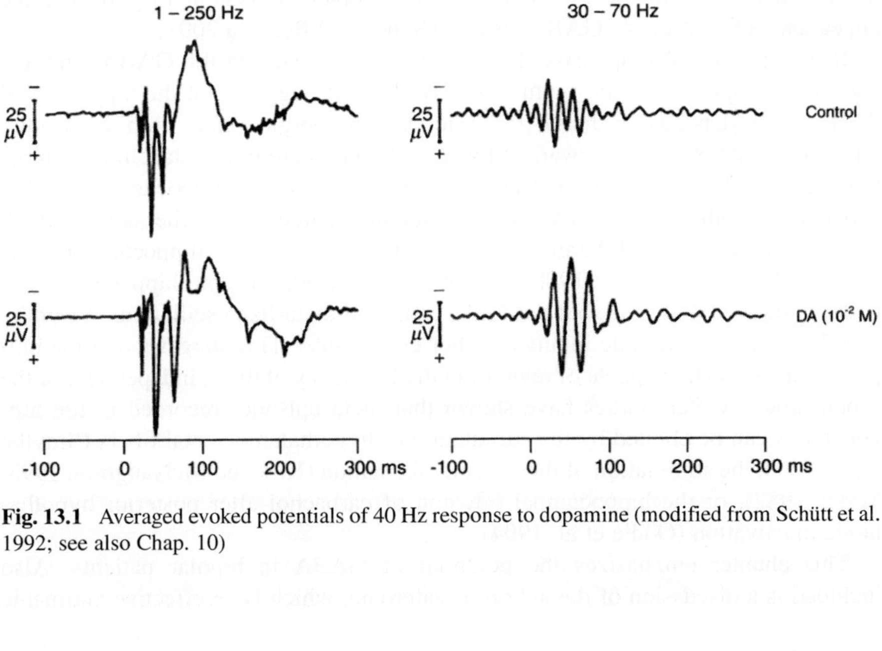

Fig. 13.1 Averaged evoked potentials of 40 Hz response to dopamine (modified from Schütt et al. 1992; see also Chap. 10)

On the left side of Fig. 13.1, the evoked potentials of the helix are shown in a wide frequency band. On the right, the filtered evoked potential in the frequency window of 30–70 Hz is illustrated. After the administration of dopamine, the gamma-evoked response reached values >200% in comparison with the control response. The basic results obtained from the study of invertebrate ganglia indicating increased activity in the 40 Hz range after the application of dopamine allows the tentative assumption that the gamma increase induced by dopamine is a universal process recorded in most low level species. Section 13.6 explains that when patients with schizophrenia are treated with dopamine they show increased gamma activity. Therefore, it would be interesting to examine patients before and after application of dopamine.

GABA (gamma-amino butyric acid) is an inhibitory neurotransmitter that is widely distributed in the neurons of the cortex. GABA contributes to motor control, vision, and many other cortical functions. GABAergic interneurons, which are the core component of cortico-limbic circuitry, were found to be defective in the cerebral cortex of bipolar patients (Benes and Berretta 2001). GABA spreads in neural networks involved in cognitive and emotional processing and modulates noradrenergic, dopaminergic, and serotonergic local neural circuitry (Brambilla et al. 2003). Several studies revealed low plasma (Berettini et al. 1983; Kaiya et al. 1982) or cortical GABA activity (Bhagwagar et al. 2007) or altered genetic expression of GABA (Guidotti et al. 2000; Heckers et al. 2002) in bipolar disorder. Low GABA activity was thought to be a genetically determined trait creating a vulnerability that, with the contribution of environmental factors, can lead to the development of either mania or depression.

It is also important to note that GABAergic activity is reciprocally regulated by dopamine, the hyperactivity of which also plays a role in mania (Yatham et al. 2002). Alterations in the modulation of the dopamine system may trigger the appearance of a defective GABA system (Benes and Berretta 2001).

It is important to emphasize the web of theta activity on the GABAergic and cholinergic inputs from the septum. In vivo studies suggest that the hippocampal theta rhythm depends on the GABAergic and cholinergic inputs from the septum (Brazhnik and Fox 1997; Stewart and Fox 1990) and requires an intact hippocampal CA3 region (Wiig et al. 1994). The cholinergic inputs to the hippocampus are distributed on both the pyramidal and interneuronal cells (Frotscher and Leranth 1985), whereas the GABAergic inputs selectively contact the hippocampal interneurons (Freund and Antal 1988). Recent in vitro work on septo-hippocampal co-cultures showed that CA3, but not CA1, exhibited theta-like oscillations driven by septal muscarinic synaptic inputs (Fischer et al. 1999). This suggests that the hippocampus is locally capable of regulating the frequency of theta, independent of the septal inputs. Other studies have shown that theta episodes recorded in the hippocampus can be elicited by the stimulation of hypothalamo-septal fibers (Smythe et al. 1991), the stimulation of the reticular formation (Kirk and McNaughton 1993; Vertes 1982), or the hippocampal infusion of carbachol after posterior hypothalamic inactivation (Oddie et al. 1994).

This chapter emphasizes the potential of GABA in bipolar patients. Also included is a discussion of the action of valproate, which is an effective antimanic agent (Bowden 2003). There is some evidence to support the role of valproate in elevating the levels of GABA within the brain (O'Donnell et al. 2003). Valproate was shown to augment the ability of atypical antipsychotic medications to increase dopamine (DA) and acetylcholine (ACh) efflux in the rat hippocampus and medial prefrontal cortex (Huang et al. 2006). It was also shown to lead to a significant reduction in presynaptic dopamine function in manic patients. This was thought to be related to improvement in manic symptoms (Yatham et al. 2002), because it regulates cell survival pathways such as the cAMP-responsive element binding protein (CREB), brain derived neurotropic factor (BDNF), bcl-2, and mitogen-activated protein kinases (MAP), which may underlie its neuroprotective and neurotropic effects (Löscher 2002; Özerdem et al. 2008; Xiaohua et al. 2002).

GABAergic interneurons and pyramidal cells were found to build and maintain complex interconnections that lead to large-scale network oscillations, such as theta, gamma (40–100 Hz), and ultrafast (200 Hz) frequency bands (Benes and Berretta 2001).

Glutamate is a major excitatory neurotransmitter that is associated with learning and memory and is also thought to be associated with Alzheimer's disease, whose first symptoms include memory malfunctions. GABA and glutamate as neurotransmitters are used by >80% of the neurons in the brain and constitute the most important inhibition. The glutamate action, together with GABA and dopamine, are presented later in this chapter.

13.4 Some Relevant Experiments Related to Oscillations and Transmitters

13.4.1 Theta Oscillations

The theta rhythm has been implicated in several brain functions, including sensory processing, memory, and the control of voluntary movement (Bland 1986; Bland and Colom 1993; Vinogradova 1995). In the freely moving rat, three types of hippocampal oscillatory activity have been observed that vary according to the behavior of the animal (Leung et al. 1982). The neuronal mechanisms underlying these oscillations are still largely unknown, but are likely to involve the complex interplay between intrinsic cellular and synaptic hippocampal properties, and external rhythmical inputs from the subcortical areas. Fellous and Sejnowski (2000) focused on the intrinsic hippocampal circuitry and showed that rhythms in these three frequency ranges may be observed in an in vitro slice preparation with different concentrations of carbachol present. According to the authors, the rhythmic activity observed in vitro might be supported *in part* by the intrinsic cellular and elementary network properties that are preserved in vitro (Konopacki 1998).

Gallinat et al. (2006) surveyed theta oscillation as follows: Synchronous oscillations at distinct frequency ranges are viewed as an important mechanism linking single-neuron activity to behavior and mental disorders (Başar-Eroğlu et al. 1992;

Gallinat et al. 2004). The crucial role of hippocampal theta oscillations (4–12 Hz) in mnemonic processes (Givens 1996; Miller 1989) is increasingly targeted in the accumulating body of literature. In the hippocampus, theta activity was suggested as being a major operational mode of grouping and segregating neuronal assemblies; the activity assigns computational tasks to these assemblies (Buzsaki 2002). In human investigations, theta activity was recorded during resting conditions and was shown to dramatically increase during mental tasks such as memory paradigms in the working memory (Gevins et al. 1997; Onton et al. 2005), and experiments using the oddball paradigm (Başar-Eroğlu et al. 1992; Röschke and Fell 1997; Yordanova et al. 2003). The most consistent increase in oscillatory theta activity was found in the frontal and central midline electrodes (Gevins et al. 1997).

Furthermore, it was argued that theta oscillations represent a basic resonance phenomenon mediating the information flow through the hippocampus, thalamus, and frontal/temporal neocortex (Başar 1980; Kirk and Mackay 2003; Miller 1991). Accordingly, paradigms eliciting hippocampal theta oscillations (e.g., oddball and visual maze) in humans (Fell et al. 2004, 2005) were shown to induce cortical theta oscillations. Moreover, there is also strong evidence that the glutamate system is involved in mediating hippocampal theta activity. For instance, the removal or isolation of glutamate-containing pathways eliminates hippocampal theta activity (Buzsaki 2002; Vanderwolf and Leung 1983). Compare also the results of Özerdem et al. (2008).

Gallinat et al. (2006) demonstrated an association between the glutamate concentration of the hippocampal region and the theta oscillations over the frontal scalp areas during auditory signal processing in humans. This association was observed for oscillations in the theta frequency range but not for other frequency ranges, suggesting that theta activity mediates neuronal coupling between the (frontal) cortex and the hippocampus via glutamatergic neurotransmission.

Further studies have also observed theta band activity across sensory modalities (Başar et al. 2001a and b; Gevins et al. 1997). This suggests that the theta band activity plays a fundamental role in stimulus evaluation and memory processing. Furthermore, a negative correlation between the RTs and the P300 amplitude elicited in an oddball paradigm was reported by Ford (1999). Theta activation was observed in or near the hippocampus, as revealed by MEG recordings and source reconstruction of working memory responses in humans (Tesche and Karhu 2000). This supports earlier findings of intra-cranially recorded hippocampal theta activity modulated by behavioral and perceptive conditions in epileptic patients (Meador et al. 1991).

13.4.2 Gamma Oscillations

According to Traub et al. (1996), in vitro models of gamma oscillations demonstrate two forms of oscillations: (a) one occurring transiently and driven by discrete afferent input and (b) the other occurring persistently, in response to activation of excitatory metabotropic receptors. The mechanism underlying persistent gamma oscillations has been suggested to involve gap-junctional communication between the axons of principal neurons, but the precise relationship between this neuronal activity and the gamma oscillation has remained elusive. According to their results, Traub et al. (1996) assumed that high-frequency oscillations occurred as a consequence of random activity within the axonal plexus. The authors further discuss that interneurons provide a mechanism by which this random activity is both amplified and organized into a coherent network rhythm.

13.5 Alzheimer's Disease and Mild Cognitive Impairment

Alzheimer's disease (AD) is a neurodegenerative disease that, in its most common form, is generally found in people aged over 65. Approximately 24 million people worldwide have dementia, of which the majority (approximately 60%) is the result of Alzheimer's disease (Ferri et al. 2005). Clinical signs of Alzheimer's disease are characterized by progressive cognitive deterioration, together with declining activities in daily life and neuropsychiatric symptoms or behavioral changes. The ultimate cause of AD is unknown. However, genetic factors are clearly indicated, as dominant mutations in three different genes have been identified that account for the small number of cases of familial, early-onset AD (Waldemar et al. 2007). Although acetylcholinesterase inhibitors appear to moderate symptoms, they do not permanently alter the course of the underlying dementing process (Bonte et al. 2006; Dougall et al. 2004; Marksteiner et al. 2007).

13.5.1 Oscillatory Responses in Delta, Theta, and Alpha Bands

Event-related oscillation studies in AD patients are one of the emerging areas in pathology (Güntekin et al. 2008; Hogan et al. 2003; Karrasch et al. 2006; Yener et al. 2007, 2008). The most commonly investigated parameters were related to spontaneous EEG analysis or spontaneous EEG coherence analysis in AD patients. There are few studies analyzing the evoked, event-related oscillations and evoked, event-related coherences.

Yener et al. (2007) investigated the theta responses of 22 mild probable AD subjects (11 non-treated, 11 treated by cholinesterase inhibitors), and 20 healthy elderly controls by using the conventional visual oddball paradigm. The authors aimed to compare theta responses of the three groups in a range between 4 and 7 Hz at the frontal electrodes. At the F_3 location, theta responses of healthy subjects were phase locked to stimulation and theta oscillatory responses of non-treated Alzheimer's patients showed weaker phase-locking, i.e., the average of Z-transformed means of correlation coefficients between single trials was closer to zero. In treated AD patients, phase-locking following target stimulation was two times higher in comparison with the responses of non-treated patients. The results indicated that the phase-locking of theta oscillations at F_3 in the treated patients is as strong as the control subjects. The F_4 theta responses were not significantly different between control and AD groups. The findings imply that the theta responses at F_3 location are highly unstable in comparison with F_4 in non-treated mild AD patients, and that cholinergic agents may modulate event-related theta oscillatory activities in the frontal regions.

Figure 13.2 illustrates the grand averages of theta response (4–7 Hz) at the F_3 location for treated AD, untreated AD and healthy subjects elicited by target stimuli. The thick black line indicates the grand average of theta response of each group; the thin gray lines show averages of single sweeps from each subject.

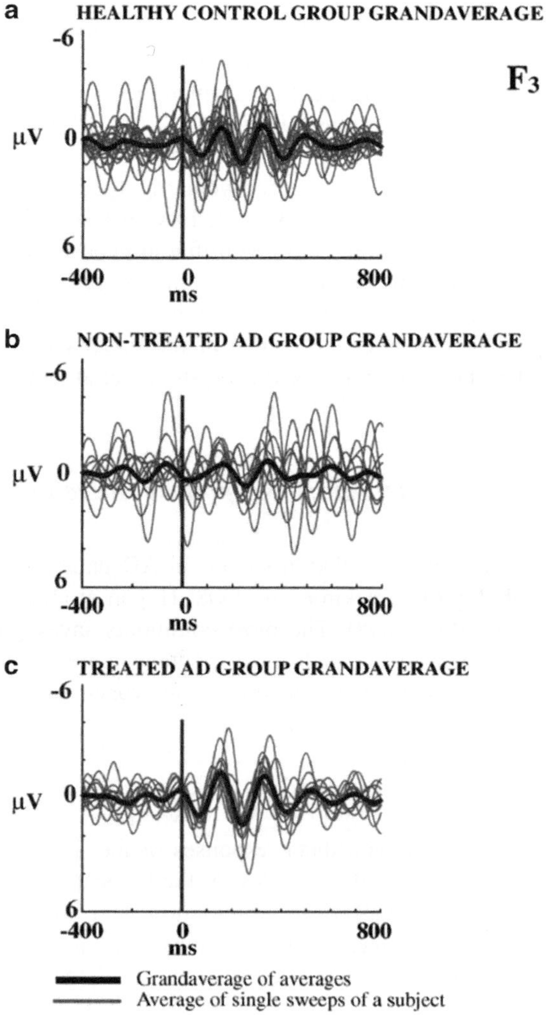

Fig. 13.2 Grand averages of group responses to the target stimuli elicited by a classical visual oddball paradigm recorded from scalp electrode site F_3. The *thick black line* indicates the grand average of each group, and the *thin gray lines* show averages of single sweeps from each subject. (**a**) Healthy elderly control group ($n = 20$). (**b**) Non-treated Alzheimer group ($n = 11$). (**c**) Treated (cholinesterase inhibitor) AD group ($n = 11$) (modified from Yener et al. 2007)

In the same group of patients (11 non-treated, 11 treated by cholinesterase inhibitors), and healthy subjects (20 healthy elderly subjects), Yener et al. (2008) investigated the amplitude measures of filtered oscillatory responses in delta (0.5–3.5 Hz), theta (4–7 Hz), alpha (8–13 Hz), and beta (15–30 Hz) on application of visual target stimuli. The authors indicated significant differences between healthy controls and the two groups of AD subjects in delta oscillatory responses, irrespective of cholinergic medication. This difference was steady at electrodes sites C_3 and C_z in the two AD groups in comparison with the control group. The findings of Yener et al. (2008) imply that the delta oscillatory responses at central locations are highly unstable in mild probable AD patients, regardless of treatment, when compared with healthy elderly controls.

Figure 13.3 illustrates examples from each group, showing single sweeps in delta oscillatory frequency range, in response to the target stimuli elicited by a classical visual oddball paradigm recorded from the scalp electrode of C_z. The thick black line indicates the average of single sweeps, and the thin gray lines show each single sweep for the subject. Figure 13.4 illustrates the grand averages of delta response on application of target stimuli for treated AD, untreated AD, and healthy subjects. In the grand averages of delta oscillatory responses at C_z, it is shown that the control group has larger amplitude than either treated or untreated AD groups. Peak-to-peak amplitudes of the control group are 7.25 (3.15) and 8.38 (3.38) μV in C_3 and C_z locations, showing a regular oscillatory pattern. In contrast, treated and untreated AD subjects have smaller amplitudes with an irregular shape.

Figure 13.4 describes the decrease of delta response to target auditory stimuli in an oddball paradigm in treated and untreated AD patients in comparison with a control group. The illustration shows that the delta response of AD patients are highly decreased, in C_z location the medication (cholinesterase inhibitor) slightly increases the delta response.

Babiloni et al. (2006, 2007, 2009) published core results on EEG rhythms in mild cognitive impairment (MCI) patients. Resting, eyes-closed EEG data were recorded in 34 MCI and 65 AD subjects by Babiloni et al. (2006). The EEG rhythms of interest were delta (2–4 Hz), theta (4–8 Hz), alpha 1 (8–10.5 Hz), alpha 2 (10.5–13 Hz), beta 1 (13–20 Hz), and beta 2 (20–30 Hz). The EEG cortical sources were estimated using low-resolution brain electromagnetic tomography (LORETA). Cortical EEG sources were correlated with MR-based measurements of the lobar brain volume (white and gray matter). A negative correlation was observed between the frontal white matter and the amplitude of the frontal delta sources (2–4 Hz) across the MCI and AD subjects. Babiloni et al. (2007) tested the hypothesis that EEG rhythms are correlated with memory and attention in the continuum from MCI through AD. Resting, eyes-closed EEG data were recorded in 34 MCI and 53 AD subjects. The EEG rhythms of interest were delta (2–4 Hz), theta (4–8 Hz), alpha 1 (8–10.5 Hz), alpha 2 (10.5–13 Hz), beta 1 (13–20 Hz), and beta 2 (20–30 Hz). The EEG cortical sources were estimated using LORETA. The results suggest that the cortical sources of resting delta and alpha rhythms correlate with neuropsychological measures of immediate memory, based on focused attention in the continuum of MCI and AD subjects.

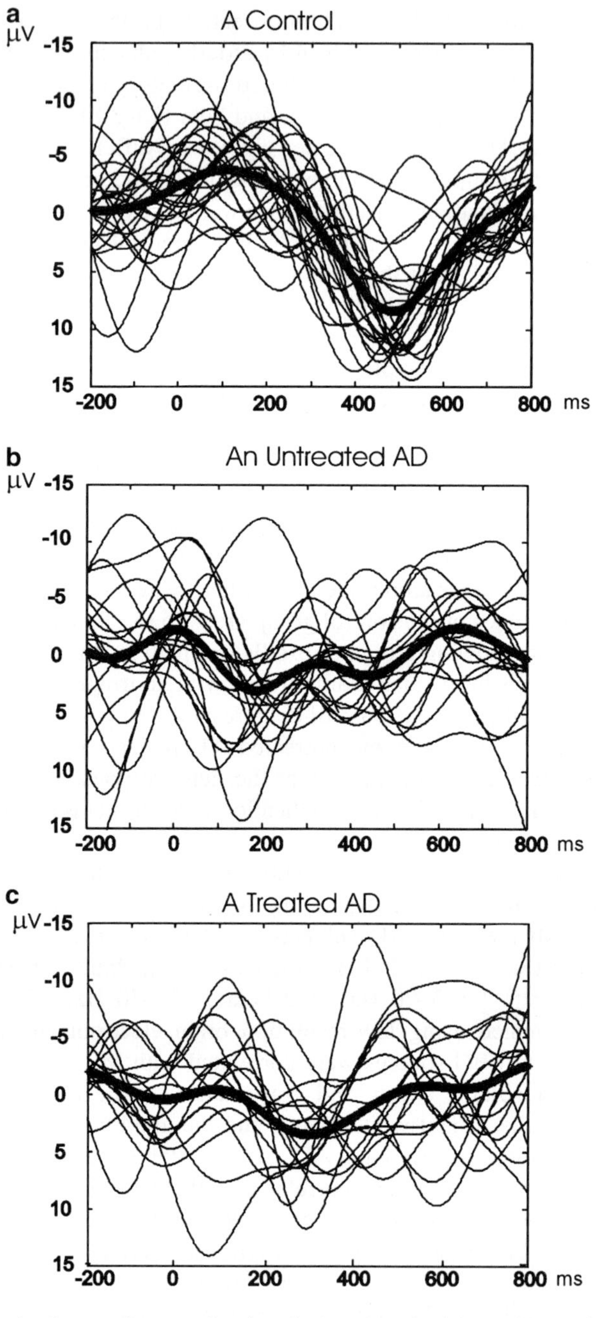

Fig. 13.3 Examples from each group showing single sweeps in delta oscillatory frequency range, in response to the target stimuli elicited by a classical visual oddball paradigm recorded from the scalp electrode of C_z. The *thick black line* indicates the average of single sweeps, and the *thin gray lines* show each single sweep for the subject. (**a**) Elderly healthy control. (**b**) Untreated AD subject. (**c**) Treated (cholinesterase inhibitor) AD subject. (modified from Yener et al. 2008)

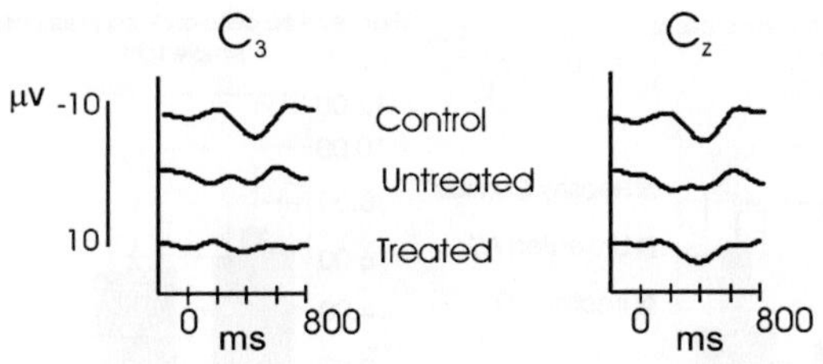

Fig. 13.4 Grand averages of delta oscillatory response of each group to the target stimuli elicited by a classical visual oddball paradigm recorded from electrodes of C_3 and C_z. (**a**) Healthy elderly control group ($n = 20$). (**b**) Untreated AD group ($n = 11$). (**c**) Treated (cholinesterase inhibitor) AD group ($n = 11$) (modified from Yener et al. 2008)

13.5.2 Comparison of Sensory-Evoked and Event-Related Oscillations

Yener et al. (2009) compared visual evoked oscillatory responses of subjects with Alzheimer's disease (AD) ($n = 22$) to healthy elderly controls ($n = 19$) elicited by simple light stimuli. The visual evoked oscillatory responses in AD subjects without cholinergic treatment ($n = 11$) showed significant differences from the controls and the AD subjects treated with a cholinesterase inhibitor ($n = 11$). Higher theta oscillatory responses in untreated AD subjects are seen on the electrode locations over bi-parietal and right occipital regions after simple light stimuli with less, if any, cognitive load. These changes were restricted to the theta frequency range only and are related to location, frequency bands, and drug effects. In their previous work Yener and colleagues observed that visual event-related oscillations elicited after the visual stimuli with a higher cognitive load, i.e., an oddball target, display lower amplitudes: between controls and AD subjects in delta frequency band without a drug effect; and over the left and mid-central region (Yener et al. 2008). These differences between the visual evoked oscillations and visual event-related oscillations imply that at least two different cognitive circuits are activated on visual stimuli in AD patients

Figure 13.5 illustrates the means and standard deviations of amplitudes of visual-evoked and visual event-related theta and delta oscillatory responses at C_z, C_3, P_3, P_4, and O_2 electrodes for treated AD, untreated AD, and healthy subjects.

13.5.3 Sensory-Evoked and Event-Related Coherences in Alzheimer's Disease

Many studies reported the successful use of EEG coherence to measure functional connectivity (Lopes da Silva et al. 1980; Petsche and Etlinger 1998; Rappelsberger et al. 1982). Accordingly, EEG coherence may be considered to be an important large-scale measure of functional relationships between pairs of cortical regions (Nunez 1997). Because coherence, in essence, is a correlation coefficient per

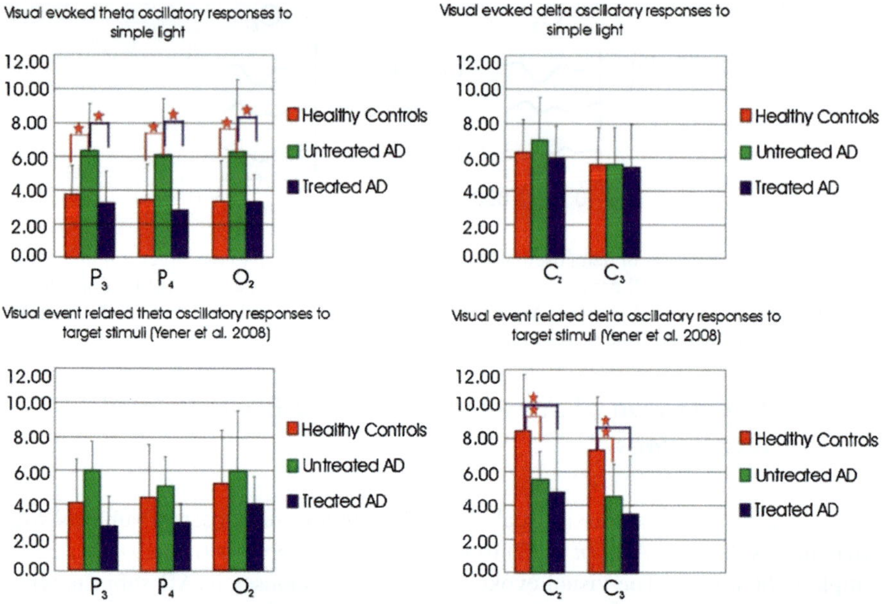

Fig. 13.5 Means and standard deviations of amplitudes of visual evoked and visual event-related theta and delta oscillatory responses at C_z, C_3, P_3, P_4, and O_2 electrodes in controls, untreated AD [T(-) AD], and treated AD [T(+) AD] groups. Significant differences between the groups were shown by an *asterisk* (modified from Yener et al. 2009)

frequency band, it is used to describe the coupling or relationship between signals for a given frequency band.

In clinical studies the measure of coherence has been used by several authors to investigate the connectivity between the various cortical areas of AD patients in resting conditions. The most common finding of these studies is that of reduced alpha and beta band coherences between distant structures (Adler et al. 2003; Besthorn et al. 1994; Dunkin et al. 1994; Leuchter et al. 1987; Locatelli et al. 1998).

Güntekin et al. (2008) investigated event-related coherences among patients with AD forms of dementia using a visual oddball paradigm. A total of 21 mild, probable AD subjects were compared with a group of 19 healthy controls. The AD group was divided into the untreated ($n = 10$) and those treated with a cholinesterase inhibitor ($n = 11$). The authors found that the control group showed higher values of evoked coherence in the delta, theta, and alpha bands in the left fronto-parietal electrode pairs vs. the untreated AD group. The control group showed higher values of evoked coherence in the left fronto-parietal electrode pair in the theta frequency band and higher values of evoked coherence in the right fronto-parietal electrode pair in the delta band when compared with the treated AD group. The only significant difference between the treated and untreated AD groups was in the alpha band; the treated AD group showed higher values of evoked coherence

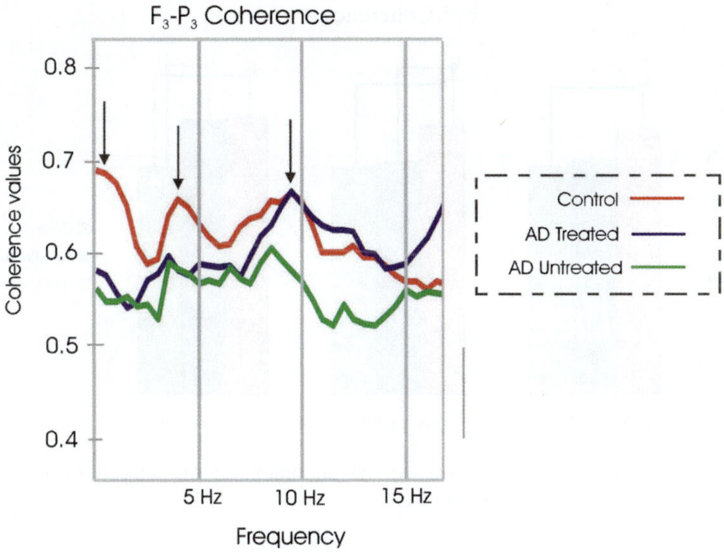

Fig. 13.6 Grand averages of evoked coherence for delta, theta, and alpha frequency bands for the F_3P_3 electrode pair for the control, treated AD, and untreated AD groups

in the left fronto-parietal electrode pair in the alpha band when compared with the untreated AD group. According to Güntekin et al. (2008), the results emphasized that the left fronto-parietal connections are highly affected by AD pathology occurring primarily within the parietal regions during the early stages of the disease.

Figure 13.6 shows the grand averages of evoked coherence for delta, theta, and alpha frequency bands for the F_3P_3 electrode pair for the control, treated AD, and untreated AD groups. In the delta frequency band (1–3.5 Hz) the coherence reached a value of 0.68 for the control group. This value was lower in both AD groups, at 0.58 in the treated and 0.56 in the untreated group. In the theta frequency band (4–7 Hz), coherence reached a value of 0.65 for the control group; however, in both AD groups this was found to be lower at 0.59. In the alpha frequency band (8–13 Hz), the coherence reached a value of 0.66 for the control and the treated AD groups, but was lower, at 0.60, in the untreated AD group. The grand averages of delta and theta coherences of the healthy controls were higher than those of both the AD groups, whereas only the alpha coherence in the untreated AD group was lower than both of the healthy controls and treated AD subjects.

Figure 13.7 shows histograms of mean *Z* values of control, treated AD, and untreated AD subjects for F_3P_3 electrode pair.

Zheng-yan (2005) stated that during photic stimulation, the inter- and intra-hemispheric EEG coherences of AD patients were at lower values in the alpha (9.5–10.5 Hz) band than those of the control group. The author reported that during

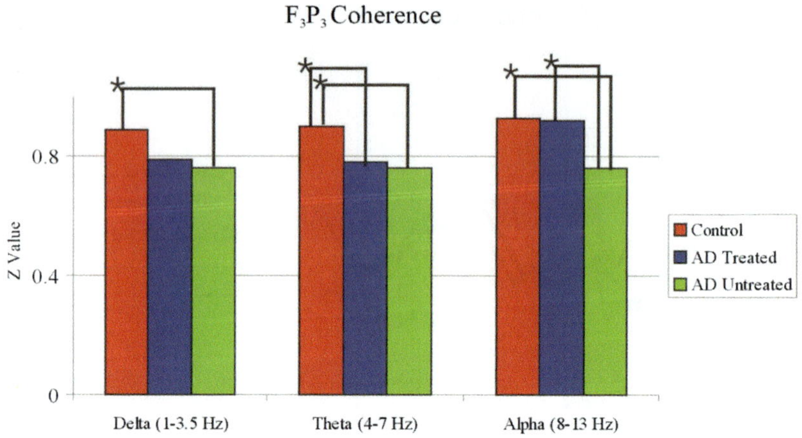

Fig. 13.7 Mean *Z* values of control, treated AD, and untreated AD subjects for F_3P_3 electrode pair. The *asterisk* represents significant results

a 5 Hz photic stimulation, the AD patients had significantly lower values of intra-hemispheric coherence in the C_4–P_4 and C_3–O_1 electrode pairs for theta band, in the C_3–P_3, C_3–O_1, and T_6–O_2 electrode pairs for alpha band, and the P_3–O_1, P_4–O_2, C_3–O_1, C_4–O_2, and T_6–O_2 electrode pairs for beta band oscillations.

Hogan et al. (2003) examined memory-related EEG power and coherence over temporal and central recording sites in patients with early AD and a normal control group. Although the behavioral performance of patients with very mild AD did not differ significantly from that of normal controls, the AD patients had comparatively reduced upper alpha coherence between the central and right temporal cortex. The findings of Güntekin and Başar (2009) have some parallels with Hogan's work. However, it is important to mention that medication has effects on long-distance evoked alpha coherence.

Zheng et al. (2007) investigated the functional relationship between calculated alpha band spectral power and inter- and intra-hemispheric coherence during a three-level working memory task undertaken by patients with mild cognitive impairment (MCI). The inter-hemisphere EEG coherence in frontal (F_3–F_4), central (C_3–C_4), parietal (P_3–P_4), temporal (T_5–T_6), and occipital (O_1–O_2) regions in MCI patients was compared with that in normal controls. The inter- and intra-hemispheric coherence during working memory tasks showed a "drop to rise" tendency compared with that at rest condition. The coherence in MCI patients was significantly higher than in the controls. The results of Zheng et al. (2007) indicate that the alpha frequency band may be the characteristic band in distinguishing MCI patients from normal controls during working memory tasks. MCI patients exhibit larger inter-hemispheric connectivity than intra-hemispheric connectivity with increased memory demand.

Evoked coherence studies (Hogan et al. 2003) showed reduced evoked coherence in AD patients between central and right temporal electrodes, whereas

Güntekin et al. (2008) showed reduced evoked coherence in the fronto-parietal recording sites. However, it is also important to emphasize the drug effects on EROs in AD patients. Drugs have been reported to take local effects on theta phase synchrony in the left frontal areas and have long-range connection effects on the alpha evoked coherence in the left fronto-parietal electrode pairs.

13.5.4 Comparison of Sensory-Evoked and Event-Related Oscillations

Several research groups have already published a number of studies related to analysis of oscillatory dynamics in MCI and AD patients. Babiloni et al. (2006, 2007, 2009) published core results on EEG rhythms in MCI patients. Zheng-yan (2005), Hogan et al. (2003), Güntekin et al. (2008), Yener et al. (2007, 2008, 2009), and Dauwels et al. (2010) published results on Alzheimer patients. At this point, it is vital to emphasize that there are important functional differences among EEG coherence, evoked coherence, and cognitive response coherence. In the EEG analysis, only sporadically occurring coherences from hidden sources can be measured. Sensory-evoked coherences reflect the property of sensory networks activated by a sensory stimulation. Event-related (or cognitive) coherences manifest coherent activity of sensory and cognitive networks triggered by a cognitive task. Accordingly, the cognitive response coherences comprehend activation of a greater number of neural networks that are most possibly not activated, or less activated, in the EEG and sensory-evoked coherences. Therefore, event-related coherence merits special attention. Particularly in AD patients with strong cognitive impairment, it is relevant to analyze whether medical treatment (drug application) selectively acts on sensory and cognitive networks manifested in topologically different areas and different frequency windows. Such an observation may provide, in future, a deeper understanding of the physiology of distributed functional networks and, in turn, the possibility of determination of biomarkers for medical treatment.

The sensory-evoked coherence and event-related target coherences were analyzed for delta (1–3.5 Hz), theta (4–7 Hz), alpha (8–13 Hz), beta (15–30 Hz), and gamma (28–48 Hz) frequency ranges for long-range intra-hemispheric (F_3–P_3, F_4–P_4, F_3–T_5, F_4–T_6, F_3–O_1, F_4–O_2) electrode pairs (Güntekin et al. 2008). The healthy control group showed significantly higher values of event-related coherence in delta, theta, and alpha bands in comparison with the de novo and medicated AD groups ($p < 0.01$ for the delta, theta, and alpha) on application of a target stimuli. In contrast, almost no changes in event-related coherences were observed in beta and gamma frequency bands. Furthermore, no differences were recorded between healthy and AD groups on application of simple light stimuli. Besides this, coherence values on application of target stimuli were higher than sensory-evoked coherence in all groups and all frequency bands ($p < 0.01$).

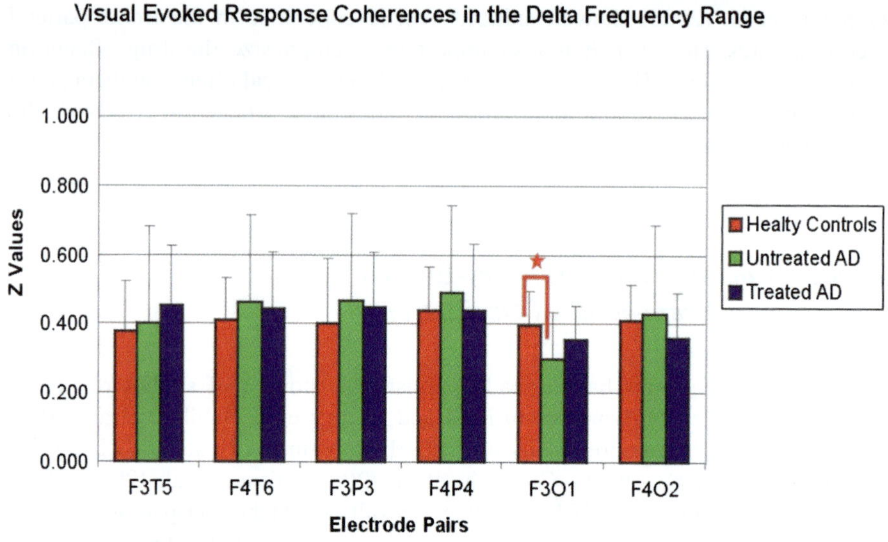

Fig. 13.8 Mean *Z* values of healthy control, treated AD, and untreated AD subjects for delta frequency range on simple light stimuli. The *asterisk* represents $p < 0.01$

Figure 13.8 illustrates the histogram of mean *Z* values for delta frequency range on application of "simple light" stimuli for all electrode pairs. Figure 13.9 illustrates the histogram of mean *Z* values for delta frequency range on application of "target" stimuli for all electrode pairs. In both figures, red bars represent the mean *Z* values for healthy subjects, green bars represent the mean *Z* values for untreated AD subjects, and blue bars represent the mean *Z* values for treated AD subjects. Figure 13.9 shows that the healthy subjects had higher delta response coherence compared with both untreated and treated AD subjects on application of target stimuli for all electrode pairs. The mean *Z* value of healthy subjects is 40–50% higher than AD patients in most of the electrode pairs on application of "target" stimuli. As Fig. 13.8 illustrates, the mean *Z* values on application of simple light are between 0.3 and 0.49, whereas on application of target stimuli, the mean *Z* values are as high as 0.9 (Fig. 13.9). The comparison of Figs. 13.8 and 13.9 shows that the evoked delta coherence on "simple light" is not high, and no difference was recorded between healthy controls and AD subjects.

Figure 13.10 shows mean *Z* values for theta frequency range on application of "simple light" stimuli for all electrode pairs. Figure 13.11 shows mean *Z* values for theta frequency range on application of target stimuli for all electrode pairs. In both figures, red bars represent the mean *Z* values for healthy subjects, green bars represent the mean *Z* values for untreated AD subjects, and blue bars represent the mean *Z* values for treated AD subjects. Figure 13.11 shows that the healthy subjects had higher theta response coherence compared with both untreated and treated AD subjects on application of target stimuli for all electrode pairs. The mean *Z* value of healthy subjects is 30–40% higher than AD patients in most of the electrode pairs

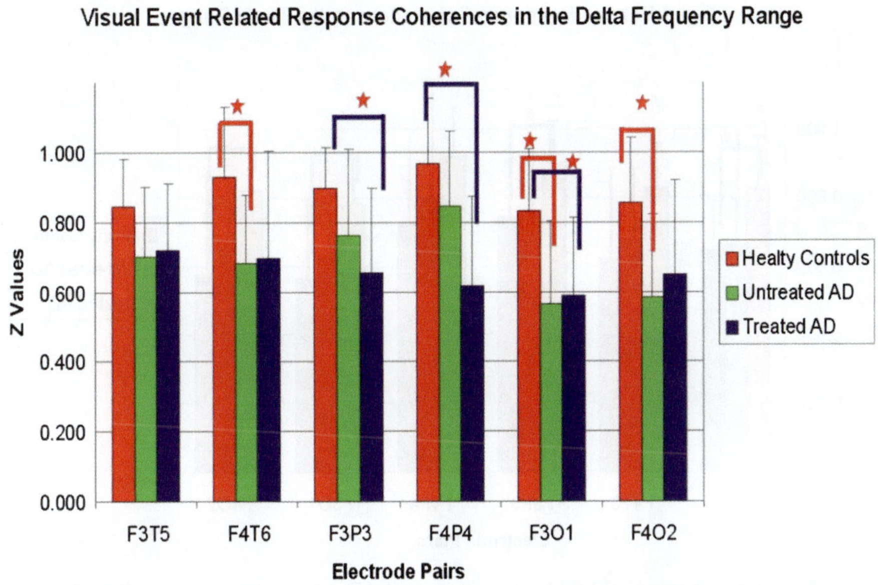

Fig. 13.9 Mean *Z* values of healthy control, treated AD, and untreated AD subjects for delta frequency range on target stimuli. The *asterisk* represents $p < 0.01$

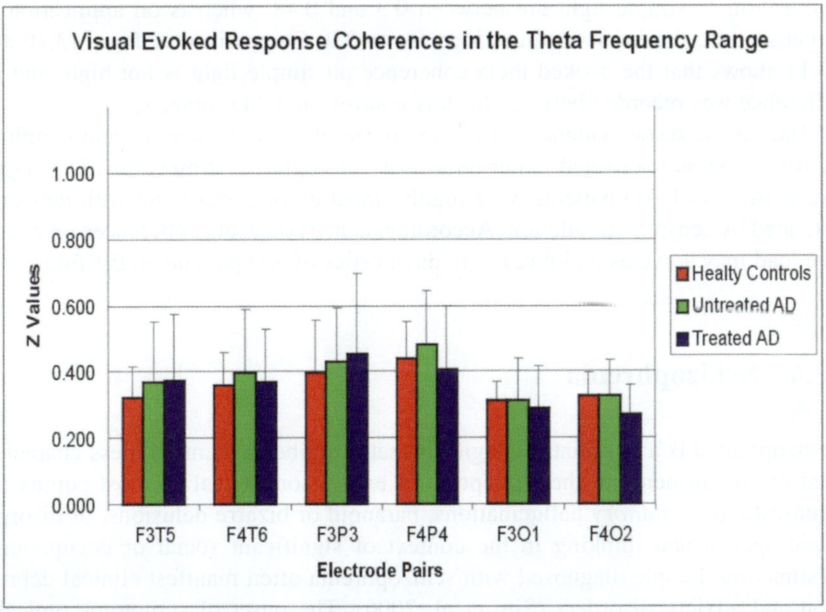

Fig. 13.10 Mean *Z* values of healthy control, treated AD, and untreated AD subjects for theta frequency range upon simple light stimuli. The *asterisk* represents $p < 0.01$

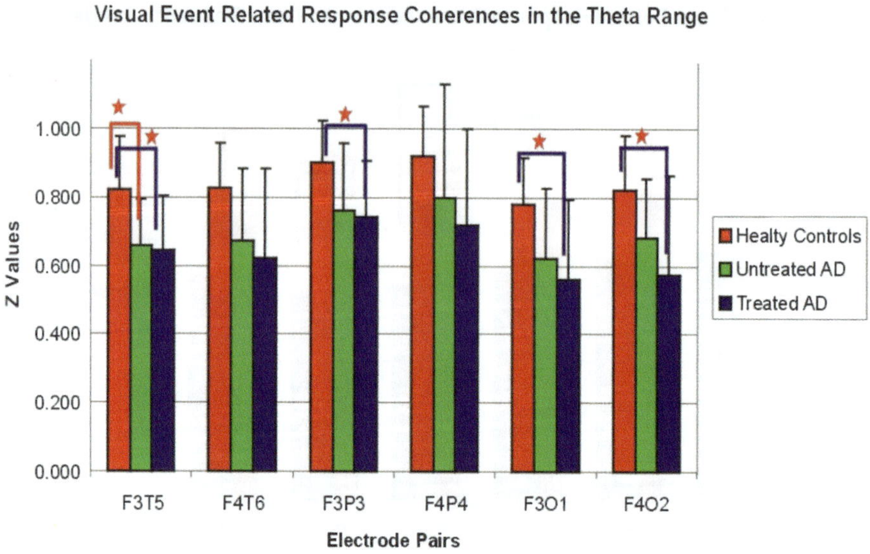

Fig. 13.11 Mean Z values of healthy control, treated AD, and untreated AD subjects for theta frequency range on target stimuli. The *asterisk* represents $p < 0.01$

upon application of target stimuli. As Fig. 13.10 illustrates, the mean Z values on application of simple light are between 0.3 and 0.48, whereas on application of target stimuli, the mean Z values increase to 0.9. Comparison of Figs. 13.10 and 13.11 shows that the evoked theta coherence on simple light is not high, and no difference was recorded between healthy controls and AD subjects.

The results show evidence for the existence of separate sensory and cognitive networks that are activated either on sensory or cognitive stimulation. The cognitive networks of AD patients were highly impaired in comparison with networks activated by sensory stimulation. Accordingly, analysis using coherences on cognitive load may serve as a biomarker in diagnostics of AD patients in the future.

13.6 Schizophrenia

Schizophrenia is a psychiatric diagnosis that describes a mental illness characterized by impairments in the perception or expression of reality, most commonly manifesting as *auditory* hallucinations, paranoid or bizarre delusions, or disorganized speech and thinking in the context of significant social or occupational dysfunction. People diagnosed with schizophrenia often manifest clinical depression and anxiety disorders (Sim et al. 2006). The onset of symptoms typically occurs in young adulthood (Castle et al. 1991). Studies suggest that genetics, early environmental, neurobiological, psychological, and social processes are important contributory factors. The current psychiatric research is focused on the role of neurobiology, but a clear organic cause has not been found. Because of the many possible combinations of symptoms, there is a continuing debate about whether the diagnosis represents a single disorder or a number of discrete syndromes.

Increased dopaminergic activity in the mesolimbic pathway of the brain is a consistent finding. The mainstay of treatment is pharmacotherapy with antipsychotic medication; these primarily work by suppressing dopamine activity (for references, see Sect. 13.2).

The disorder is primarily thought to affect cognition, but it also usually contributes to chronic problems with behavior and emotion.

Haig et al. (2000) studied gamma activity induced in response to task-relevant and irrelevant auditory oddball stimuli in 35 medicated patients with schizophrenia and 35 normal controls. The study pioneered a form of measurements utilizing a moving Welch window with short time FFT to examine the time course of the gamma amplitude. The group of patients with schizophrenia showed a significant decrease in the post-stimulus gamma response amplitude in the left hemisphere and frontal sites and an increase in the right hemisphere and parieto-occipital sites. In the non-targets (at a different latency), the schizophrenic patients showed a widespread gamma decrease. Haig et al. (2000) concluded that the gamma findings in non-targets might reflect an abnormality in appropriately processing irrelevant stimuli.

Wynn et al. (2005) analyzed gamma activity in medicated schizophrenic patients on application of a backward mask task. These authors found that medicated schizophrenic patients had significantly reduced gamma activity during the backward-masking task in comparison with healthy control subjects. The control subjects showed a significantly larger gamma activity in the right hemisphere, whereas schizophrenic patients did not show this pattern of lateralization. For the unmasked target, there was no group effect and no significant interactions in the gamma-band responses. Overall, schizophrenic patients showed less gamma activity and failed to show a lateralization of activity in the right hemisphere during the masked task, but showed comparable levels of gamma activity in relation to unmasked stimuli.

Yeragani et al. (2006) compared coherence between the first episode with schizophrenia ($n = 8$) and age and sex-matched normal controls ($n = 8$). The coherence was obtained using a cross-spectral analysis with beta (15.25–24.75 Hz) and gamma (25–44.75 Hz) frequency bands. For the analyses, non-rapid eye movement (NREM) and rapid eye movement (REM) sleep periods were used. The results showed a significant decrease in coherence in both beta and gamma frequency bands in schizophrenia parents. Post hoc *t*-tests revealed a significantly lower coherence among patients with schizophrenia only within the beta and gamma frequency bands during the waking stage.

Kwon and colleagues (1999) demonstrated that schizophrenic patients had selectively reduced, averaged evoked EEG power to 40-Hz auditory stimulation, but normal power to 20 and 30 Hz stimulation. Subsequently, Light et al. (2006) analyzed schizophrenic patients ($n = 100$) and non-psychiatric subjects ($n = 80$) undergoing auditory steady-state event-related potential testing. These authors also found that patients had reductions in both evoked power and phase synchronization in response to 30 and 40 Hz stimulation but a normal response to 20 Hz stimulation. Light et al. (2006) concluded that schizophrenic patients have frequency-specific deficits in the generation and maintenance of coherent gamma-range oscillations, reflecting a fundamental degradation of the basic integrated neural network activity.

Spencer et al. (2003) used measures of phase locking and phase coherence in the EEG to examine the synchronization of neural circuits in schizophrenic patients. Compared with matched control subjects, schizophrenic patients demonstrated: (1) an absence of the posterior component of the early visual gamma band response to Gestalt stimuli; (2) abnormalities in the topography, latency, and frequency of the anterior component of this response; (3) a delayed onset of phase coherence changes; and (4) a decrease in inter hemispheric coherence.

Spencer et al. (2007) investigated early visual-evoked gamma oscillation and early auditory-evoked gamma oscillation in chronic schizophrenic and matched healthy control subjects on the application of visual and auditory oddball tasks. These authors found that visual-evoked gamma oscillation phase locking at occipital electrodes was reduced in chronic schizophrenic patients compared with healthy control subjects. In contrast, auditory-evoked gamma phase locking and evoked power did not differ between groups. According to their results, Spencer et al. (2007) concluded that a visual-evoked gamma oscillation deficit may be a general phenomenon in schizophrenia, whereas the auditory-evoked gamma evoked by simple tone stimuli does not appear to be abnormal in chronic, medicated schizophrenic patients.

Ford et al. (2007) studied 24 patients with schizophrenia or schizoaffective disorder and 25 healthy comparison subjects. They related pre-speech neural synchrony to subsequent auditory cortical responsiveness to the spoken sound, compared pre-speech neural synchrony in schizophrenic patients and healthy subjects, and related pre-speech neural synchrony to the severity of auditory hallucination in the group of schizophrenic patients. To assess neural synchrony, the phase coherence of single-trial EEG preceding speech was calculated at a single site across repeated trials. To assess auditory cortical suppression, the N1 event-related brain potentials to speech sound onset during speaking and listening were compared. In comparison with healthy subjects, pre-speech neural synchrony was related to the subsequent suppression of responsiveness to the spoken sound, as reflected in the reduction of N1 during speaking relative to listening. There was greater pre-speech synchrony in the healthy subjects compared with the schizophrenic patients, especially those with severe auditory hallucinations. According to Ford et al. (2007), these data suggest that EEG synchrony preceding speech reflects the action of a forward model system, which dampens the auditory responsiveness to self-generated speech and is deficient in patients who hallucinate.

Başar-Eroğlu et al. (2007) investigated the modulation of event-related gamma responses in tasks varying the working memory (WM) load in schizophrenic patients and healthy controls. Gamma amplitude values were obtained for a simple choice reaction task, a low WM demand task, and a high WM demand task.

A gradual increase of gamma amplitudes after stimulus onset was associated with an increase of WM load in controls. In contrast, high amplitude gamma oscillations remained constant in patients, regardless of task difficulty. According to their results, the authors concluded that healthy subjects used various cognitive strategies depending on the task difficulty, whereas schizophrenic patients needed to initiate complex cognitive processes, similar to those used during processing of novel contexts or stimuli, even for the simple choice reaction task with a low cognitive demand.

Schmiedt et al. (2005) focused on event-related changes in post-stimulus theta oscillatory activity during varying cognitive and WM demand in healthy controls and schizophrenic patients. The results showed significant WM load and rule-switching-related increases of post-stimulus theta amplitude at fronto-central locations in controls. In patients with schizophrenia, there was no such modulation but, apart from an increased early theta at left temporal locations, there were generally reduced late theta responses in all tasks and at all locations.

Herrmann and Demiralp (2005) reviewed the literature on the alterations of gamma oscillations (30–80 Hz) during the course of neuropsychiatric disorders. Based on a study by Lee et al. (2003a and b), Hermann and Demiralp suggested that, in schizophrenic patients, negative symptoms correlate with a decrease of gamma responses, whereas a significant increase in gamma amplitudes is observed during positive symptoms such as hallucinations.

Ford et al. (2008) showed that P300 amplitude and delta and theta synchrony were reduced in schizophrenic patients; delta power and synchrony was better distinguished between groups than the P300 amplitude. Gamma synchrony was at the predicted P300 amplitude in healthy controls, but not in the patient group.

Jeon and Polich (2003) found that the P300 component in the average ERP is reduced in amplitude in schizophrenic patients and suggested that there is a deficit in the power and/or trial-to-trial synchrony of the neural activity generating the average P300.

In a single trial analysis of P300, Ford et al. (1994) used a 2 Hz half sine wave as a "P300 template" and fitted it to the EEG following a target tone. They found that patients with schizophrenia showed greater latency variability from trial to trial and had smaller amplitudes in each trial (Ford et al. 1994).

Bucci et al. (2007) investigated evoked and induced 40-Hz gamma power as well as frontal-parietal and fronto-temporal event-related coherence in patients with deficit and non-deficit schizophrenia and in matched healthy controls. In patients, correlations between gamma oscillations and psychopathological dimensions were also investigated. A reduction of both induced gamma power and event-related coherence was observed in patients with non-deficit schizophrenia, but not in those with deficit schizophrenia.

Symond et al. (2005) used a conventional auditory oddball paradigm to study 40 first-episode schizophrenic patients and 40 age- and sex-matched healthy control subjects. The authors then examined the magnitude and latency of both early (gamma 1: -150 to 150 ms post stimulus) and late (gamma 2: 200–550 ms post stimulus) synchrony with a multiple analysis of variance. First episode schizophrenic patients showed a decreased magnitude and delayed latency for global gamma 1 synchrony in relation to the healthy subjects. In contrast, there were no group differences in gamma 2 synchrony.

The first investigation of oscillatory activations in the gamma-band in an auditory oddball paradigm in schizophrenic patients was carried out by Gallinat et al. (2004). The EEG gamma-band responses of 15 drug-free schizophrenic patients and 15 age- and gender-matched healthy controls were compared. A wavelet transform based on Morlet wavelets was used to calculate the oscillatory gamma-band responses. The authors found that, in response to standard stimuli, early evoked gamma-band responses (20–100 ms) did not show significant group differences. However, the schizophrenic patients showed reduced evoked gamma-band responses in a late latency range (220–350 ms), particularly after target stimuli. This deficit occurred over the right frontal scalp regions. Furthermore, significant correlations were observed between oscillatory gamma-band responses and clinical parameters in schizophrenic patients. Gallinat et al. (2004) suggested that their results were consistent with a relatively preserved stimulus processing in the auditory cortex, as reflected by the early gamma-band responses.

According to these studies, the gamma response in schizophrenic patients is weak, even in response to simple auditory and visual stimulation (Brenner et al. 2003; Haig et al. 2000; Kwon et al. 1999; Light et al. 2006; Wynn et al. 2005). According to Spencer et al. (2003), during the visual oddball paradigm the gamma phase coherence is low. Furthermore, in general, the gamma power is also weak following all types of visual stimulation. However, the responses to auditory stimulation did not produce consistent results. According to Spencer et al. (2007), no differences were seen in the auditory oddball paradigm. According to Light et al. (2006) and Brenner et al. (2003), there are differences in the simple auditory evoked potentials. Gallinat et al. (2004) reported that differences in oddball paradigm were seen only in the second time window (220–350 ms). However, Symond et al. (2005) found a low gamma response in the first time window (0–150 ms); no differences were observed in the second time window. Spencer et al. (2007) subsequently suggested that this visual gamma oscillatory deficit may be a general phenomenon in schizophrenia, independent of task and stimulus type.

Spencer et al. (2007) comment that because their study and that of Gallinat et al. (2004) used oddball tasks with simple tone stimuli, it remains to be determined whether the absence of an auditory gamma response deficit is specific to this particular type of task or stimuli. In analyzing P300 animal studies, Başar-Eroğlu and Başar (1991) demonstrated that a gamma response followed the omitted stimuli, at a latency of 300 ms. These results show evidence of an existing gamma response as a pure cognitive response in the second time window. Several authors discuss the apparently larger influence of the second time window of the gamma response. According to Light et al. (2006), even in the simple auditory stimuli, the second window gamma response is more affected. Haig et al. (2000) and Gallinat et al. (2004) found that, in oddball paradigms, only the second time window gamma response is affected.

A precise comparison between medicated and unmedicated subjects is not encountered in the literature. This makes an assessment of the effectiveness of medication in schizophrenic patients difficult. However, in all studies on medicated and non-medicated subjects, the gamma response is low compared with healthy subjects (Gallinat et al. 2004; Haig et al. 2000; Kwon et al. 1999; Light et al. 2006; Wynn et al. 2005). It is important to note that the coherence in healthy subjects is high, whereas in schizophrenic patients it is low in spontaneous activity as well as in evoked coherence (Light et al. 2006; Spencer et al. 2003; Yeragani et al. 2006).

Although most authors have published event-related oscillations (ERO) results and the coherences in beta and gamma windows, Schmiedt et al. (2005) and Ford et al. (2008) recently published findings showing decreases in delta and theta responses. These findings are possibly the only reports showing a decrease of delta and theta in schizophrenic patients. This is significant because the most fundamental components of the oddball paradigms are delta and theta responses. In explanation, it is possible that all the other authors focused their analysis on higher-frequency windows, rather than the slow components.

It is also important to note that schizophrenic patients with hallucinations showed increased high-frequency oscillations (Baldeweg et al. 1998; Spencer et al. 2004). In addition, the results of Başar et al. (2007) are important, because they conclude that during a paradigm with working memory load, schizophrenic patients also showed an increased gamma response. All these results clearly show that the gamma responses in schizophrenic patients are not necessarily weakened; and, depending on the status of the schizophrenic behavior (negative or positive symptoms), and also depending on the difficulty of the applied paradigm, an increase of gamma activity also may be observed. In conclusion, it can be said that the oscillatory dynamics in schizophrenia also depict the unstable behavior of electrophysiology in this disease.

13.7 Bipolar Disorders

Bipolar disorder, originally described by Hippocrates and Aretaeus, is not a single disorder, but a category of mood disorders defined by the presence of one or more episodes of abnormally elevated mood, clinically referred to as mania. Individuals who experience manic episodes also commonly experience depressive episodes or symptoms, or mixed episodes that present the features of both mania and depression (Bowden 2007).

Mania is the core feature of the illness which gives rise to definitive diagnosis (DSM-IV, 1994). The manic state is characterized by increased energy and motor activity, a decreased need for sleep, distractibility with a strong element of pleasure seeking, and impulsive behavior. Manic patients also display signs of dysfunction in attentional measures, complex processing, and memory (Özerdem et al. 2008). It is suggested that having an acute episode of mania or depression can cause damage to learning and memory systems (Bearden et al. 2001). Cortical inhibitory deficits were thought to provide neurophysiological evidence for an association between bipolar disorder and disrupted cortical gamma amino butyric acid (GABA) related inhibitory neurotransmission (Levinson et al. 2007).

Diagnosis of bipolar patients is based on the individual's self-reported experiences, as well as observed behavior. The onset of symptoms generally occurs in early adulthood. Studies suggest that genetics, early environmental, neurobiological, psychological, and social processes are important contributory factors. Episodes of illness are associated with distress and disruption, and a relatively high risk of suicide (Ösby et al. 2001). Bipolar disorder is usually treated with a number of drugs termed "mood stabilizers," in particular lithium and sodium valproate. (See Sect. 13.1 for the effects of valproate.) Antipsychotic medication, sometimes called neuroleptics, is used in the stabilization of manic episodes and the maintenance phase of treatment.

Although some studies have suggested a significant correlation between creativity and bipolar disorder, this relationship is still unclear (Nowakowska et al. 2005; Rihmer et al. 2006; Santosa et al. 2007). One study indicated that bipolar patients can exhibit increased striving for, and sometimes attaining of, goals and achievements (Johnson 2005).

Patients with bipolar disorder show cognitive deficits and disorganized behavior, which may reflect a disturbance in neural synchronization. O'Donnell et al. (2004) tested whether EEG measures of auditory neural synchronization were abnormal in bipolar disorder. They evaluated 19 symptomatic patients with bipolar disorder and 32 non-psychiatric control subjects. Click trains (500 ms duration) presented at 20, 30, 40, and 50 Hz were used to evoke EEG synchronization. Patients with bipolar disorder showed reduced power across the frequencies of stimulation. Phase-locking across trials was also disturbed in bipolar disorder, and this is consistent with poor stimulus-related phase synchronization in EEG. Abnormal high-frequency neural synchronization may contribute to cognitive deficits in bipolar disorder.

Bowden (2007, 2008) emphasized that the scientific unveiling of lithium, valproate, and lamotrigine as specifically efficacious treatments for bipolar disorder is instructive in the effective application of inductive logic that has largely been absent in most other drug development efforts for bipolar disorder at the turn of the twenty-first century.

According to Bowden and Karren (2006), valproate is an effective treatment for mania, alone or in combination, but has limited benefit in bipolar depression. Carbamazepine is effective in mania, whereas lamotrigine has benefits for bipolar depression and maintenance therapy, but not for manic episodes. Other anticonvulsant drugs have been used to treat mania with mixed results. Bowden and Karren (2006) conclude that valproate, lamotrigine, and carbamazepine have a valuable place in the management of bipolar disorder. For a discussion of the use of valproate in electrophysiological studies, see Özerdem et al. (2008).

It has been reported that bilateral intra-hemispheric coherences in alpha and beta frequency bands were increased in both long-term abstinent and non-abstinent alcoholics compared with control subjects (Winterer et al. 2003a). Akiskal et al. (2001) stated that manic patients lack insight; they are generally considered unreliable observers of their own psychopathology. Self-assessment of this disorder appears feasible and potentially useful in practice. Lack of insight, poor judgment, and distractibility obviously require assessment by a clinician. Although the data from these authors are correlational and require prospective validation, they nonetheless suggest that (1) activation should be raised to the status of the stem criterion for mania and (2) mood should be specified as elated, depressive, anxious, or irritable.

According to Özerdem et al. (2008), bipolar disorder involves various cognitive dysfunctions, even in the euthymic phase of the illness. Dysfunction in GABA/glutamatergic systems and neural circuits that regulate cognitive processing seem to be involved in the underlying pathology (see Sect. 13.2). The aim of the studies carried out by Özerdem et al. (2007, 2008) was to detect differential oscillatory responses to visual target stimuli during an oddball paradigm in patients with bipolar disorder; and determine how responses change after valproate monotherapy. Event-related oscillations (EROs) by means of visual oddball paradigm in euthymic and manic, medication-free, DSM-IV bipolar patients were measured before and after 6 weeks of valproate monotherapy. At the baseline, euthymic patients showed significantly higher delta responses to target stimuli at all EEG channel positions in comparison with healthy controls, noticeably at the left frontal (F_3) and left temporal channel positions. Patients in the manic phase had significantly higher beta and lower alpha responses in the occipital area compared with the healthy controls. Six weeks of valproate monotherapy produced a significant reduction in the high delta responses at F_z, F_3, and temporal at T_3, T_4, and T_5 compared with the baseline in the euthymic group. In the manic group, following valproate monotherapy, the high occipital beta responses were significantly reduced and became similar to those of the normal controls. Valproate seems to have a selectively normalizing effect on the altered electrical activity of left frontal and bilateral anterior temporal areas in euthymic patients and the occipital area in manic patients.

This differentiation in topology is very intriguing and is a very important finding, which addresses a possible disruption in the brain's integrative working mechanisms in the manic phase of bipolar disorder. There is strong evidence for the possibility that the diminished occipital alpha response elicited by the cognitive input (P300 paradigm) is an indicator that, in mania, there is an inefficient communication between different structures of the brain (Fig. 13.12). Furthermore, Özerdem et al. (2008) propose that significantly increased occipital beta activity in the bipolar patient group may be compensatory to the presumed disrupted connectivity in the brain's integrative functioning, as indicated by the decreased alpha activity (Fig. 13.13).

Low amplitude fast oscillations such as beta and gamma (15–80 Hz) were shown to be specifically related to the brain's effort to screen and establish synchronous communication between brain regions (Munk et al. 1996). It also should be noted that the low-voltage alpha (LVA) EEG that has been measured in mania (Özerdem et al. 2008) has also been associated with a subtype of alcoholism related to anxiety disorder (Enoch et al. 1999).

Increased spontaneous beta power has been reported in frontal regions in diagnosed patients with antisocial personality disorder (ASP) (Bauer and Hesselbrock 1993). Rangaswamy proposes that the beta power increase may, inter alia, be associated with increased vulnerability, because female high-risk subjects with a larger

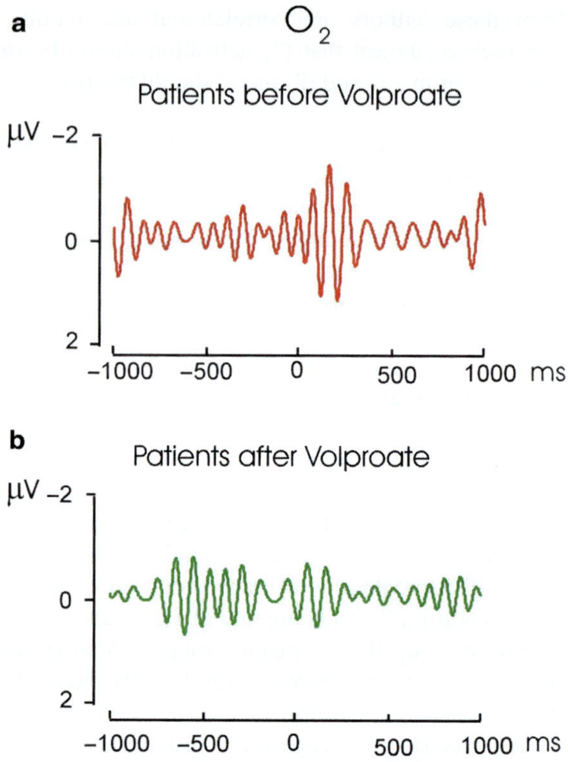

Fig. 13.12 Right occipital (O_2) alpha (8–13 Hz) response in manic patients before (**a**) and after (**b**) treatment with valproate. Graph A represents the grand average of the averages from ten patients, graph B from nine patients. Post-treatment occipital alpha response is significantly lower than the unmedicated condition

number of affected first-degree relatives displayed significantly elevated beta power compared with those with just one affected parent (Rangaswamy et al. 2002).

Interpretation of the results of the mentioned study on euthymic bipolar patients by Özerdem et al. (2007) suggests that different phases of the illness in bipolar disorder show differences in oscillatory responses: Increased delta as representative of disturbance in focused attention and decision making in euthymia; and reduced occipital alpha and increased occipital beta as a sign of more severe disturbance in the integrative cognitive processing of the brain within the core feature (mania) of the illness.

13.8 A Comparative Analysis of Alzheimer's Disease and Bipolar Disorder in the Gamma Frequency Range

The selectivity in parallel processing in the brain is produced by variations in the degree of spatial coherences that occur over long distances between brain structures/neural assemblies (Başar et al. 1999a, b; Kocsis et al. 2001; Miltner et al. 1999; Schürmann et al. 2000).

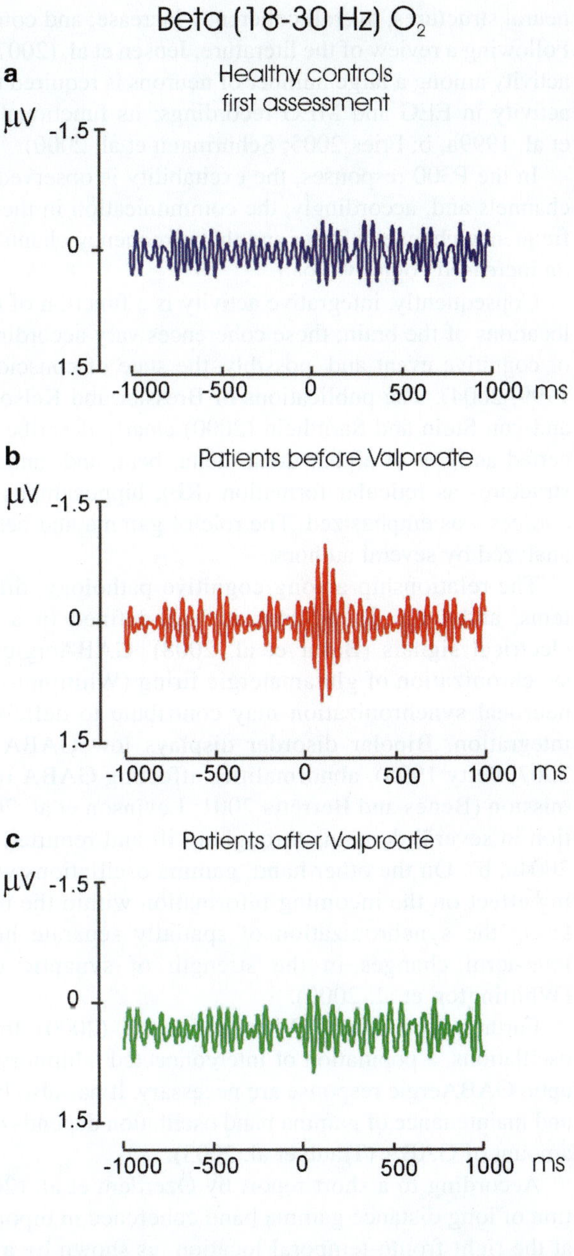

Fig. 13.13 Right occipital (O_2) beta (18–30 Hz) responses in three different conditions: **(a)** Healthy controls ($n = 10$). **(b)** Patients before treatment ($n = 10$). **(c)** Patients after treatment ($n = 9$). Graphs represent grand averages of subjects in each group. Note the significant decrease in the patients' response after treatment in comparison with its baseline and normal controls

Fries (2005) hypothesized that neural communication is mechanistically subserved by neural coherence. Activated neural groups oscillate and thereby undergo rhythmic excitability fluctuations that produce temporal windows for communication. Only coherently activated neural groups can interact effectively. This hypothesis is in agreement with the general hypothesis on excitability of neural structures, overall coherency increase, and complex matching (Başar 2004). Following a review of the literature, Jensen et al. (2007) summarize that coordinated activity among a large number of neurons is required to produce oscillatory gamma activity in EEG and MEG recordings; its functional role remains unclear (Başar et al. 1999a, b; Fries 2005; Schürmann et al. 2000).

In the P300 responses, the excitability is observed in delta and theta frequency channels and, accordingly, the communication in the brain occurs mostly in those frequency channels. Delta and theta frequency channels show selective coherences on increased cognitive load.

Consequently, integrative activity is a function of the coherences among spatial locations of the brain; these coherences vary according to the type of sensory and/or cognitive event and, possibly, the state of consciousness of the species (Başar 1999, 2004). The publications of Bressler and Kelso (2001), Varela et al. (2001), and von Stein and Sarnthein (2000) clearly describe this trend, in which the concerted activity of alpha, theta, delta, beta, and gamma oscillations in distributed structures as reticular formation (RF), hippocampus (HI), thalamus, and sensory cortices was emphasized. The role of gamma and beta oscillations was previously analyzed by several authors.

The relationship among cognitive pathology, different neurotransmitter systems, and brain oscillations is well-defined in a multidimensional model of electrical signals (Başar et al. 2008). GABAergic modulation is required for synchronization of glutamatergic firing (Whittington et al. 2000), and abnormal neuronal synchronization may contribute to deficits in cognitive and affective integration. Bipolar disorder displays low GABA activity (Bhagwagar et al. 2007; Petty 1995), abnormalities affecting GABA related inhibitory neurotransmission (Benes and Berretta 2001; Levinson et al. 2007), and cognitive dysfunction in several domains throughout ill and remitted states (Martinez-Aran et al. 2004a, b). On the other hand, gamma oscillations were suggested to have a gating effect on the incoming information within the temporal domain, thus facilitating the synchronization of spatially separate brain regions and leading to long-term changes in the strength of synaptic connectivity between areas (Whittington et al. 2000).

Further, according to Whittington et al. (2000), for an agent to manipulate fast oscillations, a population of interconnected inhibitory cells and sufficient postsynaptic GABAergic response are necessary. It has also been proposed that generation and maintenance of gamma band oscillation depends on the presence of a sufficient amount of GABA (Traub et al. 2003).

According to a short report by Özerdem et al. (2010), there is a clear disruption of long distance gamma band coherence in bipolar manic patients, especially at the right fronto-temporal location, as shown by a 35% lower coherence value among manic patients compared with healthy controls. This may correspond to a functional fronto-temporal connectivity problem in bipolar disorder. Figure 13.14 shows the grand average of coherence function between F_4–T_6 locations in the patients before (red line) and after medication (green line) in comparison with the controls (blue line). The frequency window is 28–48 Hz. At baseline, patients

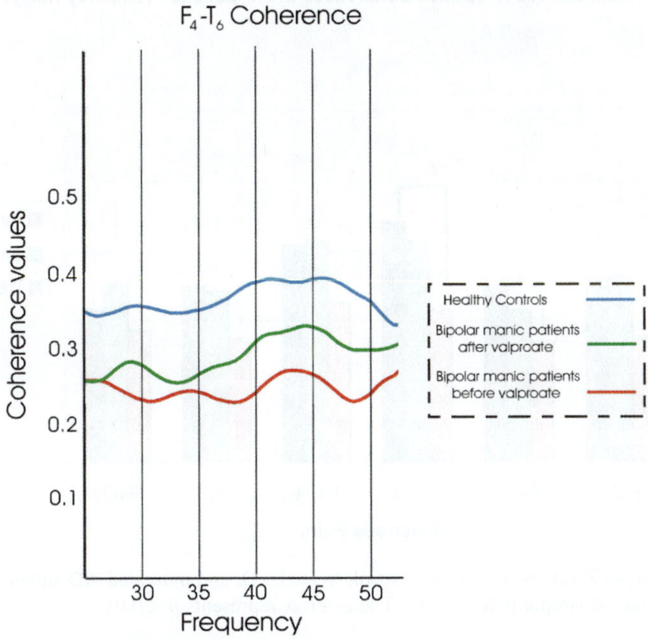

Fig. 13.14 Grand average of coherence function between F_4–T_6 locations in patients before (*red line*) and after (*green line*) medication in comparison with controls (*blue line*). The frequency window is 28–48 Hz (From Özerdem et al. 2010)

showed a 35.41% decrease in the coherence function compared with healthy controls. The post-treatment coherence function increased by 26.79% in the patient group.

It is hoped that the comparison of event-related coherences between all frequency windows and between various diseases may provide important information related to the genesis of event-related oscillations that are anchored with transmitter release.

Figure 13.15 shows a histogram of mean *Z* values for the gamma frequency range upon application of "simple light" stimuli for all electrode pairs for Alzheimer's patients. Figure 13.16 shows a histogram of mean *Z* values for the gamma frequency range upon application of "target" stimuli for all electrode pairs. In both figures, red bars represent the mean *Z* values for healthy subjects, green bars represent the mean *Z* value for untreated AD subjects, and blue bars represent the mean *Z* values for treated AD subjects. The significant differences observed between healthy subjects and AD patients for delta, theta, and alpha frequency ranges were not observed for the gamma frequency range upon application of a cognitive paradigm. The AD subjects had higher gamma response coherences when compared with healthy subjects upon application of "simple light" in a small number of electrode pairs (Fig. 13.15).

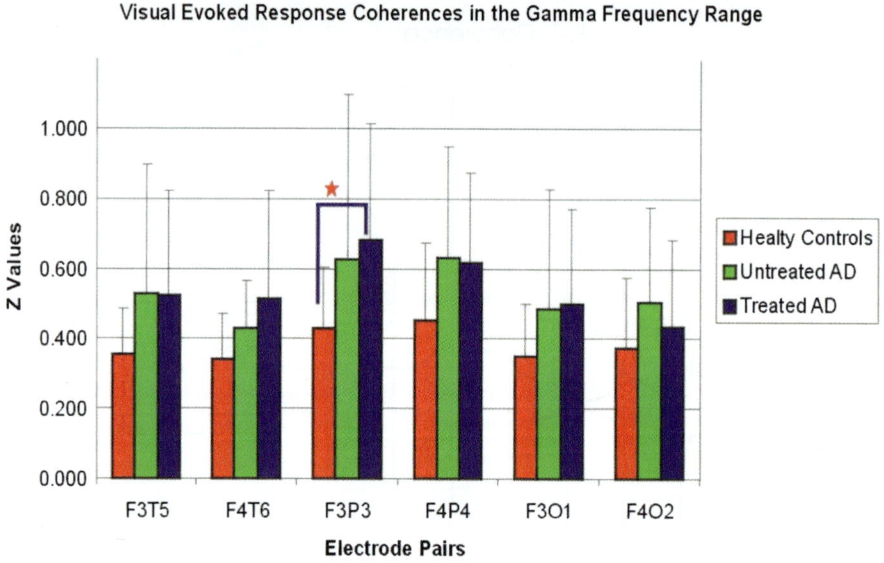

Fig. 13.15 Mean *Z* values of healthy control, treated AD, and untreated AD subjects for gamma frequency range on simple light stimuli. The *asterisk* represents $p < 0.01$

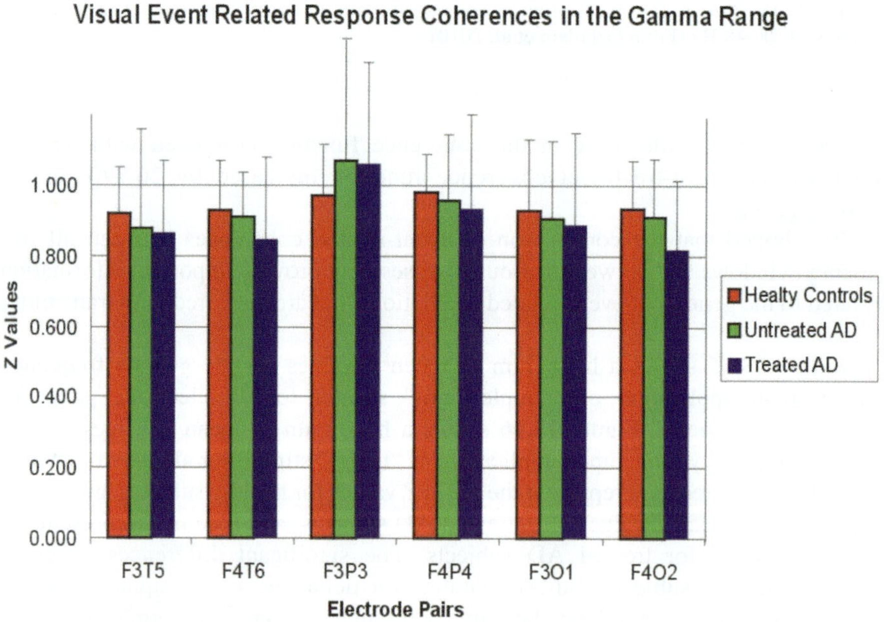

Fig. 13.16 Mean *Z* values of healthy control, treated AD, and untreated AD subjects for gamma frequency range on target stimuli

13.9 Pathologies with Special Analysis of Genetic Methods: Genetic Disorders

The research groups led by Begleiter, Porjesz, and Rangaswamy have utilized heritable neurophysiological features (i.e., brain oscillations) as endophenotypes, making it possible to identify particular genes as indicators of susceptibility, which may otherwise be difficult to detect with diagnosis alone. The review by Porjesz et al. (2005) attempted to differentiate neuroelectrical measures [electroencephalogram (EEG), event-related potentials (ERPs), and event-related oscillations (EROs)] related to acute and chronic effects of alcohol on the brain from those that reflect underlying deficits related to the predisposition to develop alcoholism and related disorders. These papers review evidence of genetic findings related to these electrophysiological measures and their relationship to clinical diagnosis. The present section is explained in detail in the work of Rangaswamy and Porjesz (2008). Many of these abnormal neuroelectric measures are under genetic control; may precede the development of alcoholism; and may be markers of a predisposition toward the development of a spectrum of disinhibitory conditions, including alcoholism. Genetic loci underlying some neuroelectric measures that involve the neurotransmitter systems of the brain have been identified. Porjesz and Rangaswamy (2007) found significant linkage and association with GABRA2 and inter-hemispheric theta coherence. They also reported significant linkage and linkage disequilibrium between the theta and delta event-related oscillations underlying P3 to target stimuli and CHRM2, a cholinergic muscarinic receptor gene on chromosome 7, which is also associated with diagnosis of alcohol dependence and related disorders. The authors' important conclusion is that quantitative neuroelectric measures (EEG, ERPs, EROs) provide valuable endophenotypes in studying the genetic risk of developing alcoholism and related disorders. It is also important to note that these delta and theta responses need to be compared with the results of Yener et al. (2008) for a more in-depth discussion of the cholinergic influences in AD patients.

In a visual oddball paradigm, alcoholics manifest significantly lower evoked theta and delta ERO amplitudes while processing the target stimuli (Jones et al. 2006). These findings are most significant anteriorly for theta, and posteriorly for delta. For the latter, it is useful to compare the results with those obtained by Özerdem et al. (2008).

EEG coherence has also been reported to be heritable, with estimates between 50 and 70% in twin populations (Stassen et al. 1988; Van Baal et al. 1998; Van Beijsterveldt and Boomsma 1994; Van Beijsterveldt et al. 1998). It has been demonstrated that the beta rhythm is generated while maintaining the balance in networks of excitatory pyramidal cells and inhibitory interneurons, involving $GABA_A$ action as the pacemaker (Whittington et al. 2000). Fast synaptic inhibition in the mammalian central nervous system is largely mediated by the activation of $GABA_A$ receptors (Tobler et al. 2001). $GABA_A$ actions are a fundamental requirement for both gamma (30–80 Hz) and beta oscillations to occur, and blockage of these receptors results in a loss of synchronization (Haenschel et al. 2000).

Some authors have suggested that mechanisms associated with thalamocortical cell populations may form the basis of theta (2–7 Hz) rhythms (Hughes and Crunelli 2005). Interestingly, Winterer et al. (2003b) reported significant association between the exon 7 variant of the $GABA_B$ receptor gene and EEG alpha voltage (classified as LVA or normal) for control subjects but not for alcoholic subjects.

The work of Porjesz et al. (2005) related to Alzheimer's and CHRM2 is important; therefore, a thorough review of this work is given in the following. These findings suggest the possible role of *CHRM2* in the generation and modulation of evoked oscillations. Theta and delta EROs depend on the level of acetylcholine (muscarinic activation). M2 receptors inhibit presynaptic release of acetylcholine, leading to inhibition of irrelevant networks. Muscarinic receptors are particularly concentrated in the forebrain and possibly serve to maintain the effective balance of relevant/irrelevant networks, and hence have a direct influence on P3 generation (Frodl-Bauch et al. 1999).

According to the work of the Porjesz group, the results with the *CHRM2* gene and brain oscillations strongly support the role of acetylcholine in the generation of N2 (theta oscillations) and in the P3 component (delta and theta oscillations). The function of acetylcholine has been demonstrated with regard to stimulus significance (Perry et al. 1999), selective attention (Mitrofanis and Guillery 1993), and P3 generation (Callaway 1983). Thus, genes are important for the expression of the endophenotype (brain oscillations) and help in the identification of genes that increase the propensity to develop alcohol dependence and related disorders (Begleiter and Porjesz 2006; Dick et al. 2006).

From the summary of the work of Begleiter and Porjesz and their research teams, it can be clearly stated that studies of *neuroelectric endophenotypes* offer a powerful strategy for identifying the genes that may be linked to susceptibility for developing psychiatric illness.

After reviewing the published literature, the integration of multiple levels of cognition, including aging and pathological cases, seems to be a necessary and useful strategy. According to the integration surveys of Lenz et al. (2008), Klimesch et al. (2008), and Porjesz and Rangaswamy (2007), it is tentatively suggested that an extended model should be used to approach the brain-body-mind problem.

13.10 Essential Changes in Brain Oscillation Lead to Essential Changes of the Mind in Pathological Brains

In the present chapter, the modification of brain oscillations in different pathologies with regard to various forms of medication were described in relation to schizophrenia, AD, and bipolar disorders; and also in those suffering from alcoholism. The tables comparatively summarize the results from different frequency windows.

13.10.1 Delta

Table 13.1 describes changes in delta response in different pathologies. In most cases, the oddball paradigm was used. Apart from bipolar euthymic patients, reduced delta amplitudes and decreased coherences were observed in all cases. Only in euthymic patients was an increase of the delta response observed before medication; then, following an application of valproate, the delta response was reduced.

13.10.2 Theta

Table 13.2 describes changes in the theta frequency window in schizophrenia, AD, and alcoholism. In all cases, the theta responses, theta synchrony, or theta evoked coherence were diminished.

13.10.3 Alpha

Table 13.3 compares AD, bipolar manic patients, and alcoholics in the alpha frequency range. In all cases, reductions in alpha responses or alpha coherences were observed.

13.10.4 Beta

Table 13.4 includes a comparison of beta responses in AD, bipolar manic, schizophrenics, and alcoholics reveal an interesting outcome. Although, in most cases, the beta coherence and beta response were reduced, both bipolar manic patients and alcoholic patients displayed increased beta response, and beta coherence was observed in bipolar manic and alcoholic groups. When these results are compared with the alpha frequency window (see Table 13.3), a similarity was observed between anxiety disorder among alcoholics and the bipolar manic patients. In both groups, the alpha responses were decreased, whereas beta responses were increased. This alpha–beta reversal in mania and alcoholism merits important consideration. Furthermore, it should be noted that the delta increase in euthymic patients is not in accord with the delta decrease in AD patients.

13.10.5 Gamma

Table 13.5 mostly presents results from schizophrenic patients, but also includes two reports on working memory demand and patients with hallucinations in whom

Table 13.1 Changes in delta response in different pathologies

	Pathology	Modality and paradigms	Methods	Results
Ford et al. (2008)	Schizophrenia	Auditory oddball	Evaluation of synchrony	Reduced delta synchrony in patients
Yener et al. (2008)	Alzheimer	Visual oddball	Amplitude of EROs	Reduced delta amplitudes in medicated and unmedicated patients
Güntekin et al. (2008)	Alzheimer	Visual oddball	Evoked coherence	Reduced delta evoked coherence in medicated (cholinesterase inhibitors) and unmedicated patients
Özerdem et al. (2007)	Bipolar euthymic	Visual oddball	Amplitude of EROs	Increased delta amplitudes in unmedicated patients. Reduced delta amplitudes after medication (valproate)
Jones et al. (2006)	Alcoholics	Visual oddball	Amplitude of EROs	Reduced delta amplitudes in alcoholics

Table 13.2 Changes in theta response in different pathologies

	Pathology	Modality and paradigms	Methods	Results
Schmiedt et al. (2005)	Schizophrenia	Cognitive and working memory demand	EROs amplitude	Reduced late theta response in all tasks
Ford et al. (2008)	Schizophrenia	Auditory oddball	Evaluation of synchrony	Reduced theta synchrony in patients
Yener et al. (2008)	Alzheimer	Visual oddball	Evaluation of synchrony	Reduced theta synchrony in unmedicated patients, medicated (cholinesterase inhibitors) patients have similar theta synchrony to that of healthy controls
Güntekin et al. (2008)	Alzheimer	Visual oddball	Evoked coherence	Reduced theta evoked coherence in medicated (cholinesterase inhibitors) and unmedicated patients
Zheng-yan (2005)	Alzheimer	Photic stimulation	Evoked coherence	Reduced theta evoked coherence
Jones et al. (2006)	Alcoholics	Visual oddball	Amplitude of EROs	Reduced theta amplitudes in alcoholics

Table 13.3 Changes in alpha response in different pathologies

	Pathology	Modality and paradigms	Methods	Results
Adler et al. (2003), Besthorn et al. (1994), Dunkin et al. (1994), Leuchter et al. (1987), Locatelli et al. (1998)	Alzheimer	Spontaneous EEG	Coherence	Reduced alpha coherence
Güntekin et al. (2008)	Alzheimer	Visual oddball	Evoked coherence	Reduced alpha evoked coherence in unmedicated patients; medicated [cholinesterase inhibitors (AChEI)] patients have similar evoked alpha response as that of healthy controls
Zheng-yan (2005)	Alzheimer	Photic Stimulation	Evoked coherence	Reduced alpha evoked coherence
Hogan et al. (2003)	Alzheimer	Memory paradigm	Evoked coherence	Reduced alpha evoked coherence in patients
Özerdem et al. (2008)	Bipolar Manic	Visual oddball	Amplitude of EROs	Lower alpha response in bipolar manic patients
Winterer et al. (2003a)	Alcoholics	Spontaneous EEG	Coherence	Increased alpha coherence
Ba ar-Ero lu et al. (2008)	Schizophrenia	Visual oddball	Evoked power and phase locking	Patients had better phase locking at frontal and central locations
Enoch et al. (1999)	Alcoholism related to anxiety disorder	Spontaneous EEG	Alpha EEG power	Low-voltage alpha EEG variant

Table 13.4 Changes in beta response in different pathologies

	Pathology	Modality and Paradigms	Methods	Results
Besthorn et al. (1994)	Alzheimer	Spontaneous EEG	Coherence	Reduced beta coherence
Zheng-yan (2005)	Alzheimer	Photic stimulation	Evoked coherence	Reduced beta evoked coherence
Özerdem et al. (2008)	Bipolar manic	Visual oddball	Amplitude of EROs	Higher beta response in bipolar manic patients; reduced beta amplitudes after medication (valproate)
Yeragani et al. (2006)	Schizophrenia	Spontaneous EEG	Coherence	Decreased beta coherence
Winterer et al. (2003a)	Alcoholics	Spontaneous EEG	Coherence	Increased beta coherence

the gamma amplitudes, phase synchrony, and coherences were reduced. It is also important to note that the two time-windows (early and late) have to be considered separately, because important differences were observed among them.

It is difficult and too early to draw significant conclusions regarding oscillatory behavior in different pathologies; the picture is too complex to define exact cognitive correlates in pathologies; the effect of medication and, accordingly, the effects upon the neurotransmitters. However, the important conclusion from these five tables is the immense differentiation related to the cognitive components detected in pathological cases by means of oscillatory dynamics. Because this chapter intends to briefly describe oscillations in pathology and transmitters, only a limited synopsis is provided in chapter 22 related to oscillations and neurotransmitters as quasi-invariants.

Table 13.5 Changes in gamma response in different pathologies

	Pathology	Modality and paradigms	Methods	Results
Kwon et al. (1999)	Schizophrenia	Auditory stimulation	Evoked EEG power	Schizophrenic patient[illegible] reduced, averaged [illegible] 40-Hz auditory sti[illegible]
Haig et al. (2000)	Schizophrenia	Auditory oddball	Amplitude of EROs	For targets: reduced [illegible] hemisphere and fr[illegible] gamma response i[illegible] parieto-occipital si[illegible] reduced gamma res[illegible]
Spencer et al. (2003)	Schizophrenia	Gestalt stimuli	Phase locking and phase coherence	Absence of the posterio[illegible] early visual gamma [illegible] hemispheric coheren[illegible] in patients
Baldeweg et al. (1998), Spencer et al. (2004)	Schizophrenia	Spontaneous EEG Gestalt stimuli	Spectral analysis Wavelet transform	Negative symptoms correlate with reduced gamma responses, whereas a significant increase in gamma amplitudes is observed during positive symptoms such as hallucinations
Gallinat et al. (2004)	Schizophrenia	Auditory oddball	Wavelet transform	In response to standard stimuli, early evoked gamma-band responses (20–100 ms) did not show significant group differences. Schizophrenic patients showed reduced evoked gamma-band responses in a late latency range (220–350 ms), particularly after target stimuli.
Wynn et al. (2005)	Schizophrenia	Backward masking task	Evoked EEG power	Patients showed less overall gamma activity
Symond et al. (2005)	Schizophrenia	Auditory oddball	Synchrony	Schizophrenic patients showed a decreased magnitude and delayed latency for global gamma 1 (0–150 ms) synchrony in relation to healthy comparison subjects. By contrast, there were no group differences in gamma 2 (200–550 ms) synchrony

Yeragani et al. (2006)	Schizophrenia	Spontaneous EEG	Coherence	Decreased gamma coherence
Light et al. (2006)	Schizophrenia	Auditory steady-state event-related potential	Evoked power and phase synchronization	Reduced evoked power and phase synchronization in response to 30–40 Hz stimulation
Spencer et al. (2007)	Schizophrenia	Auditory oddball Visual oddball	Phase locking	Visual-evoked gamma oscillation phase locking at occipital electrodes was reduced in chronic schizophrenic patients compared with healthy control subjects. In contrast, auditory-evoked gamma phase locking and evoked power did not differ between groups
Bucci et al. (2007)	Schizophrenia		Gamma power event-related coherence	Induced gamma power and event-related coherence was observed in patients with non-deficit schizophrenia, but not in those with deficit schizophrenia
Ba ar-Ero lu et al. (2007)	Schizophrenia	Cognitive and working memory demand	EROs amplitude	High amplitude gamma oscillations remained constant in patients, regardless of task difficulty
Yordanova et al. (2001)	ADHD	Auditory task	Phase locking	ADHD children produced larger and more strongly phase-locked gamma band response than controls
Lenz et al. (2008)	ADHD	Visual memory paradigm	Evoked gamma-band responses	Analysis of evoked gamma-band responses during stimulus encoding revealed a strong task-related enhancement for ADHD patients
Padmanabhapillai et al. (2006)	Alcoholics	Visual oddball task	Evoked gamma-band responses	Abstinent alcoholics manifest significantly less early (1–150 ms) gamma band response (28–45 Hz) in the frontal region during target processing

Part IV
The Physics-Biology Interface: A Cartesian System for the Twenty-First Century

Prelude to Part IV

Why is a new unified framework needed to explain the brain-body-mind relationship?

The word *mind* is the abstract and invisible notion of physicists and even biological scientists. Currently, physicists and biologists are approaching the boundaries of the processes related to mind-brain incorporation. They study cognitive processes such as attention and remembering; they measure the electrophysiological activity at the absolute threshold of perception; they can predict dream onset; they can differentiate the electrical responses of the brain on known and unknown faces. Accumulating knowledge, progress and refinements in science and technology have helped to make processes that were within the realm of philosophy (for physicists, within the realm of metaphysics) observable, measurable, and testable. Mind and cognition have been studied by the positivistic sciences for over a century, first in psychology and later in the physical/biological sciences.

The coordinate system and the analytical geometry created by René Descartes contributed enormously to the development of natural science; the concept and research based on the Cartesian system governed the positive sciences until the beginning of the twentieth century. Physicists modified their working frame by introducing Einstein's moving coordinate system and Heisenberg's uncertainty framework. In the second half of the twentieth century, brilliant proposals were offered by Norbert Wiener (1948), René Thom (1975), I. Prigogine (1980), Herman Haken (1977), and, last but not least, the nonlinear mathematics and chaos theory introduced by, for example, Abraham and Shaw (1983), Grassberger and Procaccia (1983), and Lorenz (1963). However, despite the usefulness of these new concepts, none of the contemporary multidisciplinary approaches could attain the glory of the Descartes Cartesian system. What are the reasons for this great deficit? In the twentieth century, most of the scientists who presented theories to unify the sciences were mathematicians, physicists, and theoretical scientists; thus, they did not have direct empirical experience in the realm of biology. Although N. Wiener conducted experiments with frogs, he could not thoroughly study the outcome of the biological processes. Another important criticism of the unifying trends is the fact that none of these scientists made a major step to incorporate the concepts of other contemporary pioneers to achieve a more comprehensive theory.

Besides Cybernetics, Dissipative Structures, Catastrophe Theory, and Synergetics, in the world of theoretical physics a highly influential trend is based on string theory. Although this theory is very popular among physicists trying to unify the existing theories, brain scientists have not yet considered this theory as an important metaphor for oscillatory brain dynamics and cognitive processes. This trend is mentioned briefly in Part IV and explained in more detail in Chap. 24.

Does the Cartesian System Need a Major Extension for Studies in Biology?

According to Capra (1984), the Cartesian model needs a major revolution. In his words:

> Transcending the Cartesian model will amount to a major revolution in medical science, and since current medical research is closely linked to research in biology – both conceptually and in its organization – such a revolution is bound to have a strong impact on further development of biology. To see where this development may lead, it is useful to review the evolution of the Cartesian model in the history of biology. Such a historical perspective also shows that the association of biology with medicine is not something new but goes back to ancient times and has been an important factor throughout its history.

What could a major revolution in the description of brain-body-mind integration imply? By taking advantage of the accumulated data on oscillatory brain dynamics, dynamics of the circulatory system, and the dynamics of the overall myogenic systems, a major change is proposed in the classical Cartesian system considering the increasing needs in medicine and biology. Can this new Cartesian system, named the *nebulous Cartesian system,* have a reasonable chance of being more efficient than other, earlier proposals? In the time of the evolution of brain dynamics, concepts and methods were applied which were derived from the disciplines of cybernetics, quantum dynamics, chaotic dynamics, catastrophe theory; and the author was involved in a large number of experiments in conventional physiology and brain research. All relevant concepts of these important trends were used in trying to grasp their relevant parts (Başar 1980, 1998, 1999). The theory of oscillatory brain dynamics is now completely recognized by a great number of neuroscientists; and this field is growing enormously, as this book points out.

Chapter 14
Chaos and Quantum Approach: Gateway to a Twenty-First Century Cartesian System

14.1 Introduction

Three fundamental steps were undertaken in the early chapters of this book: (1) a brief description of the fundamental work of scientists and philosophers, who opened the way to modern science based on philosophical fundaments; (2) a description of how some common principles and rules, method of thoughts, and "invariants" in various branches of science have been established. (An explanation is also given that the establishing of fundaments, general principles, and new frameworks opened the way to major discoveries in the mathematical and technical sciences.); and (3) a global description is given of the machinery of the human body and the brain, and chapters are introduced related to brain–body function in an integrative way, which is seldom encountered in the literature of neuroscience and vegetative physiology.

One of the aims of this book is also to partly make clear the factors that make the human mind distinctive. The human being is unique, because human beings do not behave like automatons, as do low level animals that are guided only by instincts and or phyletic memory (see also Chap. 17). Therefore, it is not difficult to foresee that the human being has its own type of causalities and quasi-deterministic behavior. Accordingly the present chapter provides some essential features of modern concepts in physics as a prerequisite for opening the path to an ambitious step: the proposition of a Cartesian system in hyperspace. The brain's quasi-deterministic behavior has conceptual similarities to chaotic systems and also to quantum dynamics. However, this is biology and is different from the machineries of physical systems that are incorporated into chaotic and quantum deterministic mechanisms.

By transferring the concepts of the chaos and the quantum approach to brain/body it cannot be said that the brain's behavior is completely guided and enslaved by the rules of physics. On the contrary, brain dynamics offer a type of "strangeness" not observed in physical systems. In 1980 EEG brain dynamics incorporated special features, such as particular essentials of quantum dynamics. However, the very expression *brain dynamics* denoted an amalgamation of several processes common in the physical world. Now, 25 years later, the situation is that all predicted steps in brain function have gained in importance, although pronounced similarities

E. Başar, *Brain-Body-Mind in the Nebulous Cartesian System: A Holistic Approach by Oscillations*, DOI 10.1007/978-1-4419-6136-5_14,

are observed in physical systems. Brain dynamics is different from the dynamics observed in physical, chemical, and all technical systems. With the description introduced in the present chapter these approaches are considered to be examples or metaphors; but the dynamics of brain–body–mind incorporation are new, and possibly superior to all other dynamic systems.

14.2 Some Definitions Related to Brain Dynamics

Brain dynamics has features of physical systems dynamics. It has philosophical parallels to quantum dynamics and chaotic system dynamics. It is impossible to isolate the brain from the mathematical frames of physical systems existing in nature. The mathematical tools used in computer algorithms do belong also to the arsenal of mathematicians and physicists; however, a number of observations throughout this book lead us to formulate brain dynamics as a dominant machinery of nature, which bridges Newton dynamics, quantum dynamics, and chaotic dynamics. Furthermore, it is tentatively assumed that brain dynamics is a more comprehensive system of nature superior to Newtonian, quantum theoretical, and chaotic dynamics. Some principles of brain dynamics follow:

1. *Inertia*: The brain keeps its spontaneous behavior reflected in spontaneous electrical oscillations if no external excitations are applied to the brain directly or to its peripheral organs.
   (a) Excitations to the brain can be endogenous auto-excitations from the brain–body system, including the spinal cord and the vegetative system.
   (b) Excitations can come from external stimuli, such as sensory stimulations (light, sound, heat, touch).
   (c) The brain steadily observes the changes caused by internal and external excitations.
2. In the *action and reaction* to every excitation, the brain shows probabilistic reactions as a result of multiple causalities in the brain–body system.
3. *The entropy* of the shape and electrical manifestations is increased during evolution and maturation of the brain. This is what makes the brain and brain dynamics different from the dynamics of physical systems, including Newtonian and quantum dynamics.

14.3 Chaos in Brain Function

14.3.1 Deterministic Chaos

In the present context chaos can be understood to refer to irregular fluctuation, which is described by deterministic equations, as distinct from the indeterminate fluctuation that obeys the definitions of randomness (Fig. 14.1). The problem of

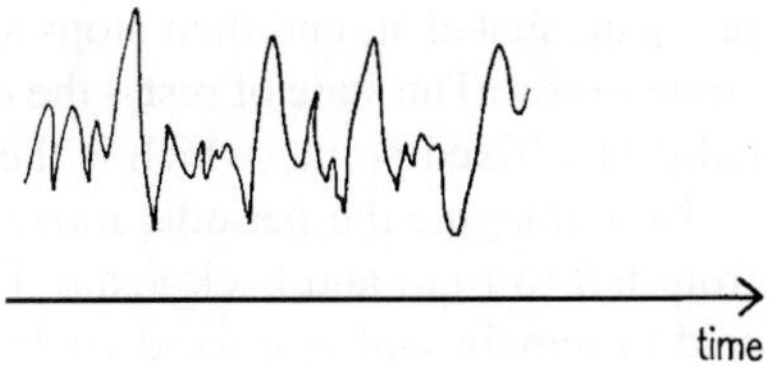

Fig. 14.1 Irregular motion (Başar 1990)

nonlinear dynamics originates in planetary motions. Henri Poincaré was the first to investigate the complex behavior of simple mathematical systems. He analyzed topological structures in phase space and discovered that the equation for the motion of planets could display an irregular or chaotic motion (Poincaré 1892). A mathematical basis for this behavior was later given by Birkhoff (1932).

Lorenz (1963), in a model of boundary layer convection, discovered that a system of three first-order nonlinear differential equations can exhibit chaotic behavior. Contrary to Poincaré's example, Lorenz discovered deterministic chaos in dissipative systems. The analysis of deterministic chaos is currently an active field in many branches of research. Mathematically, all nonlinear dynamic systems with more than two degrees of freedom can generate chaos, becoming unpredictable over a longer time scale.

In a popular book on chaos, Gleick (1987) the advocate of the new science goes as far as to say that "twentieth-century science will be remembered for just three things: 'relativity,' 'quantum mechanisms,' and 'chaos'." In other words, chaos has become the century's third great revolution in the physical sciences. Like the first two revolutions, chaos cuts away at the tenets of Newtonian physics.

The brain is a *nonlinear system* par excellence. Accordingly, in the last 25 years, the concepts of chaotic dynamics have found an important application in research on the compound electrical activity of the brain. This chapter covers most of the relevant concepts and results in the field of chaotic attractors in the brain.

14.3.1.1 Chaos in Everyday Experience

A simple example of chaos in nature is described by Hooper (1983):

> Suppose you are sitting beside a waterfall watching a cascade of white water flow regularly over jagged rocks, when suddenly a jet of cold water splashes you in the face. The rocks have not moved, nothing has disrupted the water, and presumably no evil sprites inhibit the waterfall. So why does the water suddenly "decide" to splash you?

Physicists studying fluid turbulence have wondered about this kind of event for several hundred years, and only recently have they arrived at conclusions that seem to solve the problem, at least in part. The waterfall's sudden random splashes do not come from some "imperceptible jiggle," but from the inner dynamics of the system itself. Behind the chaotic flow of turbulent fluids of the shifting cloud formations that shape the weather lies an abstract descriptor that physicists call a "strange attractor." What is an "attractor" and what makes it "strange"? It is possible to describe this, again by using Hooper's explanation (1983): "Suppose one puts water in a pan, shakes it, and then stops shaking it; after a time it will stop whirling and come to rest. The state of rest – the equilibrium state – can be described mathematically as a 'fixed point,' which is the simplest kind of attractor."

Now imagine the periodic movement of a metronome or a pendulum swinging from left to right and back again. From the viewpoint of geometry, this motion is said to remain within a fixed cycle forever. This is a second kind of attractor, the limit cycle. All the various types of *limit cycles* share one important characteristic: *regular, predictable motion*. The third variety, the *strange attractor*, is *irregular, unpredictable, or simply strange*. For example, when a heated or moving fluid moves from a smooth or laminar flow to wild turbulence, it becomes a strange attractor.

Chaotic behavior in deterministic systems usually occurs through a transition from an orderly state when an external parameter is changed. In studies of this system, particular attention has been devoted to the question of the route by which the chaotic state is approached. An increasing body of experimental evidence supports the belief that apparently random behavior observed in a wide variety of physical systems is caused by the underlying deterministic dynamics of a low-dimensional chaotic (strange) attractor. The behavior exhibited by a chaotic attractor is predictable in short time scales and unpredictable (random) on long time scales.

14.3.2 The EEG Has Strange Attractors: The EEG Is Not Noise

In the years following the first measurement of human EEG by Hans Berger and important developments by Lord Adrian and later by Grey Walter, pure EEG research remained somewhat in the shadow of new discoveries based on single-neuron recordings. From the beginning of the 1960s the use of signal averagers enabled EEG research scientists to extract the evoked potentials from the so-called random-noise EEG. In this context, the event-related potentials that contributed greatly to the understanding of cognitive functions and clinical diagnostics were considered deterministic signals, whereas EEG was considered mostly to be pure noise.

In 1980 Başar stated strongly that EEG should not always be considered as a background noise. On the contrary, in the mentioned approach on field potentials, it was assumed that the EEG must be considered as one of the most important oscillations affecting the production and conduction of signaling in the brain (Başar 1980). The view was based on experiments with compound potentials from cat and human brains. In this approach some EEG fragments were considered as internal evoked potentials coming from yet unknown or hidden sources. Further, evoked potentials were considered as forced (or evoked) oscillations following known and deterministic inputs. Later, it was tentatively assumed that the EEG has a strange attractor (Başar 1983a,b; Babloyantz 1986).

The new trend in brain research could be initiated in an appropriate manner by the evaluation of the correlation dimension D_2 to the brain's EEG during slow wave sleep (SWS) by Babloyantz et al. (1985) and then by application of the same algorithm to pathological cases. Following the most important pioneering work by Babloyantz and co-workers, Röschke and Başar (1985) published results on the strange attractors in several intracranial structures of a cat brain generally and during SWS, confirming the results of Babloyanz et al. (1985). Further, Layne et al. (1986) and Rapp et al. (1985a,b) interpreted the waking EEG as chaotic behavior.

The statement "The EEG is not noise, but a quasi-deterministic signal" results from a need to interpret experiments by many neurophysiologists. The new development in the research of chaos in brain function is fascinating. However, this field cannot be considered an isolated research field. In the period 1985–1989, during which chaotic EEG results were substantially developed, other noteworthy progress in the study of oscillatory phenomena and neural network resonance at the cellular level was also achieved.

14.3.2.1 Preliminary Remarks on the Nonlinear Approach to EEG and Brain Function

The conceptual renaissance implicit in the slogan "EEG is not simple noise" is already a gigantic step or watershed. At this point it seems useful to mention a paradoxical aspect of chaotic neuroscience: Neuroscientists have not yet discovered how to apply the Newtonian approach to physics or quantum statistics to brain science. This difficulty has been mentioned by Rosen (1969), among many other authors. The usefulness of predictability and the mathematical description of brain behavior has not been in the mainstream of brain research, with a few exceptions. Therefore, the "big bang" of applying chaotic dynamics to brain activity has struck brain scientists at a still unprepared and inappropriate stage.

In the new branch of nonlinear dynamics, new concepts, new expressions such as attractors, and new methods such as correlation dimension are used. Some of these newly introduced expressions and methods are included in Appendix C to assist the reader to understand these steps.

14.3.3 Typical Examples of Chaotic EEG Behavior

14.3.3.1 Results During Slow-Wave Sleep: Cat Cortex and Hippocampus

To analyze the dimensionality of field potentials during slow-wave sleep (SWS), cats with chronically implanted electrodes were studied. The chronic electrodes were needles of small exposed tips, implanted cortically (subdural) in the auditory cortex (GEA), and subcortically in the hippocampus (HI), and reticular formation

(RF). The 15 experimental trials were recorded during SWS. The intracranial EEG signals were low-passed to 50 Hz and then digitized using a 12-bit AD converter and stored in the memory of a computer. The sampling frequency was $f_s = 100$ Hz for all trials. Dimensions of the EEG signals were evaluated over a time period of about 20 s (n=2,048) and 40 s (n = 4,096). Details of the software are described elsewhere (Röschke 1986; Röschke and Başar 1988).

The computation of the correlation dimension for the auditory cortex is shown in Fig. 14.2, which represents an evaluation of the activity from the cat named Toni in SWS. One can detect that the slopes of the curves by plotting $N(r)$ vs. log r converge against a saturation value of about $D_2 = 5.00$. The time lag τ was $\tau = 50$ ms. Identical results are obtained, if a time lag between 20 and 80 ms is chosen.

The power spectrum of the EEG is illustrated to underline the advantage of this procedure (Fig. 14.3). This spectrum resembles that of colored noise. On the contrary, the case of the phase-space description and evaluation of the correlation dimension lead to a convergence. Such saturation is not reached in the case of a noisy signal. This is the difference between noise and strange attractors.

The mean values from 15 trials (five cats), each of 40 s duration and measured in the auditory cortex, reticular formation, and hippocampus during SWS, are shown in Table 14.1. The standard deviations lie within the same range and are, in all cases, smaller than 10% of the mean. Significant differences between cortical and subcortical structures of the cat brain are demonstrated. Detailed information on the results from five cats has been presented by Röschke and Başar (1988).

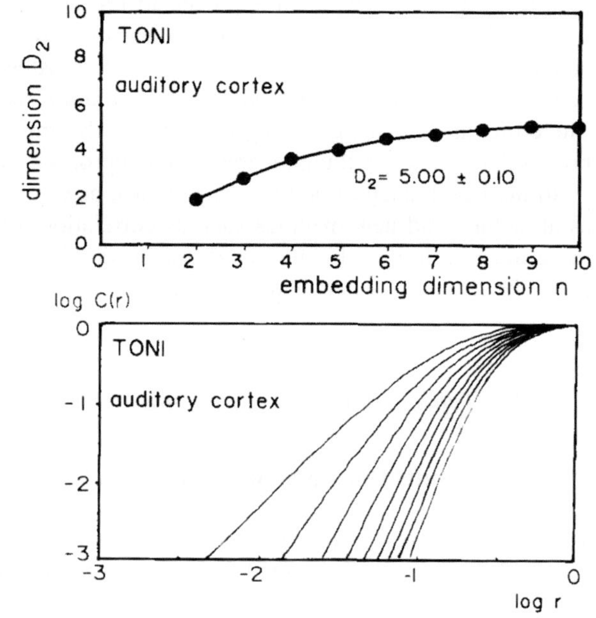

Fig. 14.2 The plot of log $C(r)$ vs. Log r (*at the bottom*) leads to a convergence of the slopes of the curves with a saturation value of $D_2 = 5.00$ (*at the top*); 4,096 data points, t = 50 ms, cat Toni, SWS, auditory cortex (from Röschke and Başar 1989)

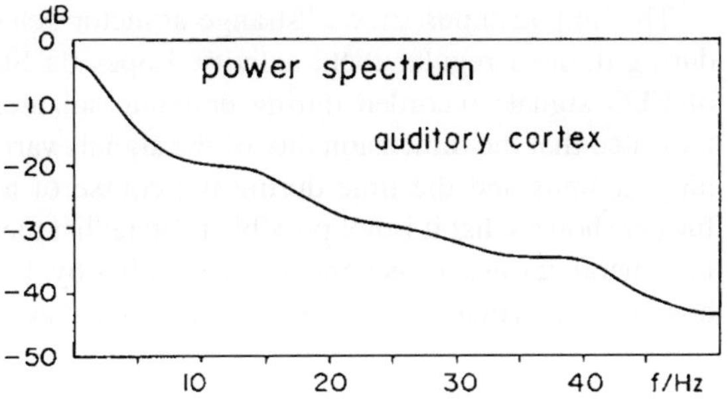

Fig. 14.3 Power spectrum of EEG epochs analyzed over a period of 40 s (analyzed in Fig.9.14). The spectrum was smoothed by means of a previously published linear prediction method (Rösche 1986; Röschke and Başar 1989)

Table 14.1 Mean values of 15 trials (five cats) from auditory cortex, reticular formation, and hippocampus

$D_{GEA} = 5.06 \pm 0.31$
$D_{RF} = 4.58 \pm 0.38$
$D_{HI} = 4.37 \pm 0.36$
From Röschke and Başar (1989)

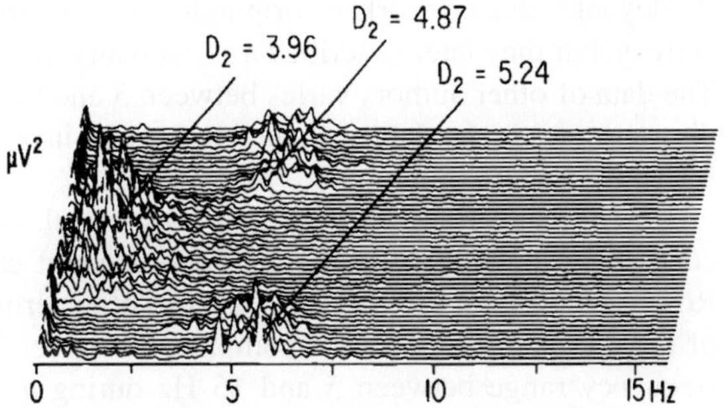

Fig. 14.4 Power spectra (5-s epochs) of the hippocampus (cat Jenny) and correlation dimension D_2 (for 40-s epochs) during theta activity (waking stage) and slow-wave sleep) (from Röschke and Başar 1989)

It should be emphasized that the dimension D_2 of the investigated system, in this case, the brain, varies from 4.37 in the hippocampus to nearly 5.00 in the cortex and that the dimensions of the different attractors are also relatively stable in experiments with individual cats. Moreover, in nearly 90% of the trials the maximal dimension is detected in the auditory cortex.

14.3.3.2 Hippocampus Theta Activity: Transitions

Figure 14.4 shows the power spectra of the cat hippocampus with a dominant theta activity during the waking stage. The results indicate that if the cat brain state has a transition from waking stage to SWS stage, the maxima of power falls from about 5 Hz to between 1 and 3 Hz with several minor peaks. In the waking stage, the cat again showed a marked theta activity, and in this stage, D_2 was relatively high ($D_2 = 5.24$). During the SWS stage, the dimension was reduced with a value of $D_2 = 3.96$. Following the SWS stage D_2 again reached higher values of around 5.

The hippocampus shows "strange attractor behavior" and not "pure noise" only during defined regular theta activity. Lopes da Silva et al. (1990) studied the D_2 of EEG signals recorded during epileptic seizures in the rat brain. Their results indicated that the dimensionality of the signals varies as a function of the part of the hippocampus and the time during the course of a kindled epileptic seizure. They further showed that it is not possible to state that the EEG signals can be represented, in general terms, as generated necessarily by low-dimensional chaotic systems; however, this could be the case during given states of epileptiform discharges.

14.3.3.3 Correlation Dimension of Alpha Activity: Brain Alpha Attractor

Since the discovery of the alpha rhythms by Hans Berger, one of the greatest puzzles in electroencephalography has been the physiological understanding of the origin of 10 Hz activities, their relation to sensory and cognitive functions of the brain, and not least, their interactions as an indicator of the brain state. In several studies, the correlation dimension was used to investigate the alpha rhythm. Babloyantz and co-workers originally did not find any saturation during alpha activity, but they later described a D_2 as being approximately 6 (Babloyantz 1988). The data of other authors varies between 3 and 8. It is not surprising that there are discrepancies in the data evaluated by several investigators (see later in this chapter for an overview).

Röschke and Başar (1989), and Başar et al. (1989b, 1990) took another step concerning alpha activity. To take into account changes in the brain alpha state, Röschke and Başar (1990) began a series of experiments in which the computations of the correlation dimension comprised long sessions. By filtering the EEG in a frequency range between 5 and 15 Hz during a state of high alpha activity and performing measurements on 3-min segments from 30-min records for a number of subjects, they found that, under such measuring conditions, the D_2 of the alpha activity fluctuated between 5 and 8 (Table 14.2), even though immobility seemed quite good over 30 min.

The simultaneously evaluated power spectra (Figs. 14.5 and 14.6) also differ. There is usually no saturation if the digital filters are not applied. On the other hand, in the measurements of the same subject there are time periods during which there is no saturation in any of the seven electrode sites. In some subjects fractal dimensions between 5 and 8 are usually seen, but in some locations (e.g., frontal) almost never is a 3-min segment with good saturation observed. Table 14.3 shows the large differences between the D_2 of occipital and frontal recordings.

Figure 14.5 shows the power spectra that were evaluated from four different locations simultaneously (vertex, parietal, occipital, and frontal) during the waking stage of a subject with eyes closed. During the analysis of such compressed arrays of power spectra, the following observation typically was made: In the central electrode (vertex), the power was usually centered between 7 and 10 Hz with large peaks in frequencies <10 Hz. In occipital locations, the subjects usually showed high amplitude alpha activity centered at 10–12 Hz. In frontal electrodes the 10 Hz component usually had lower amplitudes; high amplitude activities were mostly

Table 14.2 Changes in D_2 over 30 min: values of one subject on successive 3-min samples from a 30-min EEG during good alpha, after filtering the EEG through a 5- to 15-Hz digital bandpass filter (no phase shift)

Location	D_2 in 10 successive 3-min samples										Mean
(a) O_1	7.0	7.0	–	7.5	7.8	5.5	6.5	6.0	6.5	6.0	6.6±0.7
O_2	6.2	6.6	–	6.5	7.1	6.1	6.0	6.2	7.5	7.1	6.5±0.5
P_3	6.2	6.6	–	6.5	6.6	5.8	7.5	5.9	6.0	–	6.3±0.5
P_4	7.0	6.2	–	6.2	7.2	5.8	6.9	6.3	6.0	6.5	6.4±0.4
C_z	7.5	6.9	–	6.6	7.1	6.9	–	6.0	5.9	6.2	6.6±0.5
F_3	6.6	6.6	–	6.1	7.1	5.0	–	5.5	6.0	–	6.1±0.7
F_4	6.6	–	–	7.5	7.1	6.9	–	6.6	6.6	6.6	6.8±0.3
(b) O_1	6.9	6.8	6.6	6.7	6.5	5.9	6.4	6.0	6.5	6.5	6.4±0.3
O_2	7.1	6.9	6.8	6.5	5.9	6.1	6.2	6.1	6.3	6.6	6.4±0.4
P_3	7.0	6.9	–	6.5	6.3	6.2	6.2	6.2	6.5	6.7	6.5±0.3
P_4	7.5	–	–	6.5	6.0	6.5	7.1	6.2	6.7	–	6.6±0.5
C_z	7.3	6.9	–	6.6	6.5	6.6	6.4	7.0	7.3	6.4	6.7±0.3
F_3	6.1	6.1	6.2	5.9	6.0	5.9	5.9	6.0	6.2	6.1	6.0±0.1
F_4	–	6.9	–	–	–	6.9	–	–	–	–	6.9

–, No saturation.

From Başar et al. (1989b)

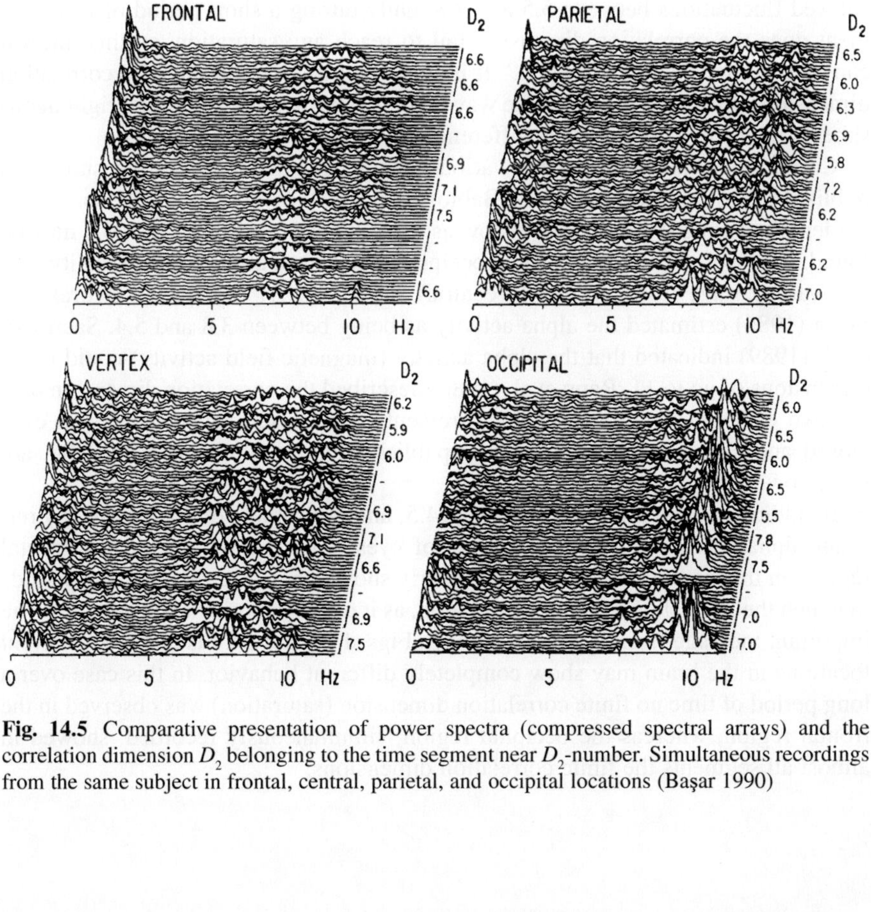

Fig. 14.5 Comparative presentation of power spectra (compressed spectral arrays) and the correlation dimension D_2 belonging to the time segment near D_2-number. Simultaneous recordings from the same subject in frontal, central, parietal, and occipital locations (Başar 1990)

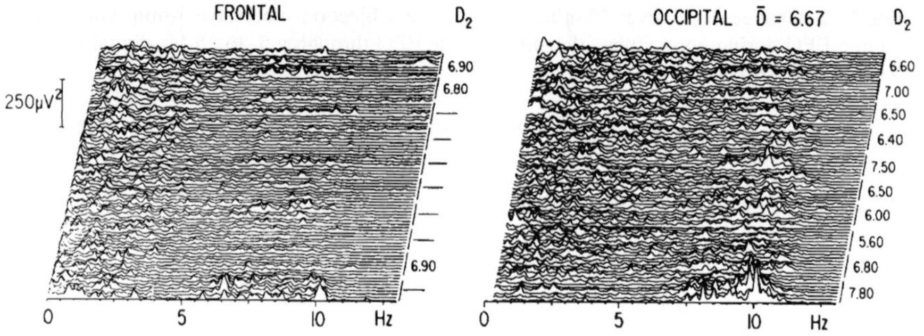

Fig. 14.6 Comparative presentation of power spectra and D_2 for frontal and occipital location of another subject (Başar 1990)

centered to lower frequencies, including the theta band. Before computing the correlation dimension of the EEG, the data were digitally filtered in the frequency range between 5 and 15 Hz.

In Fig. 14.6 the correlation dimension D_2 is also given during corresponding time segments. For evaluation of D_2, 3-min EEG segments (number of points $n = 16{,}384$) were used. The sampling frequency was $f_s = 100$ Hz and the frequency resolution was $D_f = 0.006$ Hz. The correlation dimension D_2 of the occipital region showed fluctuations between 5.5 and 7.8. Only during a short period of measurement does the correlation dimension fail to reach any saturation, so this interval can be distinguished from "noise." It is important to note that: (1) the correlation dimensions do not vary in the same way in all locations, and (2) left and right hemispheres may show considerable differences.

Correlation dimensions during various stages have already been measured and published by several investigators. Babloyantz (1989) mentioned that according to the length of time series, values as low as $D_2 = 2.6$ and as high as $D_2 = 6.6$ may be found. Layne et al. (1986) estimated occipital and central alpha dimensionality during waking stage from 5.5 to 6.6 (central) and 6.5 to 7.7 (occipital). Dvorak and Siska (1986) estimated the alpha activity as being between 3.8 and 5.4. Saermark et al. (1989) indicated that the alpha activity (magnetic field activity) could reach dimensions of up to 11. Rapp et al. (1986) described the correlation dimension during two different conditions of measurement, eyes closed and relaxed, and eyes closed and chaotic; furthermore, they published correlation dimensions that are much lower, ranging from 2.4 to 3.

In addition to the illustration in Fig. 14.5, another subject did not show any relevant alpha activity during the periods of eyes closed, especially in the frontal region. In this case, as the illustration clearly shows, there are seldom time periods in which the EEG shows a finite correlation, as it usually shows noisy behavior. The important message from the illustrations in Figs. 14.5 and 14.6 is that two different locations in the brain may show completely different behavior. In this case over a long period of time no finite correlation dimension (saturation) was observed in the frontal region, whereas the occipital region, simultaneously recorded, showed in almost all segments the finite correlation dimensions.

Table 14.3 Human EEG/MEG data

Reference	Parameters	Results
Babloyantz et al. (1985)	$\Delta t = 10$ ms $N = 4000$ $\tau = 20$ ms EEG	Sleep stage 2: $D_2 = 5.03$ Sleep stage 4: $D_2 = 4.0$–4.4 Awake, alpha activity: $D_2 = 6.1$ Beta: D_2 – no saturation
Rapp et al. (1986)	$\Delta t = 2$ ms $N = 1000$–4000 $\tau = 10$–20 ms EEG	Eyes closed, relaxed: $D_2 = 2.4$ (N = 1000) $D_2 = 2.6$ (N = 4000) Eyes closed, counting: $D_2 = 3.0$ (N = 4000)
Layne et al. (1986)	$\Delta t = 2$ ms $N = 1000$–15000 $\tau = 20$ ms (occipital) $\tau = 40$ ms (vertex) EEG	Awake, occipital: $D_2 = 5.5$–6.6 Awake, vertex: $D_2 = 6.5$–7.7
Başar et al. (1989b; 1990; this volume)	$\Delta t = 30$ ms $N = 16384$ points (segments of 3 min) sampling frequency: $\Delta f = 100$ Hz EEG	Eyes closed, occipital/vertex/parietal/frontal: $D_2 = 5.5$–8 (Finite dimension only when data prefiltered between 5 and 15 Hz)
Dvorak and Siska (1986)	$\Delta t = 5$ ms $N = 1000$–12000 $\tau = 40$ ms EEG	Eyes closed: $D_2 = 3.8$–5.4 (N = 1000) $D_2 = 8$–10 (N = 12000)
Van Erp et al. (1988)	$\Delta t = 5$–10 ms $N = 1000$–10000 $\tau = 15$–75 ms EEG	Alpha rhythm: $D_2 = 5$–6 (N = 1000) $D_2 = 7$–8 (N = 10000) Beta rhythm: D_2 – no saturation
Babloyantz et al. (1986, this volume)	$\Delta t = 0.83$ ms $N = 6000$ $\tau = 16$–60 ms EEG	Creutzfeldt-Jakob disease: $D_2 = 3.7$–5.4 Epileptic attack: $D_2 = 2.05$
Saermark et al. (1989; personal communication)	$\Delta t = 10$ ms $N = 4000$–8000 $\tau = 100$ ms MEG	Healthy subject: $D_2 = 11$ Epilepsy (2 patients): $D_2 = 7$ Epilepsy (2 patients): D_2 – no saturation

D_2, Correlation dimension; *N*, number of data points; τ, time shift; Δt, sampling time; SWS, slow-wave sleep stage; REM, rapid-eye movement sleep; MEG, Magnetoencephalography.

D_2, Correlation dimension; Dt, sampling time; MEG, magneto-encephalography; *N*, number of data points; REM, rapid-eye movement sleep; SWS, slow-wave sleep stage; *t*, time shift. From Başar (1990)

These results help somewhat to account for the discrepancies in the results of several authors, not by explaining, but by confirming variance. There are large fluctuations in the dimensionality of alpha waves. This means that, in this frequency range, the brain has two types of behavior: *noisy behavior* and *strange attractor behavior*.

Also note the important results from Freeman and Skarda (1985) with repeatable 40 Hz EEG patterns from brains of rabbits expecting specific odors. It is also worth mentioning here, once more, the following statement made by Babloyantz and Kaczmarek in 1979: If deterministic chaos is detected from a single-channel recording, it indicates the presence of chaotic activity in the recorded site. Nothing guarantees that the time series from an adjacent lead will show a chaotic activity, or if it does, whether we are dealing with the previous attractor. It seems that one site in the brain (frontal) can show noise behavior over long periods, whereas another site (occipital), can show chaotic behavior at the same time under certain conditions.

14.3.3.4 An Overview of EEG Investigations by Means of the Correlation Dimension: A Limited State of the Art

In Tables 14.3 and 14.4 the values of the correlation dimension (D_2) are outlined together, as computed by several research groups under different experimental conditions. The parameters are also included in the tables on human EEG. Table 14.3 presents the results of measurements on human subjects. Table 14.4 shows the experiments with intracranial recordings of cat, rat, and rabbit brains.

The existence of attractors and strange attractors in the brain is important since it was shown that EEG is not noise. The uncertainty of spontaneous brain behavior is clearly shown with this type of chaos analysis. Several chapters of this book and the final discussion cover the brain's chaotic behavior by analyzing the multiplicities of brain–body integration (see also Chap. 16). We return to the concept of chaos in Chaps. 22–25 and the Epilogue as essential to philosophical discussion.

14.4 Remarks on Quantum Dynamics

Two important approaches of the twentieth century are described. Applying chaos theory to brain function has already had some conceptual and practical applications in further understanding brain function. However, the application of the concepts and approaches of quantum theory to brain research is completely new. If one talks about the quantum approach to understanding a physical system it usually means looking to the processes at the micro-state level. Moreover, new trends also explain that quantum theory can similarly be applied to a macroscopic system. One of the important properties of quantum systems is the uncertainty principle. There are, however, other systems that show uncertainty in the macroscopic domain. Chaotic systems also have such properties, as described in the previous sections of this chapter. From a philosophical viewpoint the uncertainties coming from chaotic behavior of the brain

Table 14.4 Intracranial EEG (animal experiments)

Reference	Results
Başar et al. (1988), Röschke and Basar (1985, 1988)	Cat, SWS, cortex (epidural): $D_2 = 5.0 \pm 0.1$ Cat, SWS, hippocampus: $D_2 = 4.0 \pm 0.07$ Cat, SWS, reticular formation (mesencephalon): $D_2 = 4.4 \pm 0.07$ (the most stable data)
Röschke and Başar (1989)	Cat, inferior colliculus: $D_2 = 6.7$ Cat, reticular formation (mesencephalon): $D_2 = 7.05$ (unstable attractor, waking state, attractor properties in only 25% of recording time, „high frequency attractor," data filtered between 100 and 1000 Hz
Röschke and Başar (1989)	Cat, waking state, hippocampus: $D_2 = 4.00$ (during synchronized hippocampal theta activity)
Lopes da Silva et al. (1990)	Rat, hippocampus: $D_2 = 2\text{–}3$ or higher (unstable depending on location and on existence of epileptic discharge)
Skinner et al. (1989; this volume)	Rabbit, olfactory bulb: $D_2 = 5\text{–}6$ (event-related shifting from 5 to 6 in evoked activities with odor targets)

From Başar (1990)

are comparable with the processes of macroscopic parallelism in quantum dynamics. The microscope model of Werner Heisenberg is explained in the following. Then, mechanisms that are conceptually similar to the Heisenberg model are explained.

14.4.1 Quantum Dynamics and Brain Oscillations

The uncertainty principle in quantum physics was formulated by Werner Heisenberg, who developed the following model of thought: If one day a microscope with very high resolution could be used, the experimenter would be able to observe the interaction of a gamma ray with an electron, which is in the aperture of the microscope. Heisenberg assumed that the gamma ray, which is used for the illumination of the electrode, would undergo an interaction with the electron (Fig. 2.3). By supplying energy to the electron, the position of the electron should be changed according to the physical motion laws. The observer will certainly fail if he or she tries to localize the position of the electrode. One would then observe, not the exact position of the electron at the moment of collision with X-ray light, but only the position of the electron following displacement. This model of thought was the subject of discussion after the development of quantum mechanics. Finally, Heisenberg's experimental requirements were fulfilled and the microscope theory was supported by new experiments (Cassidy 1999); thus, Heisenberg's prediction was realized.

Can the uncertainty principle manifested by the microscope thought experiment be translated to brain research? To attempt this, consider the experimental electro-encephalographic (EEG) recording illustrated in Fig. 2.4

When we stimulate the brain with a sequence of cognitive working memory inputs, the spontaneous activity of the brain continuously changes. The development of alpha activity (increases in amplitude), in turn has an important influence on the alpha responses. The brain is learning and goes from a "preliminary" to a "learned" state. We mentioned the same situation in the microscope analogy; at the moment of application of the cognitive input the state of the brain changes, and accordingly the exact cognitive response to cognitive inputs or those with emotional components cannot be determined.

14.4.2 How May the Brain Show Chaotic Behavior and a Quantum Type of Uncertainty?

The brain is a *nonlinear system* par excellence. Accordingly, in the last three decades, the concepts of chaotic dynamics have found an important application in research on compound electrical activity in the brain.

Why does the brain behave as a chaotic system? This can be explained on three fundamental levels. There are various groups of neural oscillators in at least five frequency channels in several structures of the brain. Therefore, the brain has a large number of degrees of freedom related to those activities. Accordingly, the dimension of possible states of substructures within or comprising the whole brain is high. In a chaotic system, small changes in initial conditions may give rise to large changes in the system's trajectory.

On the contrary, the quantum-like "uncertain" behavior of a brain structure can be observed in even a few neurons behaving similarly. However, the dimension of the studied brain structure, which consists of only a few neurons, does not show chaotic behavior. Nonetheless, upon excitation, such a structure can behave in a probabilistic or indeterminate manner; that is, the neuron (neurons) may or may not fire. Once a neuron fires the experimenter is no longer able to excite it, as at the beginning. A good example can be given by describing a session using light stimulation on a human subject. Only a unique stimulation may cause higher alpha activity and the second stimulation will not find the neuron or neuron groups at the same functional level.

A hypothetic neural network consisting of neuron populations is presented in Fig. 14.7a. The probability of firing single neurons is similar in Fig. 14.7b.

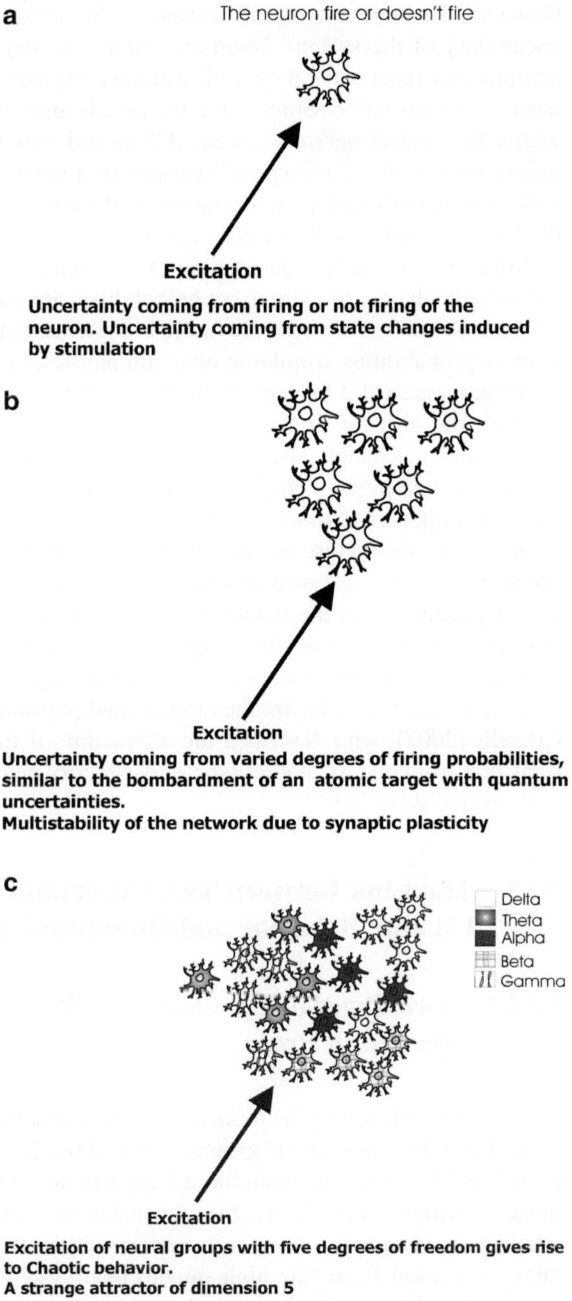

Fig. 14.7 (**a**, **b**, and **c**) Three different steps describing various possibilities of responsiveness in neural populations and their in deterministic and/or chaotic behavior

However, the large amount of neurons in the network of Fig. 14.7c increases the uncertainty of the system. Upon excitation (sensory or cognitive), the individual neurons can react separately with different degrees of probability; they can also react as an ensemble. Some neurons already may be excited by hidden sources within the central nervous system (CNS) and therefore not able to react during refractory periods. This type of reaction in a neuron ensemble can be compared with Boltzmann's statistical mechanics or the population of atoms in a target under the bombardment of a beam in an accelerator.

In Einstein's words, "Quantum physics formulates laws that govern crowds and not individuals; not properties but probabilities are described" (Einstein and Infield 1938). Laws do not disclose the future of systems, but govern the temporal changes in these probabilities. Similar to quantum physics, in cognitive processing the laws of the brain are valid for large populations of individual units. Rules for excitation are not valid only for single neurons, but also for neural populations. What applies to quantum mechanics also applies to the dynamics of chaotic systems. In both systems, *not properties, but probabilities* are described, laws disclose the change of the probabilities over time; and they are valid for congregations of units.

A given substructure of the brain or a brain tissue, in fact, shows that both properties can be observed together. Single neurons may behave with uncertainties originating from the nature of neurons similar to quantum systems and chaotic uncertainty resulting from high dimensionality. This can result from several oscillatory frequencies that give rise to higher degrees of freedom (Fig. 14.7c). We would also like to emphasize the fundamental physiological findings of Hughes and Crunelli (2007), who described the alternation of theta and alpha oscillations or transition from one to another type of oscillatory behavior in neural populations.

14.5 The Link Between Two Uncertainty States: Chaotic Behavior and Quantum-Like Behavior

14.5.1 Two Types of Uncertainties: Web of Chaotic Brain/ Quantum Brain

It is easy to explain why the brain behaves as a chaotic system. In various structures of the brain there are several groups of neural oscillators in at least seven frequency channels. Therefore, the brain has a large number of degrees of freedom related to those activities. Accordingly, the dimension of substructures of the brain or the whole brain is high. We know from earlier sections that chaotic behavior and uncertainty is caused from this high dimensional system. In a chaotic system, small changes in initial conditions may give raise to large changes in the trajectory of the system. On the contrary, the quantum-like behavior of a brain structure can be observed even by few neurons behaving similarly. Accordingly, the dimension of the studied brain structure consisting of only a few neurons does not show chaotic behavior. However, on excitation such a structure can behave in a probabilistic or non-deterministic manner: The neuron (neurons) may or may not fire. Once a neuron fires, the experimenter is no longer able to excite the neuron as at the beginning. The example was given of a session with light stimulation to a human subject. One stimulation may cause higher alpha activity, and the second stimulation will not find the neuron or neuron groups at the same functional level.

A hypothetical neural network consisting of a large number of neurons is presented in Fig. 14.7c. The probability of firing of single neurons is similar to the presentation in Fig. 14.7. However, the large number of neurons in this network raises uncertainty in the larger system. Upon an excitation (sensory or cognitive), individual neurons can react separately with different degrees of probability; they also can react as an ensemble. Some neurons are excited already from hidden sources or CNS, and are not able to react during their refractory periods. This type of reaction of the neuron ensemble can be compared with Boltzmann's statistical mechanics or the population of atoms in a target under the bombardment of a beam in an accelerator. (Compare Appendix C related to quantum resonances.)

According to Einstein, "Quantum physics formulates laws that govern crowds and not individuals; not properties but probabilities are described." Laws do not disclose the future of systems, but govern the temporal changes in these probabilities. As in quantum physics, in cognitive processing laws are valid for great congregations of individual units. They are valid, not for single neurons but for neural populations. What applies to quantum mechanics also applies to the dynamics of chaotic systems. In systems, not properties but probabilities are described, laws disclose the change of the probabilities over time, and they are valid for congregations of units.

In fact, a substructure of the brain or brain tissue shows the two properties together. Single neurons may behave with uncertainties originating from the nature of neurons and the chaotic uncertainty resulting from the existence of high dimensionality (a manifestation of the existence of several oscillation frequencies or higher degrees of freedom). A quantum-like or similar behavior to electrons in Heisenberg's microscope experiment results in an entropy change induced in the excited brain structure. The changes in entropy are explained in detail in chapter 17.

14.5.2 String Theory as a Unifying Brain Concept

As stated at the beginning of this chapter, the steps undertaken presently are those that introduce the nebulous Cartesian system that is explained in chapter 15 and chapter 16. The present chapter has given an explanation of chaos analysis and its place in brain processes, quantum theory principle and brain behavior, and finally, entropy transition as essential processes in brain function and development. These all-important parameters make the brain completely unique from other systems. Having stated these transcendent features of brain behavior, it is now necessary to show that a new reference system is needed to decipher the brain's predictability and understand more of its processes.

The basic objects in string theory are not particles, which occupy a single point in space, but one-dimensional strings. These strings may have ends or may join up with themselves in closed loops. Hawking (2001) said:

> Just like the strings on a violin, the strings in string theory support certain oscillation patterns, or resonant frequencies, whose wavelengths fit precisely between the two ends. But while the different resonant frequencies of a violin's strings give rise to different musical notes, the different oscillations of a string give rise to different masses and force charges, which are interpreted as fundamental particles. Roughly speaking: The shorter the wavelength of the oscillation on the string, the greater is the mass of the particle.

In chapter 24 a brain string theory is proposed as a model to understand the web of chaotic brain/quantum brain.

Chapter 15
The Brain in Probabilistic Hyperspace

15.1 A Unifying Step in Brain Function: The Most General Transfer Functions in the Brain According to Fessard (1961)

The transfer function describes the ability of a network to increase or impede transmission of signals in given frequency channels. The *transfer function*, represented mathematically by frequency characteristics or wavelets, constitutes the *main* framework for *signal processing and communication* (see Chaps. 6–8). The existence of general transfer functions would then be interpreted as the existence of networks distributed in the brain having similar frequency characteristics facilitating or optimizing the signal transmission in resonant frequency channels (Başar 1998). In an electrical system, an optimal transmission of signals is reached when subsystems are tuned to the same frequency range. Does the brain have such subsystems tuned to similar frequency ranges, or do common frequency modes exist in the brain?

The empirical results reviewed in this book imply a positive answer and provide a satisfactory framework to respond to Fessard's question in Chap. 2. Frequency selectivity in all brain tissues containing selectively distributed oscillatory networks (delta, theta, alpha, beta, and gamma) constitute and mathematically govern the general transfer functions of the brain. According to Fessard's prediction, all brain tissues, both mammalian and invertebrates, would have to react to sensory and cognitive inputs with oscillatory activity or similar transfer functions. The degree of synchrony, amplitudes, locations, and durations or phase lags continuously varies, but similar oscillations are most often present in the activated brain tissues (Başar 1999). As to the process of coding explained in Chap. 6, the general transfer functions of the brain manifested in oscillations strongly indicates that frequency coding is one of the major candidates for governing brain functioning.

E. Başar, *Brain-Body-Mind in the Nebulous Cartesian System: A Holistic Approach by Oscillations*, DOI 10.1007/978-1-4419-6136-5_15,

15.2 Generalization of Questions from Descartes and Fessard Concerning Brain-Body Interaction

As stated, Fessard (1961) posed a similar question to Descartes, who mentioned the possibility of the existence of some common principles and rules governing nature. In other words, are there various general principles that govern the transmission of the signals in the brain? Fessard's question related to electrical signals in the brain can be extended and generalized to the brain-body incorporation.

> Are there also general transfer functions in functional interactions of the brain with the vegetative system and biochemical pathways? Do cranial nerves and the brain stem link the vegetative system to cortex and provide cognitive/memory integration?

The autonomic nervous system is linked to the brain and regulates vital functions without our conscious control. We breathe, our hearts beat, our stomachs digest, our bladder muscles contract. Further we secrete saliva, insulin, and digestive enzymes. Our skeletal muscles are able to show vasodilatation and vasoconstriction without our conscious control. These functions are operating mainly on structures *hidden* from view. The autonomic system acts on smooth muscle in the blood vessels and intestines, cardiac muscles, and glands. The autonomic system also has afferent pathways carrying signals from various parts of the body to the brain and spinal cord. Başar and Weiss (1981), after studying common features of contractility of the organs that are under control of the autonomous nervous system, proposed general classifications of those organs with the expression the *overall myogenic system*, explained in Chaps. 4, 5 and 9.

15.2.1 *The Sympathetic System and EEG Oscillations*

Living system settings are ensembles of detectors and all types of mechanisms that allow living beings to maintain survival functions, such as normative values of blood pressure, respiratory rhythms, cardiac pacemakers, and body temperature. Such mechanisms that are important to maintain the body within the limits of healthy life should be also categorized within the level of persistent memory because damage to these settings strongly affects higher levels of nervous activity and all levels of memory activation. Gebber et al. (1995a) published a series of articles from their laboratory on the 10 Hz rhythmic sympathetic nerve discharges of cats and offered a hypothesis on its functional significance.

The question of whether they are tuning and matching effects between cognitive 10 Hz oscillations and sympathetic discharges may be answered in the future. However, the fundamental findings of Gebber's group indicate that the physiological settings of the circulatory system and phyletic memory characteristics are in the 10 Hz frequency range. A link between cognitive and vegetative processes possibly may be supplied with general 10 Hz oscillations (see Chap. 9).

Başar and Weiss (1981) measured and reviewed mechanisms of control, auto-oscillations of blood flow, contractility of the vasculature, forced oscillations in the peripheral circulatory system, spectral activity of peristaltic organs, and dynamics of the lymph nodes and lymphatic system, and found that all these subsystems showed spectral properties (oscillatory activity) in the ultra-slow frequency range of 0.01, 0.04–06, and 0.1 Hz. Başar and Weiss (1981) introduced the concept of the overall myogenic system after quantifying and analyzing the performance of smooth muscle dynamics, by considering this as the dynamics of an overall and coordinated dynamic system. Smooth muscle cells are building blocks and basic effectors of the overall myogenic system. The overall myogenic system incorporates the: (1) *vascular system*, with all the arteries, arterioles, etc., in the systemic circulation; (2) *lymphatic system*, with lymphatic vessels and nodes; and (3) *visceral system*, which performs the visceral functions of the vegetative system and peristalsis in vegetative function. Figure 15.2a and b show the organization of the overall myogenic system. Further, a holistic approach is introduced in Chap. 9. The data presented in Chap. 9 show that the coordination and communication in and between the vegetative systems and brain occur with tuned frequencies in the range of EEG oscillations and the ultra-slow frequency range.

15.3 Mutual Excitation and Overall Tuning in the Brain and the Overall Myogenic System

The approach to the brain-body-mind problem requires a deeper account of *multiple causalities* in comparison with the prerequisites of modern physics and quantum dynamics. In the analysis of the brain-body-mind-construct, multiple uncertainties or uncertain causalities must be included. These multiple causalities originate from: (1) nonlinear properties of the vegetative system, such as irregularities in biochemical transmitters, cardiac output, turbulences in the vascular system, respiratory apnea, and nonlinear oscillatory interactions in peristalsis (see Fig. 15.1); (2) nonlinear behavior of neuronal electricity, for example, chaotic behavior of EEG (see Fig. 14.9); (3) genetic modulations; and (4) the nonlinear proprieties of the body's physical processes. The tuning and connections among, brain, circulatory and respiratory system, spinal cord, and the overall myogenic system are schematically illustrated in Fig. 15.2. All these organs and/or systems are embedded in biochemical pathways functioning with the release of neurotransmitters.

15.4 What Are Multiple Causalities? What Is a "Hyper-Probabilistic Cartesian System?"

Newton's concept fits perfectly with the Cartesian system. The establishment of a more efficient new Cartesian system in the brain-body-mind incorporation is a most fundamental and difficult step. The search for probabilistic causal factors in brain

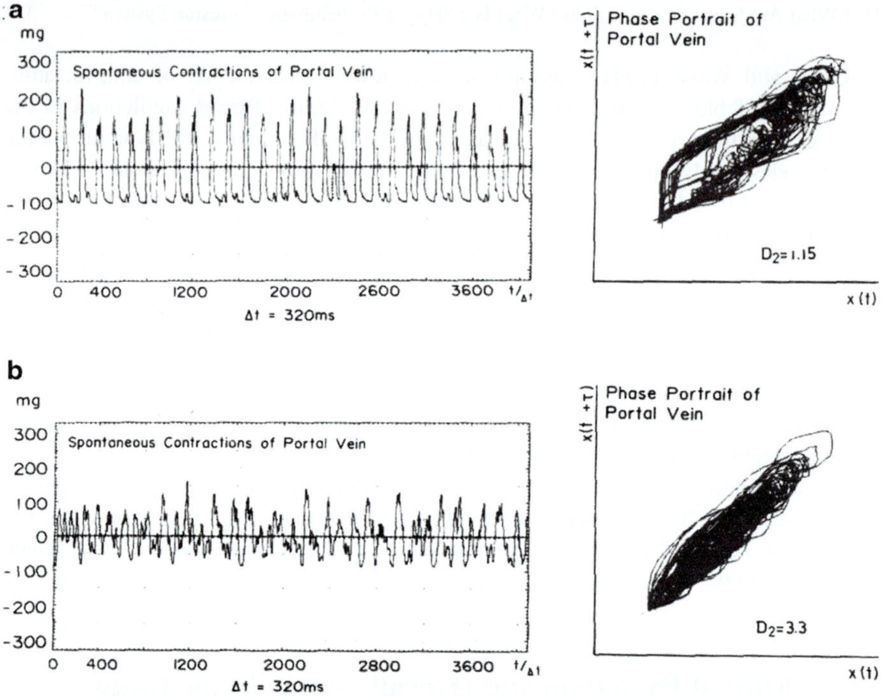

Fig. 15.1 Spontaneous mechanical contractions of the portal vein. *Left:* time series; *right*: phase portrait. (**a**) Example of the "minute rhythm" activity. (**b**) A period in which the smooth muscle presents a mixed contractile activity. A comparison of the changes in the dimensionalities in (**a**) and (**b**) (modified from Başar et al. 1990)

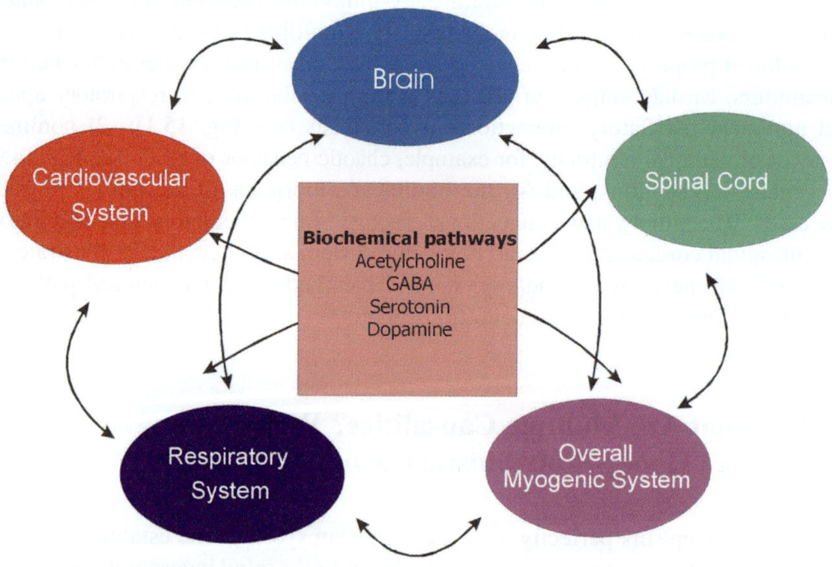

Fig. 15.2 Schematic illustration of the autonomous nervous system, biochemical pathways that are interconnected and interactive in functioning

research, similar to the task of von Weizsäcker, is more difficult in comparison with physical sciences. In the 1930s Heisenberg's theory was classified to be in the realm of metaphysics (Popper 1935). Today, his theory of quantum mechanics is a dominating branch of physics. The concept of "probabilistic causality in the brain-body-mind incorporation" in this book differs from the quantum theoretical causality. This results from multiple causalities as consequence of the richness of the interacting biological components in the phenomena or processes of psychophysiology. Accordingly, a formulation of a concept of "probabilistic causalities in biology" is attempted. This is still partly a metaphysical question. According to the coordination and/or tuning of oscillatory activity in communication within the brain and its link to the vegetative system and spinal cord, the most general transfer functions of brain-body-mind incorporation seem to be in concert (see Fig. 15.2). However, multiple uncertainties and nonlinear interactions have probabilistic reactions. These general transfer functions are also influenced by activation of biochemical pathways (see the Alzheimer's study by Yener et al. 2007).

According to the developments of the positive sciences, four steps are tentatively formulated here, to generate the new proposal or the new Cartesian system.

First Step: The Cartesian system created by Descartes was the first fundamental concept and analytical framework related to, and interwoven with, the concept and applications of Newtonian dynamics. Although, in physics, the conventional and classical Cartesian system made gigantic steps up to the beginning of the twentieth century, this Cartesian system could not respond to new developments because of the probabilistic nature of quantum dynamics and the moving reference systems of the theory of relativity. Capra (1984) explained that already in the twentieth century this reference system was far from responsive to the needs emanating from progress in biology.

Second Step: The second step includes the developments in quantum dynamics at the beginning of the twentieth century. To analyze quantum processes, physicists moved to a probabilistic Cartesian system in which the causality principle became probabilistic. This means that the trajectories of particles, obeying quantum rules, can be described only with the concept of cloudy wave packets. Although the processes of quantum dynamics were not classified under the title of the probabilistic Cartesian system, this step has been mentioned from the perspective of historical development.

Third Step: Next, a hyperprobabilistic Cartesian system is proposed to describe the processes in the brain-body system, i.e. a framework with a nonlinear linking of the central nervous system and autonomous control system performing the vegetative functions (see Fig. 15.2). This reference system is named hyperprobabilistic because subsystems of brain-body incorporation depict all probabilistic causalities as described.

15.5 A Commentary on Some Philosophical Thoughts Related to the Brain in Probabilistic Hyperspace

It is clear that physicists were forced to introduce a whole-view conceptual schema called quantum mechanics. Quite unexpectedly they found that only quantum laws were capable of resolving a number of puzzles and explaining a variety of data acquired from the realm of atomic physics. Greene (2004) stated that:

> According to quantum laws, even if you make the most perfect measurements possible or how things are today, the best you can ever hope to do is predict the *probability* that things will be one way or another at some chosen time in the future, or that things will have been one way or another at some chosen time in the past.

The universe according to quantum mechanics is not etched into the present; in fact it participates in a *game of chance.*

Is every movement, every function, every process of thinking completely dominated by chance and possibly the target of unexpected influences so that the brain functioning could one day attain a state of complete unpredictability? Greene (2004) answers this question by putting together concepts of entropy and gravity. According to the theory of gas dynamics, it is established that a uniformly dispersed gas has high entropy. This is also true in the classical rules of statistical mechanics. However, the story is very different when gravity is taken into account. Gravity is a universally attractive force; therefore, if there is a large enough mass of gas, every region of gas will pull on every other, and this will cause the gas to fragment into clumps.

The explanation is that the second law of thermodynamics states that in the formation of order there is generally more than a compensating generation of disorder. One of the fundamental forces of nature is that which exploits this future entropy tally to the hilt. Because gravity operates across vast distances and is universally attractive, it induces the formation of ordered clumps; the stars seen on a clear night are all in keeping with the net balance of entropy increase.

Can Greene's explanation be used to understand brain dynamics? In the physical universe, the electromagnetic forces of gravity act as huge long distance generators to change the entropy; however, this entropy, in turn, also has feedback and great influence on changing the constitution of stars and black holes, and can balance them. What are the correlates of gravity and electromagnetic forces in the human body system? Upon sensory and cognitive stimulation of the brain, the firing of neural populations goes from a disordered to an ordered state. Chap. 17 relates to evolution, and describes how neurons exist in less ordered clumps in the beginning of Aplysia or Helix ganglia species evolution. As the brain evolves, more ordered configurations can be seen in the low vertebrates, then later in mammals, and finally in human brains. Ordered structures such as the cerebellar cortex, the six layers of the cortex and neuronal formation of the hippocampus, do not exist in the Aplysia ganglion. The highly ordered neuronal clumps in the late stages of evolution indicate that the shape of neural formations reaches low entropy configurations in parallel with the huge volumes of gas, according to Greene (2004). Further, according to Greene, gravity is one of the causal factors giving rise to the constitution of the galaxy and stars. What place is there for the role of gravity in the brain-body-mind universe? There is a tentative argument that memory (phyletic memory) and cognitive processes constitute the core causality to change the entropy of neural populations during evolution, and this can be defined as a macro evolution. Upon stimulation of the brain with sensory and cognitive inputs, the brain's electrical activity (oscillations) goes from a disordered to an ordered state, as shown in Chap. 6. It can also be stated that, in parallel with the physical universe, entropy and the acting forces on the brain (sensory and cognitive inputs) give rise to the balancing effects of entropy. We go even a step further and assume that quantum-like probabilistic processes also matter in the functional processing of the brain, as described in Chap. 14. However, from all the processes described in this chapter it is very difficult to find a solution or determine a clear hypothesis. The brain is not a gas, but it seemingly obeys the rules of statistical mechanics, and changes in entropy play a major role. In the major gameplay of balances in the physical universe of elementary particles, strong interactions among elementary particles, gravitational forces, and electromagnetic forces are very important. However, order is a function of invariance in the physical universe, and these invariants are the constitutions of elementary particles. There are oscillatory activities of the brain-body in the frequency range, including ultra-slow oscillations (Chap. 9) and EEG oscillations. These types of oscillatory properties could be also named *eigenvalues*. All reactions of the brain, ranging from the simplest such as a simple flashing light to the most developed complex reactions, which occur in the recognition of faces, and the facial expressions of love or hate, are governed by these eigenvalues, which produce various types of configurations and superpositions.

15.6 From the Cartesian System in Probabilistic Space to a Nebulous Cartesian System

The aim here is to establish a new Cartesian system. Chap. 14 deals with the chaotic and quantum approaches that led to the need of a probabilistic hyperspace. In Chap. 16 the Cartesian system in the probabilistic hyperspace is defined. The proposal aims to define a common matrix to study several parameters together, thus integrating brain, body, central nervous system, and transmitters in a multidimensional Cartesian system. Such a Cartesian system can be used in several laboratories working on integrative brain function. Concrete solutions can be found, provided that scientists take care to consider all the influences emanating from a great number of functional processes. It is assumed that the use of such a Cartesian system is feasible; however, it requires hard work, and a large number of computers and data bases. As in the first Cartesian system, research scientists must evaluate all possibilities to avoid errors as the rule of R. Descartes explained in Chap. 2. The concept of the previous chapter shows that in spite of all precautions, brain-body integration can be approached only in probabilistic hyperspace because of multiple causalities. Now the fourth step introduces the processes of the unconscious brain that are described in Chap. 19. Although the unconscious brain is an important part of a number of phenomena in the brain, including creative states, we do not as yet have precise parameters to include them in a Cartesian system. The parameters and all variables are hidden, coming from unknown mechanisms. However, even if it is not possible to concretize these processes, a new type of Cartesian system is needed to understand all the processes in brain-body-mind integration. In the seventeenth century, when Blaise Pascal described two types of reasoning, it was not yet possible to measure thought processes. Now it is possible to partly measure cognitive processes. Thus, in the future new techniques could become available to more concretely approach the metaphysics of the brain and in turn its unconscious processes. As science develops, the metaphysical processes could be understood step by step and become real processes.

Fourth Step: Descartes, although undertaking a brilliant mental leap in his time, did not possess today's knowledge of physiology and physics. However, now the time has come to consider a new Cartesian system to understand the brain-body-

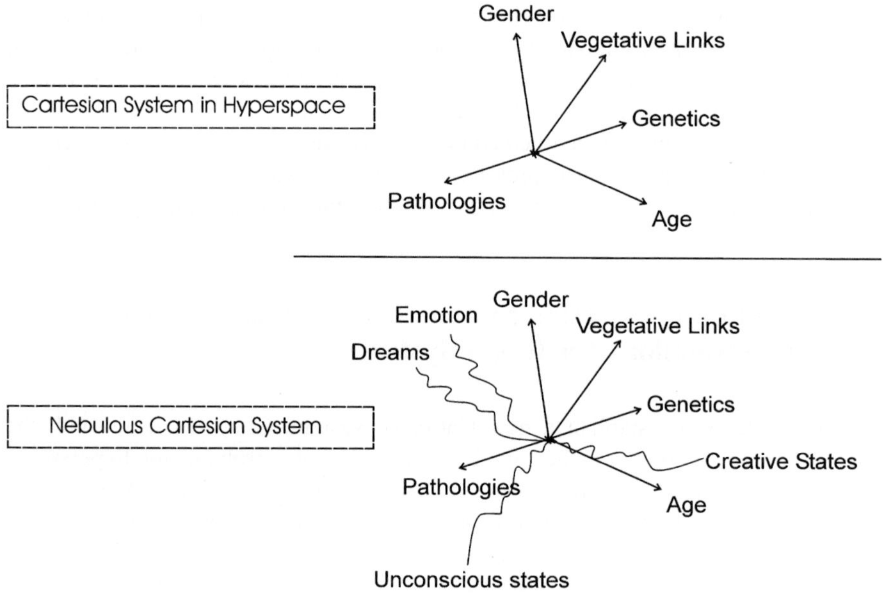

Fig. 15.3 In the *upper part* of the illustration a Cartesian system with *multiple coordinate axes* is presented. The *coordinate axes* contain information related to gender differences, vegetative links, genetics, age, and pathologies. This means that hyperspace is a constellation that embraces all possible influences on brain responses. The *bottom* of the illustration shows an example of the nebulous Cartesian system. In addition to the Cartesian system in hyperspace, there are further *coordinate axes* related to dreams, creative states, unconscious states, and emotion. Because there is not as yet any reliable information on these states, these *axes* are not given *rigid lines* but rather *irregular lines*. The time of unconscious states, dreams, and creative states are inhomogeneous, subjective, and immeasurable with physical instruments. Bergson's concept of duration is represented with the irregularity of these *coordinate axes*

mind relationship. In the description of brain-body-mind integration, it is necessary to include some processes that belong to the study of brain metaphysics, such as dreams, intuition, creativity, and unconscious states, including unconscious learning and problem solving. This means that the hyperprobabilistic Cartesian system is not a sufficiently complete framework or reference system. Accordingly, additions must be made to this hyperprobabilistic Cartesian system, which include sentiment, emotion, and creativity, so as to complete the spectrum of processes comprising both conscious and unconscious domains. This is the fourth step, after which the hyperprobabilistic system will be transformed to a nebulous system, because the future can only be predicted in a nebulous way through unconscious states such as dreams and creativity. The intuitions that are necessary for creativity, according to Henri Bergson (1907) take place in an inhomogeneous time space called duration, which is not measurable with conventional physical clocks (Chaps. 2 and 18). Accordingly, as the fourth step it is tentatively assumed that the processes or mechanisms of the brain-body-mind system can be analyzed and predicted similar to the metaphor of "finding the pathway a cloudy or foggy day." This is the new nebulous Cartesian system, as illustrated in Fig. 15.3.

Chapter 16
Quantum Brain and the Nebulous Cartesian System

16.1 Possible Ways to Approach Functioning of the Brain-Body-Mind Incorporation in the Framework of a "Nebulous Cartesian System"

By proposing the nebulous Cartesian system it is necessary to accumulate knowledge on electrophysiology, anatomy, learning processes, and all physiological settings and store them on several levels of a multiple coordinate system of the nebulous Cartesian system. During the work of the brain several parameters and entities related to brain functioning are usually operating in parallel, linked to all parameters of subsystems of brain-body-mind integration. When considering the brain as a probabilistic (nebulous) working and adaptive machine, one would expect to be able to approximately predict the next steps of this machine. It is possible to say that this is an attempt to describe the working principles and time course of these machines starting with the initial conditions. From this construct one must take into account all the histories; then there is the problem of determining the integral overall histories. This is the path-integral, and from this viewpoint it is necessary to recognize the need to use Feynman diagrams or Heisenberg's S-matrix, both of which are interrelated. The possibility of using this method in brain research has been referred to in earlier publications (Başar 1983a; Başar and Güntekin 2007).

16.1.1 *S-matrix Formulation of Heisenberg, Brain Dynamics, and Physical Causality*

In 1943 Werner Heisenberg formulated the S-matrix theory of particle interactions, in which Heisenberg (1961) tried to include in the theory only those concepts that have a clear operational significance. The theory is concerned only with the outcomes of scattering or collision processes and not with the detailed sequence of events taking place during the process, as in the earlier approach of quantum mechanics. The basic quantities of interest in high-energy physics, and more particularly in the study of strong interactions, are the collision, or scattering, amplitudes

E. Başar, *Brain-Body-Mind in the Nebulous Cartesian System: A Holistic Approach by Oscillations*, DOI 10.1007/978-1-4419-6136-5_16,

between sets of initial and final particles, the collection of which is the S-matrix (Barut 1967; Feynman 1962; Heisenberg 1961; Iagolnitzer and Barut 1967). The basic assumption of the S-matrix formalism is that each physical system, considered with “all its evolution,” can be represented before interactions by a well-determined ray I_{on} (or collection of vectors) in a Hilbert space (I_n) of “incoming” or “initial” free-particle states and after interactions, by a well-determined ray $Ø_{out}$ in a Hilbert (H_{out}) of “outgoing” of “final” free-particle states. The S-matrix should determine the cross-section for the production or annihilation of particles. The S-matrix can also be considered as a pure function that transforms all the momenta before the collision to the momenta after the collision (Iagolnitzer and Barut 1967); accordingly:

$$\langle | \mathrm{S} | \rangle = \langle \mathrm{out} \,|\, \mathrm{in} \rangle$$

Başar (1983a) presented the proposal that the brain response with the same formalism introduces matrices already denoted as brain matrices. This formalism should again present a metaphor to the S-matrix that predicts cross-sections of the production of elementary particles. Currently, the tentative assumption is that it is possible to approach the task of understanding brain function and many predict responses by using a large number of indicators.

Conceptual work and experimental designs lead to essential steps in brain research. When designing an experiment, the EEG should not be considered a non-dynamic or passive background during a cognitive process. According to this new type of experimental design, it appears that for the comprehension of event-related potentials (ERPs), a new set of parameters in our work on the paramount EEG must be considered, tentatively called brain indicators. The following is a provisional list of indicators:

1. The nonlinear correlation dimension that influences the degree of order in the spontaneous activity
2. Phase angle of the brain waves and their amplitude modulation envelopes (Bullock and Başar 1988), rms
3. Values of various EEG frequencies
4. Coherence in space and coherence in time for each frequency.

Using these indicators, it is now possible to move on to a new type of file, called the brain state matrix. Earlier publications attempted to create a picture of this matrix by stating that the brain state could be described by several measures of instantaneously defined EEG properties, outlined as indicators, during a given short period of approximately 0.05–1 s (Başar 1983a, b). The knowledge of parameters in such a matrix enables the experimenter to roughly predict the shape, amplitude, and frequency content of the ERP. The amplitude of an evoked response [evoked potential (EP)] often strongly depends on the amplitude of the ongoing activity (Rahn and Başar 1993a, b). These experiments showed that if the subject’s EEG is already in a coherent state, the physical sensory stimulation does not create a new, more coherent state. There is no EP to a physical stimulus in such a coherent state of the brain.

The results can be extended with nonlinear descriptors by stating that, if the dimension of the EEG is low, the transition of the EEG to a lower dimension is not possible or difficult. In other words, as explained, if the brain's electrical activity shows a low entropy state (high-order), the transition to a lower entropy state is difficult (Başar 1980). The amplitude and shape of the outgoing response or activity (evoked activity) are inversely correlated with the ongoing activity. The outgoing response is a function of the ongoing activity. The expression of ongoing and outgoing activities is used here in reference to an important analogy in elementary particle physics. The S-matrix introduced by Heisenberg was applied to elementary particle physics and nuclear reactions by considering ongoing and outgoing waves.

16.2 Feynman Diagrams

The development of an ensemble of rules has already been proposed, similar to those of the *Feynman diagrams* for the interaction of elementary particles, and these are used as a tool to understand the reaction function of the brain. This should be considered a tentative step for combining several simultaneous measurements of brain processes and approach to question of how brains work in a global manner. The rules will be completely different from the physical Feynman diagrams. Nevertheless, a similar way of thinking may be used, in the hope that this step can be enhanced by developing a "brain dynamics study program," including the characterization of the brain's pathological states, as well as the brains of lower vertebrates and invertebrates. The Feynman diagrams that are used in elementary particle physics have been developed to describe and predict the electromagnetic processes whereby electrons and photons interact. The interactions are indeed complicated. There is a type of "grammar" to these diagrams, which allows only certain configurations to be realized in nature. This grammar results from the basic laws of physics, such as conservation of energy and electric charge.

Particle physicists have found that this complexity should be handled in a reduced form; and to understand the behavior of electrons and photons, approximations are used that neglect all but the simple Feynman diagrams. By considering roughly the hundred simple diagrams for certain processes, physicists have been able to precisely predict important relations. As Başar and co-workers have shown (Başar 1980, 1983b, 1988; Başar et al. 1987), there are several alloyed and unalloyed transitions of the EEG following stimulation. For example, if a subject emits abundant high amplitude alpha waves before the application of sensory stimulation, usually no enhancement of that frequency is seen in the response encountered. On the contrary, an alpha blocking is observed and the same rule is also true for 40 Hz activity (Başar et al. 1987).

Furthermore, if the overall coherence among various structures of the brain is high, then again, the enhancement of EEG activity is low or vanishes. Additionally, a coupling between frequency components and amplitudes of various EEG components among different brain structures exists; for example, there is an important coupling or similarity between 10 Hz activities of the reticular formation and the thalamus (Başar 1983a, b). Starting with a brain state matrix and developing new rules step by step (which should be experimentally evident and allow the facilitation and prohibition of several transitions of the brain rhythms) it could be possible to predict a large number of brain reactions that are analyzed as the brain's compound response potentials.

This way of thinking maintains that the EEG is not only an activity that reflects some brain state, but also one that anticipates reactive mechanisms and controls the response to stimulation. Accordingly, the introduction of this new type of "grammar" may also serve to assist in the design of experiments that will contribute to the understanding of a large number of cognitive processes. It is further suggested that the brain obeys the same dynamic laws or rules that govern the control of the brain's excitability as described in quantum mechanics. If there is an excited state in an atom it is very difficult to increase the energy output of that atom. The brain behaves similarly – if a neuronal population is in an excited state, cognitive or sensory stimulation cannot excite this population any further. Some rhythms or patterns in natural phenomena can be explained and/or predicted by the powerful Feynman diagrams.

16.2.1 Brain-Body Feynman Diagrams

What will be analyzed with brain–body Feynman diagrams? The aim is to try to insert the entire history of EEG activity combined with physiological settings before stimulation. Chapter 7 explains that the memory is strongly influenced by physiological settings such as blood pressure and respiratory cycles. Fig. 15.2 illustrated the interaction of the CNS with subsystems of the vegetative system. Because all functions of the brain are mostly in concerted action, the same chain of reasoning is also valid for integrative brain functions. If the electrophysiological responses of the brain depend on changes of cardiovascular input, such as blood pressure, respiratory cycles, and the level of cholinergic or adrenalin secretion, then Feynman diagrams that predict the brain responses must also incorporate these physiological parameters.

These physiological parameters are also extended and/or influenced by emotional states. The emotional states can directly influence brain responses; however, emotional states can affect cardiovascular responsiveness as well, and this in turn may modify the electrical brain response. Figure 16.1 illustrates several factors to be evaluated when building brain Feynman diagrams. The most adequate way to start is to consider separate simple Feynman diagrams to describe all these different psychophysiological events. A partial Feynman diagram could be developed to show the influence of emotions directly on the brain or one developed to act on the cardiac output as well as the influences of the cardiac output to the brain. The final Feynman diagram, including all histories or physiological settings, is constituted from a chain of partial Feynman diagrams as a large tree with several branches.

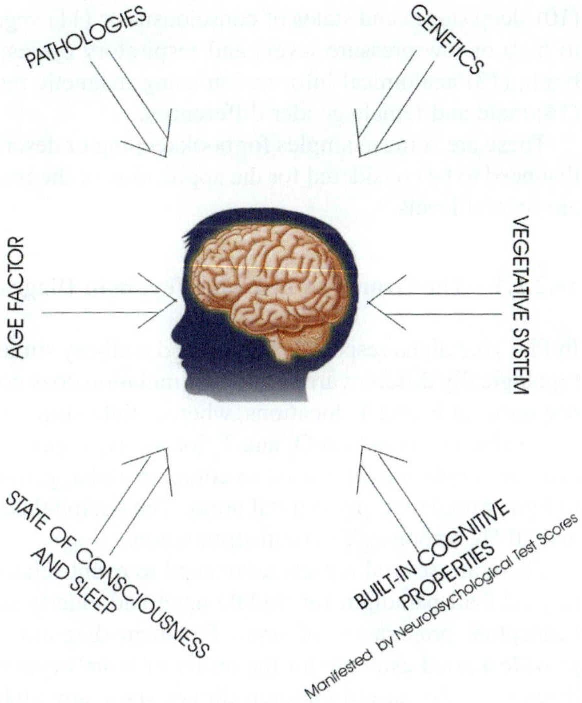

Fig. 16.1 Explanation of several factors influencing brain responsiveness. The same factors about which empirical weights are known will also be used in constructing Feynman diagrams

16.2.1.1 Factors Shaping the Computing of Brain-Body Feynman Diagrams

A Feynman diagram is a bookkeeping device for performing calculations in quantum field theory. In physics, the interaction between two particles is quantified by the cross-section corresponding to their collision. This cross-section, or more precisely the corresponding time evolution operator, propagator or S-matrix, can be explained as a sum of terms. What needs to be considered in the interaction of stimulations with the neural populations in the brain? As a first step here is a short story in time.

A sensory or cognitive stimulus to the brain evokes or induces oscillations. For general bookkeeping the following processes are to be considered: (1) activation of a given brain area with superposition of oscillations in alpha, beta, gamma theta, and delta; (2) phase re-ordering and phase-locking of the ongoing activity; (3) topology dependent oscillatory response; (4) blockings or enhancements in several frequencies depending on the level of pre-stimulus activity; (5) coherences between the studied structures that have an influence on the response; (6) the age factor, which plays an important role (shifting of alpha frequency from occipital to frontal areas; (7) genetic factors, which play an important role in oscillatory responses (Porjesz and Begleiter 1996, 2003; Porjesz et al. 1998, 2002); (8) neurological test scores (Doppelmayr et al. 2005; Karakaş et al. 2003; Klimesch et al. 1997a); (9) health conditions, pathologies as Alzheimer's disease or multiple sclerosis (Başar Eroğlu et al. 2007; Tass and Hauptmann 2007; Yener et al. 2007);

(10) sleep stages and states of consciousness; (11) vegetative system factors related to high or low pressure levels and respiratory cycles; (12) emotional input to the brain; (13) anatomical information using magnetic resonance imaging (MRI); and (14) male and female gender differences.

These are some examples for bookkeeping or describing the evolution of signals that need to be considered for the application of the brain Feynman diagrams. There are several levels.

16.2.1.2 The Grammar of Brain Feynman Diagrams

In Fig. 16.2 alpha responses to light and auditory stimulation are illustrated for four topologically different areas. Light stimulation does not evoke any significant alpha responses at F_4 and T_4 locations, whereas light stimulation and auditory stimulation evoke alpha responses at O_2 and T_4 locations, respectively. From this simple model, another simple model is used to compare alpha, gamma, theta, and beta responses to light stimulation in occipital areas. The occipital cortex responds with 10, 4, 40, and 20 Hz responses to visual stimulations.

There are several reports associated to event-related oscillations by using auditory oddball paradigm for middle-aged and elderly subjects. We also think that a conceptual preparation of brain Feynman diagrams for 3-year-old children can provide a good example for the utility of brain Feynman diagrams. The Feynmann diagram of 3-year-old children do not show any alpha response, as illustrated in Fig. 16.3 (compare also with results in chapter 11).

What are the practical advantages of trying to develop such diagrams? Before such an experiment is planned the analyzer should consider that no alpha responses would be recorded and in search of interpretations and pathological deviations this main property of alpha knowledge can be an important step to orient the research scientist. Next, the responses of cerebral ganglia in Aplysia are considered.

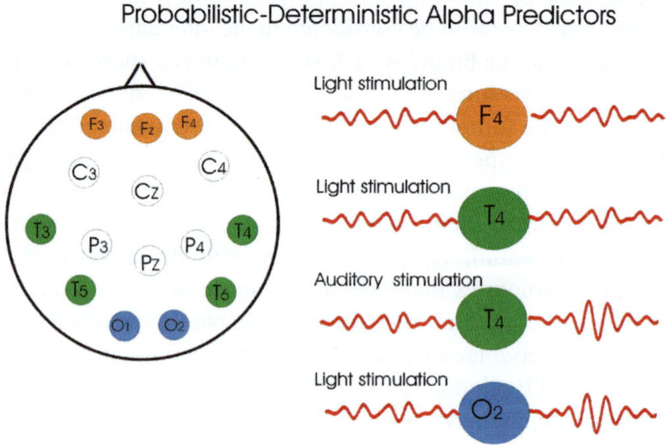

Fig. 16.2 Feynman diagrams for alpha responses in different areas of the cortex

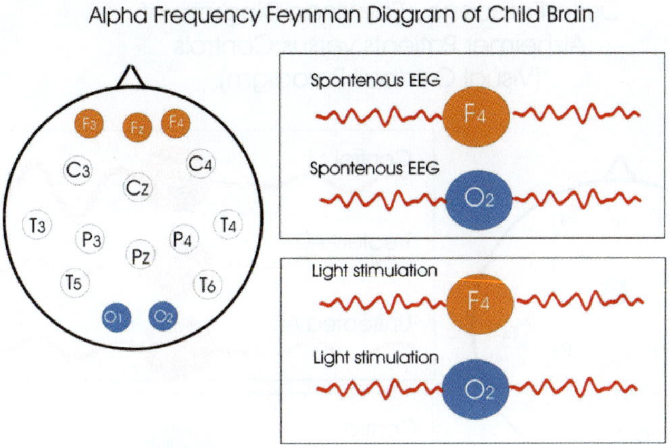

Fig. 16.3 Feynman diagram of the child brain in the alpha frequency range

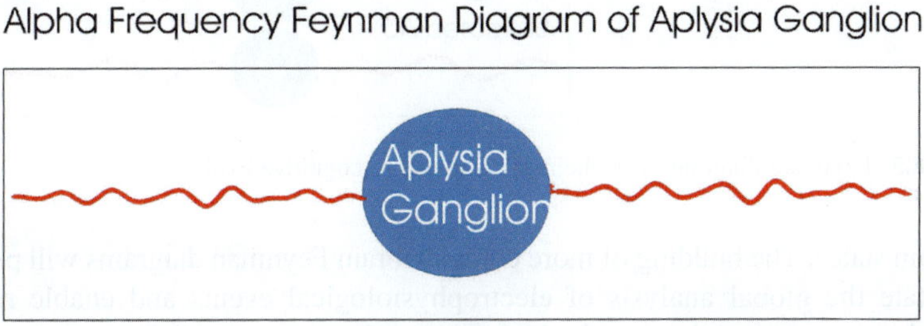

*

Frequency band (Hz)	1–8	8–15	15–30	30–48	52–100
Species	Maxima in frequency characteristics (Hz)				
Aplysia					
Group I	2	to	20	40	100
Group II	3	–	25	-	60

Fig. 16.4 Feynman diagram of the Aplysia ganglion

Also, there are almost no alpha responses in the Aplysia ganglion, as indicated by the results shown in chapter 10 (Fig. 16.4).

Although in healthy subjects there are ample delta responses to light stimulation, in Alzheimer patients highly reduced delta responses are recorded, as indicated in the Feynman diagram of Fig. 16.5 (see chapter 13).

These few simple analyzed diagrams can give important insight into a comparative analysis. These types of simple rules can be extended for functional and comparative studies, including diverse types of brain states, as well as coherence measures as descriptors of connectivity or correlation dimensions as descriptors

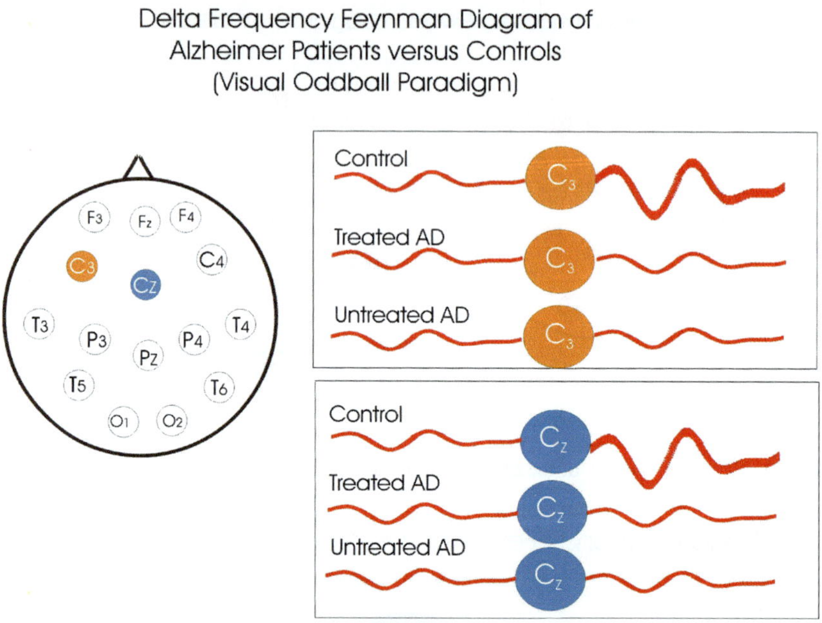

Fig. 16.5 Feynman diagram of Alzheimer's patients to cognitive load

of brain states. The building of more complex brain Feynman diagrams will probably facilitate the global analysis of electrophysiological events and enable research scientists to gain insights into brain functions that are more difficult to understand using detailed analytical research.

16.2.2 Computing of Brain-Body Feynman Diagrams

It is more appropriate to use the term *quantum computing* to refer to any use of the effects considered quantum mechanical rather than classical. Today, quantum parallelism is of utmost interest. As a metaphor in brain theory, the term *uncertainty of brain reactions* is used instead of *quantum parallelism*. According to David Deutsch (2003), this "parallelism" can be understood as an extension of the Feynman path integral approach to quantum mechanics, in which the probability of a physical system for transition from state A to state B can be statistically modeled. In quantum computing, the computer evolves along all possible paths from its initial state, and the probability of any particular final state is given by a sum of all paths that lead to that state. Another way of describing this is to say that the computer evolves along an exponentially growing number of paths (multiplying with each step), and in the final step all of these parallel computations interfere with each other to determine the probabilities of various outcomes. Feynman suggested the possible relationship between quantum computing and nanotechnology as early as 1959. He also pointed out the fact that quantum computing is potentially more powerful than classical computing, because classical computers cannot simulate quantum mechanics efficiently, whereas quantum computers should be able to do so.

In the previous section the approximate steps were described in order to put together several experiments to predict brain responses by considering all histories and the evolution of processes in the whole brain before stimulation. However, the other relevant processes of vegetative and biochemical processes in the whole body have been described, which can or may strongly influence brain responsiveness (see Fig. 15.2). As also considered in the following sections, the computation or prediction of brain responses with Feynman diagrams is difficult. Therefore, powerful supercomputers (quantum computers) should be used to evaluate all possible combinations and interactions in the brain and CNS, and the following step is proposed, first taking into account all possible processes in the brain and attempting to roughly predict the brain responses of a given subject depending on the age, pathological states, and possibly emotional behavior. After doing this the corrections can be added to the computations stemming from changes in vegetative parameters, such as the effect of the increase or decrease of arterial pressure on alpha, theta, or gamma responses; how diseases in the gastrointestinal system accompanied by increased or decreased peristalsis affect the measured responses; and the influence of changes in the balance of the lymphatic system on the brain's responsiveness. Most of the changes in brain oscillatory responses on these physiological changes cannot be found in the neurophysiology literature. However, by continuing to use the brain oscillatory approaches it will only be a matter of time before sufficient empirical results are collected to describe modifications of brain oscillatory responses in all these non-physiological or pathological changes.

16.2.3 Possible Advantages of "Brain-Body Feynman" Diagrams

The simple brain Feynman diagrams that have been presented are easy to understand. However, it is possible to progressively build hundreds of such grammar rules in the study of brain responses. Comparative studies and steps to create archives for all functions can be extremely complex and difficult. However, a simple visual analysis of such basic Feynman diagrams can allow the brain investigator to predict more complicated functions by an analysis of reduction. For example, the frontal brain alpha responses usually have low amplitudes and show no phase locking. On the contrary, frontal theta responses have large amplitudes and strong phase locking. If a great deviation from the simpler Feynman diagrams is observed in a brain response, the new added values or differences could make a rapid understanding of brain function possible and help to interpret new results.

The aim here is not to survey a vast literature on the computation of Feynman diagrams, but rather to indicate their relevance. Furthermore, Feynman diagrams could also be considered a schematic approach parallel to the Heisenberg S-matrix.

According to the literature the evaluation of Feynman diagrams can be performed by Monte Carlo calculations, which is a kind of experimental mathematics. If, in future, it would be possible to use quantum computers, the probabilistic nature of quantum devices can also be considered to solve brain-body Feynman diagrams. Then it would be appropriate to assess whether the Monte Carlo method would be adequate for the modeling of brain processes. When considering the schematic illustration of the Feynman diagrams, the problem of multiple causalities appears (see also Fig. 16.6). The prediction of the occurrence of a brain response would depend on various types of initial conditions, meaning initial brain states and a great number of factors from body and environment converging as multiple inputs to the brain. A great number of initial conditions have to be considered for brain processes. During signal processing of the brain, several hidden variables and/or parameters influence the brain processing, and accordingly the manifestations of oscillatory activity. This means that several main processes and subprocesses are in play and several links have to be considered. These multiple processes, which occur in series and in parallel, can be computed as multiple trials by using random trials

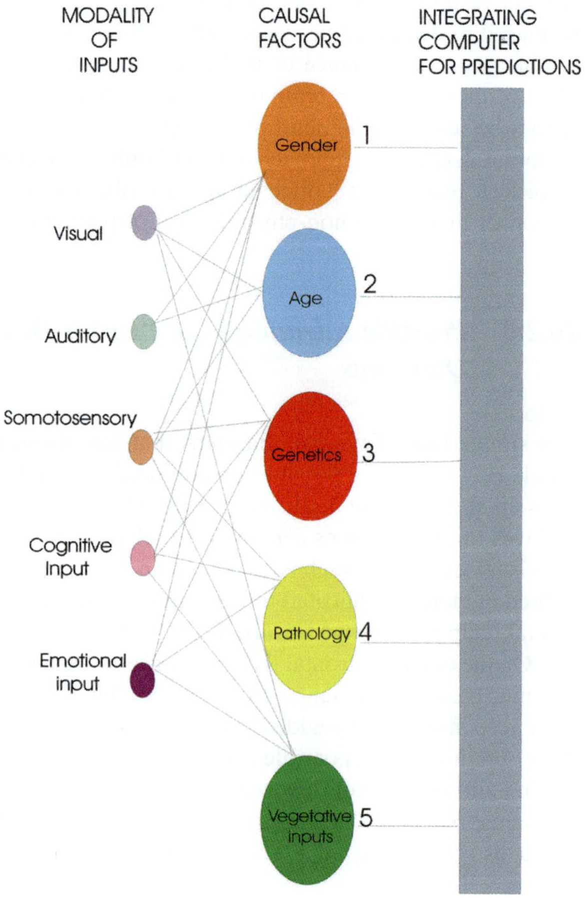

Fig. 16.6 Several factors influencing brain responsiveness will be embedded in fast computers in parallel or series so as to compute brain Feynman diagrams. Evaluation of large statistics, including the Monte Carlo method, could be used for predictions of brain responses in subjects

generated by a computer billions of times (see Fig. 16.6). In this way, brain reactions could be described and/or predicted within limits of probabilistic windows. This is the essence of the Monte Carlo method, which is used mainly to describe life histories of neutrons in nuclear reactions. Therefore, the Monte Carlo method seems to be a suitable approach for modeling brain processes.

16.3 Does the Language of the Brain-Body-Mind Need the Evolution of a New Discipline? Parallels to Quantum Theory, String Theory, and Chaos

In several sections of the present chapter and various papers, a number of experimental approaches for the analysis of experimental results were supported by the main idea frameworks initiated by Norbert Wiener, The Copenhagen School, Hermann Haken, Donald Hebb, Charles Darwin, and F.A. von Hayek. Much has been learned from the work of outstanding mathematicians, system scientists, and theoretical physicists. This has given rise to the multidisciplinary framework presented in this book and the development of oscillatory brain dynamics. However, a higher-order way of thinking is that all these pathway-opening scientists came from a mathematical background and tried to describe rules of the brain and mechanisms of thinking. Although enormous credit must be given to the development of cybernetics, dissipative structures, catastrophe theory, and synergetics, there is a strong argument for the necessity of direct knowledge from the brain (i.e. empirical results from the brain) so as to be able to understand the *principia* of brain working and the *principia of thinking*. This means learning from experimental results of the brain, and from all that learning establishing a theory containing a series of rules on brain functioning (Başar 2006). When trying to understand the brain, the goal is to develop a physical-physiological and philosophical construct. The starting point is not with mathematics to derive definitions; first the *principia-mathematica* of the brain must be decoded. The hypothesis is that this way of thinking is adequate. The human brain is the most complex structure known to us in the universe. Accordingly, a framework, which could enable an understanding of the brain, should be derived only from the language of the brain. Certainly, the search for a framework or description of the dynamics of the brain cannot solve all the problems related to brain-body-mind integration. However, the developed framework may help for recurrent measurements and computations to further understand brain-body-mind functioning.

Considering the descriptions in this book, three important features should be underlined: (1) The brain is a learning system, its ability to react to external or internal inputs changes with time. The reactions of the learning brain can be completely different compared with the reactions of the emotional brain (Güntekin and Başar 2007; chapter 11). (2) The brain's reactions change during evolution of species. (3) Responses also change in the maturing brain from the early days of childhood to the adult brain. (4) Creativity and related states of intuition cannot be explained by the earlier frameworks developed by mathematicians and theoretical physicists. By the application of oscillatory brain dynamics an area is approached in which an attempt can be made to measure all the four features mentioned. It is important to emphasize that the nebulous Cartesian system is a construct, a work in progress; in turn, this work may open new ways to deepen the understanding of brain function and possibly the metaphysics of the brain. This is in keeping with the opinion of John von Neumann and Arthur W. Burks (1966), who stated that: "[…] logics and mathematics in the central nervous system, when viewed as languages, must be structurally essentially different from those languages to which our common experience refers."

We believe that our proposal to establish the nebulous Cartesian system is in the sense of Capra (1984; see also Sect. 16.1) and that such a proposal is needed in neuroscience to initiate a breakthrough after the accumulation of a large amount of data derived from the evolution of species, the learning brain, and creative brains. This volume was based on the evaluation of papers on the frontier of emerging investigations into oscillatory dynamics in neuroscience. Brain oscillations are tuned like the strings of a violin during cognitive processes; accordingly, the string theory of quantum physics is a metaphor for the functional processes in the brain. (See also chapter 24 for a string model of brain-body-mind interaction.) EEG oscillations are the "natural frequencies" of the brain, and the essence of oscillatory brain dynamics are based on EEG resonances, as the resonances of strings that make the basic tones in music similar to the ground mechanisms of the brain.

John Carew Eccles suggested in 1986 that the synapses in the cortex may respond in a probabilistic manner to neural excitation; a probability that, given the small dimension of synapses, could be governed by quantum uncertainty. Further, Hameroff and Penrose (1996) developed elegant working hypotheses that take into account the possible quantum nature of signal transmission at the micro-level, considering probable processes at the synaptic level. Such hypotheses will probably be more profoundly examined in the future and will need experimental extensions. On the contrary, the parallelisms of Başar that were first initiated in 1980 with quantum-like resonances and in 1983 with the S-matrix metaphor have the less ambitious goal of describing the quantum-like probabilistic behavior of brain wave responses in observed brain reactions upon sensory-cognitive excitation. In such processes hidden variables can also originate from the vegetative system, as described in previous sections. The quantum probabilistic behavior is not only found at the micro-level (synaptic level) and or the single level, but chaotic dynamics resulting from multiple processes are also crucial entities, as described in chapter 15. In pure physics, the dualism between wave and particles has a quasi-metaphysical role; however, the brain is much more complex. Brain reaction susceptibility has multiple causalities. This can simply be called the brain's "multiplicities," and these will be dealt with in parallel.

It is hoped that the proposals in this work may motivate a number of young neuroscientists to jointly evaluate results from various types of measurements and accumulate them in the S-matrix as well as series of Feynman diagrams.

How far will this go?

Time will tell.

Part V
Metaphysics of the Brain: The Cutting Edge of Philosophy

Prelude to Part V

Stefan Zweig (1932), who had a long friendship and longstanding and interesting correspondence with Sigmund Freud, commented on the task of Freud in search of the unconscient world as follows: "A table in the middle of a dark room is invisible; but this does not mean that the yet invisible table does not exist. So is the unconscient world of Sigmund Freud."

Richard Panek (2004) has documented a stunning and intriguing look at how Einstein and Freud shaped the world we live in and our perception of it in his book, The Invisible Century. They were men of genius: Einstein did for the field of cosmology what Freud did for psychoanalysis, both courageously going where no minds had gone before and achieving groundbreaking results in their respective pursuits to understand how the world works. As Einstein came up with his theory of general relativity and challenged the existing paradigm of classical physics, Freud was struggling to make his field acceptable to the scientific community, which shunned psychoanalysis as not worthy of being called a true science simply because the analysis of a human being could not be duplicated in a lab or reduced to a mathematical equation. In his fascinating book, Panek leads us to the undeniable conclusion that the work of these two men culminated in the greatest scientific achievements of the twentieth century – the discovery of the invisible workings of both the universe and our inner minds, or the macrocosm and the microcosm of life.

The present book adds three more genies to the invisible century as the time period between 1850 and 1950: Charles Darwin, Gustav Jung, and Henri Bergson also should be regarded as discoverers of the invisible century. Darwin ventured out beyond his system so as to analyze an as-yet-invisible world. Darwin tried to understand the differentiation among species by comparing those in the Galapagos islands. The result was a model that explained theory of evolution. Similarly to Einstein, he compared different worlds: Einstein learned about the gravitation of the Earth by looking outward to other galaxies.

Bergson, too, tried to learn one of the most important abilities of the human being, that of intuition, by studying the *Overseas Voyage* of Darwin. He introduced the concept of durée (duration) so as to find the boundaries of episodic memory and

Fig. 1 Henri Bergson (October 18, 1859–January 4, 1941)

Fig. 2 Sigmund Freud (May 6, 1856–September 23, 1939)

Carl Gustav Jung

Fig. 3 Carl Gustav Jung (July 26, 1875–June 6, 1961)

creative memory, and thereby also performed a task in the sense of Eric Kandel and Nancy Andreasen, who asked questions related to the brain's ability to travel back in time or explain the meaning of creativity (see Figs. 1–3).

Chapters 17–20 deal with the work of these important men, whose work may lead to an understanding of the as-yet-unknown side of the brain-body-mind.

Can creativity and intuition be analyzed in future with more concrete steps? Is travel back in time within the mind and episodic memory interwoven and complementary?

First, we have to build a somewhat clear classification of unconscient and conscious states and also employ introspection.

Chapter 17
Darwinism, Bergsonism, Entropy, and Creative Thinking

17.1 Darwinism and L'Evolution Créatrice

17.1.1 Darwin's Theory

One of the most revolutionary developments in biological sciences was Charles Darwin's publication, *On the Origin of Species* (1859). Darwin worked within a framework of the living world as initiated earlier by Jean-Baptiste de Lamarck (1809).

Darwin's theory rests on two fundamental ideas: The first is the concept of *heritable variation*, which appears spontaneously and at random, as in individual members of a population, and is immediately transmitted to descendants. The second is the idea of *natural selection*, which results from a *struggle for life*. Only individuals whose hereditary endowments enable them to survive in a particular environment can multiply and perpetuate the species.

A review of Darwin's *On the Origin of Species* will show that the word *brain* is found only in one short paragraph. The reason is clear: Knowledge of the anatomy and physiology of the brain was very rudimentary during the 1850s. It should also be noted that Darwin did not mention anything about cognition or network abilities of the brain. This is deceptive because, in his notes, Darwin frequently referred to the brain as the organ of thought and behavior, and to heredity of behavior as being dependent on the heredity of brain structure (de Beer 1960). As Smulders indicates, additional mentions of the brain do arise in later editions of *On the Origin of Species*. The sixth edition contains a passage that explicitly states that natural selection applies to the brain as it does to all the other organs (Darwin 1872, p. 98; Smulders 2009).

Fifty years after its publication, Henri Bergson closely analyzed Darwin's theory in his work, *Creative Evolution*. Darwin's evolution theory provided an excellent model for Bergson's description of "intuitive processes and creativity." He formulated a theory of *intuition*, stating that this ability is unique among humans. Like other researchers of the time, Bergson was limited by the rudimentary level of contemporary neurophysiological and anatomical knowledge.

E. Başar, *Brain-Body-Mind in the Nebulous Cartesian System: A Holistic Approach by Oscillations*, DOI 10.1007/978-1-4419-6136-5_17,

In comparison with Bergson's time, present-day knowledge of neural networks, electrophysiology of neural populations, histology, and neural transmitters is much more advanced.

Contemporary knowledge enables us to say more: We reviewed the work of Hebb (1949), Hayek (1952), and Edelman (1978) on learning about neural networks, in which connectivity of neural networks or neurons achieve changes in morphological structures and connective abilities of the brain in a very short time (see Sects. 17.5.1–17.5.3). The theory proposed by Hebb is supported experimentally by the work of Kandel and Schwartz (1982) on the Aplysia model.

Accordingly, simple measurement strategies overwhelmed fundamental conceptual questions. Therefore a tentative attempt to reconcile the relevant natural philosophy of Bergson with contemporary knowledge is made here, in the hope of adding value to the progress of neuroscience.

Henri Bergson (1907), who studied the work of Charles Darwin, came to the conclusion that the *superiority of the human brain* in comparison with lower species is defined by the ability of "intuitive and creative thinking."

Bergson (1907) emphasized three types of mental abilities during the evolution of species: instinct, intelligence, and intuition. Instincts are observed in lower species such as invertebrates; intelligent behavior also belongs to the functional properties of lower vertebrates and mammals. Writing at the beginning of the twentieth century, Bergson did not have the opportunity to observe the electrophysiological correlates of evolution. He did not have knowledge of Hebb's theory (1949). Knowledge of the morphology of the evolving brain and maturation of the human brain were not yet established and/or discovered. Subsequent knowledge and new tools developed during the twentieth and twenty-first centuries provide an opportunity to analytically discuss Bergson's theory of intuition. The ability to interpret transitions in the evolution of brain waves presents a new and highly efficient framework to interpret higher brain activity.

Further, according to Bergson, a radical distinction exists in higher vertebrates between pure automatism (originating mainly in the spinal cord) and voluntary activity, which requires the intervention of the brain. Bergson (1907) stated that there are three types of judgments or development of skills. All lower vertebrates have instincts, which are regular and very simple, and also more complicated stereotypical behavioral patterns. A number of more developed animals resemble humans in that they display intelligence, not only automatism. In humans, parallel to almost automatic spinal cord reflexes, one observes intelligent behavior coordinated by the cerebral cortex. Humans also display a third mechanism, described as intuition.[1] Bergson assumed that the most important characteristic of humans is the ability to make use of this third type of cognition. According to Descartes (1840) and Locke (1690), this is indeed what differentiates humans from other species.

---

[1] Intuition is the act or faculty of knowing immediately, directly, and holistically without rational processes, and without being aware of how we know. It is also the channel or process through which we access realms of truth and knowledge.

To open a new gateway to analyze electrical signals during human evolution, we confine our attention to three matters. The first is the change of electrical *oscillations* through the evolution of species; the second is the measurement of *coherence* within the brain during its evolution; and the third is the analysis of *entropy*. The third is supported and extended through the work of Hebb, Kandel, and Edelman, which are related to reorganization processes in neural populations.

This chapter is based partly on a long-term collaboration with Theodore Bullock between 1984 and 2005, which was conducted in San Diego and Lübeck.

17.2 Electrical Activity from Aplysia Ganglion to Human Frontal Cortex: Possible Role of the "Alpha Oscillation" During Evolution of Species, in "Creative Evolution" and Maturing Brain

Alpha activity is possibly the most important dynamic pattern in brain–body interaction (Gebber et al. 1995a, 1999; Başar and Güntekin 2007). Very little spontaneous oscillation in the same frequency ranges of EEG and evoked oscillations were recorded in the isolated ganglia of invertebrates. Goldfish, ray, or Aplysia cannot develop spontaneous regular and large and rhythmic 10 Hz alpha activity. Animal recordings clearly indicate that entropy is high. In contrast, the alpha activity in human subjects depicts a high level of regularity, which is possibly a result of synchronization between neurons. Accordingly, the entropy of alpha is much lower in human EEG recordings. Although the frequencies of EEG oscillations seem to be somewhat invariant during evolution, in human electrical signals the degree of synchrony is much higher and the entropy is lower (Rosso et al. 2001). Different brains investigated by the groups of Başar and Bullock are presented schematically in Fig. 10.2, together with the characteristic averaged evoked potentials from different loci.

Başar et al. (1999d), Schütt et al. (1999), and Başar and Güntekin (2009) described the changing oscillations, particularly of the alpha activity, in the evolution of brains. The alpha activity was very weak in recording spontaneous and evoked activities of invertebrate ganglia (Aplysia and Helix pomatia). In Table 10.1 it can be seen clearly that there are almost no alpha responses in these species; the more highly evolved fish brain displays more efficient alpha responsiveness. Dominant and powerful alpha responses are seen, in the cat brain, although spontaneous alpha activity is scarcely recorded. Finally, in human cortical recordings there is huge alpha activity, reaching amplitudes up to 100 μV. However, human alpha activity is observed only in young and middle-aged human subjects; babies do not show alpha activity until the age of 3 years. Spectral analyses of spontaneous and evoked activities depict clear differences in the alpha activity in the maturing brain (Başar 1998; Başar et al. 1999d). Thus, alpha patterns are strongly dynamic manifestations of the evolving and maturing brain.

What may be the role of alpha activity during the evolution of species? According to the concept proposed by Bergson (1907), lower animals and plants do not have the ability to perform developed cognitive tasks. Invertebrates function instinctively, also described as the result of phyletic memory and stereotypical behavior patterns. At a higher stage of evolution, animals start to display intelligent behavior, and are able to perform some forms of cognitive tasks. This ability was described by Blaise Pascal as the "geometrical mind." Developed mammalian species such as cats certainly display this kind of performance, and are more intelligent than lower vertebrates; they are also able to develop new abilities, such as hunting and searching. Cat brains show a higher degree of spontaneous and evoked alpha. In addition, the cat brain can selectively respond to sensory stimulation and develop responses based on cognitive loading (Başar-Eroğlu and Başar 1991; Başar-Eroğlu et al. 1992). However, the cat brain lacks the abundant and high amplitude synchronized spontaneous 10 Hz activity seen in human alpha activity (Başar 1980). Humans can speak, solve geometric problems, invent new machines, and investigate and substantiate theoretical problems, such as searching for the cause of gravitation and planetary motion. This stimulates several research questions: Is the presence of alpha activity a consequence of the highly sophisticated human brain in comparison with the cat? Or do the creative evolution and creativity of the human brain trigger this alpha activity? Section 17.4 discusses the potential existence of a Maxwell Demon.

17.2.1 The Evolution of Alpha Activity During Cognitive Loading, and Its Role in the Maturing Brain

High amplitude and recurrent alpha activities are usually recorded at the end of an experimental session in which subjects are induced to develop cognitive tasks or pure thinking (Başar et al. 1989a). Also, in an experimental paradigm with cognitive tasks, the event-related oscillations in the alpha frequency range are more prolonged. Here again, the mechanisms providing high level of cognition are accompanied by increased amplitude of 10 Hz oscillations.

The increase in the amplitude of 10 Hz oscillation is accompanied, in most cases, by recurrent alpha wave packets and also regularity of the alpha oscillation shapes. The studied brain structures move from a state of disordered activity to a state of ordered activity, indicating a decrease in entropy. This point can be summarized by simply saying that the cognitive excitation of the mature brain elicits higher synchronization and a state of lower entropy. This process is explained in further detail in Fig. 17.1.

Children do not show alpha activity until the age of 3 years. The cognitive performance of children is also does not develop until the age of 3 years. Young adults have better cognitive performance and speech ability; their alpha activity is also highly increased. In other words, during the maturation of the human brain, the alpha activity also reaches a type of micro-evolution. As alpha activity reaches

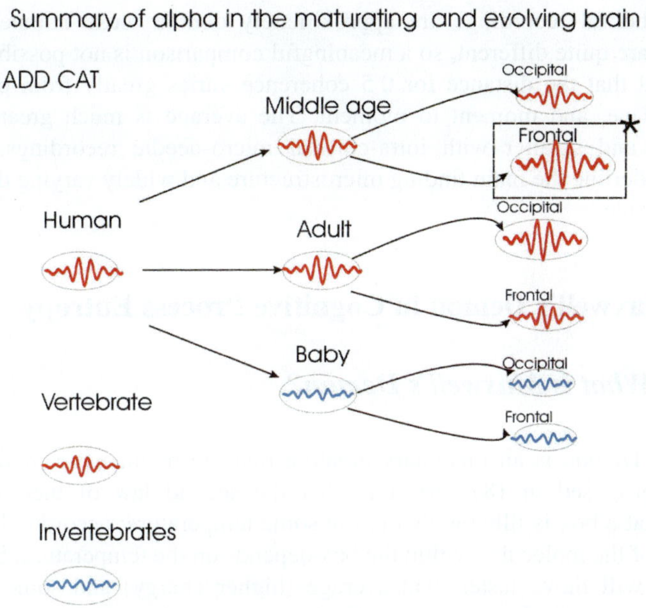

Fig. 17.1 Globally illustrated waveforms of alpha activity during evolution of species and maturation of the human brain (modified from Başar and Güntekin 2009).

higher amplitudes and an almost sinusoidal waveform, its entropy decreases. This means that the maturing brain's alpha activity is involved with the transition from a high to a low entropy state. It can be easily assumed that the higher cognitive ability of the adult brain in comparison with the child brain is accompanied by a decrease in alpha activity entropy.

17.3 The Role of Coherence in Brain Evolution

The term *coherence* refers to a pairwise measure of correlation at each frequency between two simultaneous time series. This is the best estimator for synchrony to date. It tells us that the eye is a poor estimator—confusing amplitude with synchrony (see Appendix A). According to Bullock et al. (1995b), there seems to be a difference between biological classes, with virtually no synchrony at any frequency in sea slugs (Aplysia), even at <1 mm and even for low frequencies. Bullock and Başar (1988) previously showed that fish had significant coherence at 1 or 2 mm; turtles and geckos displayed slightly more, and mammals still more. Bullock et al. (1990) found that detecting EEGs from the surface of the smooth brain of a rabbit, the average of many pairs showed coherence of approximately 50% when the electrodes were about 3–5 mm apart, falling to chance level at approximately 7–10 mm apart (Bullock 2002; Bullock and McClune 1989; Bullock et al. 1990, 1995b).

Human subdural recordings are approximately double those numbers, but the electrodes are quite different, so a meaningful comparison is not possible. Bullock emphasized that the distance for 0.5 coherence varies greatly from pair to pair, place to place, and moment to moment. The average is much greater in scalp recordings and smaller with intra-cortical micro-needle recordings. All these features underline the main finding microstructure and widely varying dynamics in time and space.

17.4 Maxwell's Demon in Cognitive Process Entropy

17.4.1 What is Maxwell's Demon?

Maxwell's Demon is an imaginary creature that the mathematician James Clerk Maxwell proposed in 1871 to contradict the second law of thermodynamics. Suppose that a box is filled with a gas at some temperature; accordingly, the average speed of the molecules within the box depends on the temperature. Some of the molecules will move faster than average (higher energy) and some will move slower than average (lower energy). Further, suppose that a wall is placed across the middle of the box, separating it into left and right sides. Both sides of the box are initially filled with the gas at the same temperature. Maxwell imagined a molecule-sized trap door in the partition, operated by the demon, which is observing the molecules. When a molecule moving faster than average approaches the door, the demon allows it to pas to the left side (by opening the door if the molecule comes from the right). When a slower than average molecule approaches the door, the demon ensures that it ends up on the right side. The result of these operations is a box in which all the faster than average (higher temperature) gas molecules are in the left side and all the slower than average (lower temperature) molecules are in the right side (Fig. 17.2).

According to several authors (Monod 1970; Prigogine 1980; Szilard 1929) this theoretical situation appeared to contradict the second law of thermodynamics. To explain this paradox it was pointed out that to realize such an outcome the demon would still need to use energy to observe the energy level of the molecules (e.g., in the form of photons). The demon itself (plus the trap door mechanism) would gain entropy from the gas as it moved the trap door. Thus, the total entropy of the system would still increase.

Fig. 17.2 Schematic figure of Maxwell's Demon

In Maxwell's thought experiment, the demon manages to *decrease* the entropy, in other words, it *increases* the amount of energy available by *increasing* its knowledge about the motion of all the molecules. Norbert Wiener (1948) referred to this as the Maxwell demon; the phrase Maxwell's Demon is also used.

Wiener (1948) further asked the question, "Why wouldn't a setup like Maxwell's demon work?" Any real demon would not be a disembodied spirit, receiving its information telepathically. To acquire information about the world it is necessary to be in physical interaction with it and, on the atomic and molecular scale, it is not possible to ignore the quantum mechanical nature of the world. For instance, to be able to observe the molecules, the demon would need to absorb whole photons at a time, and any detailed version of the thought experiment would encounter the uncertainty principle and the fact that an interacting demon would acquire the same temperature as the rest of the system.

The link between thermodynamics and quantum physics is even stronger: Macroscopic entropy can only be computed correctly from cumulative contributions from microscopic states.

17.4.2 Does a Maxwell's Demon Exist During Some Cognitive Processes?

In several publications on the learning brain, the author has published EEG records showing a transition of alpha activity from a disordered state to an ordered state as cognitive loading on the brain and the degree of working memory during the learning process increase. In one experimental scenario, the subject pays attention to three auditory or visual signals occurring at regular intervals. The task of the subject is the mental detection of the time of the fourth signal, which is omitted. At the beginning of the experiment, the subject's pre-stimulus alpha activity has low amplitudes and the 10 Hz oscillatory shapes are less regular. However, in the second stage of the experiment, the subject's pre-stimulus alpha manifests high amplitudes with regular, almost sinusoidal shapes, and a very good phase locking before the omitted stimulation. This means that the 10 Hz activity goes from a disordered state to an ordered one, i.e., from a *high to a low entropy state* (Fig. 17.3; see also Figs. 7.10–7.13).

In these experiments no physical input was applied to the brain before the omitted stimuli. However, during that thought process, the brain does decrease the entropy of the electrical activity. Visual observation of a recording of the alpha activity in the EEG indicates that regular alpha activity was observed at the end of the experiment. In other words, during thought processes the brain's electrical activity exhibits a potential form of the Maxwell Demon scenario, which triggers the brain to violate the second law of thermodynamics. Let us assume that, during problem solving or a high level of attention, processes leading to creative evolution the entropy law would be changed. If this assumption is tenable, the creative evolution within the brain would put the universe of human beings in a new dimensional state.

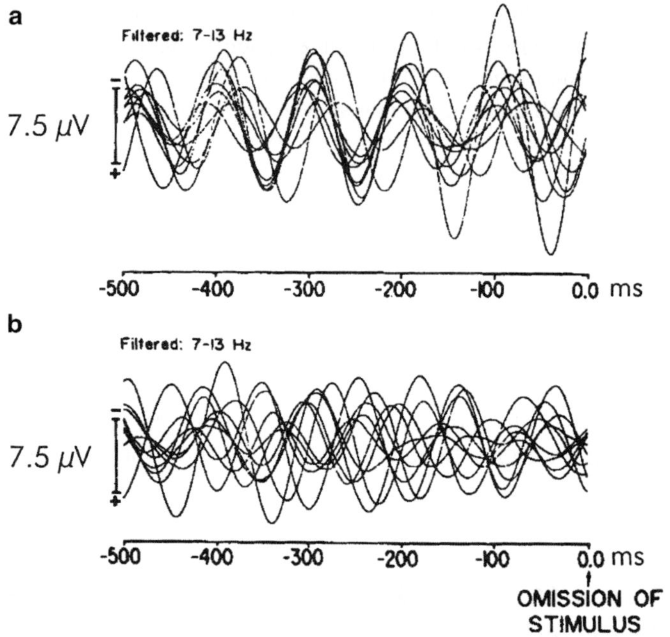

Fig. 17.3 Vertex recordings in a "repeatable pattern formation" experiment (**a**) Approximately ten pre-stimulus EEG sweeps at the end of the experiment (filter: 7–13 Hz). (**b**) Approximately ten pre-stimulus EEG sweeps at the beginning of the experiment (filter: 7–13 Hz) (modified from Başar et al. 1989)

Higher mental activity would be a key to change not only the structural form of oscillatory activity but, as a consequence, also the morphology of a number of brain structures, according to the rules of Hebb (1949).

The second law of thermodynamics is valid for closed systems. The question now is whether the brain can be considered a closed system despite the great number of stimuli from the exterior milieu. The same question is also valid for the entropy measurements described in the decrease of alpha in Sect. 17.2 and Fig. 17.1.

To address this question, the same cognitive paradigm was repeated with Alzheimer patients who were unable to perform difficult cognitive tasks. These subjects did not show a similar entropy decrease in alpha to that observed in healthy subjects, although the exterior milieu was the same.

In conclusion, it can be said that there are three types of entropy decreases in the 10 Hz activity in various species and the human brain. All these three types of developments of alpha activities have a common dominator: In all three cases, the augmentation in the ability of thinking or cognitive processes is in accord with the entropy decrease. Also a parallel process with the decrease of anatomical entropy in all the studied brains is observed. According to these empirical results, we pose the following relevant questions: (1) Do brains that begin thought processes also induce entropy decreases in electrical signals and anatomical shape, or does the decrease of entropy in anatomical structure result in a concomitant decrease in electrical entropy, and elicit intuition in the human brain during cognitive processes? (2) Are these three types of processes (entropy of electricity, entropy of anatomical organization, and degree of thinking processes) in mutual interaction? Do they occur in parallel? It is not possible to identify a clear causality for their origin, but it is possible that all three types of processes present a recurrent form of development. If brain dissipative structures are considered, the start oscillations may also result in structural changes (Prigogine 1980).

17.5 Hebb, Kandel, and Edelman: Entropy Changes

17.5.1 Hebb's Theory: Growth of Neural Assemblies

In 1949 Hebb published a theory of perception and learning in which a rapprochement was proposed among (1) perceptual generalizations; (2) the permanence of learning; and (3) attention. It was proposed that a repeated stimulation of specific receptors will slowly lead to the formation of an "assembly" of association area cells that can act briefly as a closed system after stimulation has ceased. This process prolongs the time during which the structural changes of learning can occur and constitutes the simplest instance of a representative process (image or idea). Furthermore, according to Hebb, the interrelationships between cell assemblies are the basis of temporal associations of temporal organization in central processes (e.g., attention, attitude, thought).

Does some kind of modification in neurons or connections between neurons take place as a result of learning (Hebb 1949)? For example, when we learn to associate two stimuli (e.g., an unconditioned stimulus and conditioned stimulus, as in classical conditioning), what exactly happens in the brain to support this process (Tranel and Damasio 1995)? Hebb (1949) proposed that the co-activation of connected cells would result in a modification of weights, so that when the presynaptic cell fires, the probability of the postsynaptic cell firing is increased. Hebb stated

This principle did not describe what was meant by "growths" or "metabolic changes." However, it served as a useful pioneering idea, and has become one of the widely cited concepts for neurobiological investigations of learning and memory. *It is clear that increased connectivity between neural assemblies increases synchrony and regularity of responses, thus manifesting a decrease of entropy.*

17.5.2 Fundamental Results by Kandel Support Hebb's Theory

Important advances in the understanding of learning and memory at the molecular level have come from the work of Eric Kandel and his colleagues (Hawkins et al. 1983; Kandel and Schwartz 1982). Much of this work was conducted using the marine mollusc *Aplysia californica*, which has a simple nervous system composed of approximately 10,000 neurons. The neurons are unusually large and easily identifiable, making Aplysia far more convenient for cellular-level studies than are vertebrates with infinitely more complex systems.

Research by Kandel and colleagues provided the first steps for direct evidence that alterations of synaptic efficacy play a causal role in learning. Specifically, they discovered that behavioral habituation of the gill and siphon withdrawal reflex, a staple behavioral preparation in Aplysia, was mediated by a reduction in transmitter release at a defined synaptic locus (Castellucci and Kandel 1974; Pinsker et al. 1970). In turn, these results supported the principle of Hebb. Later, Bailey and Chen (1983) showed that habituation was accompanied by alterations in the morphology of electrophysiologically identified synapses. These findings provided direct evidence for forms of synaptic plasticity that could afford the cellular and molecular basis for at least some forms of learning and memory.

17.5.3 Re-entrant Signaling: A Theory of Higher Brain Function (Scope of G.M. Edelman)

Edelman (1978) raised the following important questions:

> "Does the brain operate according to a single principle in carrying out its high-order cognitive functions? That is, despite of the manifold differences in brain subsystems and the particularities of their connections, can one discern a *general mechanism* or *principle* that is required for the realization of cognitive facilities? If so, at what level does the mechanism operate- cells, molecules, or circuits of cells?

According to Mountcastle (1976), the central problem for brain physiology is "how to understand the actions of large populations of neurons, actions that may not be wholly predictable from properties of subsets." The central problem of the intrinsic neurophysiology of the cerebral cortex is to discover the nature of neuronal processing with the translaminar chains of interconnected cells (in columns). Edelman (1978) transformed these statements as follows: "The main problem of brain physiology is to understand the nature of repertoire building by population of cell groups." It has been stated elsewhere that EEG oscillations should be considered as building blocks or as belonging to primitive repertoires of brain physiology.

Edelman (1987) developed a theory of neuronal group selection. This theory assumes a genetic endowment of neuronal groups, such as the columnar modules of the cortex, with an inherent degree of *variability* and *plasticity* in their connections. They constitute the units of selection of the primary repertoire. By exposure to external stimulation and a Hebbian mechanism, certain groups of cells, which tend to fire together, are selected by stimuli insofar as groups respond to them, and their connection is thus strengthened. Some of those connections make *recurrent* or *re-entrant* circuits that are an essential feature of the model and of its theoretical and computational elaborations (Tononi et al. 1992). Groups not selected are crowded out by the competition.

According to Edelman (1978), *re-entry is dynamic* and can take place via multiple parallel and reciprocal connections between maps. Re-entry takes place between populations of neurons, rather than between single units. Neurons within a group tend to be strongly connected. At a higher level, the integration of perceptual and conceptual components is required to categorize objects.

As described in previous sections, the theories of Hebb, Kandel, and Edelman are interrelated, and each indicates a transition from irregular states to regular states during sensory-cognitive processes. These interrelated scopes have been interpreted as a decrease of entropy during learning process, as hypothesized in this section.

17.6 Is Hawking's Scope on Entropy Also Needed in Brain Research?

In the nineteenth century, the French physicist Pierre Laplace (1878) suggested that if the position and motion of all the particles in the universe were known, physics could predict the evolution of the universe into the future. Laplace formulated the ultimate version of classical determinism: that the behavior of a system depends on the behavior of its parts, and its parts obey deterministic law of physics.

According to Heisenberg's uncertainty principle, the more accurately one tries to measure the position of a particle, the less accurately one can measure its speed and vice versa. The uncertainty principle had profound implications for the way in which we view the world.

The uncertainty principle signaled the refutation of Laplace's dream of a theory of science, a model of the universe that would be completely deterministic: One certainly cannot predict future events if one cannot even measure the present state of the universe precisely. This view of Hawking was a discussion point in the understanding of the nebulous Cartesian system (NCS) described by Başar and Güntekin (2007). The NCS presents a parallel situation to quantum mechanics, which does not predict a single defined result for an observation. Quantum theory introduces an unavoidable element of unpredictability or randomness into science. The wave particle processes may be neatly represented by the so-called "sum over histories" introduced by Richard Feynman. In this approach, the particle is supposed not to have a single history or path in space-time, as it would in a classical, non-quantum theory. Instead it is supposed to go from A to B by every possible path. The probability of going from A to B is found by adding up the waves for all the paths. This view has an important parallelism in studying brain waves. One has to study all the time histories or sum over histories of brain waves and of all settings in the vegetative system, which is linked to the brain by the cranial nerves. Again, this indicates that predictions on brain–body incorporation are probabilistic in nature (Başar and Güntekin 2009; see also Chap. 18).

The laws of science do not distinguish between the past and the future. In any close system, disorder (entropy) always increases with time; things always tend to go "wrong." The increase of disorder, or entropy, with time is an example of what is called an arrow of time – something that distinguishes the past from the future, giving a direction to time. There are at least three different arrows (directions) of time in which disorder or entropy increases. There is also a psychological arrow of time; this is the direction in which we feel time passes, the direction in which we remember the past but not the future. Finally, there is the cosmological arrow of time; this is the direction of time in which the universe is expanding rather than contracting (Hawking 2001).

We mention here the similarity or parallels with the concept of Bergson, who introduced the inhomogeneity of subjective time (Başar and Güntekin 2009).

17.7 Conclusions and a Tentative Synthesis

The frequency code in brain oscillations remained almost unchanged during the evolution of species, as shown in Table 10.2 and indicated in several other publications. This poses the question, "What did change during evolution of species in electrical activity by assuming that the frequency codes were kept almost invariant?" As described by Başar and Güntekin (2009), four types of differentiations were observed despite similar frequencies:

1. The amplitudes of slow EEG oscillations were increased from invertebrates toward the human brain, especially that of alpha activity (Sect. 17.2).
2. The synchrony of oscillations within a neural population also increased during evolution.
3. According to Bullock et al. (1995b) and Bullock and Başar (1988), no significant coherences were found in Aplysia neural networks, whereas large coherences between distant structures are recorded in mammalian and human brains.
4. The entropy of electrical oscillations also diminishes during evolution. Strong evidence to support this can be observed in 10 Hz oscillations. (There are almost no 10 Hz oscillations observed in land snails [Helix pomatia] and sea slugs [Aplysia], whereas there are coherent and very regular 10 Hz oscillations in the human cortex). From these four important differentiations, the decrease of entropy seems to have been a crucial factor during evolution from invertebrate ganglia to the human brain. Further, according to Hebb (1949), perceptual inputs or thoughts would increase the connectivity between neural assemblies. According to Edelman (1978), this can open the way to selective neural groups. Increased connectivity and building of selective neural groups would possibly induce a greater order in neural assemblies. In fact, the hippocampus, cerebral cortex, and cerebellar cortex of mammals demonstrate regular cell orders that are not encountered in invertebrates and lower vertebrates. For this reason, one can speak of a *decrease of entropy* in the shape of anatomical structures. On the other hand, the lower entropy of (at least 10 Hz) electrical oscillations shows a parallel to the development of lower entropy of neural structures. This means

that during the evolution of species, the entropy of structural organization and electrical oscillations is decreased. However, the second law of thermodynamics suggests the opposite; is the second law of thermodynamics thereby violated? During evolution and thought process only the existence of a Maxwell Demon can trigger the change of entropy. What might this Maxwell Demon be? Can thinking processes be considered to function like a Maxwell Demon? Humans are capable of displaying the highest levels of mental performance, including intuition, which leads to new discoveries or inventions. How might this decrease of entropy during evolution open the way to mental processes, and vice versa, are mental processes able to increase connectivity between neural populations, leading accordingly to modifications in ordered neural structures and regular alpha activity? Both are possible. One may even propose that both provide a recurrent and reciprocal activation of the incorporation of low entropy and higher nervous activity.

Consider again Bergson's "creative evolution" at the turn of the twentieth century (1907). Electrical activity of the brain, especially EEG oscillations and anatomical knowledge of neural structures, had not yet been discovered. Moreover, the theory of Hebbian neurons or the results of human brain maturation were unknown. We must acknowledge the importance of Bergson's attempt to find parallels between the evolution of species and that of thought. It also should be emphasized that Bergson's association between the human brain and its capacity for intuitive solutions or creation is highly significant. Presently, there is biological and electrophysiological evidence that associates the human brain with intuition. Also, the creative mind described by Blaise Pascal (1657) can be considered to be a parallel concept to Bergson's theory of intuition.

The present chapter is focused on the search for biological and electrophysiological correlates in evident changes of cognitive processes during evolution. There are at least three important pieces of evidence that are physiologically founded: the increase of alpha activity, the increased efficiency of cognitive processes, and the decrease of entropy. Two more findings support this strong statement: Before the age of 3 years, children's brains do not display either high cognitive processes or alpha activity (see Chap. 11). The mature brain has the capacity for higher cognitive processes and also displays alpha activity. During an experiment with cognitive loading, subjects exhibited low entropy and high amplitude alpha activity (see Fig. 17.3).

By viewing the chain of reasoning provided in this way, the boundaries of the brain's meta-processes are somewhat forced. Modern methods can probably open the way to attack the nature of intuition. As well, quantum theory may lead to determining the nature of intuition.

Başar (2009) also surveyed the existence of 10 Hz activity in the spinal cord and the sympathetic nerves of the heart. Accordingly the "alpha" was found to be not only a correlate of thought processes, but also of fundamental functional body processes (see also Chap. 9). This is the case as well because the visceral ganglia of invertebrates also have less regular and low amplitude 10 Hz oscillations. In other words, 10 Hz oscillation is a building block in all species for performing tasks and forms a fundamental part of higher nervous function.

17.7.1 From Body–Brain to Mind

Başar and Güntekin (2007) and Başar (2008) assumed that the entire responsiveness of the brain is highly influenced by changes in the vegetative system, which is also embedded in biochemical pathways (see Fig. 7 from Başar and Güntekin, 2007; see Fig. 15.2) in this volume). The questions, "What is mind?" and "What is conscious experiment?" can be approached only by considering multiplicities in a hyperspace of measured parameters.

Therefore, Başar and Güntekin (2007, 2009; see also Chap. 18) argued that, to understand "mind" it is necessary to determine functions defined as instinct, intelligence, and intuition. The work of philosophers such as Aristotle to Descartes, Pascal, John Locke, Kant, Spencer, and Bergson are extremely useful and relevant. However, in trying to describe what the mind does, these prominent philosophers all lacked the benefit of current empirical, physiological, and anatomical knowledge. Significant advances in these fields were made during the twentieth century, and the level of scientific knowledge available in the twenty-first century now permits radically different approaches to the investigation of such questions. The possibility of defining the "mind" is one step toward a global understanding of the brain, incorporating both instincts and phyletic memory.

According to Bergson, intuition is required to understand a problem and solve it, which is often not possible by means of intelligence. Furthermore, the creativity for which intuition is required cannot be achieved only with simple intuitive behavior. Creativity can emerge only by the combined effort of brain–body integration, including memory, the capacity for focused attention, knowledge, and rich episodic and semantic memories. The ability of association and speed of attention are also prerequisites for creativity.

17.7.2 What Is the Place of Bergson's Work in Memory and Quantum Brain?

In evolution theory one is confronted with the concept of *diminishing entropy*, which in turn leads us to propose the existence of a Maxwell Demon.

The relationship between quantum physics and Bergson's theory has been addressed by preeminent authors (Papanicolaou and Gunter 1987). However, biologists and memory research scientists have yet to fully consider such issues within their fields. Chapter 18 proposes initiating a debate that considers Bergson's theories as a trailblazer in search of memory function, evolution of species, and the evolution of cognition and intuition. Processing the brain's alpha activity is a tenable example, and other empirical evidences may follow.

Chapter 18
Bergson's Intuition Memory and Episodic Memory

18.1 The Importance of Bergson's Philosophy in the Era of New Physics and Contemporary Biology

At the beginning of the twentieth century quantum theory and later quantum mechanics replaced the classical Newtonian view. Einstein's theory of relativity changed the idea of absolute time. Henri Bergson had a strong knowledge of physics and mathematics and was able to interpret Einstein's theory of relativity, although Einstein was not in agreement with Bergson's concept of relativity. Bergson somewhat changed his interpretation and also created ideas related to the indeterministic world.

Several authors (Beauregard 1987; Stapp 1987) recognized that Bergson's interpretation of a new type of philosophy, in essence, was similar to Heisenberg's work on the S-matrix. These thoughts and interpretations are described by Papanicolaou and Gunter (1987) in their book (Beauregard 1987; Stapp 1987). Because Başar (1983a) argued that a possible interpretation of brain processes could be reached by Heisenberg's S-matrix theory and later by Feynman diagrams, it is important to mention Bergson's work in relation to time and intuition.

Bergson's interpretation of Darwin's evolution theory is also relevant to some new models. The present chapter describes the essence of Henri Bergson's work on time and intuition. This description is a prerequisite for understanding Chap. 17, which investigates new avenues of evolution theory, intelligence, and the electrophysiology of the brain.

According to this reasoning, the aim of this chapter is to provide a new perspective on Henri Bergson's core ideas of duration *(La dureé)* and intuition. Bergson's philosophy was also a candidate to provide a bridge between quantum theory and the general framework of philosophy.

E. Başar, *Brain-Body-Mind in the Nebulous Cartesian System: A Holistic Approach by Oscillations*, DOI 10.1007/978-1-4419-6136-5_18,

18.2 Essentials of Intuition

Intuition is the human ability to perform creative thinking and apply these processes to produce inventions or new discoveries. This ability is a property of human beings and is not seen in animals. Cats, dogs, and primates can display a relatively sophisticated level of intelligence and emotional behavior. However, they cannot make discoveries; they cannot develop instrumentation, wheels, or vehicles. In other words, they cannot perform synthesis on accumulation of their knowledge and facts from the environment. We understand this superiority of the human brain as intuition.

The next step in understanding the dynamics of intuition consists of identifying a way to measure the duration of these dynamic intuitive processes. According to Bergson, the dimension of time in the process of intuition is not real time; it is not measurable with conventional physical external clocks. The timescale of intuition is an internal time and is inhomogeneous. We additionally understand that the time of intuition is Bergson's concept of duration.

18.3 S-matrix

Heisenberg (1943) pointed out that the full empirical dynamical content of relativistic quantum theory is contained in a certain function of momentum vectors called S-matrix (see also Chaps. 15 and 16). This quantity involves neither time nor its dual energy variable that distinguishes before from after. In a recent publication, we discussed the importance of Heisenberg's S-matrix and the use of Feynman diagrams in brain research (Başar 1980, 1983a, 2009; Başar and Güntekin 2007).

In classical theories of dynamics, the momentum distribution function is independent of the spatial distribution function. However, in quantum theory, all of the necessary information may be obtained from either of the two functions. Thus, one can represent the full mathematical content of quantum theory without explicitly referring to the special continuum, simply by expressing everything in terms of momentum variables (Schmidtke 1987).

An important aspect of S-matrix representation in quantum physics is that the equations that determine the S-matrix can be expressed in terms of the S-matrix alone, i.e., without reference to the continuous time variable or its dual energy variable (Schmidtke 1987).

The content of physics, as presented in the S-matrix, is expressed in terms of inertial states and a before-and-after variable that accommodates discontinuous jumps between these inertial states. This conceptual framework appears to conform reasonably well to Bergson's concept of intuitions. According to this view, Başar (2009) proposed the use of Feynman diagrams to present complex processes in the brain. This is proposed in the following way: The S-matrix consists of functions corresponding to the different possible combinations of initial and final systems.

These are systems that, if left to themselves, would endure, unchanging, for all eternity. Each of these functions is a function of the momentum of each of the initial of final stable systems.

18.4 Bergsonism: Material and Memory and Creative Evolution

Several authors point out that Bergson's best known work is his 1907 book, *Creative Evolution*, in which the concept of vital force[1] is put forth to explain why living beings are fundamentally different from inanimate matter. Further, Bergson states that the "superiority of human beings" to all other living beings is because of the human brain's intuition. By using a terminology of non-linear systems, one may say that this is the most important bifurcation that occurred in nature to make the human mind unique.

Jacques Monod (1971) commented that Bergson's poetry was not able to establish a durable theory. According to Deleuze (1966), there were no solid links among the different books of Bergson, which accounted for the declining influence of Bergsonism in the second half of the twentieth century. By reading the original text of *Matter and Memory* (1896) and *Creative Evolution* (1907), it is possible to discover the existence of several leitmotivs in Bergson's work. In the author's opinion, the existence of two types of memories, the concept of duration, and the description of human intuition inspired by Darwin's theory on the evolution of species (1859) constitute a solid chain of ideas that were described in an era of research in which electronic instrumentation and diverse methods of analyzing anatomy and electrophysiology were not available. The following statement should be emphasized: Had Bergson been able to draw on the experiences of modern neurophysiology and brain oscillation theory, he would certainly have been able to produce a much more profound description of creative evolution. His idea to use the Darwinian concepts to analyze the creative evolution of the human brain was inspired, as Chap. 17 explains. The following text presents Bergson's essential ideas related to the two types of memory.

Bergson said that if a perception occurs within a very short time it always needs certain duration and as a consequence of the effort of the memory, "Perception and memory are not separable." Our percepts mostly reflect images in our environment; the first component is the sensation. However, the memory also embraces a realm of remembering linked to our immediate perception. A multiplicity of moments constitutes the essential core of subjective perception. Bergson's *Matter and Memory* (1896) distinguishes three types of memory: pure memory, image memory, and perception. None of these components can occur separately. Perception is not characterized by simple contact of the mind with the present object. Rather, perception is embedded in the memory of images that, in turn, participate in pure remembering.

---

[1]Elan vital

The views of Hayek (1952), Fuster (1995b), Baddeley (1996), Desimone (1996), and Başar (2004) strongly assume that perception and memory are inseparable. In other words, these authors support once more the unification of perception and memory first introduced by Bergson.

18.5 What Is Time? What Is Duration?

Does Bergson's concept provide a gateway to a transcendental philosophy and metaphysics of the brain?

Figure 18.1 offers a schematic description of two different time positions, homogeneous time and inhomogeneous time.

Homogenous time is the traditional "real-time" measurement with physical clocks. In inhomogeneous time, it is not possible to measure real time, because during sleep and episodic memory the brain can condense time periods of many years to a fraction of second. Long episodes also occur within short periods of time in dreams.

According to Deleuze (1966), Bergson commented that science seeks prediction and utilizes concepts and intellect in its methodology. Scientists need not claim that reality is deterministic in any metaphysical fashion to pursue their goals. Bergson recognized the utility of prediction in an intellectual order, but suggested that the vitality *(élan)* of life is understood better in terms of duration and recognition of its fundamental unforeseeability. Past-Present-Future are not three separate points or

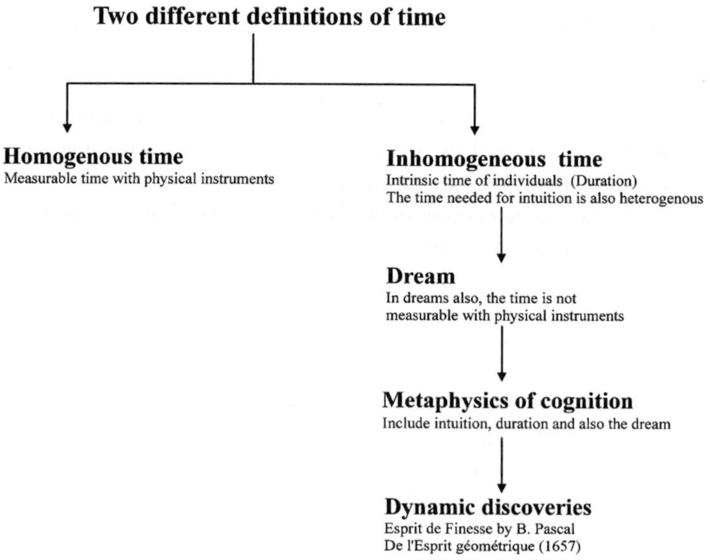

Fig. 18.1 Two different definition of time

areas on a line; rather, for Bergson, the past real memory is flowing through consciousness; the present is continuous perception with its characteristic *durée;* and the future is the creation, newness, and the unforeseeability of experience. *Durée*, thus, is best understood intuitively. We must be careful not to replace the uniqueness, continuity, and flowing quality of our temporal experience with quantitative, objective, instants of time:

> "…as soon as one aspect of it *(durée)* is treated in isolation – and this happens almost inevitably when we begin to talk about it – such treatment tends to be misleading; we unconsciously confer the discontinuity of the discourse into the complex indivisibility of the referent" (Deleuze 1966).

Is the duration of thought and Alpha correlated?

According to Bergson, reality is mutable, continuous, heterogeneous, and indivisible; moreover, it is evolving and creative. We can understand this reality intuitively and thus grasp its dynamic, creative, changing character, or we can conceptualize it intellectually and deal with it objectively or quantitatively. Bergson's opinion is that science and metaphysics are products of intuition. "A thoroughly intuitive philosophy would realize the much desired union of science and metaphysics." The following illustration is modified from Charles Schmidtke (1987) (Fig. 18.2).

In Fig. 18.2, past memory and anticipation of the future are connected with the *duration.* Further, the important feature of the diagram is that duration is the linking element between past and future. According to Schmidtke (1987), this means that, although we are participating agents, we are involved in a continuous process of creating and retaining memories, as well as using and recalling memories. The meaningfulness of our memories requires that they remain within a context and remain dynamic. To this consideration, two important remarks must be added.

1. Our memory is not considered as a static function, stored somewhere in the brain and is used when we try to remember the past. When thinking or during processes of creativity, the memory is interwoven with or embedded in the duration of the creativity. This means that the process of memory evolving is included in the duration and is a completely dynamic process.
2. The memory is acting and performs functions, because it is not separable from all other brain functions. Memory and duration are interwoven (Schmidtke 1987).

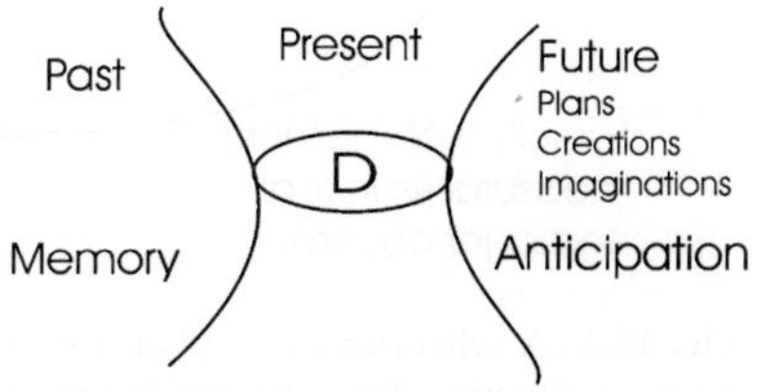

Fig. 18.2 For the explanation see the text (modified from Charles Schmidtke 1987)

18.6 A New Interpretation of Intuition and Duration in Relation to Creative Processes

This concept is somewhat different from Bergson's intuition and duration. During an intuitive process, our brains require high-speed access to past memories, meaning episodic and semantic memories. Further, the intuitive human brain must be able to transform events in the episodic and semantic memory to a type of intermediary virtual memory, which extends past events to some future projection. This may be explained using a simple example. We can recall the figure of our partner and imagine her or him wearing new clothes that we have just seen in a department store. According to Bergson, this is pure imagination. Certainly, this is a very simple example, which is suitable to explain the core of the creative memory or the necessary way of thinking. To perform this process in reality might require several hours. However, this type of creative synthesis can be performed in our creative memory in a fraction of a second. In the author's view, this is what duration means in the processes of discoveries in the creative mind (Fig. 18.3).

A more complicated or difficult example would be the application of a mathematical equation to an observed phenomenon; again, an experimenter or mathematical scientist can foresee results that are not measurable with external clocks in a very short time (Penrose 1989). Note the famous story of Henri Poincarré, who proposed a mathematical solution (which he intuitively felt to be correct) to a difficult mechanical problem that could be solved later with a number of mathematical steps

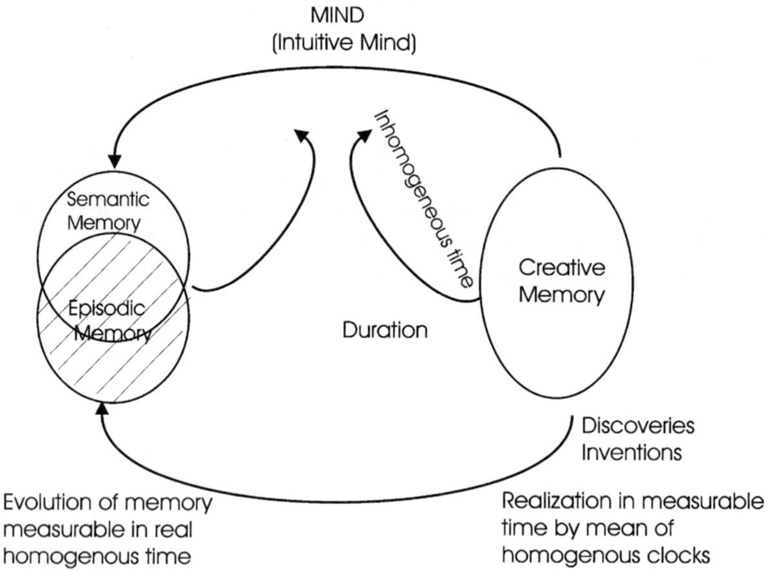

Fig. 18.3 A schematically explanations of link between episodic memory, creative memory by means of duration, for more detailed information see text

that certainly took longer when measured with external clocks (See Chap. 20). At this point, we should make two comments:

1. Such an intuitive process is dynamic, in essence, non-linear dynamic. We say non-linear dynamic, because the time axis is inhomogeneous; possibly even a nebulous time axis, as described by Başar and Güntekin (2007).
2. The second point is the importance of memory during the duration and occurrence of intuition. Without memory it is impossible to transform past memory into a creative memory (see Fig. 18.3). If we accept this chain of reasoning, it is easy to recognize that memory, duration, and intuition are inseparable. This is similar to the space-time-matter continuum in physics. Our creative brain or thinking brain moves in a universe that consists of memories of real life and virtual life created from memory. Additionally, this universe is not separable from our body.
3. This is also similar to what is observed about the processes of dreams. The time in dreams may virtually encompass years, hours, or minutes, whereas real time may span only a few minutes (see Fig. 18.1). Is the ability of our brains to be aware of very long processes in short physical time measured by clocks in creative thinking? Do creative thinking and dreaming require similar inhomogeneous time? Is the "zipping" of time during creative evolution the basic property of the brain, or the essential key to understanding metaphysics of the brain? The elapsed time during creative processes and dreams may extend to years. Moreover, is this time evolution a macro-process of time expansion in Einsteinian trips to the galaxy?

18.7 Lessons from Bergsonism

Why is Bergson's concept important to a theory of brain-body-mind integration? There are several reasons:

1. Chap. 17 describes an electrophysiological synthesis of creative evolution and Darwin's theory on the evolution of species. The contents and implications of this chapter are possibly the first step toward unifying a biological empirical theory (Darwinian evolution) and a philosophical interpretation of Darwinism together with new electrophysiological measurements that are useful to amalgamate these three items as a tentative step toward a new type of model. The establishment of such a concept required the synthesis of ideas from several fields of physics, particularly thermodynamics. Whether this synthesis is reliable may be tested in the future via experimental studies. Here the lesson from Bergson is that a philosopher may develop a theory by closely analyzing empirical results. In developing the present chapter, the author studied the original text of Darwin's *On The Origin of Species* as well as Bergson's *L'Evolution Créatrice*. The author discussed electrophysiological parameters according to Bergson's question, "What is different in the brains of humans in comparison to other lower

species?" Synchronization, amplitude, and entropy of alpha activity were identified as relevant parameters for comparison. Further investigation indicated that human alpha activity is a potential candidate to explain the electrical behavior related to intuition. For philosophers to develop new theories and, accordingly, greater understanding or new discoveries, it is necessary to ask the appropriate initial questions. From the time of Descartes, Pascal, Hume, and Locke, the world of science has witnessed great discoveries. However, at the beginning of the twentieth century almost no philosophers studied Darwin, James, Stuart Hill, and Albert Einstein to ask new questions following major advances in scientific understanding. Following from the path-breaking work of Bergson, we have an opportunity to perform more empirical analysis to further the understanding of the mind.

2. Bergson introduced the gateway to new definitions of memory and duration. As stated in previous sections, duration is a method of explaining inhomogeneous time. Inhomogeneous time is also interwoven into the essence of episodic memory. We do not yet have the means to measure inhomogeneous time. However, neither did Descartes, Pascal, and Locke have the means of measuring thoughts, nor as we call them today, cognitive processes. By asking how to measure inhomogeneous time, one day it may become a possibility.
3. From this chapter as well as Chap. 17, it should be evident that philosophy can serve as a means to understand processes in physics and biology. Carl Friedrich von Weizsäcker (1985) extended the classical causality principle to the modern probabilistic causality concept applied in quantum physics. Multiple causalities in brain-body-mind integration are also probabilistic in their nature. The theory of the probabilistic brain or quantum brain also derives from relevant philosophical questions. The question of whether brain-body-mind integration can be better understood by means of the S-matrix or Feynman diagrams cannot yet be answered completely. If these problems can find relevant application in future, it will demonstrate the importance of this method of philosophy.

Chapter 19
Towards Metaphysics: Conscient and Unconscient States

19.1 A Short Physiological and Psychological Classification of Unconscious States

Why does this chapter start with a classification of unconscious states? Its thesis emphasizes that conscious states can be understood only if we try to also learn processes occurring in unconscious states.

The study of consciousness is an extremely difficult task, which in the past often led to erroneous statements.[1] However, in a book on the machinery of the mind, it is almost obligatory to say a few words about this concept.

A classification of various states of unconsciousness is given in Fig. 19.1. This illustration does not reflect a complete view of unconscious states. However, according to several results, an objective definition of mind should encompass all physiological and psychological fundaments and processes. Several chapters (4 through 6 and 13 through 15) explain that changes in blood pressure, respiratory cycles, and release of transmitters are processes influencing or contributing to the functioning of the mind. The examples given in Chap. 20 show the necessity of considering unconscious states as processes that are part of real life, and that unconscious processes are interwoven with conscious processes. These processes are not separable, as the examples of Poincaré, Mozart, Proust, and Loewi in Chap. 20 demonstrate. Accordingly, this book teaches that a description of consciousness is only possible by incorporating conscious and unconscious states.

[1] Roy John organized a successful meeting Havana in 1989. This meeting was followed by a 3-day symposium. A session on consciousness was planned on the first day; however, the event was chaotic and consequently the discussions did not come forward. At the end of the first day Roy John said, "Erol, you have organized and published two successful conferences in Berlin, and have sufficient experience. Why don't you take over the discussions the following day?" I accepted with the condition that discussions of consciousness would be stopped. We then had very fruitful discussions over the following days. I remember that Karl Pribram, Hermann Haken, Walter Freeman, and Christophe Deecke were also present at the symposium.

E. Başar, *Brain-Body-Mind in the Nebulous Cartesian System: A Holistic Approach by Oscillations*, DOI 10.1007/978-1-4419-6136-5_19,

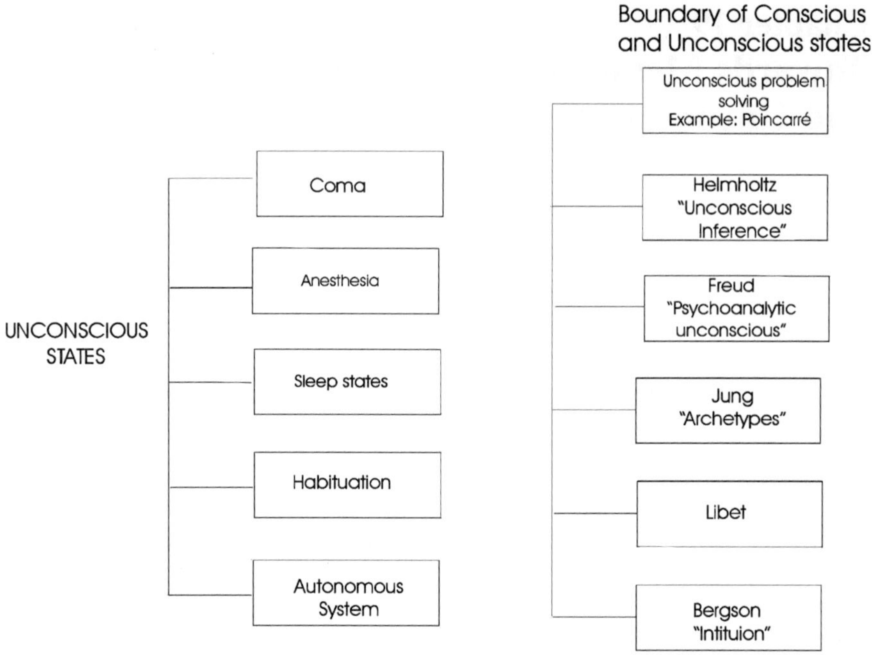

Fig. 19.1 Unconscious states and boundaries of conscious and unconscious states

19.2 Physiologic Unconsciousness

Coma is a profound state of unconsciousness. A comatose patient cannot be awakened, fails to respond normally to pain or light, does not have sleep–wake cycles, and does not take voluntary actions. Coma may result from a variety of conditions, including intoxication, metabolic abnormalities, central nervous system (CNS) diseases, acute neurologic injuries such as stroke, and hypoxia. It may also be deliberately induced by pharmaceutical agents to preserve higher brain function after another form of brain trauma has occurred.

19.2.1 Anesthesia

In modern medical practice, general anesthesia is a state of total unconsciousness resulting from general anesthetic drugs. A variety of drugs are applied to the patient that have different effects, with the overall aim of ensuring unconsciousness, amnesia, and analgesia. Inhalational anesthetic substances are either volatile liquids or gases; they are usually delivered using an anesthesia machine. An anesthesia machine composes a mixture of oxygen, anesthetics, and ambient air, delivering it to the patient and monitoring patient and machine parameters. Liquid anesthetics are vaporized in the machine.

19.2.2 Dream States

According to Fuster (1995a), dreaming is consistent with the activation of cortical memory networks, refashioned in the absence of guiding reality or critical judgment. However, the memory activated in dreams is not only distorted and elaborated, but it is often beyond the recognition of the awakened dreamer.

The dream, although replete with past experience, is also experienced in the present, and it lacks the phenomenal attributes of past and future (Fuster 1995a).

According to Hobson, dreaming has its origins in the same region of the pons that generates rapid eye movement (REM) sleep, which is said to produce relatively chaotic signals that activate the forebrain and force out the noisy input it is receiving from various intermittent events, primarily in the form of ponto-geniculo-occipital (PGO) spikes (Hobson and McCarley 1977). This noisy input is a candidate to be the key reason why dreams are bizarre and disjointed. Recently, Hobson confirmed that there is a greater degree of forebrain control of the REM mechanism than in his original theory, especially through the hypothalamus (Hobson et al. 2000b).

On the contrary, Solms (1997) asserted that his research with brain-injured patients proves that dreaming is possible without REM. In conjunction with other research, Solms' results led him to locate the origins of dreaming in the ventral tegmental area of the midbrain, a few centimeters from Hobson's point of origin (Solms 2000a).

A critique of the Hobson versus Solms debate was published by Domhoff (2005).[2]

19.2.2.1 Blaise Pascal's "Mythos on Dreams"

Blaise Pascal described two dreaming persons in his famous book, *Pensées:* "A king has dreamed during every night that he was a simple poor worker, whereas the poor worker has dreamed every night that he was living as a king." Pascal asks the question whether the king and the worker would have in this case a similar type of life. Happiness and unhappiness is mixed during day and night. The crucial questions are, "Are dreams as important and real as our everyday conscious life?" and "Where is the boundary between real life and unconscious life?"

19.2.2.2 Dream of Otto Loewi

Otto Loewi (1921) discovered the important transmitter acetylcholine as described in Chaps. 2 and 3. The discovery of this transmitter was crucial; but the dream of Loewi preceding the experiments was also extremely thought-provoking.

---

[2] I invited Mark Solms to the World Meeting of Psychophysiology in 2006 in Istanbul. I had a very long and productive discussion with him and Bahar Güntekin during a dinner on the Bosporus. His new Freudian view may add important insights to the theory of unconscient states, emotions, and sleep.

This famous dream follows:

On Easter Saturday 1923, Loewi dreamed of an experiment that would prove once and for all that transmission of nerve impulses was chemical, not electrical. He woke up, scribbled the experiment onto a scrap of paper on his night-stand, and went back to sleep. The second night he had the same dream. This time, he got up, went to the lab, did the experiment, and by dawn, knew that there was a Nobel in his future. Fourteen years later, there was.

19.2.3 Autonomous System

The autonomous system is described in Chap. 4. The autonomous system also works in the absence of electrical signals from the brain; accordingly, its functions can be classified as unconscient bodily work.

19.3 Psychological Unconsciousness

19.3.1 Habitual Behavior

Procedural Memory is a kind of "bodily" memory. It is memory for habitual motor skills, or as Solms and Turnbull (2002) state, general perceptuo-motor or ideomotor skills. This type of memory allows us to learn skills and abilities. Constant repetition in the learning phase is especially important for procedural memory. All levels of ideomotor ability, from walking to playing the piano, are skills that are learned gradually. Skills such as riding a bicycle are also extremely resistant to decay with time. Procedural skills are hard to learn, but also hard to forget (Solms and Turnbull 2002). The literature teaches us that cortical motor structures in the parietal and frontal lobes are engaged in procedural learning. However, once a skill becomes habitual, the motor program representing it is progressively consolidated into non-cortical structures mainly in the basal ganglia and cerebellum. In this functional process there is a switch of the function from cortex to deeper brain areas (see Fig. 5.5, Solms and Turnbull 2002, p 158). We may further say that the voluntary learning process becomes an unconscious functioning process. This means in turn that a change occurs in the brain between involved functional structures as the brain under goes a transition from conscious to unconscious. An important feature of procedural memory is its *implicit functioning*. Habitual behavior is executed automatically and therefore unconsciously.

Further, it can be said that this automatic functioning has parallels with the autonomous functions of the CNS. The difference is that the *autonomous system is genetically wired*. Habitual behavior includes the effects of continuous ongoing processes that we almost do not perceive after a while: If we live near a waterfall, with time we no longer hear the noise of the water. If we are near a railway station, with time we do not mark the passing of the trains. Such events were described by I. Kant.

19.4 Sigmund Freud and Gustav Jung

19.4.1 Sigmund Freud and the Unconscient

The theories of Sigmund Freud have again become a focal point among several neuroscientists. As initiators of this new stream we first mention Mark Solms and Eric Kandel. The modern view of the scope of Sigmund Freud is described mainly by Mark Solms; the following briefly summarizes the modern interpretation of Freud's hypothesis. Freud's theory of the mind was based on the concept of the human organism as a complex biological machine. According to Freud, psychological processes are deeply rooted in the body's physiology and biochemistry and follow the principles of Newtonian mechanics (meaning they are not probabilistic). Mental life in health and illness reflect the interplay of instinctual forces within the organism and their clashes with the external world. This volume also searches the roots of the mind in physiological and biochemical mechanisms as well as deeper brain structures such as the brain stem. However, we introduce physical concepts and mathematical tools for the interpretation of the empirical results.

The most significant contribution Freud made to neuroscience was his argument for the *existence of an unconscious mind.* During the nineteenth century the dominant trend in Western science was positivism. Freud, however, proposed the concept of unconsciousness by assuming that awareness exists in layers and that there are thoughts occurring below the surface. The basic proposition of Freud was that our motivations remain mostly hidden in our unconscious minds. Moreover, they are actively withheld from consciousness by a repressive force. The executive apparatus of the mind (the Ego) rejects any unconscious drives (the Id) that prompt behavior that are incompatible with our civilized conception of our self.

According to Mark Solms *(Scientific American)*, Freud is back in fashion with his theory of the unconscious. Further, Eric Kandel states that psycho-analysis is "still the most coherent and intellectually satisfying view of the mind." Freud introduces the central notion that most mental processes that determine our everyday thoughts, feelings, and volitions occur unconsciously. Further, analysis of the behavior of patients who are unable to consciously remember events that occurred after the damage of memory and coding structures show that their brains are clearly influenced by the forgotten events. Different memory systems process information consciously or unconsciously (implicitly and explicitly). Neuroscience has shown that the major brain structures essential for forming conscious (explicit) memories are not functional during the first 2 years of life, providing an elegant explanation of what Freud called infantile amnesia. As Freud surmised, it is not that we forget our earliest memories; we simply cannot recall them to consciousness.

19.4.2 Gustav Jung and Archetypes

Freud never abandoned the basic Cartesian orientation of his theory. Gustav Jung, by contrast, was not so much interested in explaining psychological phenomena in terms of specific mechanisms, but rather attempted to understand the psyche in its totality and was particularly concerned with its relations to the wider environment. The key difference between the psychologies of Freud and Jung is their views of the *unconscious*. For Freud, the unconscious was predominantly personal in nature, containing elements that had never been conscious and others that had been forgotten or repressed. Jung acknowledged those aspects, but he believed that the unconscious was much more. Jung's concept of the collective unconscious distinguishes his scope on psychology not only form Freud's but from all others. Jung saw the *unconscious* processes involving "collectively present dynamic patterns" that he called *archetypes*. These patterns, formed by the remote experience of humanity, are reflected in dreams as well as the universal motives found in myths and fairy tales around the world.

Archetypes, according to Jung, are "forms without content," representing merely the possibility of a certain type of perception and action. As described in Chap. 15, our concept of the new Cartesian system or nebulous Cartesian system aligns better with the more extended view of Jung. Further, the modern view of *genetics* reconciles well with the theory of archetypes. Jung was attracted to quantum mechanics as a result of his longstanding interaction with physicists, especially Wolfgang Pauli. He described in *Aion* that we may have to consider transcendental concepts such as quantum theory to understand the psychology of the brain. We are in good agreement with this idea as it relates to the quantum brain, described in Chaps. 14 through 16.

19.5 A General Scheme Jointly Analyzing Unconscious Conscious and Preconscious States

The previous sections discussed levels of consciousness by giving a detailed account of them. After surveying the literature and considering the explanations given in the previous sections, we then formulated a more abstract scheme. What is new in our classification?

1. We separate physiologically induced consciousness from the unconsciousness described by Freud. However, the unconsciousness, which also takes into account sleep and cognitive latent states, is more comprehensive that the unconsciousness described by Freud. Unconscious behavior can be also extended by the archetypes of Gustav Jung (Fig. 19.2).
2. The preconsciousness described here also has an extended context in comparison with Freud's description. Freud described a transition state. However, we have to add discoveries that take place during vigilance and also immediately upon waking to this transition state. Possibly here the topological activation of

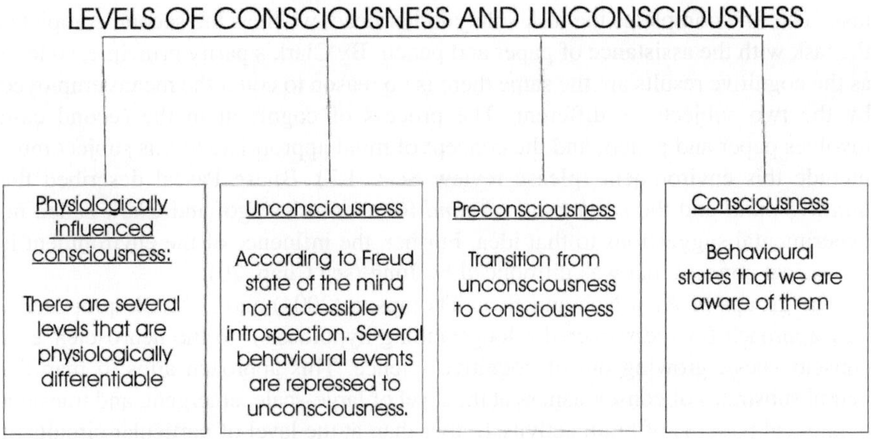

Fig. 19.2 Levels of consciousness and unconsciousness

the brain is similar to the configuration proposed by Mark Solms and Oliver Turnbull (2002, Fig. 5.5, p 158). Also on the contrary discoveries can take place during the transition from an unconscient state. Do those transitions correlate with intuitive behavior as well? Do learning states also belong to preconscious states? The end of this chapter discusses these questions.

3. The interference of memory is almost obligatory in the given description. The interaction between memory-intuition and thoughts deliberated during the preconscious state are described at the end of the chapter, which also emphasizes the view of Marcel Proust related to episodic memory.

19.6 Do New Integrative Trends of Mind and Consciousness Exist?

Philosophy is an extremely respectable branch of science; however, it should have a mathematical, experimental, and strong scientific base.

To talk about mind and consciousness, one needs strong empirical evidence as well as insight into this evidence, which is one of the claims of this book.

Sitting in a comfortable chair, with books on the desk and a sharp pencil in the hand, may lead to interesting ideas, but also to erroneous avenues.

When this book was almost finished, I considered philosophical trends and the work of several authors, which will be described in the following.

According to Andy Clark, the dynamic loops through which mind and world interact are not merely instrumental. The cycle of activity that runs from brain through body and world and back again actually constitutes cognition. The mind, by this account, is not bounded by the biological organism, but extends into the environment of that organism. Consider two subjects carrying out a mathematical task. The first completes the task solely in her head, whereas the second completes the task with the assistance of paper and pencil. By Clark's parity principle, so long as the cognitive results are the same there is no reason to count the means employed by the two subjects as different. The process of cognition in the second case involves paper and pencil, and the concept of mind appropriate to this subject must include this environment (please review Sect. 1.7). Blaise Pascal described the *intuitive mind* and the *mathematical mind* four centuries ago, and Clark added no experimental suggestions to that idea. Further, the influence of the environment is a classical concept that was introduced by Jung (see Chap. 20).

The groups of Alva Noe and Evan Thompson (2004) and F. Varela proposed a new approach (or rediscovered a longstanding hypothesis)[3] to the neuroscience of consciousness, growing out of cognitive science. This approach aims to map the neural substrates of consciousness at the level of large-scale, emergent, and transient dynamical patterns of brain activity (rather than at the level of particular circuits or classes of neurons), and it suggests that the processes crucial for consciousness cut across brain–body-world divisions, rather than being brain-bound neural events. Resonant neural assemblies and large-scale brain integration have also been articulated by this group.

In this hypothesis, no proposals are made about how the brain and body are linked. For large-scale brain work the reader is referred to the work of E. Başar (especially *EEG-Brain-Dynamics* [1980] and Chaps. 6 through 9 of this volume).

[3] In all the works of the groups of Varela, Lachaux, Thomson, etc. no unique citation is recorded about Başar's considerable publications over the last 30–40 years.

Chapter 20
Mysteries of the Mind: Conscient and Unconscient States in Creativity and Sleep

20.1 Thoughts on the General Scope of Metaphysics

Patricia Churchland (2002) used the expression "pure metaphysics" to refer to the school of thought that assumes that metaphysical answers are beyond the reach of scientific methods and discoveries, and that the job of metaphysics is to lay the absolute foundation for all sciences. Churchland said that Aristotle's rather innocent expression "first philosophy" came to acquire a more self-important significance associated with supra-scientific methods and principles.

According to Max Tegmark (2003), the borderline between physics and metaphysics is defined by whether a theory can be tested experimentally, not by whether it appears strange or involves unobservable entities. The frontiers of physics have gradually expanded to incorporate ever more abstract (and once metaphysical) concepts such as a round Earth, invisible electromagnetic fields, the slowing of time at high speeds, quantum superposition, curved space, and black holes. Unusual progress in physics at the turn of this century surprised conventional scientists, especially biologists, suddenly and in an unexpected way. In the three centuries following Newton's mechanics, the Newtonian–Cartesian paradigm opened major doors for scientists to understand conventional physical phenomena at medium scales, not in macro-space or within a microscopic framework. According to Newton's framework, engineers could easily calculate the trajectory of a launched projectile.

This is impossible in quantum mechanics.

Tunnel effect, time reversal, fuzzy systems, and the generation of supernovae manifesting huge energy transfers in the galactic system are strange phenomena for biological scientists; before the beginning of the twentieth century they were also strange phenomena for physicists.

Progress within the last few decades means we can partly measure attention, some remembering processes, sensations at the perception threshold, and predict the occurrence of dreams. We can measure how the brain's electrical response differentiates known and unknown faces when shown to patients. In other words, progress and refinements have already penetrated the difficult barrier to the realm

E. Başar, *Brain-Body-Mind in the Nebulous Cartesian System: A Holistic Approach by Oscillations*, DOI 10.1007/978-1-4419-6136-5_20,

of biological sciences. Such progress means that meta- processes in the brain become testable processes.

The Cartesian system and the Newtonian–Cartesian way of thinking have created wonders. Can we now develop a type of nebulous Cartesian system, in which tunnel effects or unusual high resonance phenomena can also be discussed in the framework of theoretical brain dynamics? Are the concepts of unconsciousness described by Freud (a neurologists and psychologist) and unconscious inference described by Helmholtz (a biophysicist) somewhat comparable?

What was the evolution of concepts from Mesmer to Charcot to Freud? What are hidden variables in quantum physics? Do we have hidden sources in the brain as most EEG manifestations; and do they have a hidden, chaotic nature?

This chapter contains measurable ideas, measured and tested ideas, and also includes non-testable predictions. Philosophy and utopia have parallels. The utopian descriptions of several philosophers have been realized. For example, thoughts or thinking processes described by Blaise Pascal, and episodic memory described by Marcel Proust or Henri Bergson, can be partially measured now. The *Brave New World* described by Aldous Huxley (1932) could be partly realized, at least, with the cloning of animals or plants (although I strongly hope that this will not occur with human beings!).

The author (E. Başar) believes that imagination and utopian thinking provide important meat for science. He prefers also to start with measurements or experiments, by keeping in mind this sentence by Newton: *Hypothesis non fingo* (I feign no hypotheses). This is particularly the case in the book, *EEG-Brain Dynamics* (Başar 1980), which contains descriptions on oscillatory phenomena including the almost completely unknown gamma band activity of the brain. At the time of publication, this book was regarded as "interesting, but pure fantasy" by most neuroscientists. It took only 10 years to see the great change in neuroscience, as stated by Vernon Mountcastle.[1]

The present chapter describes a chain of ideas combined with experiments or strongly anchored in empirical foundations. It is attempted to separate reality from fantasy; however, the combination of science and meta-science may provide a useful synthesis to open the way for new measurements. The Gedanken (thought) experiment of the microscope by Heisenberg, which was described to explain the uncertainty principle, was recently realized (see Chaps. 2 and 15).

20.2 Web of Metaphysics and Unconsciousness: Consciousness Versus Unconsciousness

This chapter now studies the metaphysics of the brain, the *quasi-functions* of the brain that are not understandable by means of well-known and accepted concepts and tools. Processes that take place during unconscious states are not yet directly measurable. Nevertheless, these mechanisms belong to hidden but vital properties of the brain and

---

[1]Mountcastle: Preface in E. Başar and T. H. Bullock (1992).

possibly to the most important ones for creativity and discovery, such as *intuition*. According to Henri Bergson, intuition is the most important mental ability of humans, one that is not seen in any other species (see Chaps. 17 and 18).

Before dealing with the concept of intuition, let us classify conscious and unconscious states of the brain. Although states of consciousness and unconsciousness are linked or interwoven, we first prefer to show them separately and schematically.

20.2.1 Is the "Esprit De Finesse"[2] of Blaise Pascal Now Measurable?

In Chap. 2, the *intuitive mind* of Blaise Pascal is described in detail. A person with an intuitive mind must have solid knowledge of the problems that he or she wants to understand or solve. Further, to solve a difficult problem one requires focused attention, good working memory, association, and fast perception processes. As cognitive science made progress, the measuring of attention, association, and working memory became feasible. At the time of Pascal, no one could imagine that measurement of cognitive processes would be realized one day. Chapter 17 offered a speculative theory about the link of the brain's alpha activity and intuitive behavior. Is this a good theory, or is it mostly speculation? It can be assumed that this approach is at least semi-empirical, and that it can become a good testable hypothesis. The measurement of cognitive processes is a typical example of the way in which scientific progress means that metaphysical processes can become a reality or can be approached with measuring devices.

20.2.2 Is "Le Temps Perdu" of Marcel Proust a Pioneering Description of Episodic Memory?

The following description by Marcel Proust comprises an interesting essay written during his lifetime:

> One day in winter, on my return home, my mother, seeing that I was cold offered me some tea, a thing I do not ordinarily take. I declined at first, and then for no particular reason, changed my mind. She sent one of these squat, plump little cakes called "petites Madeleines", which looked as though they had been molded in the fluted valve of a scallop shell. And soon, mechanically, dispirited after a dreary day with the prospect of a dreary morrow, I raised to my lips a spoonful of tea in which I had soak a morsel of the cake. No sooner had the warm liquid with the crumbs touched my lips an extraordinary feeling was happening to me. An exquisite pleasure had invaded my senses, something isolated, detached, with no suggestion of its origin. And at once the vicissitudes of life had become indifferent to me, its disaster innocuous, its brevity illusory – this new sensation having had

[2] Intuitive mind.

> on me the effect which love has of filling me with a precious essence; or rather this essence was not in me, it was me.

Without being aware of it, Marcel Proust may have described what neuroscientists later called "episodic memory." (See Chap. 7 and references for the definition of episodic memory.) It can be also tentatively proposed that Bergson's concept of *duration*, discussed in Chap. 18, is also a part of the process of episodic memory.[3] To strongly argue for this, it is easy to consider an episodic memory process. If one remembers an important day in one's childhood, such events are usually recollected through a life story of 20 or 30 years. A person who remembers his or her own past needs only a fraction of a second to finish the remembering task. However, in this fraction of a second this person condenses a life story of the last 20 or 30 years. The fraction of time needed to experience an episodic memory event is not congruent with the remembered life story. This is not measurable with devices that gauge physical time. Accordingly, it can be proposed that this type of inhomogeneous time has parallels with the concept of "duration," as proposed by Bergson. In other words, as a next step it can be assumed that episodic memory and intuitive behavior have a common component: *inhomogeneous and undefined time.*[4] Do the time reversal in quantum processes and the backward voyage in time during the process of episodic memory have parallels in different metrics?

20.3 Unconscious Problem Solving

20.3.1 Poincaré

To provide some historical context to the power of intuition described in the previous chapter and its time duration, this section describes two examples that are explained by Penrose (1989, pp. 144, 175). Poincaré's narration is as follows:

> In order to participate to a geological research project I started a trip from my town Caen to Countances. During the excursion, due to a number of social events, I had forgotten my mathematical work. As we reached the city of Countances, we continued our trip with a bus. As I stepped up into the bus, I suddenly had the impression that the transforms I had used in order to describe the Fuchs functions could be identical to those in non-Euclidian geometries. At that moment I could not check this thought mathematically, since I did not have time, as I was participating in other talks. As soon as I returned to Caen as usually, I finished the mathematical operations in order to verify this idea (Poincaré).

The striking point in this example is this: As Poincaré oriented his conscious thinking to completely different areas, he suddenly found a solution for a complex judgment, which was later confirmed mathematically. This sudden emerging idea cannot be easily explained with words. Penrose says that Poincaré would need at least a seminar of 1 h to explain this idea to the experts. As Poincaré entered the

[3] As the manuscript of the present book was almost in print we encountered a book by Jonah Lehrer entitled, *Proust Was a Neuroscientist* (2007). It is a very interesting read.

[4] Marcel Proust and H. Bergson were relatives.

bus, the mathematical solution appeared in an unexpected way and as a homogeneous and clear idea. Further, the most interesting point is that Poincaré did not feel any suspicion about the correctness of this idea. This type of thinking is also not unusual for Penrose and a number of other scientists. The following also mentions the example of Sherlock Holmes, as described by Einstein.

20.3.2 Conan Doyle and Einstein

In describing the way in which Sir Arthur Conan Doyle's fictional detective Sherlock Holmes solves problems, Einstein pointed out the following:

> The great detective, however, realizes that no further investigation is needed at the moment, and that only pure thinking will show the pattern of relation between the collected facts. So he plays his violin, or lounges in his armchair, enjoying a pipe, when suddenly, by Jove, he has it! Not only does he have an explanation for the clues at hand, but he knows that certain other events must have happened. Since he now knows exactly where to look for it, he may go out, if he likes, to collect further confirmation on his theory.

For some thoughts of Einstein pertinent to a theory of brain-mind, see Sect. 2.7, in Chap. 2.

20.3.3 The Dream of Otto Loewi

Otto Loewi (1921) discovered the important transmitter acetylcholine, as described in Sect. 8.13 in Chap. 2. The discovery of this transmitter is crucial, but Loewi's dream preceding the experiments is extremely thought provoking as well.

This famous dream follows:

> On Easter Saturday 1923, Loewi dreamed of an experiment that would prove, once and for all, that transmission of nerve impulses was chemical, not electrical. He woke up, scribbled the experiment onto a scrap of paper on his night-stand, and went back to sleep. The second night, he had the same dream. This time, he got up, went to the lab. He performed the experiment and, by dawn, knew that there was a Nobel in his future. Fourteen years later, there was.

20.4 How Is Creativity Elicited?

20.4.1 Mozart

It has been said that Mozart had the ability to look through an entire piece of music with all the essential motifs. However, if Mozart attempted to perceive consciously, he would need a time interval equal to the performance of the entire piece of music. On the other hand, if Mozart started to read all the individual notes of this piece of music, it would be very difficult to determine the length of time needed to do this. This means the *metaphysical time* needed by Mozart to go through this piece of music would take much longer than was apparent to someone observing his rapid way of looking through of the entire piece of music. These examples indicate one thing: There is a difference between physical time or external time, measured by physical conventional synchronized working clocks, and time within our unconsciousness, non-measurable with physical clocks, during which we solve problems, register whole passages of music, or have the best ideas for new inventions or discoveries. The time measured or lived in our unconscient is probably the duration, i.e., metaphysical clock of Henri Bergson. The difference between the measured time in the physical world and the time of our inner world could also be approached with the process of accessing our episodic memory. By trying to remember an event that occurred a long time ago, we can rapidly go backward through time and access perceived or lived events.

20.4.2 Balzac's Description of Stefan Zweig (1932)

A similar example to Mozart is the creativity of the famous French novelist Honoré de Balzac, who described in his novels a great population of people with different names, abilities, and actions. Balzac had extraordinary powers of memory and could remember everything in his manuscript, including the names of characters, their abilities, their friends, etc. He also had the ability not only to remember but also to project in his mind events that were to come. Therefore, by writing the novels in his imaginary world, he could visualize all the scenes, dialogues, or important events that were going to be described. In other words his vision enabled him to see future spectacles that would be written later. By the time he was 12 years old, he could make a clear picture of the events in the books from which he read. In his words:

> As I was reading about the battle of Austerlitz, I saw all things that occurred. The bang of the cannon shots, the shout of soldiers penetrating my ears. I smelled the powder; I heard the tapping noise of the horses.

If this behavior is compared with the ability of Mozart to look through an entire piece of music with all of the essential motifs, one can easily recognize that Balzac had the ability to look through an entire section of a novel with all of its details. Accordingly, it can be stated that both geniuses, Mozart and Balzac, showed immense memory power and ability of association. One can also raise the question of whether creativity needs a huge ability of remembering and associating.

20.4.3 Hermann Hesse's Essay: How to Start Writing a Book

Hermann Hesse, one of the most gifted writers of the early twentieth century, was awarded the Nobel Prize in Literature in 1946. In one of his novels, Hesse described how he usually started to write a book. In his words:

> Finally, on a Saturday evening I was ready to start writing. When I plan to write a new essay or novel I cannot start immediately with the writing. I do not make a rigid schedule and plan; I am waiting for the moment I would collect my ideas and focus my mind to the process of writing. I cannot explain why I have chosen this moment at the Saturday evening; this is a feeling which started outside of my head and I cannot describe how this moment was coming.

In this description, we again witness a process that occurs in the unconscient world of a famous writer.

20.5 The Important View of Eric Kandel on the "New Science of Mind"

Eric Kandel (2006) raises the following question: "Where is the new science of mind heading in the years ahead?" He goes on to say that we have some understanding of the cellular and molecular mechanisms of memory storage, but we need to move from these mechanisms to the systems properties of memory. What neural circuits are important for various types of memory? How are internal representations of a *face, a scene, a melody*, or *an experience* encoded in the brain?

Further, Kandel (2006) says that major conceptual shifts must take place in how we study the brain so as to cross the threshold from where we are to where we want to be:

1. With regard to Kandel's proposal, one such shift will come from studying elementary processes – single proteins, single genes, and single cells – to studying systems properties – mechanisms made up of many proteins, complex systems of nerve cells, the functioning of whole organisms, and the interaction of groups of organisms.
2. Further, cellular and molecular approaches will certainly continue to yield important information in the future, but they cannot by themselves unravel the secrets of internal representations in neural circuits or the interactions of circuits – the key steps linking cellular and molecular neuroscience to cognitive neuroscience.
3. Kandel recommend the following steps: To develop an approach that can relate neural systems to complex cognitive functions, we will have to move to the level of the neural circuit, and we will have to determine *how patterns of activity in different neural circuits are brought together into a coherent representation.* To study how we perceive and recall complex experiences, we will need to determine how neural networks are organized and how *attention* and conscious awareness regulate and reconfigure the actions of the neurons in those networks. Therefore biology will have to focus more on nonhuman primates and human beings as the model systems of choice. For this, we will need imaging techniques that can resolve the activity of individual neurons and neuronal networks.

Kandel's view – as primarily a molecular scientist and psychiatrist creating a new trend – is of basic importance, as it suggests that only then can we address, in biologically meaningful terms, the theories of conscious and unconscious conflicts and memory first proposed by Freud in 1900. Kandel likes to develop a reductionist approach to the problem of attention by focusing on how *place cells in the hippocampus* create an enduring spatial map only when an organism is paying attention to its surroundings. What is the nature of this spotlight of attention? How does it enable the initial encoding of the memory throughout the neural circuitry that is involved in spatial memory? What other modulatory systems in the brain beside dopamine are recruited when an animal pays attention, and how are they recruited?

Do they use a prion-like mechanism to stabilize place cells and long-term memory? Obviously it would be beneficial to extend such studies to people. "How does attention allow me to embark on my mental time travel to our little apartment in Vienna?"[5]

Freud has added the interesting idea that, although we are not aware of most instances of mental processing, we can gain conscious access to many of them by paying attention. From this perspective, to which most neural scientists now subscribe, most of our mental life is unconscious; it becomes conscious only as words and images. Brain imaging could be used to connect psychoanalysis to brain anatomy and neural function by determining how these unconscious processes are altered in disease states and how they might be reconfigured by psychotherapy. Given the importance of unconscious psychic processes, it is reassuring to think that biology can now teach us a good bit about them.

20.5.1 A Remark on Hawking's Arrow of Time[6]

In physics, *space-time* is usually interpreted with space being three-dimensional and time playing the role of a fourth dimension that is of a different sort from the spatial dimensions. According to certain Euclidean space perceptions, the universe has three dimensions of space and one dimension of time. By combining space and time into a single manifold, physicists have significantly simplified a large number of physical theories, as well as described in a more uniform way the workings of the universe at both the galactic and subatomic levels.

Hawking (1988) explained that the laws of science do not distinguish between forward and backward directions of time. He further stated that there are at least three arrows of time that distinguish the past from the future. They are the thermodynamic arrow, the direction of time in which disorder increases; the psychological arrow, the direction of time in which we remember the past and not the future; and

[5] Eric Kandel was born in Vienna; he later emigrated to the United States.

[6] Hawking's theory is now often strongly criticized. However, Hawking talks about physico-biological metaphors that can be highly attractive to forebrain modelists and those wishing to model the forebrain.

the cosmological arrow, the direction of time in which the universe expands rather than contracts.

20.6 Episodic Memory and Search for "Temps Perdu:" Traveling Back to our Apartment in Istanbul In 1944

E. Kandel asks, "How does attention allow me to embark on my mental time travel to our little apartment in Vienna?" This time-space travel is, without any question, the sudden travel back in time through the process of episodic memory, which is triggered by an episode (or event) in search of temps passé, or time perdu,[7] as Proust expressed it.

The author had a similar experience, as described in the following:

> I left my home city, Istanbul, in 1958 as a 20 year old student, to study in Germany. Later, I went to New York as a postdoctoral fellow; then I came back to Ankara, Turkey 11 years later as a member of Hacettepe University. In 1980, I returned to Germany for more than 20 years; and finally I came to Istanbul as a member of Kultur University 4 years ago. In other words, from 1958 to 2006 almost 50 years passed with only short visiting sojourns in Istanbul.

Interesting enough, following my return to Istanbul, I started to remember detailed episodes covering my childhood and time at school. An important example is related to a book entitled, *Beethoven* by the famous French Nobel Prize winner Romain Roland. This book was brought to our home by my father in the winter of 1944–1945. Since I was not able to read such a book during that infamously cold winter, my father read the book aloud for my brother and me. Later, during my studies at university, I read the original French version of the book because of my deep interest in the story. The French edition is still on my bookshelf; however, I do not pay special attention to this one. Around 20 years ago I visited Beethoven's house in Bonn and I was not overly impressed. In contrast, several months ago I found a 1944 copy of that Beethoven book in an antique book store – the same edition, with blue and red pictures related to relics from Beethoven's life. This book triggered in me the deepest interest and joy. Figure 20.1 illustrates one of the blue and red photos from the book. For me, this book produced a much greater impression than Beethoven's actual house in Bonn. It seems that this special book on Beethoven, seen in early childhood and within my own family home, provides a special cue to travel back through time and memory, similar to the case of Marcel Proust. In that case, the real Beethoven's house was not important; the old color pictures shown in Fig. 20.1 are more essential. Moreover, it is most possible that place cells offer an important contribution to these types of episodic (and emotional) memories.

As I held the old book in my hands, hundreds of episodes from my childhood between 1945 and 1947 reappeared in my memory. I saw our apartment in Taksim

[7] The lost time.

Fig. 20.1 Beethoven and Beethoven's house, illustrated in a book on Beethoven by Romain Roland, printed in 1944

in Istanbul, with old paintings, my parents' gramophone, my toys, and a number of other things. I heard Beethoven's Kreutzer Sonata, interpreted by Hubermann.[8] I saw the garden house in Kalamış (Istanbul), where two White Russian musicians invited by my family, played the Kreutzer Sonata on a summer night in 1947. I relived numerous melodies, visions, toys, and individual episodes within only fractions of a second.

What mechanism in my mind was triggered that allowed me to travel back in time to our home in Istanbul? What are the roles of place cells in relation to episodic memory, as Kandel also wondered (see Sect. 20.6)?

The remark related to Hawking's arrow of time merits important consideration when developing an opinion on traveling backward in time. Hawking's book is recommended reading.

20.6.1 Measurement of Episodic Memories and Emotions Can Be Crucial to Interpret the Creative Mind

Now, the more important question is whether manifestation of such an emotional episodic memory is measurable?

I think the answer is yes.

We described episodic/emotional memory experiments in Chap. 12. One could also design an experimental setup similar to that in Sects. 2.1 and 2.2 and replace the grandmother picture with the old book about Beethoven and the anonymous

[8]I later earned a 1934 version in CD form.

face within the cover page of an unknown book. In that case, it is conceivable to register increased frontal theta responses to Beethoven's book in comparison with an unknown, anonymous book.

What do all these things mean?

1. We tentatively propose that it may be possible to measure processes within the brain related to traveling back in time.
2. Once such things can be measured, a greater number of measurements can be extended, similar to Newtonian pioneering experiments that became increasingly refined in the seventeenth and eighteenth centuries.
3. Accordingly, electrophysiological research on episodic memory may provide a key for measuring the possibilities of heterogonous time during episodic remembering, which is not possible to measure with regular physical clocks.
4. Another type of indirect measurement related to memory and creative evolution is described in an extended way in Chap. 17. This is a comparative study during the evolution of species. A comparison between different type of subjects can be also achieved with various levels of subjects, as will be mentioned in Chap. 25. (At least, one may possibly find differences in event-related oscillations between the brains of creative artists or scientists.)

20.7 What Is the Nature of Creativity in the View of N. Andreasen (2005)?

A study by Andreasen (2005) concludes that "extraordinary creativity" is the result of neural processes that "differ qualitatively as well as quantitatively" from those of other people. The author's admiration for creative genius and the arts lead her to look for evidence of such great minds as Mozart, da Vinci, Michelangelo, Tchaikovsky, mathematician Henri Poincaré, and chemist Friedrich Kekulé.

What conditions further creativity? What is going on inside the brain of a Mozart or a Shakespeare during the creative process? And is there a relationship between creativity and mental illness, as often posited? Nancy Andreasen (who tends to draw conclusions primarily from her own work) explains in her book that she is performing a study to see if any such tendency exists among especially creative scientists. Despite the paucity of evidence, she suggests that creativity arises largely from the "association cortex" – parts of the frontal, parietal, and temporal lobes that integrate sensory and other information. This idea, however, has just begun to be researched. Andreasen, again, relies heavily on her own study performed with positron emission tomography (PET). Free association is an instance of episodic memory – a type of autobiographical memory that recollects the information linked to a person's experience – but, in this particular case, according to Andreasen, it is "more mysterious," because it "is clearly less sequential and time-linked (and) may be the repository of information that is stored deeply and is therefore sometimes less consciously 'accessible'." What Andreasen's experiment reveals is that the area of the brain that registers activity during free association is the association cortex. This cortex is what gathers and links information from various other areas in the brain, and (here is the interesting part) "in potentially novel ways" (p. 71). Therefore, the claim is that the genesis of new ideas and concepts is attributable to this neural process, which links information in the subject's brain in novel ways. However, what makes these discoveries fascinating within the study of creativity is that:

1. Much of this linking process occurs in what we refer to as "the unconscious mind," and
2. This capacity uses the parts of the brain that are its "most human and complex parts."
3. According to Andreasen, there is a distinction to be made between ordinary creativity (creating sentences in conversation) and extraordinary creativity (composing symphonies), and she connects the empirical evidence back to the introspective accounts presented to the reader earlier. What this link successfully demonstrates is that the creative processes in the instances of people such as Mozart and Tchaikovsky are extraordinary and characterized by a unique thought process, which in turn must (although Andreasen is careful to say presumably), be caused by a unique neural process. In essence, the claim is that the type of creativity we are interested in, the type that produces paintings like the Mona Lisa, is a distinct type of neural activity that can be distinguished from other types of brain activity. Furthermore, it appears to be something that occurs in the unconscious mind, via a process of free association. Andreasen herself puts it as follows:

> I would hypothesize that during the creative process the brain begins by disorganizing, making links between shadowy forms of objects or symbols or words or remembered experiences that have not previously been linked. Out of this disorganization, self-organization eventually emerges and takes over in the brain. The result is a completely new and original thing: a mathematical function, a symphony, or a poem.

Here, we repeat several remarks from Chap. 18, in which we extended the view of Henri Bergson on *Creativity.* For the occurrence of an intuitive process, our brains require high-speed access to past memories, meaning *episodic* and semantic memory. Further, the intuitive human brain must be able to transform events in the "episodic and semantic memory" to a type of intermediary virtual memory, which extends past events to some future projections.

We also strongly emphasized the importance of memory during the duration and occurrence of intuition. Without memory, it is impossible to transform the past memory to a creative memory (see Figs. 18.2 and 18.3). If we accept the preceding chain of reasoning, it is easy to recognize that *memory, duration, and intuition are inseparable*. This is similar to the space-time-matter continuum in physics. Our creative brain or thinking brain moves in a universe that consists of memories of real life and virtual life created on the existence of memory. This universe is also inseparable from our body (see Sects 2 and 3 in Chap. 18).

The remarks related to the importance of intuition and episodic memory are in accordance with the pioneering work of Andreasen; however, we have a different interpretation of the role of the association cortices. Chaps. 8 and 12 show that episodic memory also requires functional connectivity with other areas, for example, the occipital cortex.

Concluding thoughts are reached in Chap. 25, which aims to integrate metaphysics of the brain and the described events of the mysterious brain. It is a difficult mission to approach the mysterious brain or invent a type of measurement to explain all that, but one must start with attempts to explain these abilities of the brain with tenable biophysical and biochemical methods. Possibly, one has to develop new paradigms. Measuring episodic memory and facial expressions enabling "travel to the past" may possibly open a new era in our understanding of brain-body-mind integration.

Part VI
Essentials and Unifying Trends in Brain Body Mind

Prelude to Part VI

Part VI aims to summarize and draw conclusions from the important impact of various chapters of the book. Further, after surveying all data chapters, philosophy chapters, review chapters, and concept chapters, we aim to arrive at some synthesis or to establish a tentative new trend. Most of the chapters within Part VI are very short, so as to highlight the most relevant ideas and essence of the book. Accordingly, the reader may also start to read the book by beginning with this last part. After this, the reader can return to appropriate chapters to grasp the basic leitmotifs of the book. At the end of this part, the most general and global conclusions are described.

The Curious Story of These Blackboard Drawings

Before presenting the leitmotifs and unifying trends leading to three adjacently described and interwoven models in Chaps. 24–26 it seems to be interesting to describe a type of "brainstorming" process employed in the first days of planning this book. After finishing this final part, I found three neglected and somewhat forgotten photos in my book archive.

The cardinal step to develop these semi-empirical Gedankenmodels in Part VI was explained 4 years ago with to my coworker B.G.,[1] during a 30-min graduate course.

---

[1]Dr. Bahar Güntekin told me later that the language of my short talk was very difficult to understand. My talk was in Turkish; however, I seemingly used a number of words, mostly in English, but also frequently in German and French. Therefore, she took pictures and we discussed the matter in more expanded sessions during the following week. I remember, according to this event, the words of Henri Bergson: In reality, the past is preserved by itself automatically. In its entirety, probably, it follows us at every instant; all that we have felt, thought and willed from our earliest infancy is there, leaning over the present which is about to join it, pressing against the portals of consciousness that would fain leave it outside. (From Bergson, 1907, *Creative Evolution.*) Possibly, I lived in the past by compressing the time and work of the last 40 years into only 30 min.

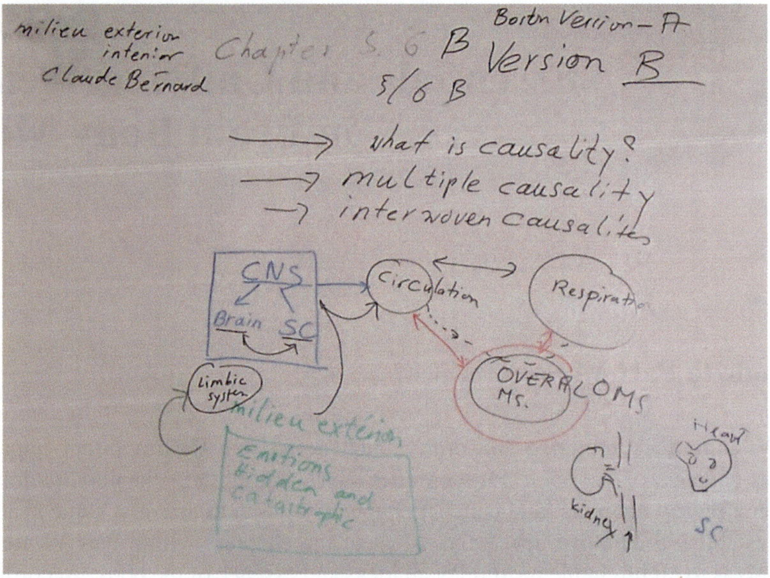

Fig. 1 Description of multiple causalities and web of brain – CNS, OMS (overall myogenic system), SC (spinal cord) and organs as kidneys and heart

Figures 1–3 show a collection of findings and ideas, illustrated as blackboard notes, which opened the way to the writing of the present book. To explain the essence of the machineries of mind (as I see them) I needed only half an hour; completion of the present book took more than 4 years.

Figure 1 presents a number of initial ideas in the form of these immediate spontaneous drawings, briefly described in the following:

1. The mind is under the influence of "Milieu Intérieur" and "Milieu Extérieur."
2. There are multiple causalities in the machineries of brain-mind: The overall myogenic system, the vegetative system (including the heart and kidneys) are interwoven and lead to multiple causalities (see also Chaps. 4, 5, and 9).
3. The central nervous system (CNS), the overall myogenic system (OMS), and the organs of the vegetative system (including the respiratory system and circulation) show mutual excitability.
4. The CNS, OMS, and vegetative system (comprising circulation, respiration, etc.) are all embedded or interact with the biochemical pathways (transmitters).
5. All the physiologic systems are governed with invariants that are natural frequencies of brain-body functioning.
6. Emotions from the milieu extérieur may have catastrophic consequences.

At the beginning of the book we presented preliminary trends to define "What is Mind?" by starting with the pioneering work of Renaissance philosophers. Earlier thoughts on mind and the popular definition of mind are also presented in Chap. 1.

Four centuries have passed since the time of earlier definitions of mind. However, even in the twenty-first century there are some explanations of the mind as a deep version of cognition, i.e., thinking process. According to the standard definition presented in encyclopedias, *mind* refers to the collective aspects of *intellect* and consciousness, which are manifest in some combination of *thought*, *perception*, *emotion*, *will*, and *imagination*.

After presenting considerable empirical evidence on the psycho-physiologic basis of the brain, vegetative system, and spinal cord, we are confronted with a new strategy of how to approach the mind, or observations to understand the mind. This task is advanced as follows.

Chapters 21 and 22 bring together all evident empirical findings that are described in earlier chapters of the book. As presented in these chapters, there is concrete empirical evidence to describe the machinery of the mind and common phenomena in nature.

Furthermore, several schools of thought made considerable progress toward understanding the machinery of brain-mind. The question of whether new schools are needed is treated in Chap. 26.

However, there is also some empirical evidence that is impossible to explain analytically. For example, work on the creative mind, unconscious problem solving, and intuitive mind provide empirical evidence that is currently impossible to explain analytically. They are processes that we cannot physically or chemically localize, whether inside the brain or in the vegetative system. However, such processes are observable and are constituents of mind.

According to these results, the need is clear to develop models that could bring together all measurable unifying concepts, hypotheses based on these and also tentative Gedankenmodels.[2] Such models do not clarify precisely the related phenomena; they are, however, useful to plan further experiments and identify the shortcomings in future planning. The Gedankenmodels in Chaps. 22–25 are based on the most important message: The mind cannot be considered a unique entity originated or located only in the brain. Only the web of brain-body can provide a framework for understanding the mind.

It is evident that, as yet, it is impossible to develop a clear definition and model of the mind. On the other hand, it is also evident that oscillations, neurotransmitters (biochemical pathways in Fig. 2), and the evident links described between brain and vegetative system (body) provide crucial keys to understanding the web of

[2]Thought models (from the German).

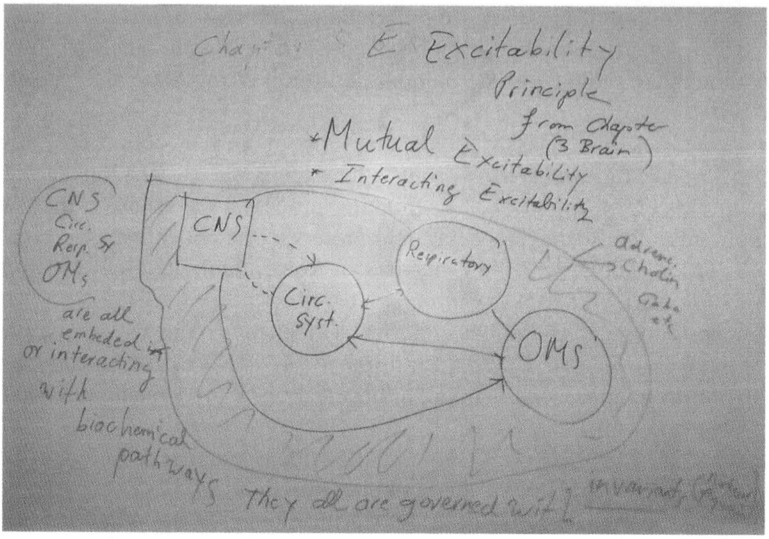

Fig. 2 *Blue shaded* areas correspond to biochemical pathways; CNS and OMS, etc. are schematically embedded in biochemical pathways. Do not forget that this is a very rough explanation. All three whiteboards were filled in about 30 min.

brain-body-mind. The titles of Chap. 23–25 reflect the steps to the elevation[3] of ideas.

Part III discusses the necessity of analyses using a new nebulous Cartesian system to integrate multiple causalities. This was explained in the photo of the blackboard shown in Fig. 3.

Chapter 23 includes a quasi-deterministic model, mostly an engineering approach to explain coupled oscillators in the brain, spinal cord, and organs of the vegetative system. Although individual oscillators can be described as being predictable, less predictable non-deterministic resonance phenomena can arise from multiple couplings or a multiplicity of subgroups.

Chapter 24 is a consequence of steps undertaken in Chaps. 14, 15 and 16 related to the quantum brain. Although the oscillators in these cases are also globally coupled, the individual oscillators are highly non-predictable. This model is inspired by the unifying string theory in physics.

Chapter 25 provides a most tentative (extravagant) metaphysical argument that is based on globally coupled oscillators and the string model, and also describes the influence of intuition and creativity, as interpreted through the views of Bergson, Kandel, and Andreasen. Chapter 25 also includes the reasoning's on brain-body-mind that can be considered as a final essential of the book.[1]

---

[3]The reader is advised to read the poem *Elevation* by the French poet Charles Baudelaire, inspired by the Prelude to Lohengrin, by Richard Wagner. An English translation can be found via Google.

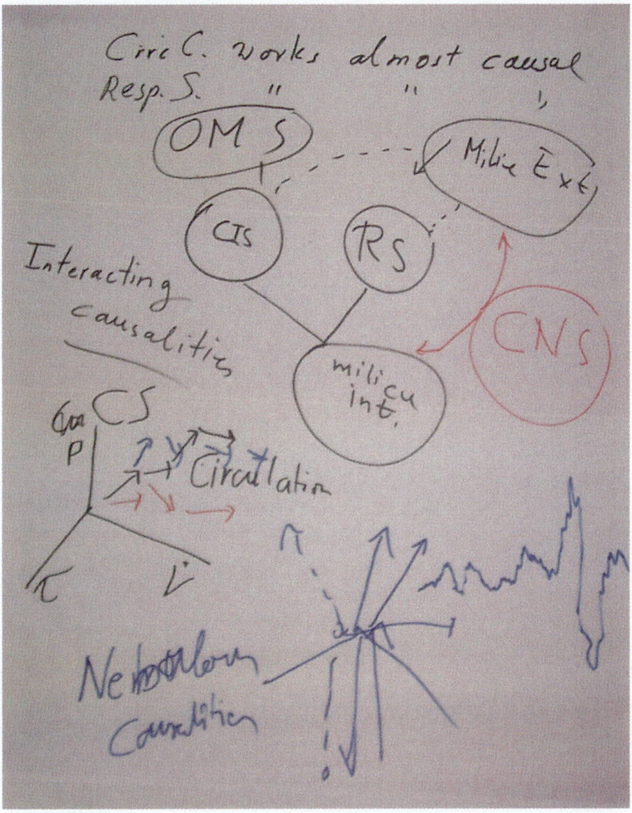

Fig. 3 From causalities to a nebulous Cartesian system: Note the irregular and multiple axes in the unusual and hypothetical coordinate system.

Chapter 21
Leitmotifs and Common Concepts: An Interim Description

21.1 Introduction

This chapter provides a type of interim summary of the foregoing parts, and it is planned to guide the reader to a more profound synthesis in Chaps. 22–26. Some of the concepts and results will be more accentuated in the last chapters.

This book covers only a small area of general sciences. Nevertheless, the brain is an extremely complex system, and by analyzing brain mechanisms it is possible to discover some overlapping concepts in neuroscience and general sciences. Consideration of problems of the brain machinery, including the vegetative system (i.e., circulation, overall myogenic system, and respiration) indicates the importance of electrical links with neurotransmitters. We described concepts and methods from general systems theory, Fourier theory, quantum theory, and chaos theory. As a consequence, it is tentatively suggested to look to physics and astrophysics, namely string theory. Reference was made to the essence of Darwin's theory, which is one of the most fundamental frameworks in modern biology. The present chapter outlines three fundamental steps to describe general and universal trends. These are as follows:

1. There are interactive processes in the brain-body-mind integration.
2. Common concepts and principles emerging from various processes of both biological and physical sciences are to be jointly interpreted.
3. Chapter 2 of this book described the important progress in general science achieved by a few outstanding philosophers, who established fundamental models to approach the brain-mind concept. Furthermore, the importance of the Copenhagen school, cybernetics, synergetics, dissipative structures, and catastrophe theory are emphasized. Further, as an important conclusion of the present book these frameworks are amalgamated and extended to a more comprehensive framework.

E. Başar, *Brain-Body-Mind in the Nebulous Cartesian System: A Holistic Approach by Oscillations*, DOI 10.1007/978-1-4419-6136-5_21,

21.2 General Principles of the Brain-Body-Mind in the Web of Biology/Physics

21.2.1 Overall Frequency Tuning in Brain-Body Functioning

Broadcasting systems and radio devices in our homes communicate through similar discrete frequency channels by means of electromagnetic signals. Oscillatory signals sent from antennas in broadcasting companies are received at distant locations by tuners (also called radio devices). Most importantly, the receiving devices and the antennas' radiation have to be tuned to the same frequency range to achieve good communication. One can now ask the following question: Are the selectively distributed brain networks eliciting alpha, beta, gamma, theta, and delta oscillations also tuned to achieve good communication in the brain? Chapters 6–8 showed that the frequency tuning properties of neural populations are similar despite neuronal and anatomical differences of distant brain areas. The neuronal network of the cerebellum-cortex is dominated by Purkinje cells, the cerebral cortex has six layers dominated by large pyramidal cells, and the hippocampus also has a different neuro-anatomical constitution. Despite these considerable structural differences the most general network properties of the cerebral cortex, cerebellum, and hippocampus are controlled by a *general tuning*. In other words, most general transfer functions of the brain are dominated by common network properties: EEG-oscillations. Again, a simple metaphor can be provided: There are several types of radio devices ranging from those containing transistors to others with triodes and all types of old-fashioned devices using valves. Structurally these devices are different, but they communicate using the same oscillatory dynamics.

This principle of general and common tuning can be found, not only in the brain but also in the brain-body integration. According to the work of Gebber et al. (1995a, b), Aladjalova (1957), and Ruskin et al. (1999a) structures in the vegetative system are also tuned to the same frequencies. Transmitters such as acetylcholine and norepinephrine are also excellent vehicles for general tuning in brain-body interaction. The body and the brain use the same transmitters (see Fig. 3.10 and Chap. 13) and the same frequencies for general tuning for brain-body interaction. Here the ensemble of these phenomena is tentatively called *the overall tuning in brain-body interaction* (see also Chap. 23). It is also necessary to note that the principles of the synchrony of oscillators and overall tuning somewhat overlap. Furthermore, those conventional electronic networks can have similar frequency tuning regardless of their constitution with capacitor resistances or inductances, and this is similar to the networking properties of the brain.

These thoughts will be explored in greater depth in the next section, with Einstein's view showing the parallel with physics.

21.2.2 Synchrony of Clocks According to Einstein: Good and Bad Clocks in the Brain

Einstein asks the fundamental question, "What is a clock?" The primitive subjective feeling of time flow enables us to determine that one event happened earlier than another. However, for estimating the exact time interval between two events, a "clock" is needed, and thus the time concept becomes objective. Any physical phenomenon may be used as a clock. By taking the time interval between the beginning and end of an oscillation, arbitrary time intervals may be measured by using oscillators. *All clocks, simple hourglasses as well as the most refined instruments, are based on this idea*. The neural oscillators in the brain work in a similar way to clocks.

In physics, two perfect clocks at two distant points should show exactly the same time. However, there are, according to Einstein, *good clocks* and *bad clocks*. "Good" clocks are synchronized and always show exactly the same time even when they are spatially separated. In the brain various clocks are oscillating with several discrete frequencies. They are synchronized or partially synchronized by executing diverse types of brain functions. In Chap.13 the insufficient synchronization between oscillators or "clocks" of different neural populations in neuropsychiatric patients was demonstrated. Metaphorically speaking, this means that clocks in the pathological brain are often bad clocks – they are not synchronized. In the theory of relativity, asynchrony of clocks plays an important role in processes observed between places separated by a long distance. The role of brain oscillators is fundamental for the functioning of the brain, but the clocks in the brain are not such homogenous time indicators as those in physical systems (see also Chap. 23). However, *delays and prolongations* are also functionally important.

Subjective time in the brain-body, in Henri Bergson's concept, called *duration*,[1] are governed by neither good, nor bad clocks. Bergson's duration cannot be measured by means of physical homogenous clocks. Rather, a new type of time dimension should be included in all types of implications in the metaphysics of the brain. This new way of thinking was revealed by Marcel Proust and also significantly discussed by Roger Penrose (1989) (see also Chaps. 18–20 related to consciousness and intuition processes). Accordingly, the addition of thought systems in living beings add a greater importance to clocks in biology. Despite these ambiguities between physics and thought processes, the existence of synchronized clocks in nature also belongs to the most important common processes because a clock usually regulates functionality; this is also true in the brain-mind.

[1]See Chaps. 18 and 19 for the definition of duration.

21.2.3 The Concept of Resonance in the Brain and in Nature

In Appendix C, oscillatory processes and resonance phenomena in physical systems and nature are discussed. The usefulness of resonance phenomena in biological systems is also discussed in Part II (Chaps. 6–9). It is clear that resonance phenomena play a major role in the communication processes of the brain, in the vegetative and myogenic systems. There are common and possibly interactive resonances among the brain and vegetative system.

21.2.4 The Fundamental Role of Causality in Classical Mechanics, Quantum Mechanics, Brain, and Evolution Theory

In Newton's concept of mechanics the trajectories of particles can be exactly predicted when the knowledge of the coordinates and the initial forces acting on the particle are known. This is in accordance with the classical view of causality by David Hume (see Chap. 1). In atomic physics, Carl Friedrich von Weizsäcker introduced a new concept of *quantum causality*. The exact position of an atom can only be predicted in a vague way, by means of a probabilistic wave packet (see Chaps. 14–16).

In Darwin's theory there are multiple causalities for evolution, including the causalities dominated by the environment of a living being and its efforts for survival. The mutations of a living mechanism cannot be predicted by starting from initial conditions. Furthermore, mutations do not always occur in an ideal manner. The evolution cannot be predicted because chance and necessity govern this process. Jacques Monod (1971), who described the rules of evolution in accordance with findings in molecular biology, has emphasized this view, which was introduced by Darwin's findings.

Chapter 16 contains the description of several types of causalities with regard to brain-body-mind integration. The superposition of these causalities has indicated the necessity of using the expression *the quantum brain*. Furthermore, it is proposed that brain-body-mind integration can be described in a Cartesian system in hyperspace. For this case it is necessary to apply the concept of statistical mechanics.

Summarizing and reinterpreting the described facts on the causality principle it can be said that the brain-body-mind integration has a *probabilistic causality*. However, this type of causality is different from the causality described in quantum processes. Because, thinking processes and the existence of intuition is much more difficult to predict and require another concept of relative times.

Chapter 17 describes the changes of alpha activity during the evolution of species and offers a tentative proposal that the evolution of alpha activity is correlated with the development of intuition in humans. The development of intuition led to new discoveries and the invention of machines. All intellectual efforts achieved by humans created a new type of survival in nature and also the possibility of changing some components of nature. In our world, trees grow spontaneously according to the rules of the environment, but human beings are able to influence and change the flora in their environment. These changes are not part of the concept of chance and necessity; the self-organization of the developed human being highly diminished the effects of chance and necessity.

21.2.5 Statistical Mechanics and the Quantum Brain

Einstein and Infeld (1938) summarized the processes of quantum physics and statistical mechanics in a brilliant manner.

> The rich variety of facts in the realm of atomic phenomena forces us to invent new physical concepts. Quantum physics formulates laws governing "crowds and not individuals." Not properties but probabilities are described, not laws disclosing the future of systems are formulated, but laws governing the changes in time of the probabilities and relating to great congregations of individuals. The new quantum physics removes us still further from the old mechanical view and the retried to the former position seems, more than ever, unlikely.

In the brain, individual neurons are the basic building blocks for signaling and communication between brain structures. If a neuron is excited at a higher-than-threshold level, the individual neuron will certainly fire in a similar way to the predicted motion of a particle in Newtonian physics. However, if we excite a neuronal population it is only possible to predict the "probability of the response" of that population. This is similar to the analogy between Newtonian physics and quantum physics. In Appendix B on resonance phenomena, the similarity was shown between an oscillatory cord the membrane of a drum, a wind instrument or any other electrical instrument on the one hand, and a radiating atom on the other. There is also some similarity between the mathematical equations governing an acoustical problem and those governing quantum physics; but again, the physical interpretation of the quantities determined in these two cases is quite different. The physical quantities describing the oscillating cord and the radiating atom have quite a different meaning, despite some formal likeness in the equation.

Einstein further states:

> The deviation from the normal can thus be calculated for any other moment from the mathematical equations for the oscillating cord. The fact that some definite deviation form the normal position corresponds to every point of the cord is expressed more rigorously as follows: for any instant the deviation from the normal value is a function of the coordinates of the cord. All of the cord from a one-dimensional continuum, and the deviation is a function defined in this one-dimensional continuum, to be calculated from the equations of the oscillating cord.

Analogously, in the case of an electron, a certain function is determined for any point in space and any moment. Physicists call this function the probability wave. In the current analogy, the probability wave corresponds to the deviation from the normal position in an acoustical problem. At a given instant, the probability wave is a function of a three-dimensional continuum, whereas at a given moment, the deviation of a cord is a function of the one-dimensional continuum. The probability wave forms the catalogue of our knowledge of the quantum system under consideration, and enables us to answer all sensible statistical equations concerning that system.

There is now a third type of interaction, namely, the behavior of the brain-body-mind system. Because the reaction of neural populations has a probabilistic nature, it is possible to talk about a similar behavior that could be expressed as a *quantum brain*, or better as a *hyper-probabilistic brain*. Again, direct structural or functional similarities with physics do not exist here. However, the description of the responsiveness of the brain has fundamental conceptual similarities to the events in statistical mechanics or quantum physics.

It is important to emphasize that neither episodic memory nor intuition can be measured with physical clocks. As Bergson explained, both processes, as known by subjective experience, occur in nonmeasurable, heterogeneous time-space.

21.2.6 Chaos

The application of *chaos theory* to physical systems, especially its applications in meteorology and to all systems susceptible to develop turbulence, has led to valuable results and interpretations. Chap. 14 confronts the chaotic processes in the brain with a basic question: "Do the chaotic brain and the quantum brain obey some common principles?"

Brain responses to sensory and cognitive input evoke *multiple causalities* in the brain that can be the reason for chaotic behavior. It is also possible to state that the brain behaves with uncertain and multiple causalities similar to the events in processes of quantum theory. At this point it is proposed that brain function can be described only in a probabilistic hyperspace and with both quantum indeterminism and the concepts of chaos theory (see Chaps. 15 and 16). This is a dualism, which is possibly found in all types of nonlinear systems.

21.2.7 String Theory

Considering the phenomena of oscillations resonance, the conclusion that string theory (or M-theory) could also be used as a metaphor to describe brain functions (see the tentative scope in Chaps. 24–26).

21.2.8 Brownian Motion and Einstein's Concept in Search of Hidden (or Invisible) Processes

What is Brownian motion? In liquid, a suspended particle is constantly and randomly bombarded from all sides by other molecules. If the particle is very small, the hits it takes from one side will be stronger than the bumps from the other side, which causes it to jump. These small random jumps comprise Brownian motion. The first mathematical theory of Brownian motion was developed by Einstein in 1905 (Einstein and Infeld 1938).

Einstein and Infeld (1938) described tracks of molecules according to Brownian motion. However, Einstein did not describe only these tracks, but also analyzed the causes of Brownian motion. In his search for causes of gravitation, Einstein wished to understand the reasons for the dissipation of energy. Attempting to establish what happens in the galactic system, he predicted black holes. He integrated the existing knowledge about the motion of stars into one theory using the laws of physics. Thus, Einstein's work in extraterrestrial physics is similar to Darwin accomplishment in comparing species' differences in the Oversea Voyage.

One of the aims of electroencephalographic (EEG) research is to explore brain functions. Accordingly, the analysis of Brownian motion trajectories initiated by Einstein (Einstein and Infeld 1938) is an excellent theoretical model or metaphor in the search for brain functions. The key to these functions is embedded in the puzzling engrams of EEG-oscillations. A number of explanatory formulations were given in a special issue of the *International Journal of Psychophysiology* (2006) – the trajectories of EEG-oscillations are used to discover their hidden origins. These formulations show the immense usefulness of a function-oriented investigation of the brain. As Einstein's fundamental model shows, signal analysis alone can never be sufficient. Signal analysis must be used with an ensemble of strategies introduced in this book as evolving brain, maturating brain, emotional brain, and pathologic brain. The comparison of such strategies helps considerably in understanding functionalities.

21.3 Common Concepts

21.3.1 Entropy in Brain Structures

The concept of *entropy* was introduced in this book as an important entity to describe transition phenomena. Because entropy is a measure of disorder in a given system or signal, this concept is extremely suitable to describe dynamic processes in the brain. The entropy concept explained in this book occurs at three different levels that are not independent of each other.

First, there is synergetics. For example, when energy is pumped into a laser, the oscillating atoms change from a disordered system to an ordered system. In this way phase locking and frequency locking phenomena are observed. When the energy of the laser device is increased, the laser light becomes more focused. The use of lasers is well-known in several branches of science, industry, and medicine. However, according to Haken (1977), the process of frequency locking in the laser is just one example of a general phenomenon and can be conceptually used to understand synergy effects in physics, biology, and social sciences (see Fig. 2.2). In the brain it is also possible to observe phase locking and frequency locking after sensory and/or cognitive stimulation (see Fig. 6.4; Gönder and Başar 1978). Second, as Rosso et al. (2001) have shown, EEG activity shows a decrease of wavelet entropy during the transition of a resting state to sensory-cognitive stimulation. Comparable with what happens with the laser, brain signals go from a disordered to an ordered state. Third, the concept of entropy during electrophysiological evolution of species and maturating of the brain is analyzed. This concept is detailed in the following.

21.3.2 Entropy During Evolution and Maturation

The entropy of the electrical brain oscillations in the alpha range (10 Hz) is considered to be a marker for the evolution of species (see Chap. 17). The amplitudes of the 10 Hz activity increase during the evolution from Aplysia to human being. Even more important is the fact that alpha (10 Hz) oscillations attain the most regular, ordered oscillations in the human brain. *In other words, evolution leads to a decrease in the entropy of alpha (10 Hz) activity*. Anatomical observation is also important. The morphology of the human brain shows a higher degree of regularization than the brains of simpler animals; thus depicting the lowest entropy in the human brain. Furthermore, the entropy of alpha activity in children below the age of three is high. The entropy of the alpha activity decreases considerably during the maturation of human brain (see Chap. 11). Accordingly, the measure and changes of entropy are considered to be one of the most important control parameters in the investigation of brain-body-mind integration.

21.3.3 Going Out of the System and Darwinism

Darwin's evolution theory is considered to be one of the most important revolutions and theoretical frameworks in biology. Darwin used the method of *going out of the system* by comparing the anatomical structures of several species in Europe, America, and finally the Galapagos Islands. Darwin also studied plants so as to develop his theory of natural selectivity with the aim of survival (Darwin 1859).

Accordingly, it can be said that Darwin used the method of going out of the system when he compared different living beings, as well as when he compared the same living beings in different geographical areas.

The method of going out of the system began to be used in 1976 in the analysis of the process of autoregulation in the circulatory system, the coronary system of the heart, and also the dynamics of the smooth muscles. The processes of autoregulation in the circulatory organs are dominated by the dynamics of smooth muscle contractions, leading to the conclusion that smooth muscles are effectors of the flow regulation in autoregulating organs. Later, research was extended to describe the contractile properties of peristalsis organs (e.g., stomach, intestines, and uterus) and it can be shown that even the lymph nodes depict the same dynamic properties that are also dominated by the smooth muscle (see Fig. 5.1). These facts led to the opinion that the overall myogenic coordination system and the brain are governed by an ensemble of common transfer functions manifested in electrical or muscle contraction, and this idea gives an important insight into the understanding of brain-body behavior.

Furthermore, the concept of whole brain dynamics has been formulated and has shown that the principles of selective distribution of oscillations and superposition of oscillations give important hints to the understanding of several processes of brain functioning. As mentioned, this interest in the evolution of species also led to the use of the method of going out of the system. It can be demonstrated that similar EEG oscillations are found in species ranging from Aplysia to the human brain. Moreover, in Chap. 17 a tentative proposition is made that human intuitive behavior is manifested in alpha activity. Again, studies of the maturing brain revealed that alpha activity is also strongly correlated with brain maturation. No alpha activity is observed in children until the age of 3 years; in the brains of adults alpha activity has highest amplitudes in the posterior areas. However, during maturing in the brain of elder subjects, the frontal alpha activity and the alpha response gain important weight in comparison with posterior alpha activities. In relation to the process of intuition, it is important to refer to Gustav Jung (1959), who saw the *unconscious* processes as involving "collectively present dynamic patterns," called archetypes. These patterns formed by the remote experience of humanity, are reflected in dreams as well as the universal motives found in myths and fairy tales around the world. *Archetypes*, according to Jung, are "forms without content," representing the mere possibility of a certain type of perception and action.

21.3.4 Micro-Darwinism in Brain-Body Interaction

Einstein analyzed gravitation phenomena in different galaxies and predicted the existence of black holes. Therefore, he followed the same principle as Darwin. The gravity of the earth cannot be understood without considering other galactic systems in parallel. Another example is the discovery of the periodical table of elements. To understand properties of the Fe- or Na-atom, it is important to know the system that includes all the elements.

In the field of brain functions and all other physiological functions, in the twenty-first century, research scientists often narrow their work down to restricted areas. There are specialists of the hippocampus or the cerebellum; there are specialists of the kidney or the heart. Only seldom do scientific approaches consider the *whole brain* and the *whole body*. Metaphors of comparisons from physics, astrophysics, or general systems theory are generally ignored. Here, it is suggested that comparative approaches are useful, even if one tries to understand only a small area of the brain. Narrow approaches will not lead to a proper understanding of brain functions.

Micro-Darwinism in brain-body interaction means the following: To learn about the functioning of the coronary system of the heart, it is necessary to compare the properties of the vascular system, kidneys, and peristaltic organs. Oscillatory processes in brain stem neurons are also similarly tuned with oscillations of the overall myogenic system. As stated in chapter 9, the existence of similar oscillatory components is a manifestation of the integrative coordination of the brain-body: 10 Hz oscillations can be observed in the neurons of the sympathetic system of the heart as a response to sensory stimulation as well as during memory, attention, perception, and learning in the brain. Joint analysis of these facts constitutes one of the core themes of this book. Without the micro-Darwinian concept, that is, without comparing properties of several distant organs in the body, it is not possible to understand the homogeneity/similarity of all these components. This concept could be called micro-Darwinism, because it is related to the body and not to evolution or the evolution of species.

At a more general level, does the use of the brain's S-matrix or brain's Feynman diagrams also provide a type of Darwinism? The sum of time histories or the so-called path-integral takes all possible mechanisms into consideration. Unfortunately, no such comparative effort has yet been undertaken by brain scientists; therefore, one of the most important messages of this book is to encourage scientists to apply such principles.

The holistic approach in Chap. 9 is an application of such a step and opens the way for unifying models linking the brain, vegetative system, and spinal cord (see Chaps. 22–25).

21.4 Interaction of Schools in Search of Brain-Body-Mind

Norbert Wiener's concepts were used in this volume, as described in his renowned book, *Cybernetics*, as preliminary steps in the analysis of brain-body functioning. The first application was the analysis of circulatory dynamics (see Chap. 4). For this purpose, general systems theory methods (evaluation of frequency characteristics) were used. Later, the same approach was applied to event-related potentials from the brain of cats and human beings. The analysis of the power spectra of brain signals has been a common approach since Norbert Wiener and Mary Brazier's group's first work. The adaptive digital filtering method was applied for the first time by Başar and Ungan (1973). Thus, the first important results in brain dynamics were achieved within the scope of the Norbert Wiener School. However, at the beginning of the 1980s two new approaches began; the consideration of nonlinearities in brain functions inspired by the work of Erich von Holtz (Başar 1976), and the catastrophe theory of René Thom and its interpretation by Zeeman (1977) related to the concept of alpha attractors (Başar 1980). Parallel to this development, brain responses were compared with the concepts of high energy elementary particle physics and nuclear resonance phenomena. The terms *strong resonance* and *weak resonance* were used to describe event-related oscillations in an analogy to elementary particle physics. Başar (1983a) suggested the use of S-matrix for visual quantification of brain processes (see also Appendix B). A few years later (Başar 1989) brain Feynman diagrams were proposed. Currently there are a number of proposals related to the quantum brain, and the number of such papers is increasing; the probabilistic interpretation of EROs finds more and more interest in neuroscience. Chap. 16 shows that this approach provides one of the major components of the new Cartesian system (Fig. 21.1).

21.5 Common Codes, Principles, and Rules in Brain-Body Integration

The essential messages emerging from this book are dominated by the operating principles of neural populations. The essences of these principles described in Chaps. 7–9 are briefly outlined in the following:

1. Although the basic signaling elements of the central nervous system (CNS) are neurons, higher mental activity is achieved by an ensemble of all neural populations in the whole brain.
2. During functioning of brain-body distributed neural populations are activated within the frequency channels of natural frequencies of the brain (EEG-oscillations) and ultra-slow oscillations linking the brain and vegetative system.
3. Neural populations work with the principle of the superposition of oscillations in several frequency channels. The topological distribution of the super-imposed oscillations is of great importance for brain-body integration.

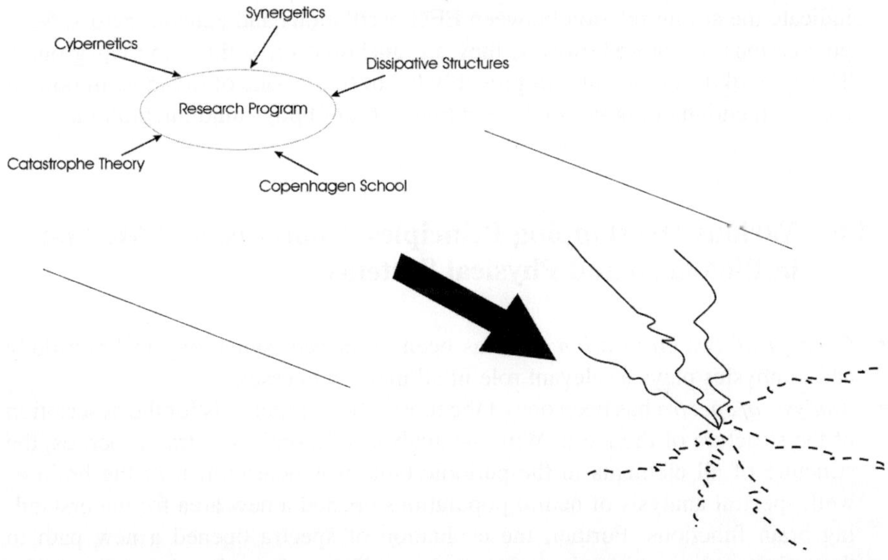

Fig. 21.1 Illustrative explanation showing the transition from research schools to new Cartesian systems as an allegory. (These concepts and frameworks are explained in greater detail in Chap. 26)

4. Although there is emphasis on the selective distribution of oscillations in diverse areas of the brain, the realization of different major brain mechanisms is achieved by the whole-body subsystems that are working in synergy. Accordingly, long distance coherences are observed in the brain. Not only neurons, but often neural populations are connected by manifesting common and time coherent behavior (Chaps. 7 and 8).

Sensory cognitive functions, memory, emotion, and change of cognitive processes in pathologies are governed by oscillatory activities in delta, theta, alpha, beta, and gamma frequency windows (Başar et al. 2000, 2001a).

5. The natural frequencies are not only common building blocks of the brain-body. Natural frequencies also serve as *common codes* during evolution of the species.

Ultra-slow oscillations and brain oscillations belong to common functional building blocks for communication in the brain, vegetative system, and spinal cord.

6. Similar to invariant frequency codes and time constants, neural transmitters display *quasi-invariant* properties. *Neurotransmitters* acting as biochemical pathways influence both the brain and the vegetative system. Therefore, neurotransmitters provide another type of coding for the evolution of species. Chaps. 10 and 13 give examples of the action of transmitters on the Helix pomatia ganglia and the human brain as well as their influence on neural pathologies.
7. The most crucial question is, Are genetic codes, EEG codes, and transmitters interrelated, or do they include common mechanisms? (See also the final chapters.) Relevant publications of the Begleiter and Porjesz (2006) research group indicate the strong relation between EEG oscillations and genetic factors. Such publications are new; however, they are anchored on solid empirical ground. This type of development will probably be, in future, one of the most important factors in enlightening our understanding of brain-body-mind integration.

21.6 Various Overlapping Principles, Concepts, and Methods in Biological and Physical Systems

- *Concept of synchronization:* As has been described, synchrony in brain-body and in physics plays a relevant role in all major processes.
- *Analysis of spectra* has been one of the most efficient methods for the description of the structure of the atom. With the analysis of atomic spectra properties, the structure of all elements in the periodic table was determined. In the brain as well, spectral analysis of neural populations opened a new area for understanding brain functions. Further, the evaluation of spectra opened a new path to the understanding of invariants during the evolution of species. According to the spectra of neural populations in evolution, it is possible to assign properties of "coding" to EEG oscillations.

- *The principle of causality* is one of the most governing concepts in the understanding of the general sciences. In addition to the conventional causality described by Hume and the quantum theoretical interpretation of causality by C.F. von Weizsäcker, it should be mentioned that the prediction of particle reactions in Feynman diagrams are also based on all causal factors before a nuclear interaction. The S-matrix introduced by Werner Heisenberg is also based on the causality included in the time histories of in-going particles.
- *Living systems*, living subsystems and the large number of physical systems, have type response susceptibility. Magnetic materials such as ferromagnetic or paramagnetic substances have intrinsic properties called magnetic susceptibilities. In living systems there exist brain response and contraction susceptibilities in the overall myogenic system. These types of reaction properties are mostly based on resonance properties of these systems (see Chaps. 4, 6, and 7, and Appendix C).
- *Resonance phenomena in nature* also have common properties. The resonance property of a system plays a major role in communication and amplification of weak signals. A speculative model of the brain-body system is proposed in which the resonant properties of multiple strings may play a major role.
- *Statistical mechanics.* These properties lead again to another general principle: the similar consideration of statistical mechanics in biology and physics. Irregular fluctuations of atoms in a gas and the random firing of single neurons in neural populations can be described by means of statistical mechanics.
- To return to one of the most essential principles of Descartes, who described the rules for and principles of research in his book, *Discourse de la méthode*. To solve a problem in the most satisfactory way, the research scientist must collect all the acts and data related to the system or process under investigation. In Descartes' time research scientists were not in possession of computers and sophisticated measuring instruments. We are now in possession of such enormous facilities. Therefore, this can be the beginning of a new Cartesian area.
- Finally, the nebulous Cartesian system is a framework that takes all types of structural causalities, time histories, genetic causalities, and resonance properties of the brain-body system into account.

21.7 The Importance of Brain Metaphysics

Metaphysics is often replaced by concrete theories after accumulating new measurable results. Often, metaphysical thoughts lead to well-founded and solid scientific disciplines. Some metaphysical trends for approaching brain work were included herein with the hope that one day the described processes can be illuminated or better understood (see Chap. 20). One example follows. At the time of Blaise Pascal, *thinking* (La pensée, 1660) could not be measured or evaluated; in no way was it possible to classify different type of spirits or thought processes (see Chap. 1). Today, several elements of cognitive processes can be measured. This is one of the reasons why this book contains a chapter on metaphysics. It is hoped that one day utopian ideas can be combined with new methods to measure "reality." Also keep in mind that Heisenberg's microscope thought experiment was realized some 70 years later.

21.8 Frameworks

After so many years of collaborative research with several research groups, the author tentatively proposes a new framework basing on several approaches: cybernetics, concepts of dissipative structures (prigogine), the catastrophe theory (René Thom), the new psychology by F.A. Hayek, and synergetics (Herman Haken). All these schools of thought are very creative and have been extremely useful. Despite criticism, Wiener's cybernetics approach was the most dominant and important stream in interdisciplinary sciences. The streams created by Prigogine and Rene Thom were brilliant steps from a mathematical perspective.

However, all of the mentioned multidisciplinary trends have two common shortcomings. First, none has considered the foregoing approaches or new parallel developments. Second, all these theories were derived from mathematical concepts or originated from mathematicians and theoretical physicists. Accordingly, these trends were not necessarily based on empirical biology.

The aim of the proposal in chapter 26 is different:

1. All useful and important features of the mentioned multidisciplinary approaches are accepted as most useful steps; and these approaches will continue to be applied.
2. The new approaches originate from empirical findings on brain-body integration. It uses possible and adaptable methods from mathematics and physics. However, this mathematics does not treat biology as an application area of physics and mathematics. On the contrary, brain-body integration has its own language. Mathematical tools are applied later with more biology-oriented algorithms.

The core ideas of outstanding philosophers are described in Chaps. 1 and 2, and a number of philosophers, including Descartes, Hume, and Locke, have opened new gateways in neuroscience. However, it is Bergson who was unique in his efforts to establish the philosophy as a positive science. Although he was a mathematician, he preferred to observe the events of nature and the evolution of species described by Charles Darwin. He was different from many other philosophers, who emphasized their own theories and neglected to pay attention to the theories of other philosophers. Bergson tried to unify what other philosophers had established. The best example is his work on *creative evolution*, in which he used biological findings from the world of plants and animals and knowledge from chemistry and general biology. According to Bergson, philosophy should not be the task of an individual scientist, but can be only successful as a joint work of several philosophers. Accordingly, this book draws on Bergson's ideas by using measurements as well as theoretical discussions (see Chap. 18).

Chapter 22
Oscillations and Transmitters Are Quasi-invariants in Brain-Body-Mind Integration

22.1 Parallel Analysis of "Whole Brain Body and Brains" as Minimal Study Prerequisite

The brain is the most complex system known to us in the universe, and previous chapters have presented empirical evidence related to the dynamics of the brain, i.e., a biological system that is continuously changing. Assuming that the brain is an organ that also controls our mind and body, the understanding of brain dynamics should help in the search for the communicative processes between the body and the brain. In addition to electrical oscillations, brain-body machineries are also controlled by the release of neurotransmitters (see Chaps. 3 and 13). Feedback loops and recurrent loops also provide continuous exchange among the brain, body, and neurotransmitters. Furthermore, the oscillatory activities in the brain and body are affected by these transmitters; the electrical processes also control the release of transmitters.

Questions such as, "What is the mind?" and "What is the brain-body-mind?" can be approached only by functional and comparative analysis of the *ensemble of brains*, and links as shown in Fig. 22.1. In the upper part of Fig. 22.1 is a schematic description showing that the machinery of brain-mind cannot be understood only by analyzing processes at the level of the adult brain. Our thesis is this: To approach the brain-mind, we also have to observe the machineries of invertebrate ganglia and brains during the evolution of species. Furthermore, physiological processes and anatomical changes need to be analyzed during maturation of the brain, from infancy through adulthood to old age (see Chaps. 10 and 11).

In pathologic brains, the release of transmitters, and accordingly, oscillatory processes and control of cognitive processes, are highly altered (see Chap. 13). Therefore, the analysis steps in Fig. 22.1 that include a loop indicating the influences of pathology (Alzheimer's disease, schizophrenia, bipolar disorders) constitute a minimal analysis – prerequisite to approach the integration of brain-body-mind.

When one reads this book and learns about sensory-cognitive processes, electrophysiology in the evolution of species, differences between child and adult brains, and memory activation and emotional brain, one observes results indicating that

E. Başar, *Brain-Body-Mind in the Nebulous Cartesian System: A Holistic Approach by Oscillations*, DOI 10.1007/978-1-4419-6136-5_22,

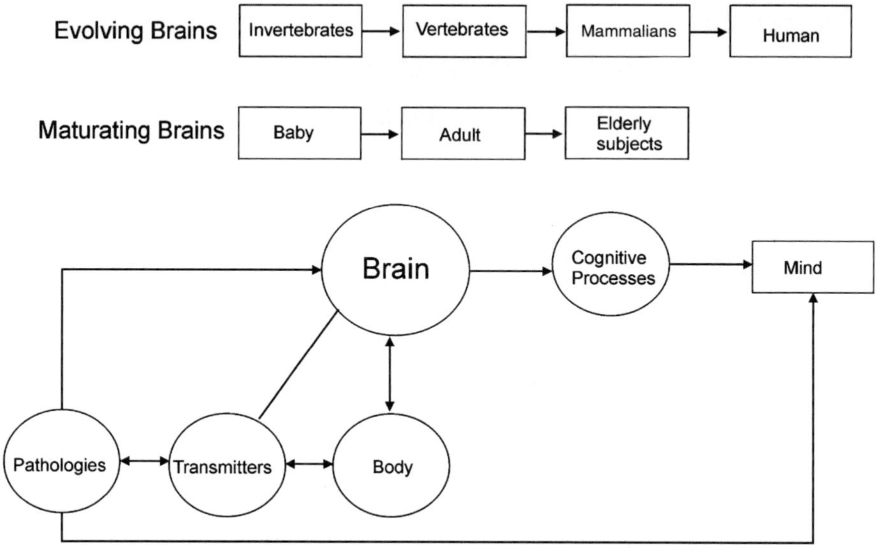

Fig. 22.1 Schematic explanation of the steps proposed for an approach to brain-body-mind

oscillatory processes and transmitters are interwoven with all types of processes in brain-body integration. This acquired empirical evidence led us to the concept of quasi-invariants as building blocks in brain-body integration.

22.2 EEG Oscillations as Quasi-Invariants and Importance of Coherences

In research undertaken 30 years ago, it was stated that EEG oscillations are invariants in the brain's sensory-cognitive processing (Başar 1980). Presently, we are inclined to slightly revise our earlier statement. This volume favors the expression quasi-invariant rather than invariant. This revision of our earlier position is based on a number of important factors:

1. When pursuing the various chapters of this book it is easy to observe that 10 Hz oscillation does not have the same shape, amplitudes, and even frequencies during evolution of species (see Chap. 10), and maturation of the human brain (see Chap. 11).
2. Moreover, in the adult brain 10 Hz oscillations depend on the related function and topology; for example, the occipital cortex shows high amplitude regular alpha activity, whereas the frontal cortex has poor alpha activity during resting conditions. Upon increase of cognitive load, 10 Hz oscillations are also observed in the frontal cortex. In Alzheimer's patients, alpha is reduced, and is almost completely absent in patients with bipolar disorder.

3. Further, delta-response has large amplitudes in frontal and parietal lobes in response to target stimuli during a working memory paradigm (see Chap. 6).
4. Several examples show that alpha, theta, delta, and gamma activities are not present at the same area during application of different task conditions in the brain. By observation of responses to various facial stimulations (angry face, happy face), i.e., during emotional processes, the activated oscillations are altered by observation of angry faces, occipital beta response increases during observation of angry faces, and occipital alpha activity is increased as well (see Chap. 12).
5. The beta response is also considerably increased during gender differentiation (Güntekin and Başar 2007a, b).
6. Is the mind of a child different from that of an adult? Is the mind of an Alzheimer's patient different from a healthy subject of the same age? There are crucial differences between the oscillatory patterns of healthy and Alzheimer's subjects as well as subjects with bipolar disorder (see Chap. 13).
7. In other words, the oscillatory processes show different amplitudes and shapes, depending on the modality of sensory cognitive input and also topology. It was even possible to show that the human behavior defined as *intuition* can be assigned to alpha activity during the evolution of species (see Chap. 17). This last statement is somewhat speculative as of yet. However, this view anchors on solid measurements; it is worthwhile to continue research in this area.

According to the factors explained in the preceding, it is simply stated that oscillatory responses are the building blocks of dynamic processes and, because of the crucial changes in amplitude, location, and dependence on modality, age, and disease, we describe them as *quasi-invariants* in brain-body-mind integration.

8. Not only the parameters of oscillatory patterns, but also the *connectivity of oscillations* among various structures of the brain is vital for brain functioning when studying coherences in the healthy adult brain and clinical disorders (see Chaps. 6, 7, 10, and 13). It is strongly recommended here to re-read Bullock's (Bullock et al. 1995a and b) comparison of coherence measurements in Chap. 10 and Table 10.3. The tissue of the human brain, which is supposed to have a different type of mind in comparison with low vertebrates and invertebrates, shows higher coherences. Accordingly, *coherences* play a major role in differentiating "minds."

22.3 Ultraslow Oscillations Are also Quasi-Invariants

The concept of functional building blocks is extended by including the ultraslow frequency range between 0.001 and 0.1 Hz, and also including the heart, kidneys, and all organs of the autonomous system (see Chaps. 4, 5, and particularly 9). Accordingly, all frequencies ranging from the minute rhythm to fast EEG rhythms can be also considered to be *quasi-invariants* in the machinery of brain-body-mind integration.

22.4 Web of Oscillations and Transmitters as Quasi-Invariants

As explained in Chaps. 3 and 13, there is an important interplay between transmitters and diseases such as Alzheimer's disease (AD) and bipolar disorders. In such diseases, the oscillations are impaired or are irregular because of the reduced release of transmitters such as ACh and GABA. We describe the complete absence of alpha and theta oscillations as "break of mind," because AD and bipolar patients show different types of mind in comparison with healthy subjects. Furthermore, breaks of coherence in the alpha and gamma frequency ranges are also accompanied by various types of pathologies. The brain work described in Chaps. 6–8 is broken down in cases of cognitive impairment, as manifested by several types of diseases.

Figure 22.2 describes the changes in oscillatory responses in alpha, beta, theta, and delta frequency ranges in AD and bipolar patients. The illustration includes findings related to patients before and after medications. Figure 22.2 is a type of short concluding scheme of limited results; it does not even comprehend coherence analyses that open deeper horizons.

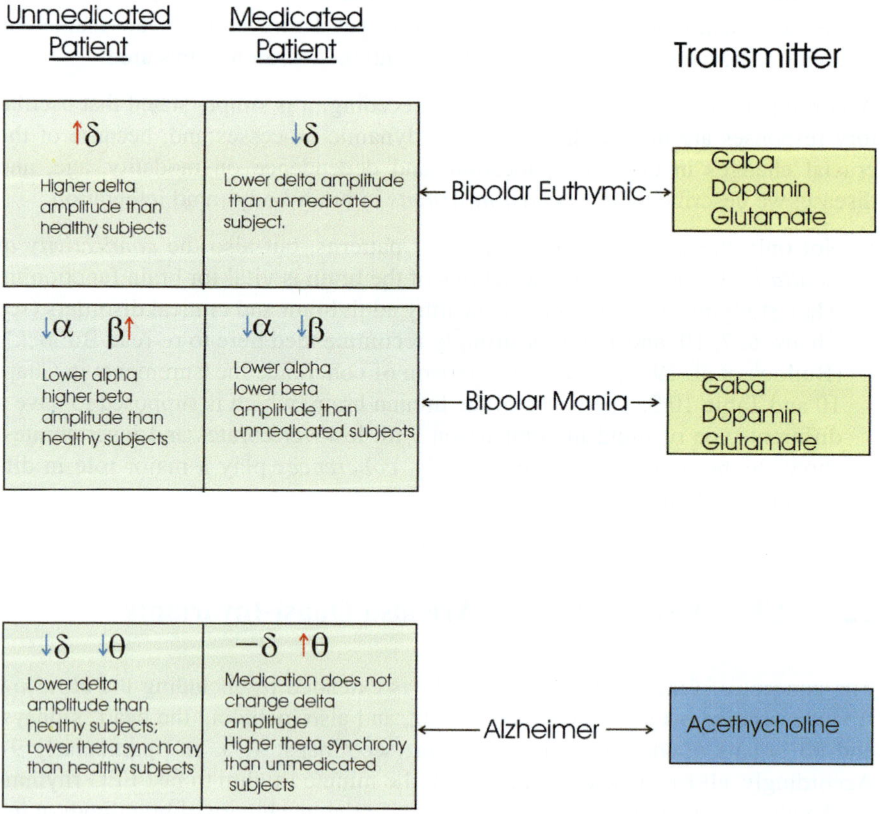

Fig. 22.2 A scheme globally describing changes in oscillatory responses in Alzheimer's and bipolar patients. Results based on the findings in Chap. 13

At the right side of the illustration the most important neurotransmitter responsible in both diseases is roughly explained.

The AD patients were medicated with cholinesterase inhibitor, and the bipolar patients with valproate.

The following description is also important: Neurotransmitters, which form the building blocks of brain-body-mind integration, are also quasi-invariants within the mechanisms of whole body integration. It is also crucial to note that there is an important *interplay between oscillations and transmitters*. This interplay could also be considered as evident when we analyze the schematic description of Fig. 15.2

22.5 A Synopsis on the Relation of Neuropathologies to Brain-Body-Mind

According to the reviewed publications, a valuable description of integrative brain-body-mind processes by means of electrical signals requires the analysis of brain responsiveness at various levels. The following steps are recommended to achieve this integrative approach (see also Fig. 22.1).

1. Stimulations in single modalities are not sufficient to determine the electrical correlates of brain functions: Sensory signals (visual, auditory, somatosensory signals) should be applied to a human subject and recorded from distributed areas of the cortex. Additionally, cognitive inputs (e.g., the P300 paradigm) and presentation of complex cognitive stimuli (figures with various shapes and face pictures) should be applied, so as to compare them with sensory signals, for a better analysis and to determine functional correlates.
2. A more advanced step consists of the evaluation of the emotional input to a human subject. According to Solms and Turnbull (2002), emotions can be considered as a sixth sense; and facial expressions (e.g., those conveying love and hate) can be used to understand the modulation of electrical signals from the brain (Başar et al. 2008; Güntekin and Başar 2007c; Zeki 2007, see also Chap. 12)
3. The mentioned steps are necessary for an integrative understanding, but are not sufficient for a complete description. In addition to the necessary application of various modalities, further basic causal factors and hidden causal factors should be considered, and jointly analyzed if possible.
   Some of the causal factors to take into account include (see Fig. 22.3):
   (a) Gender differences
   (b) Age factors (babies, adults, elderly subjects)
   (c) Genetic factors (alcoholism, ADHD, polymorphisms)
   (d) Vegetative inputs (including blood pressure, respiration). Changes in these levels also belong to the hidden casual factors
4. The model in Fig. 22.3 proposes that to understand the functional correlates of brain oscillations, it is necessary to develop an integrative research application using inputs of several different modalities, and then compare the responses to

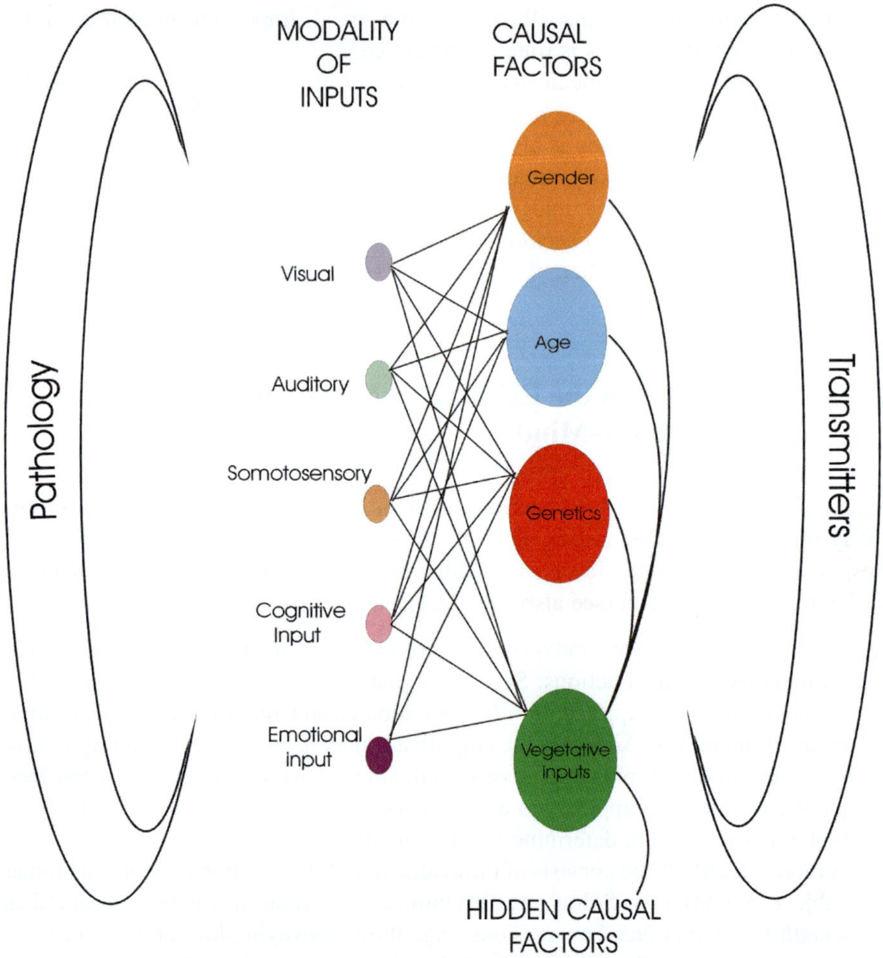

Fig. 22.3 A multidimensional model of electrical signals

visual, auditory, somatosensory, cognitive, and emotional inputs. Furthermore, causal factors such as gender, age, and genetics need to be included in this research program. Also included in Fig. 22.3 are the vegetative inputs to be considered as hidden causal factors. Spontaneous changes in blood pressure, respiratory cycles, gastrointestinal peristalsis, and changes in the noradrenaline or ACh levels in the autonomous system also belong to the causal factor group. The pathological level is also included in this model, because it can be extremely influential in modulating brain oscillations, both spontaneous and evoked. Moreover, the medication of patients by means of pharmacological agents containing various types of transmitters can help to reduce pathologies and greatly assist the oscillatory responses to return close to the level of healthy subjects (Güntekin et al. 2008; Özerdem et al. 2008; Yener et al. 2007, 2008). Moreover, as the results of Porjesz and

> Rangaswamy (2007) and Rangaswamy et al. (2002) underline the fact is that pathologies, genetic factors, and the application of transmitters all induce changes in the results in such a way that the current knowledge can be extended to the understanding of brain oscillations and cognitive processes.

After reviewing the reports cited in the present chapter, it becomes obvious that pathology (cognitive disorders, diseases) and medication (which influences the transmitter release) entirely change the understanding and big picture of cognitive processes:

1. Various pathologies cause significant and differentiated changes in the oscillatory dynamics of patients.
2. Medication (modulation of transmitters) can partly reduce the cognitive impairment of patients. Additionally, medication helps to reduce pathological deformation of electrical signals.
3. Not just oscillations in a unique frequency window, but also multiple oscillations have to be jointly analyzed for the description of pathological cases. The release of transmitters also selectively influences brain oscillations (see the first and second gamma and theta windows).
4. In addition to the analysis of amplitudes of oscillatory responses, a coherence analysis is highly recommended. The latter analysis methods help in the understanding of the connectivity or lack of connectivity between distant locations.
5. With regard to EEG coherence and phase-delays, multiple windows should be analyzed so as to determine a more accurate picture of the pathology.
6. The analysis of superposition of oscillations is also an extremely relevant tool for the description of oscillatory responses in cognitive processes, both in healthy subjects and those with cognitive disorders.
7. To obtain a true picture of the oscillation processes in pathology, frequency analysis has to be performed separately in medicated and non-medicated patients.
8. As the studies of Yener et al. (2008) and Özerdem et al. (2008) show, it is essential to perform analyses separately, before and after medication.
9. According to Bowden (2008), therapy with the application of multiple pharmacological products is useful (polypharmacy). The selectivity of the oscillatory response can assist in achieving optimal use of medication following the analysis of brain oscillations.
10. The ambiguity of gamma responses can be better understood using adequate input modalities and also early and late time windows. Accordingly, it is proposed that a standard assessment should be developed.
11. The alpha and beta frequency windows have been neglected in most studies of schizophrenia. As demonstrated by the results of Başar-Eroğlu et al. (2008), Ford et al. (2008), Özerdem et al. (2008), Fehr et al. (2001), and Fehr (2002), it is obvious that all frequency windows need to be analyzed in all cognitive disorders.
12. The link among genetics, brain oscillations, and transmitters is better understood from the work of Begleiter and Porjesz (2006). This genetics-brain oscillation concept will potentially form one of the most important windows in neuropathologies, as this short chapter shows.

Chapter 23
Unifying Trends: Globally Coupled Oscillators in Brain-Body

23.1 New Possible Models as Integration of the Brain with the Body

Some of the important new features of the present book related to physical-mathematical frameworks/models are described in Chaps. 14, 15 and 16. We propose new Cartesian frameworks in the twenty-first century. The aim of this chapter and Chap. 24 is to propose new models of thought on brain-body-mind integration. The first model is an adaptation of the model of globally coupled oscillators developed by Pikovski et al. (2001), briefly mentioned in Chap. 9 (see also the model in Fig. 9.2). A second model of thought is the brain string theory, derived from an analogy to string theory in physics and adapted to the theory of brain-body-mind integration. The second of these models also includes concepts of quantum theory; this time being more concrete than the description of the *quantum brain* explained in 1980.

Both models have common basic structures: The brain, spinal cord and organs of the vegetative system are globally (and/or diffusely) linked. They communicate through signals in similar frequency bands of EEG-oscillations and ultraslow oscillations (compare also Chaps. 4, 5, 6, 9, and 22). The first of these two models is derived by using engineering concepts and has a more or less causal structure. In contrast, the second model presents a metaphor to string theory, in which the links and communications in the incorporation of brain-body have a probabilistic nature, similar to quantum dynamics.

23.2 A Model of Globally Coupled Oscillators in Brain-Body-Mind Integration

Based on the accordance of sensory-cognitive oscillatory processes in the body, and brain and cognitive processes, try the following. The model of globally coupled oscillators described in Fig. 9.2 can be expanded by taking into account the links

E. Başar, *Brain-Body-Mind in the Nebulous Cartesian System: A Holistic Approach by Oscillations*, DOI 10.1007/978-1-4419-6136-5_23,

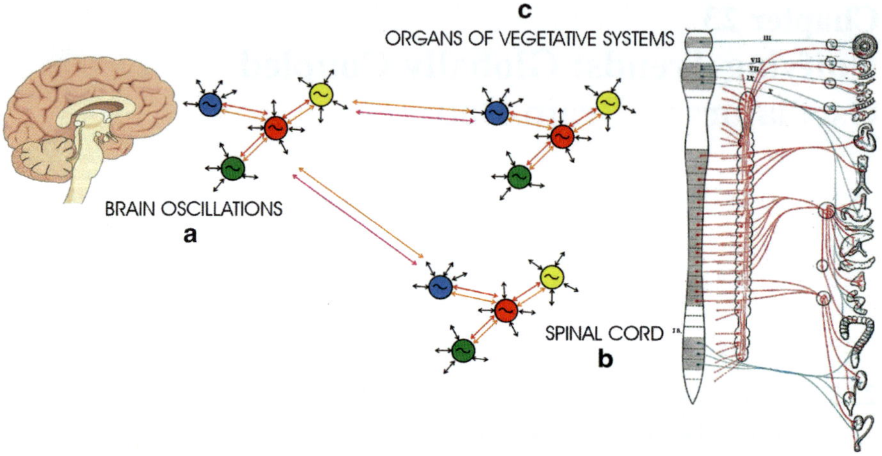

Fig. 23.1 (**a**) Oscillators in the brain (*left side*). (**b**, **c**) Spinal cord and organs of the vegetative system (*right side*)

among the brain, spinal cord, and autonomous system. In this case, it can be easily assumed that three different assemblies of globally coupled oscillators are involved in processing of mutual links and communication. Because the dynamics of cognitive processes are also governed by oscillatory dynamics and coherent oscillations (EEG and ultraslow oscillations; shown in Chaps. 4–8), it can be concluded that oscillatory processes play a key role in brain-body-mind integration. The role of transmitters in pathological situations of the mind was described by Bowden (2008), Başar and Güntekin (2008), and in Chaps. 3, 13, and 22. Therefore, transmitters that greatly influence oscillatory dynamics should be included within the main parameters of brain-body-mind integration.

Figure 23.1 (top) illustrates the adapted concept of globally coupled oscillations in brain-body-mind integration. Various structures of the brain are globally linked to the spinal cord and vegetative system organs. It is impossible to individually track the links, EEG, and ultraslow oscillatory activities in neural structures and elements of all the organs embedded in the overall myogenic system. According to the measurements described in Chap. 9, global coordination and communication among brain, spinal cord, and all organs are expected.

23.2.1 Overall and Mutual Excitability

We also assume the existence of mutual excitability and overall resonances in such a model, because the vegetative and neural elements are tuned to have the same excitability for communication. For an explanation of the excitability rule, the reader is referred to Sects. 7.3.3 and 11.3. Global resonances and global overall excitability are also controlled and altered by neurotransmitters.

Chapter 24
Unifying Concepts: Brain-Body's String Theory

We possibly have to move to transcendental theories as quantum theory to understand the psychology of the brain.

(Jung 1951a)

Quantum physics formulates laws governing crowds and not individuals. Not properties but probabilities are described.

(Einstein and Infeld 1938)

Chapter 14 explained the view of Einstein related to quantum physics. According to this view, as in quantum physics, laws in cognitive processing are valid for great congregations of individual units. They are valid not for single neurons, but for neural populations. What applies to quantum mechanics also applies to the dynamics of chaotic systems. In systems, not properties but probabilities are described; laws disclose the change in probabilities over time; and they are valid for congregations of units. We also remember the duality of chaotic and quantum behavior of the brain outlined in Chap. 14. There are also several new theoretical methodologies describing the relation between the quantum and the chaotic approach (e.g., Reichle 2004). However, these analyses are beyond the scope of the present volume.

24.1 "Brain's String Theory" in the Brain-Body Syncytium

String theory is a more transcendental configuration of the globally coupled oscillators. Both of these models are fictive models so as to reach unification in the very complex brain-body-mind system.

String theory (or M-theory) is a new and important unifying concept in physics that brings together quantum theory, relativity theory, and gravitation theory. Strings, i.e., types of macro-oscillators, are the basic elements of this theory. Within contemporary research related to general theories of the brain, physicists have tried to develop a theory of everything (M-theory). This implies that physicists are

E. Başar, *Brain-Body-Mind in the Nebulous Cartesian System: A Holistic Approach by Oscillations*, DOI 10.1007/978-1-4419-6136-5_24,

working on general theories of science. The essence of string theory is this: In all physical systems, at the microscopic level, a number of strings exist. These strings are capable of displaying resonance in 11 frequency channels. The resonance of strings and their collision are essential to create elementary processes and also, building on those, more complex phenomena.

In earlier research, and previous sections of this book, we have proposed that the investigation of oscillatory processes might help to understand unifying trends in brain-body-mind integration. We have empirically described that oscillations are candidates to be one of the major building blocks governing the integration of the brain-body-mind. Resonance principles related to oscillators are also observed in several biological systems. Accordingly, we propose the following a model (or metaphor) similar to the string theory of physics. A demonstration of this model is not yet possible. Nevertheless, if one day equations of the oscillatory processes in the brain can be derived, we think that this model essentially could be able to mathematically describe the dynamic processes in brain-body-mind integration.

We have shown that functional correlates of brain-body functioning are manifested with EEG and ultraslow oscillations (Chaps. 4, 5 and 6). The steps now undertaken are as follows:

1. First, according to the cardinal question of Fessard in Chap. 2, we looked for the most general transfer functions in the brain.
2. In a more general view we were confronted with the more general question for connecting brain-body transfer functions (see Chaps. 4, 5, and 9). The experimental conclusion is that the common or most general transfer functions in the brain are represented by oscillations. These transfer functions are characterized by brain oscillations in the EEG frequency range and ultraslow oscillations to link brain-body processes (see Chap. 9). *Resonances* and *mutual resonances* result from tuning of oscillatory information reaching the brain from the spinal cord, vegetative system, cerebellum, and also hidden sources (see also Fig. 15.2). This oscillatory information reaching the brain can survive if the elementary strings that are brain communication signals can co-exist and show resonance. *Only resonance states lead to function.*
3. Non-resonant collusion of elementary strings form noise and cannot survive as acting signals. The collision following resonance of the oscillatory strings is, accordingly, the cause for general processes in the brain-body interaction. However, one cannot explain the formation of resonance (and accordingly functions) in the conventional Cartesian system or within the scope of Newtonian dynamics. The collisions of vibrating strings should take place in a hyper-probabilistic space that obeys rules of quantum mechanics and/or statistical mechanics. This is one of the results in this book: In the study of brain dynamics, we combine rules of oscillatory processes including resonance, rules of quantum mechanics with its metaphor of the uncertainty principle, and chaos theory. The resonance ability of single strings (oscillators) leads to more complex resonance stemming from multiple strings colliding in the brain. That would also create

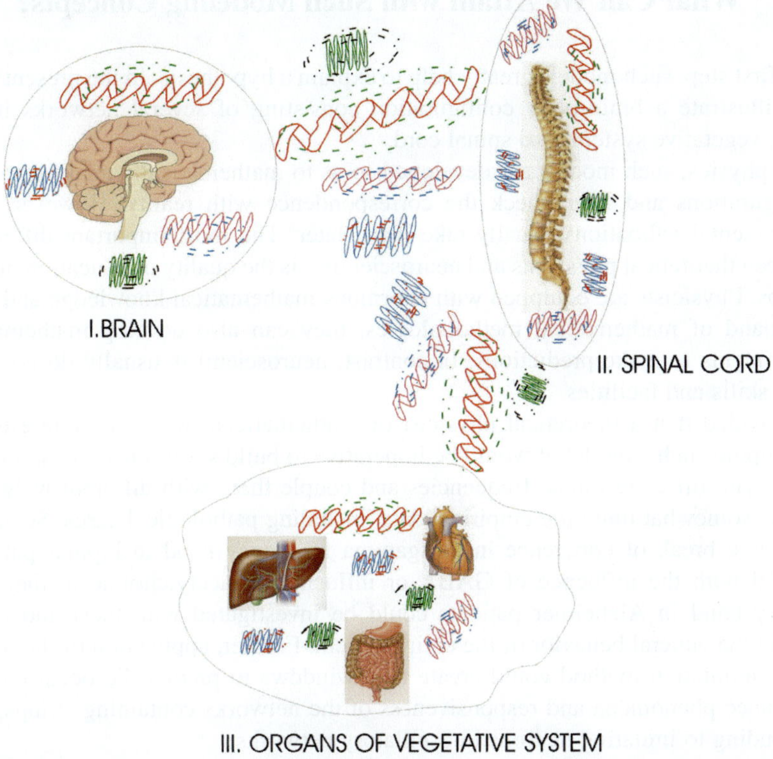

Fig. 24.1 In this model (**a**) the brain, (**b**) the spinal cord, and (**c**) the organs of the vegetative systems are illustrated as three different functional groups that are connected with strings having different innate frequencies (EEG-oscillations and ultraslow oscillations). Strings (or oscillators) of different frequencies that are presented with *different colors* bind these three groups, according to coherences described in Chapters 6 and 7. The strings also bind neural groups within the brain, spinal cord, and organs of the vegetative system. Different neurotransmitters, shown as *dashed lines in different colors*, influence the frequencies of strings and coherences among neural groups. Several types of strings are represented with *different colors*. Strings have different amplitudes and are also accompanied by (embedded in) *dots*, *points*, and *short lines with different colors*. These *lines* are either *horizontal* or *oblique* and represent different transmitters. The illustration shows three groups of body brain organs: (**a**) brain, (**b**) spinal cord, and (**c**) organs of the vegetative system. There are also strings of *different size* and *colors* between these three groups. These strings present couplings (links) among groups (i.e., coherences)

multiple coherent modes that could be described as *super-synergy* and *super resonance*. Refer to Chapter 8 for the concept of super-synergy, and Appendix III for a general explanation of resonance phenomena. Such resonating and hyper-probabilistic working machinery provide multiple brain functions. These are the elements of super string brain theory, analogous to M-theory in general physics.

The brain's string theory is illustrated in Fig. 24.1.

24.2 What Can We Attain with Such Modeling Concepts?

As a first step, such models greatly help to explain a hypothesis. In the present case, they illustrate a brain-body configuration consisting of several networks in the brain, vegetative system, and spinal cord.

In physics, such models are developed so as to mathematically analyze various configurations and later check the correspondence with reality. However, such experimental validations usually take place later. The most important difference between theoretical physicists and neuroscientists is the quality of education in both groups. Physicists are equipped with enormous mathematical knowledge and have command of mathematical methodologies; they can also develop mathematical experiments to make predictions. In contrast, neuroscientists usually do not have these skills and facilities.

Provided that a theoretical physicist or mathematician would be interested in developing such a model, it would be imperative to build several networks of oscillators with different innate frequencies and couple them with different weighting factors somewhat imitating empirical data, including pathological cases. Scenarios such as a break of coherence in the gamma frequency band in bipolar patients, parallel with the influence of GABA or influence of acetylcholine in theta frequency band in Alzheimer patients could be investigated with these models in predict the general behavior of the entire system. Further, application of the Monte Carlo simulation method could create new windows to predict the occurrence of resonance phenomena and responsiveness of the networks containing strings, corresponding to imitating neurons in oscillatory networks.

Chapter 25
Unifying Concepts: Dynamic Syncytium of Brain-Body-Mind and Intuitive Processes

> *... the mind must remain forever in a realm of its own which we can now only directly experience it, but which we shall never be able fully to explain or to "reduce" to something else.*
>
> (Hayek 1952)

25.1 Is Hayek's View Related to New Psychology a Precursor of the Integrative Mind?

C.F. von Hayek (1952) asked the question, "What is mind?" and discussed the relationship between mind and body or between mental and physical events as follows:

> What we call "mind" is a particular order of a set of events taking place in some organism and in some manner related, but not identical, to the physical order of events in the environment. (See Sec. 2.12.)

This view is in accordance with all processes described in Chaps. 9, 12, and 18–20. Hayek classifies *emotion* as a special type of disposition for a type of actions that, in the first instance, are not necessitated by a primary change in the state of the organism, but that are complexes of responses appropriate to a variety of environmental conditions. Fear and anger, sorrow, and joy are attitudes toward the environment, and particularly toward fellow members of the same species. This means that a great variety of external events, and also some condition of the organism itself, may evoke one of several patterns of attitudes or dispositions, which will affect the perception of, and responses to, any external event. Emotions may thus be described as affective qualities similar to the sensory qualities and forming part of the same comprehensive order of mental qualities. The similar views of Solms and Le Doux are discussed in Chap. 12.

Hayek further proposes that we must distinguish between two different kinds of physiological "memory" or traces left behind by the action of any stimulus: One is the semi-permanent change in the structure of connections or paths that determines the courses through which any change of impulses can run (similar to Hebb's principle).

E. Başar, *Brain-Body-Mind in the Nebulous Cartesian System: A Holistic Approach by Oscillations*, DOI 10.1007/978-1-4419-6136-5_25,

The other is the pattern of active impulses proceeding at any moment as the result of a stimuli received in the present and past and perceived also as merely part of continuous flow of impulses of central origin, which never ceases altogether, even when no external stimuli are received.

Hayek's most important conclusion is that "We shall find that the same set of external stimuli will not always produce the same responses, but also that altogether new responses will occur." This is similar to the coordinated movement of the organism, which is not determined by the movement of an individual muscle, but by the whole complex of body muscles. Following the presentation of fundamental evidence in Chaps. 14–16, we introduced the brain S-matrix, which takes into account the whole history of the organism. Hayek does not comment on the S-matrix in his 1952 book, but his concept is in accordance with the principle of application of the Brain's S-matrix and Feynman diagrams that include the history of whole brain-body organism. Hayek talks about the coordinated movement of the organism. This view is similar to the whole brain-whole body models presented in the chapters on unifying trends (see Chaps. 9 and 22–24). Within those chapters, the concepts of globally coupled oscillators and brain's string theory that comprehend *coherence-links, coordination, probabilistic coupling of oscillators*, and *neurotransmitter influences* are explained.

Even though we may indicate that the mental event experienced can be produced by the same forces that operate in the rest of nature, we shall never be able to say which particular physical events correspond to a particular mental event. The brain's string theory in Chap. 24, which is governed by the uncertainty principle, is dominated by a similar principle. All these thoughts are also in accordance with the modern causality principle of von Weizsäcker.

25.2 Intuition Revisited: Intuition and Emotion Highly Influence the Machineries of the Mind

Chapter 20 described some events or processes that led to new discoveries by outstanding scientists or the development of creative solutions to existing problems, such as Poicaré's solution for a mathematical problem, or Otto Loewi's discovery of acetylcholine. Several other examples were described by Nancy Andreasen (2005), and Roger Penrose (1989). The comments on Sherlock Holmes by Einstein, as described in Chap. 2, are also very interesting.[1] Next, von Hayek, Bergson, and Andreasen highly interesting observations related to the creative brain and psychiatric patients are reviewed. Bergson explained in detail the role of intuition in

---

[1]Einstein also played the violin. Was it a preparative stage for him to also solve problems?

creativity (see Chaps. 17 and 18). Hayek excluded the possibility of finding physical elements or forces corresponding to a particular mental event. This last view corresponds with the context of what can be called intuition. Two appropriate definitions of intuition are repeated here:

1. Intuition is the act or faculty of *knowing immediately*, directly, and holistically without rational processes and without being aware of how something is known. It is also the channel or process through which realms of truth and knowledge are accessed.
2. *Insight, sensitivity to environment, and problem-solving ability.* Once relegated to the realm of superstition, this mental capability can model non-rational aspects of reality about as well as intellect can force its user to impose a Cartesian, mechanistic view on the world.

By requiring appearances to be given in space and time, intuitions allows one to perceive particular correlations between representations, thereby limiting empirical knowledge to the sensible realm.

It is crucial to add that emotions, which are also often a product of earlier experiences, influence intuitive behavior and/or creativity. Accordingly, in crucial decision-making processes, perceptions and cognitive processes are often governed by intuitive and emotional processes. According to Fuster (2003), memory can enter consciousness in a multitude of forms and states. By definition, the recall of any memory is conscious. *Imagining* is also conscious in that it consists essentially of the conscious retrieval of long-term memories and establishes cognitive networks (*cognits*) that are assembled and reconfigured in a variety of ways. Creative intelligence is the ability to invent goals, projects, and plans or, as Fuster (2003) says, to invent the future. Creative Intelligence is served by the conscious interaction of memory and imagination. Figures 18.2 and 18.3 describe the relationship (or boundaries) of episodic memory, semantic memory, imagination, and creativity. As mentioned in Chap. 18, we are able to recall the body of our partner and imagine her or him wearing new clothes, which we have seen immediately beforehand in a department store. According to Bergson, this is *pure imagination*. Certainly, this is a very simple example, which is suitable to explain the core of creative memory or the necessary way of thinking. To perform this process in reality might require several hours; however, this type of creative synthesis can be performed in a fraction of a second in creative memory, which is able to construct past, present, or future scenarios that do not rely on reason or causalities.

Return to the tentative correlation between alpha activity and intuition, and recall that creativity and intuition are interwoven according to the empirical and philosophical scope of Bergson. In tests of remote association and alternate uses, Martindale and Hines (1975) found that alpha frequency is positively correlated with creativity.

It also should be remembered that the increase of coherence in the mammalian brain and the human brain is high in comparison with lower animals.

25.3 Intuition, Episodic Memory, and Emotion Form a Functional Syncytium

What is the role of oscillators or strings in models such as globally coupled oscillators (Chap. 23) and the brain's string model (Chap. 24) during the processes of creativity and intuition, which are almost inseparable, as explained in Chap. 18 and 20? Moreover, in those chapters, it is hypothesized that the processes of creativity and episodic memory have interwoven machineries, because both processes occur in the non-homogenous time-space or in a *time-space not measurable with physical clocks* (Figs. 18.2 and 18.3). This is because during sudden creative epochs, functional tasks needing enormous time segments occur in a fraction of a second, as in the working methods of Poincaré or Mozart.

Currently, it is possible to measure electrical changes of the brain's frontal and occipital lobes on presentation of known episodes and emotional events (see Chap. 12). Accordingly, it can be predicted that in future more refined measurements may open the way to also analyze creative periods. At least one may find differences in event-related oscillations between the brains of creative artists and scientists. The evident interplay between oscillations and transmitters was also noted, as well as the neurotransmitter-oscillation web in unifying trends for the understanding of mind (see Chap. 22).

Note the special emphasis of Nancy Andreasen, who pays attention to the creativity of neuropsychiatric patients. The results and interpretation of refined analysis of brain oscillations in Chaps. 13 and 22 present a case for studying in more detail changes of the creative mind in diseases of cognitive impairment and/or changes in brain-body-mind integration. Also, in the light of changes in neurotransmitter level among patients with schizophrenia, Alzheimer's disease, and bipolar disorder, these favored types of measurements of episodic and emotional memory may lead to a profound approach to the understanding of mind and partially answer the question posed by Eric Kandel in Chap. 20.

25.4 Reasonings on the Web of Brain-Body-Mind: Joint Interpretation of the Brain-Body's String Theory and Intuition

The statement of René Descartes, "Everything in the universe could be explained in terms of a few intelligible systems and simple approaches…" is strongly supported by results and hypotheses described herein. Oscillations, neurotransmitters, the resonance principle and entropy factor belong to governing intelligible systems in the integration of the brain-body-mind. This is similar to the physics of the earth and galaxies, in which atomic principles and quantum oscillatory processes explain a great number of phenomena.

The brain's string theory is the most developed model of thought presented in this volume, because all the thoughts and empirical fundaments are included in this model, which is based on the following results and fundaments:

1. The functional syncytium brain-body-mind replaces the concept of mind, because of the fundamental findings related to EEG oscillations, ultraslow oscillations and neurotransmitters that are quasi-invariants, as described in Chaps. 9 and 22. The overall myogenic system (OMS) and vegetative system (including the heart, kidneys, and lymphatic system) function in an interwoven way, and lead to multiple and uncertain causalities in the machineries of brain-mind (see also Chaps. 5, 7 , 9, and 22).
2. Each oscillatory activity represents multiple functions; vice versa, each function is represented by multiple oscillations (superposition principle in Table 7.1).
3. Functioning in all physiological and biochemical pathways are governed by quasi-invariant natural frequencies and neurotransmitters of brain-body functioning (see Chap. 22). This is similar to Descartes' view on the governing role of a few intelligible systems.
4. The principle of general and common tuning can be found not only in the brain, but also in brain-body integration. According to the work of Gebber et al. (1995a, b), Aladjalova (1957), and Ruskin et al. (2001a, b), structures in the vegetative system are also tuned to the same frequencies. Transmitters such as acetylcholine or norepinephrine are also excellent vehicles for general tuning in brain-body interaction. The body and the brain use the same transmitters (see Fig. 3.10 and results in Chap. 13) and the same frequencies for general tuning for brain-body interaction. Here, the ensemble of these phenomena is tentatively called the overall tuning in brain-body interaction.
5. EEG codes and natural frequencies process within the limits of the uncertainty principle. Oscillations and neurotransmitters form one combined activity. Therefore, the web of oscillations and neurotransmitters can also be jointly considered as dual building blocks for function.
6. The CNS, OMS, and organs of the vegetative system show mutual excitability and, accordingly, mutual resonances in communication. This principle supports the fundaments of brain-body-mind and leads to the brain-body string theory.
7. Spontaneous and event-related oscillations in the CNS, OMS, and vegetative organs are all embedded or interactive with the biochemical pathways (neurotransmitters). These oscillatory processes can be considered as manifestations and building units of brain-body functioning.
8. In pathology, the role of oscillatory dynamics and coherence between various parts of the brain are determined for breakdown of the mind. As the brain matures, an enormous change of alpha activity is observed. Accordingly, the mind of a child and that of a dementia subject are completely different. Creativity is not excluded in psychiatric disorders (Andreasen 2005). Good clocks in healthy subjects become bad clocks in pathology, because the synchrony decreases. (See the scope of Einstein in Chap. 2.)
9. Episodic and semantic memories, which are manifested with superposition of oscillations and "traveling back" in time are among the most important processes

of the mind: intuition and episodic memory, an emotion, form one combined entity.

10. Extended genetic analysis is essential to differentiate clinical disorders in the scope of the work of Begleiter and Porjesz (2006).
11. The *archetypes* described by Jung (1951b) also form a type of phyletic-genetic memory that also highly influences the processes of brain-mind.
12. Consciousness can be analyzed and interpreted only if it is compared with different states of unconsciousness.
13. Entropy is a useful parameter in measuring learning processes and also in the development of mind, as demonstrated by studies of electrical activity during evolution of species (see Chaps. 10 and 17).
14. The advantages of theories of oscillations: Fig. 3.11 compared several modern methods that are applied in brain research. Of these, EEG- or MEG-oscillations are possibly the most appropriate methods for measuring episodic memory and emotions because of the very short time resolution of these methods as well as the fast occurrence of the episodic process.
15. The machineries of mind can be approached only within the scope of the uncertainty principle, i.e., with the quantum causality of von Weizsäcker.
16. Bergson states that intuition requires an accumulation of knowledge and/or "transition of unconsciousness knowledge" to "conscious experience." Therefore, intuition plays a crucial role in the machineries of mind, especially creative states.
17. Measurements of episodic memory and episodic emotions can also present crucial opportunities to interpret the creative mind in those with clinical disorders.
18. It should be emphasized that neither episodic memory nor intuition can be measured with physical clocks. As Bergson explained, both processes of intuition and episodic memory (as known by subjective experience) occur in non-measurable, heterogeneous time-space.
19. Physiological and biochemical changes observed during the evolution of species will contribute considerably to further advances in the search for brain-body-mind.
20. According to these results, the mind cannot be defined with a unique sentence, as stated in Sect. 1.1. One may address the question, "How does the mind work?" but cannot yet define it: The answer to this question requires manifold functional implications in the brain and body. The brain, body, and mind are inseparable entities.

Last, but not least, according to the points presented herein, Hayek's opinion is favored as the most adequate position from which to approach the mind:

> …. the mind must remain forever in a realm of its own, which we shall never be able fully to explain. (Hayek 1952)

However, the search for mechanisms of brain-body-mind should not be considered a mission impossible. In the sixteenth and seventeenth centuries, solving the mind problem or understanding cognition could almost be termed a mission impossible.

Subsequently, the Cartesian system was developed; Newton advanced scientific understanding, and so on. Today, scientists are able to judge and positively develop research on cognitive processes. Therefore, the proposed framework for the nebulous Cartesian system or other new Cartesian systems presented by other researchers may lead to gigantic steps toward achieving a new "white revolutions" in Bullock's words (1984).

By examining all of the chapters one may find several other important issues or discover new avenues to approach brain-body-mind integration. For the time being, the essential message is included in these reasonings.

What will be the next steps in brain-body-mind integration? Wiener (1948) stated, "If a new scientific subject has real vitality, the center of interest in it must, and should, shift in the course of years." The field of oscillatory brain dynamics is now in the center of interest. I think that now is the time for a new paradigm shift, or at least a gigantic extension. This paradigm shift also should include further analysis of brain-body oscillations as a continuation of the results presented in Chap. 9 and involve research on the intuitive mind and the quantum brain in a new Cartesian system.

Chapter 26
The Need for a Paradigm Shift and a New Cartesian System

According to Capra (1984), the Cartesian model needs a major revolution. In his words:

> Transcending the Cartesian model will amount to a major revolution in medical science, and since current medical research is closely linked to research in biology – both conceptually and in its organization – such a revolution is bound to have a strong impact on further development of biology. To see where this development may lead, it is useful to review the evolution of the Cartesian model in the history of biology. Such a historical perspective also shows that the association of biology with medicine is not something new but goes back to ancient times and has been an important factor throughout its history.

26.1 The Importance of Philosophy

What is the importance of the Cartesian system in the development of science? What is the importance of philosophy in today's science? Who are the philosophers? Charles Darwin was a philosopher. Einstein was a philosopher. Norbert Wiener and Werner Heisenberg were philosophers, but they were new Cartesian philosophers. Bergson and von Weizsäcker are relevant philosophers as well. The significance of the Cartesian system is reflected in the fact that its creator, René Descartes, is considered both a founder of modern science and as well as a pioneer of modern philosophy, which is based on empirical findings.

Descartes also developed a methodological framework that formed the basis for all subsequent modern scientific investigation. Descartes was active in several fields leading to a breakthrough in scientific methods. He initiated three major steps:

1. He created the Cartesian system and analytical geometry, opening the way to Newton,[1] and later to Maxwell, and so on.

[1]In the *Principia,* Newton not only wrote the three laws of motion, but also gave a systematic mathematical framework for exploring the implications of those laws. The success and power of Newton's laws led to great optimism about the ability to predict the behavior of mechanical objects and, as a consequence, led to the huge growth in science that occurred up to the beginning of the twentieth century. The Epilogue explains more about Heisenberg's view of Newton's principles.

E. Başar, *Brain-Body-Mind in the Nebulous Cartesian System: A Holistic Approach by Oscillations*, DOI 10.1007/978-1-4419-6136-5_26,

2. He developed an important scientific methodology with his principles and methods (see Chaps. 1 and 2).
3. He said to try to omit nothing in your research before starting to judge on the "essence of problems" under study.

The present book aims to adopt a similar strategy of trying to omit nothing: The proposal of the brain dynamics research program (see Chap. 3, Fig. 3.12) aimed at considering methods and measurements across a broad spectrum and tried to omit as few items as possible.

We started with the observation, "Put everything on the table, study brain oscillations, smooth muscle oscillations, the electrophysiology of invertebrates in comparison to the fish brain and mammalians including the human brain." Further, we studied the effect of transmitters in invertebrate ganglia and human brains. By doing these things, we tried to reach conclusions about the essence of brain-body-mind structure. The question is now, "Was everything considered?" Certainly, we could not succeed in achieving perfection. However, we tried to apply Descartes' principles to the present book, and have applied them to our research program since the beginning of the 1970s.

I am not convinced by the newly emerging concept of *neurophilosophy*. This is because philosophy is science, and not a collection of knowledge in various areas of the sciences. Philosophy is *integrating* and not *separating!* Churchland (2002) performed an excellent job by emphasizing the importance of philosophy in neurosciences. Her work has been influential among neuroscientists in highlighting the importance of philosophy. However, her book neglects the importance of "nature philosophy" as a general framework to study science, particularly neuroscience. Philosophy in neuroscience cannot ignore general science, in particular the few intelligible systems found in every branch of science, as Descartes expressed.

Wiener, Eccles, and Kandel are all philosophers because these scientists have all performed experimental and theoretical research with concrete measurements and then developed new thoughts. Hume asked crucial questions, and the process of addressing his questions led to discoveries in modern sciences.

von Weizsäcker changed the causality principle of Hume by introducing non-deterministic causality, which is based on quantum theory. Bergson described the episodic memory, intuition and important steps in creative evolution for the first time (see Chaps. 1, 17, and 18). Von Hayek is a founder of a new type of psychology with his important concept of neurophysiology and psychology, as discussed by Fuster (1995a) and Başar (2004).

Haken is a philosopher. He asked crucial questions in his book, *Synergetics* (1977). The phase reordering in laser atoms, synchronization, and change of entropy in the laser have also been key ideas for both brain research and social scientists. As I was completing *EEG Brain Dynamics* (1980), I discovered Haken's book, *Synergetics* (1977). I was fascinated by this book, and emphasized the importance of this new discipline in brain studies in the Preface to *EEG-Brain Dynamics*.

Scientific advances during the twentieth century produced an enormous body of knowledge and methods, which it is not possible to include in every individual study. Accordingly, we should slightly revise Descartes' requirements:

> Try to consider everything, which is in your knowledge and the impact of your methods; consider every possibility for discussion. Thereafter, reject all junk papers and speculations of yours and of other scientists. At the beginning analyze clear and safely acquired data.

Certainly the present book cannot consider every possible method, analysis, and judgment of each topic; this is not possible in the twenty-first century. Despite this, we take Descartes' view in trying to omit the fewest number of things from the great bulk of knowledge acquired in our time.

Return to the question, "What is the importance of philosophy within a multidisciplinary science platform?" As Bergson stated, philosophy can also be interpreted as method (see Chap. 18). The important message is that every scientist must develop a conceptual framework to be able to succeed within a scientific environment. As Lord Kelvin said, "All science is measurement, but all measurement is not science."

26.2 A Synthesis from the Concepts of Wiener, Prigogine, Thom, Hayek, and Haken

These four philosopher-scientists strongly emphasize common features in interdisciplinary sciences, carefully analyzing the following ideas: the concepts of order and disorder; the second law of thermodynamics; entropy; and nonlinear phenomena. The energy input in lasers induces the transition of oscillating atoms from a disordered to an ordered state, similar to brain oscillations on sensory stimulation. According to Prigogine, the second law of thermodynamics implies that the emergence of ordered states, in terms of increasing biological complexity, is improbable. Wiener mentioned the role of a "Maxwell demon" in biological processes. As seen in Chap. 17, Monod also emphasized the importance of the Maxwell demon in the creation of life (see also Thom). These frameworks also have a common general theme: All of these scientists started from physical, mathematical, or chemical metaphors by trying to identify common abstract mechanisms or symbols to create new interdisciplinary sciences. Unfortunately, none of these frameworks has its origin in biological empiricism. One essential biological framework is Darwin's evolutionary theory, in which "natural selection" plays a major role. However, in turn a selection needs a type of transition to move to a new order. Wiener, Prigogine, Thom, and Haken also mention the importance of non-linear phenomena and, related to this, deterministic chaos. According to these philosopher-scientists, new branches of sciences need to deal with the second law of thermodynamics, equilibrium, feedback mechanisms, and communication and information processes.

The scope of the present book and its aim in proposing a new Cartesian system consists of a synthesis of the excellent ideas governing the frameworks developed during the twentieth century. The author of the present book has worked with the concepts of these four frameworks by Wiener, Haken, Prigogine, and Thom to launch the EEG-brain dynamics concept, now the prevailing approach in the research of a number of neuroscience laboratories. Therefore, the aim here is not to deny the importance of existing frameworks, but to enlarge them and incorporate the philosophical schools following the Renaissance, quantum physics, and the new results in chaotic brain dynamics. A new Cartesian system in the twenty-first century raises questions that could be answered with the help of a large number of experiments and scientists. This system will provide a working branch, as in Wiener's cybernetics, but have the additional possible advantage of incorporating the experiences of existing frameworks as well.

Figure 26.1 globally illustrates the evolution of philosophy and sciences from old Athens, via the Renaissance, to the development of physics and the new contemporary unifying schools. In the twentieth century cybernetics, quantum theory, chaos theory, dissipative structures, and synergetics provided an essential step in opening the way to brain dynamics. By application of the concepts and methods of these schools, scientists have collected a vast amount of empirical data to approach brain-body-mind integration. Furthermore, the application of these data should serve to develop new methods of evaluation, closer to the language of the brain and

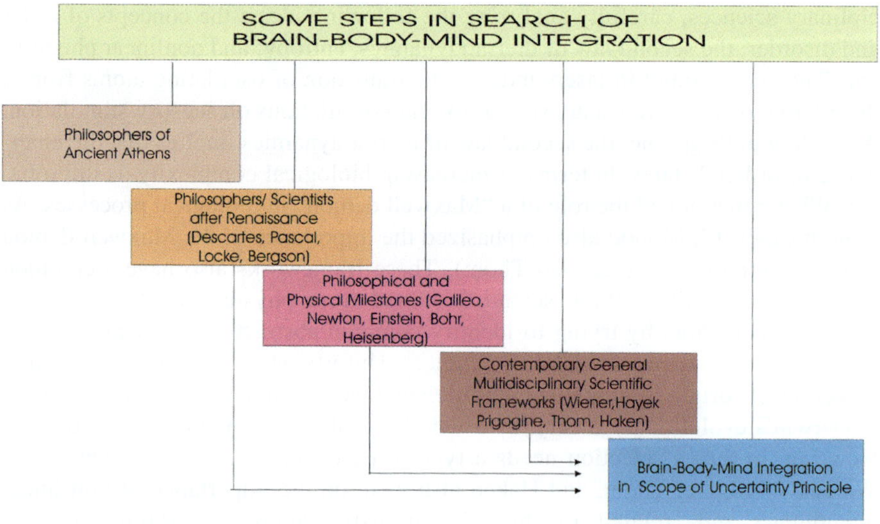

Fig. 26.1 Some fundamental approaches during the evolution of science that led to brain-body-mind integration in the scope of the uncertainty principle. This means that transcendent views should also be incorporated within research of what is called "mind"

the understanding of the brain-mind. These possibilities are outlined in Chaps. 15–17 and 22–25.

To some researchers, Norbert Wiener's approach is no longer modern, but I am not of that, because none of the general multidisciplinary frameworks offer the intellectual impact that Wiener achieved by bringing together philosophy, brain waves, computing machines, society, time series in information, and communication. To put Newton and Bergsonian time and statistical mechanics together was an initial and crucial step in preparing for the jump from Newtonian mechanics and the Cartesian view to the Bergsonian view and modern physics. Unfortunately, Wiener did not have extensive experience in biological research; therefore, his approach was strongly grounded in mathematics and engineering, without essential roots in biology. Nevertheless, Wiener is an important figure in modern science.

The present book also describes the renowned frameworks of Prigogine, Thom, and Haken. Among these scientists, it is Haken who most actively initiated ways of interacting with several areas of science, mostly by organizing symposia and editing books ranging from mathematics to cognitive processes and social sciences. In 1983, I organized a symposium on the Synergetics of the Brain with Herman Haken (Başar et al. 1983), and thus began an interesting and fruitful scientific friendship. He came with me to New York for a workshop in 1990 and later to Izmir in 2005. I profited greatly from his deep intellectual approach into the essence of synergic effects in the analysis of lasers. I also had brief discussions with Thom in Vienna and Prigogine in Austin, Texas. Wiener's approach to brain research was described to me by the brilliant brain scientist John Barlow at the Massachusetts Institute of Technology (MIT). Barlow also contributed a chapter to one of my edited books, in which he discussed the first days of his collaboration with Wiener at MIT in the 1950s.

The successes and shortcomings of the schools of cybernetics, catastrophe theory, dissipative structures, and synergetics are jointly discussed with the concept of the quantum brain as an emerging trend in neuroscience. In physical science, string theory is one of the most important unifying trends. Because string theory and oscillations are interrelated, the next question that arises is, "Can string theory also be used as a metaphor in brain research?"

The discussion of common principles and unifying trends concludes that a new Cartesian system is indeed needed to understand brain-body-mind interaction. The author proposes a nebulous Cartesian system, which could jointly encompass brain-body processes that have multiple causalities. Accordingly, the way is open for new frameworks from other scientists that can be different or better.

The time has come to begin searching for unifying trends and multiple causalities in joint processes of physics and the brain.

Hayek's (1952) view on a new psychology was a precursor to integrative mind, and is emphasized in Chap. 25. This view can also be included among the mentioned frameworks.

26.3 Possible Shift to a New Concept Based on Results of the Present Book

The present book initiates a reversal of Wiener's approach. Despite my admiration for the era of research that his work commenced, I find that Wiener somewhat lacks solid experience in biological sciences. The language of the brain is different from the language and techniques of mathematics (Von Neumann).[2] Accordingly, concepts of cybernetics, quantum theory, chaos theory, synergetics, and dissipative structures can be used only as interim approaches in forging a completely new framework that is more closely related to the brain's language.

The brain is the most complex system known in the universe, and some of the new concepts that are developed in this book are similar to a walk in a dark room. In spite of this, if several scientists could start with similar scientific élan, new realms could be reached in the coming decades.

I write this because I embrace the hope that the present book will not only appeal to scientists from the spheres of neuroscience, philosophy, psychology, and physics; but also to graduate students and research scientists who are interested in common approaches, and may find important possibilities throughout the book. When I first read Wiener's book I was a young scientist, and I certainly was not able to understand everything in it; but it was a unique experience. My work showing that the study of brain oscillations is one of the most important future fields of neuroscience can be attributed to the fact that I was able to absorb the core of Wiener's multidisciplinary philosophy. Books that reconcile philosophy, empirical neuroscience, and concepts of physics are rare. An example is the book by Scott Kelso, *The Complementary Nature* (MIT Press, 2007), which opens an important avenue in the area of multidisciplinary science.

Since Leibniz and Newton, no scientist has had a full command of the total intellectual activity of his or her day. Because measured data are needed, science has increasingly become the realm of specialists in fields that tend to grow progressively narrower. Currently, in the twenty-first century, only a limited number of scientists seem to be involved with larger study windows.

Despite this, the collaboration of scientists bridging continents has become possible. As a result of increasingly powerful computers, we have access to ever-greater memory capacity; thus, new methods and associative algorithms have become extremely rich. Feynman diagrams, Heisenberg's S-matrix, and statistical mechanics can be applied much more easily. Accordingly, the boundaries between multiple fields of science offer the richest opportunities to qualified investigators.

---

[2]...logics and mathematics in the central nervous system, when viewed as languages, must be, structurally, essentially different from those languages to which our common experience refers (Von Neumann and Burks, 1966).

This book primarily aims to promote a new understanding of the machineries of brain-mind, in solidly empirical as well as philosophical ways. Another important goal is to indicate new avenues that scientists might explore to understand the general principles of science. From Newton to Einstein and Heisenberg, from Descartes to Bergson, all science is developed by searching for general principles. Should we not perform similar steps to understand brain-mind integration?

Epilogue

What May Happen 30 Years from Now?

1 EEG-Brain Dynamics

Thirty years ago, at the end of my book, *EEG-Brain Dynamics* (1980), I included a somewhat philosophical Epilogue that predicted future developments in the field. At that time I was convinced that neuroscience sought a theory of dynamics somewhat parallel to Newtonian dynamics, which had contributed enormously to the evolution of modern sciences. However, Newtonian dynamics is governed by a rigid, almost deterministic framework that does not conform entirely to the probabilistic nature of the brain. Accordingly, I also included in that 1980 Epilogue a few sentences indicating that the statistical nature of a quantum brain approach could trigger essential progress to extend the limits of brain theories. Additionally, I was convinced that clinical applications should be sought for a deeper understanding of brain dynamics, and consequently brain functionality.

At the beginning of the 1980s only two laboratories (those of Walter Freeman and Erol Başar) resolutely pursued the concept of brain oscillations, whereas most neuroscientists remained indifferent.

2 Why the Transition from Newton's Deterministic View to Quantum Brain?

In the meantime, a great number of neurophysiologists, neurologists, and psychiatrists started to successfully apply the concept of EEG-brain dynamics. Furthermore, the concept of quantum brain, which is extensively described in Chaps. 14–16, is also increasingly utilized by research scientists seeking new models of the brain. Studies of the pathologic brain, the evolution of species, and the links between the vegetative system and the brain that are reflected in brain oscillations led to new models in the study of brain-body-mind integration (see Chaps. 9, 23, and 24).

In the present book, the new research trends of the probabilistic brain, quantum brain, pathological brain, evolving brain, and coupling of the vegetative system and the brain

E. Başar, *Brain-Body-Mind in the Nebulous Cartesian System: A Holistic Approach by Oscillations*, DOI 10.1007/978-1-4419-6136-5, constitute the primary concepts of twenty-first century neuroscience research. Additionally, the subject of brain metaphysics is discussed extensively throughout this volume.

2.1 A Lesson by Werner Heisenberg

In my third or fourth semester as a young student at the University of Munich, Professor Heisenberg asked about my preferred courses at the university. I mentioned that I found theoretical mechanics most interesting, to which Heisenberg immediately replied, "Yes, yes, this is most important! If you first understand Newton, quantum mechanics will be very easy for you." Still relying on this lesson, I tend to say that the Newtonian era of brain oscillations provided satisfactory foundational material; and the time is now ripe to consider the quantum brain with more vivacity. Accordingly, I hope that clinical studies, together with a quantum approach that comprehends the application of brain Feynman diagrams to cognitive processes, may offer a pathway to significant progression in brain studies.

3 Intuitive Brain, Philosophy, and Clinical Applications

Further, questions related to intuition and unconscious brain work are greatly emphasized in comparison with earlier studies, as explained in Chaps. 19, 20, and 25. In relation to this step, the consideration of philosophy as a strong method is re-entering the scenario of integrative research on neuroscience in search of the brain-body-mind.

There are few questions related to possible future developments:

- Can the proposed unifying models, the new Cartesian system with probabilistic nature and the quantum brain become attractive prospects in the next 30 years?
- Will the solid empirical findings related to the interplay between transmitters and oscillations be used to study impairment of mind and clinical studies, and open the way to a more precise and differentiated drug application?
- Will more progress be attained in understanding the link between the brain and vegetative organs by also considering neurons and smooth muscles forming one entity, and being controlled by oscillations and transmitters? This entity, now tentatively denoted as the brain-body-mind syncytium and also shown by means of coherent electrical activity does not coincide with the neuron doctrine by Cajal. Would the return of the expression *syncytium* be acceptable?
- Can brain states such as intuition and unconscient world be more elegantly analyzed in the next 30 years?

I think the answer to these questions is yes, because there is a need for the types of studies described in the present book, which rely on empirical evidence and large number of measurements.

4 New Land in a Few Decades?

The *Pensées* (thoughts) of Blaise Pascal was published three and half centuries ago (1670). In that famous book, Pascal elegantly described notions of mathematical mind and intuitive mind (see Chap. 1). However, at the time of Descartes, Pascal, and Locke, it was even not conceivable to measure efforts of memory or any type of cognitive processes. Even at the turn of the twenty-first century, thought processes could not be measured. Only in the last 50 years has considerable progress been made. Do we need another 350 years to read more inside the brain related to creativity, intuition, and unconscient problem solving? In the last 30 years, the progress in measuring thoughts or approaching brain-body-mind integration has been considerable; therefore, it is conceivable that we may see new land in a few decades. From 1980 to the present, in only 30 years, the field of neuroscience has witnessed transformational progress in EEG oscillations and cognitive processes. Graduate students with mathematical skills, the ability to process experimental work, and an openness to multidisciplinary research, including nature philosophy, may contribute immensely to a new era in brain research. This era will need conceptual frameworks, as was the case with physicists at the beginning of the twenty-first century, who applied metaphysics to modern physics and engineering.

Galileo Galilei (February 15, 1564–January 8, 1642)

Poetry from Paul Verlaine communicated in a letter from Henri Begleiter (September 2005) is repeated here:

L' éspoire est comme un brin de paille quit luit dans l'etable

(the hope is like a bit of straw, which glistens in a stable)

The book concludes with a quote from Roger G. Newton (2004), which has been slightly extended:

> Little could Galileo have realized that the harmonic oscillator, whose synchronisms he discovered in his youth, would turn out to be the most basic, all-pervading physical/ biological system in the world and a crucial building block for the understanding of nature.

Appendix A
Abbreviations, Glossary, and Basic Methods

A.1 Glossary and Abbreviations

AD	Alzheimer's disease
ADHD	Attention deficit hyperactivity disorder
EROs	Event-related oscillations
MCI	Mild cognitive impairment

A.1.1 Anatomical Abbreviations

CA3	Ca3 layer in hippocampus
CE	Cerebellum
GEA	Gyrus ectosylvian anterior, auditory cortex
HI	Hippocampus
IC	Inferior colliculus
LG	Laterale geniculate nucleus
MG	Medial geniculate nucleus
OC	Occipital cortex, area 17
RF	Reticular formation
SC	Superior colliculus

A.1.2 Abbreviations

3.ATT	The third attended signal in the omitted signal paradigm (last auditory stimulation before omitted one)
AFC	Amplitude frequency characteristics
EP	Evoked potential
ERO	Event-related oscillation

ERP	Event-related potential
fMRI	Functional magnetic resonance imaging
LTM	Long term memory
MEG	Magnetoencephalography
MMN	Mismatch megativity
OB	Oddball
OMS	Overall myogenic system
STM	Short term memory
TRFC	Transient response-frequency characteristics method
VEP	Visual evoked potential
WMS	Working memory system

A.1.3 Glossary

- Active memory: We modify the definition of active memory stated by Fuster (1995a) as follows: At any given time in the awake organism, a widely distributed and changing representational network is active in the whole brain. That is the *active memory manifested by the EEG-oscillations*.
- AEP: Auditory evoked potential
- AFC: Amplitude frequency characteristics. The spectra of the evoked responses in the frequency domain potentials (see Appendix B).
- Alpha response: Oscillatory component of an evoked potential in approx. 8–13 Hz frequency range (see Chap. 6)
- Alpha system (selectively distributed): See selectively distributed oscillatory systems (see Chap. 6)
- ALPHAS is an expression characterizing the ensemble of diverse 10 Hz oscillations in the brain
- CAP: Combined analysis procedure of EEG and evoked potentials
- Delta response: Oscillatory component of an evoked potential in approx. 0.5–3.5 Hz frequency range (see Chap. 6)
- EEG: Electroencephalogram
- EHF (enhancement factor): In a given experimental record of EEG-EP epochs, the enhancement factor χ is the ratio of the maximal time-locked response amplitude (max) to the rms value of the spontaneous activity just prior to the stimulus, with both signals (spontaneous and evoked activities) being filtered within the same frequency pass bands.

$$\chi = \frac{\max}{2\sqrt{2}\mathrm{rms}}$$

- EP: Evoked potential
- Episodic representations: Neural firing patterns, which encode the sequence of events that compose a unique, personal experience (Eichenbaum 2000)
- ERP: Event-related potential

- Event-related oscillation: It also includes ERP and induced ryhthms
- Evoked frequency response: Dominant maximum in AFC
- Evolving memory: The process of formation of memory, which we denote also as evolving memory, probably constitutes the most important of processes during transition from one memory state to another one
- Feature detectors: The primary features of stimuli as heat, force, light, sound, and chemical substances are selectively transduced at the peripheral ends of sets of sensory (afferent) nerve fibers. Different groups of those sensory fibers respond selectively at lover thresholds than do others to different forms of impinging energy. This "tuning" is often called *feature detection* and is accomplished during evolution of species by the development of specific transducer mechanisms for different forms of energy, either in the nerve endings themselves or in complex sensory organs in which the afferent fibers terminate. Examples are the mammalian retina, cochlea, and the pressure transducers of the primate hand (see Chap. 7, statements of Sokolov 2001 and Mountcastle 1998).
- FFT: Fast Fourier transform
- Gamma response: Oscillatory component of an evoked potential in the approx. 30–60 Hz frequency range (see Chap. 6)
- Gamma system (selectively distributed): See selectively distributed oscillatory systems (see Chap. 6)
- Habituation: A decrease in the behavioral response to a repeated stimulus (Bailey et al. 2000).
- Internal EPs: The "rule of excitability" is formulated as follows: If a brain structure has spontaneous rhythmic activity in a given frequency channel, then this structure is tuned to the same frequency, and is producing internal evoked potentials to internal afferent impulses originating in the CNS, or responding in the form of evoked potentials to external sensory stimuli with patterns similar to those of internal evoked potentials.
- Limbic structures: A collection of subcortical structures, as the prominent hippocampus and amygdala, important for processing memory and emotional information
- Longer-acting memory: In the present book the expression longer-acting memory is used instead of long-term memory to show the differentiation between the working memory and persistent memory. A new proposition in memory categorization fresh memory traces acquired in everyday experiences are temporarily stored in longer-acting memory, before reaching the persistent memory level. According to the description of memory levels introduced in Chap. 9, persistent memory combines built-in memory with physiological memory (being an ensemble of submemories as echoic memory, iconic memory, olfactory memory, etc.) and of stabilized parts of longer-acting memory acquired during life (see Fig. 9.1).
- MEF: Magnetic evoked field
- MOR (major operating rhythms): Experiments have shown that in several areas of the brain some ryhthms are more distinguished and dominant in comparison with other ones; for example, the posterior 10 Hz and frontal theta.

- Multimodal responses: Neural activity elicited by more than one sensory modality
- Phase-locked and non-phase-locked activity: Non-phase-locked activities contain evoked oscillations that are not rigidly time-locked to the moment of stimulus delivery. These are, for example, induced alpha, beta, gamma, etc. oscillations that may relate to specific aspects of information processing. In the framework of the additive model of evoked potentials, non-phase-locked activity includes the background EEG. For analysis of only non-phase-locked or both phase-locked and non-locked EEG responses, specific approaches have been used. Phase-locked activity is suggested to include all types of event-related brain potentials. For quantification of the phase-locked activity, the averaging procedure is usually applied whereby the phase-locked responses are enhanced, and the non-phase-locked ones are attenuated.
- Place cells: Hippocampal principal cells that fire selectively when an animal is in a particular location in its environment.
- Prepotent responses: Reflexive actions, either innate or well established through a great deal of experience (Miller 2000)
- Priming: The facility of recognition, reproduction, or bases in selection of stimuli that have recently been perceived
- Procedural memory: The representation of a series of actions or perceptual processing functions that occur unconsciously, and typically result in increased speed or accuracy with repetitions (Eichenbaum 2000)
- Resonance: The response that may be expected of underdamped systems when a periodic signal of a characteristic frequency is applied to the system. The response is characterized by a surprisingly large output amplitude for relatively small input amplitude (i.e., the gain is large). A translation of these comments by illustration is afforded by the annoying vibrations developed in a house when certain periodic stimuli are applied.
- RMS: Root-mean-square
- Selectively distributed oscillatory systems in the brain: By means of the application of combined analysis procedure of EEG and EPs we recently emphasized the functional importance of oscillatory responses (in the framework of brain dynamics) related to association and (long distance) communication in the brain. We assumed that alpha, theta, and gamma networks (or systems) are selectively distributed in the brain (for the delta, theta, and alpha ranges see Chap. 6). We also have tentatively assigned functional properties, namely sensory-cognitive functions, to alpha, theta, delta, and gamma resonant responses. According to this theory a sensory stimulation evokes 10 Hz enhancements in several structures of the brain, both cortical (primary auditory cortex, primary visual cortex) and subcortical (hippocampus). The synchronous occurrence of such responses in multiple brain areas hints at the existence of distributed oscillatory systems and parallel processing in the brain. Such diffuse networks facilitate the information transfer in the brain according to the general theory of resonance phenomena. Although alpha responses are observable in multiple brain areas, they are markedly dependent on the site of recording. The dependence of the alpha

response on whether or not the stimulus is adequate for the brain area under study thus hints at a special functional role of alpha responses in primary sensory processing. The term *diffuse* is used to describe the distributed nature of the frequency response in the brain. It is not yet not possible to define connections between the elements of these neuron systems by neuron tracking, or to define the directions of signal flow and exact boundaries of the neuronal populations involved. However, this description is necessary to emphasize that rhythmic phenomena in these frequency ranges are not unique features of the observed single subsystem of the brain, and that their simultaneous existence in distant brain structures may be a relevant and important point in the description of an integrative neurophysiology.

- Semantic knowledge: An organization of factual information independent of the specific episodes in which that information was acquired
- Synaptic plasticity: A change in the functional properties of a synapse as a result of use (Bailey et al. 2000)
- Theta response: Oscillatory component of an evoked potential in approx. 4–8 Hz frequency range
- Theta system (selectively distributed): See selectively distributed systems (see Chap. 6)
- Top-down: Brain signals that convey knowledge derived from prior experience rather than sensory stimulation
- TRFC method: A Fourier method that enables one to obtain the frequency characteristics from the transient response
- VEP: Visual evoked potential
- Wavelet analysis: Method of time-frequency analysis. This method can be used to search and find repeatable and phase-locked signals in a given frequency window.
- Working memory: The representation of items held in consciousness during experiences or after retrieval of memories

A.2 Relevant Mathematical Methods Used in This Book

A.2.1 Method of Transient Response Analysis

The method of transient response analysis is a common method from the general systems theory. This method studies the system's response in the time domain by application of either step or impulse functions at the input of the system. The method of transient responses has the advantage that the observer immediately obtains the responses of the system under study when sudden changes (jumps or steps) in the input function occur.

A practical method of obtaining the evoked potential (brain system transient response) is the averaging where the mean value being a function of time is taken from EEG records that follow a number of identical stimulus presentations. In such processing, the averaged values of time-unlocked activities and the noise (brain response-unrelated activities) tend toward zero, whereas the average of the evoked potential (e.g., the event-locked and repeatable signal) tends to remain constant. This approach has been most commonly used and has led to the accumulation of an enormous amount of results (for mathematical description and extended survey see e.g., Başar 1980; Regan 1989).

The greatest disadvantage of the method stems from the fact that the information about the distinct components of the system is obscure in the transient response. When two, three, or more components exist in the system response, the observer cannot distinguish these different components without further mathematical analysis. Usually, physiologists prefer the method of transient response analysis, but peak identification of distinct components (subsystems) in the time domain is often erroneous. Simple-looking system transient responses sometimes have a large number of components and vice versa, a large number of peaks in the transient response do not necessarily reveal the existence of a large number of system components. For verification of these remarks the reader should refer to the examples presented in Başar (1998, 1999).

Another important disadvantage is that by applying averaging, certain information about the dynamic properties of the brain is lost. The methods that follow in our brain dynamics research program (BDRP) are introduced in an attempt to overcome these shortcomings.

A.2.2 Frequency Characteristics and TRFC-Method

Before describing the method used we would like to explain the theoretical basis of the analyses. When the transfer properties of a system are studied, the investigator is often confronted with the resonance phenomenon. Resonance is the response that may be expected of underdamped systems when a periodic signal of a characteristic frequency is applied to the system. The response is characterized by a surprisingly large output amplitude for relatively small input amplitudes, i.e., the gain is large.

Resonance phenomena or forced oscillations can be analyzed in the direct empirical way as follows: A sinusoidal signal of a frequency f is applied to the system. After a certain period sufficient for the damping of the transient, only forced oscillations will remain, having the frequency of the input signal. Then the amplitude of the applied signal (input), the amplitude of the forced oscillations (output) and the phase difference between input and output will be measured. By gradually increasing the frequency from $f = 0$ to $f = f_0$, the output amplitude relative to the input amplitude and the phase differences will be measured as a function of frequency, called amplitude and phase characteristics, respectively.

Although this approach reveals the natural frequencies of the system, only a small number of scientists have investigated the EEG response using sinusoidally modulated light or sound signals (for details on pioneering experiments with sinusoidally modulated light, see Van der Tweel 1961). Difficulties result from the requirement to record evoked responses to sinusoidal signals of over at least 3 decades of stimulation frequencies, with the evoked responses in each stimulation frequency being averaged for at least 100 single stimulus applications. Another difficulty comes from the frequent changes in brain activity stages: they may change within a few minutes and have a limited duration, which is not sufficient for the application of many sinusoidal stimuli of different frequencies.

There is, however, another way of obtaining the frequency characteristics of a system. This is the transient response frequency characteristics (TRFC) method: According to general systems theory, all information concerning the frequency characteristics of a linear system is contained in the transient response of the system and vice versa. In other words, knowledge of the transient response of the system allows one to predict how this system would react to different stimulation frequencies, if the stimulating (input) signal was sinusoidally modulated. If the step response *c*(*t*) of the system – in our case the sensory evoked potential – is known, the frequency characteristics, $G(j\omega)$, of this system can be obtained with a Laplace transform (or one-sided Fourier transform) of the following form:

$$G(j\omega) = \int_0^{\infty} \frac{d\{c(t)\}}{dt} \exp(-j\omega t) dt$$

or,

$$G(j\omega) = \int_0^{\infty} \exp(-j\omega t) d\{c(t)\}$$

$$G(j\omega) = \int_0^{\infty} \exp(-j\omega t) \lambda(t) dt$$

where $G(j\omega)$ represents the frequency characteristics of the system; $c(t)$ = step response of the system; $\lambda(t)$ = impulse response; $\omega = 2\pi f$, the angular frequency, and f is the frequency of the input signal.

The frequency characteristics $G(j\omega)$ – including the information of amplitude changes of forced oscillations and the phase angle between output and input – is also called the *frequency response function*. It is a special case of the transfer function and is, in practice, identical with the transfer function (Bendat and Piersol 1968).

For numerical evaluation a fast Fourier transform (FFT) is used as follows: Let X_n be a discrete time series $X_n = X(n\Delta t), T = ((N-1)\Delta t)$. Then the Fourier transform Y_k of X_n is:

$$Y_k = Y(\omega_k) = \sum_{n=0}^{N-1} X_n \exp(-i2\pi N^{-1} nk); \quad \omega_k = 2\pi k T^{-1}$$

where $Y_k = a_k + ib_k$ are the complex Fourier coefficients the geometric mean of which is the amplitude spectrum.

Although this transform is valid only for linear systems, it can be applied to nonlinear systems as a first approach because the errors caused by system nonlinearities are smaller than errors resulting from the length of measurements in sinusoidal stimulation experiments given the rapid transitions of the brain's activity from one stage to another (Başar 1980).

In the mathematical literature, the TRFC-method is simply called one-sided Fourier transform or the Laplace transform. We use the expression TRFC-method to indicate that this method gives all characteristics in the time and frequency domains. In particular, the physiologist is used to observe the experimental parameters in the time domain by obtaining the transient responses of the studied system (see the preceding). Using this method, the most frequently used physiological transient responses are obtained. Moreover, it is possible to analyze the frequency content or the components in the frequency domain by computing the amplitude-frequency characteristics from the same transient response. Therefore, we find it more useful and descriptive to call this method the TRFC-method.

In view of practical application, the methodology to evaluate EPs, AFCs, and digitally filtered data has been described (e.g., Başar 1980). The essential steps are as follows:

1. *Recording of EEG-EP epochs:* With every stimulus presented a segment of EEG activity preceding and following stimulus application are digitized and recorded. This operation is repeated for each trial.
2. *Selective averaging of EPs:* The stored raw single EEG-EP or EEG-ERP epochs are selected with specified criteria after the recording session. EEG segments showing movement artifacts, sleep spindles, or slow waves were eliminated.
3. *Amplitude-frequency characteristics* (*AFC*) are computed according to the formula given in the preceding.
4. *Adaptive digital filtering* is performed, as described in the next section.

A.2.3 Response Adaptive Filtering

Response adaptive filtering is the ideal theoretical filtering of the transient response of a system in such a way that a selective blocking of one or more components (or subsystems) is obtained. Ideal filters are defined as transmission elements that, within a given frequency range, transfer the input signal without any change in amplitude and with a fixed (independent of frequency) time shift. Outside this frequency range they have zero transmission (or vice versa, depending on whether the filter has a band-pass or a band-stop characteristic). They are not physically realizable, but they should be considered as useful analytical tools when the contribution made to a signal by a frequency band is to be deduced without any distortion.

Let us assume a system, $G(j\omega)$, which should result from the interconnections of the subsystems,

$$G_1(j\omega), G_2(j\omega), G_3(j\omega), \ldots, G_K(j\omega), \ldots, G_N(j\omega),$$

in such a way that

$$G(j\omega) = G_1 G_2 G_3 \ldots G_K \ldots G_N.$$

If we already know the amplitude frequency characteristics of the system $G(j\omega)$ under study, and we further want to know how the transient system response would be affected if one or more of the components of the system were missing, we first of all determine the frequency band limits of the component to be eliminated (or the component that should be removed from the system). The procedure consists of the following steps:

1. The amplitude characteristics $|G(j\omega)|$, of the system under study are obtained by means of Laplace transform (or one-sided Fourier transform) using the transient evoked response, $c(t)$:

$$|G(j\omega)| = |L(\mathrm{d}\{c(t)\}/\mathrm{d}t)| = \left|\int_0^\infty \exp(-j\omega t)\mathrm{d}\{c(t)\}\right|$$

2. Frequency band limits of theoretical filters are determined according to the frequency and band-width of amplitude maxima in the amplitude-frequency characteristics, $|G(j\omega)|$.
3. After determination of ideal filter characteristics in the frequency domain, $G_{KF}(j\omega)$, the weighting function, $g_{KF}(t)$, of the filter is computed by means of the inverse Fourier transform:

$$g_{KF}(t) = F^{-1}\{G_{KF}(j\omega)\} = \frac{1}{2\pi}\int_{-\infty}^{+\infty} \big(|G_{KF}(j\omega)|\exp(-j\omega\tau)\big)\exp(j\omega t)\mathrm{d}\omega$$

By taking t to be equal to zero, any fixed or frequency dependent time shift (which would have been inevitable in the case of a real electrical filter) can be avoided easily.

4. The experimentally obtained transient evoked response, $c(t)$, is theoretically filtered by means of the convolution integral using the weighting function, $g_{KF}(t)$, of adequately determined ideal filter:

$$c_F(t) = g_{KF}(t) \times c(t) = \int g_{KF}(\tau)c(t-\tau)\mathrm{d}\tau$$

where $c_F(t)$ is the filtered evoked response. Because the time response is available in the form of discrete data with sampling interval of dt, the integrals in the preceding equations can be replaced with iterative summation. Evaluation of these integrals is achieved by using the FFT algorithm.

The method of response adaptive digital filtering has a very important advantage in the study of biological systems. Usually it is very difficult to remove or attenuate subsystems from the biological system under investigation; but if the frequency characteristics of the system are known, we can do it theoretically by using the theoretical isolation method. This is the theoretical version of the method of selective blocking by application of pharmacological agents or surgical ablation techniques.

Although some electronic filtering methods have already been used in the study of brain waves and evoked potentials, the theoretical isolation method presented here affords the possibility to choose amplitude and phase frequency characteristics of the filters separately. Therefore, the investigator can apply ideal filters without phase shift. It is also possible to use filters with exact characteristics and change them adequately according to the amplitude characteristic of the system under study. Therefore, the use of theoretical filters is much simpler and more flexible than the use of electronic filters. Theoretical filters are designed as digital filters. They can be applied because they introduce no phase shift in the signal (Başar 1980; Başar and Ungan 1973; Cook and Miller 1992; Farwell et al. 1993). However, filter characteristics, especially for narrow filter pass-bands as well as for abrupt amplitude changes typical for averaged EPs, should be chosen so as to avoid the production of filter-related oscillations (De Weerd 1981; Farwell et al. 1993; Wastell 1979).

We should mention here that the choice of filters used can be made independently of any frequency characteristics. This choice, however, would be arbitrary. Adaptive filtering aims at a component analysis in the study of a given brain response. Important examples of how powerful this method can be are given in Başar (1998, 1999).

A.2.4 Combined Analysis Procedure: EEG and EP Comparison

The theoretical background for developing the combined analysis procedure is the concept of the EEG as an active signal in the brain. The spontaneous EEG is regarded as a signal that determines or governs the responses of the brain. Within this framework, we need a technique providing for analysis of both the spontaneous (ongoing) and evoked EEG activity. The methodology for comparing the brain's spontaneous activity and EPs can be briefly described as follows:

1. A sample of the spontaneous activity of the studied brain structure just before the stimulus is recorded.
2. A stimulation signal is applied to the experimental subject (animal or human). Visual, acoustical, somato-sensory, etc. inputs may serve as stimulation; for example, an auditory step function in the form of a tone burst with frequency of 2,000 Hz and intensity of 80 dB SPL.
3. Single-sweep evoked response following the stimulation is recorded. As a result, the EEG activity before and after stimulation are stored together as a combined record.
4. The operations explained in the three preceding steps are repeated about 100 times. (The number of trials depends on the nature of the experiment and the behavior of the experimental subject.)

5. The stored single-sweeps are averaged using a selective averaging method (Başar 1980; Başar et al. 1975a; Ungan and Başar 1976).
6. The selectively averaged EP is transformed to the frequency domain with the Fourier transform to obtain the amplitude-frequency characteristics, $|G(j\omega)|$, of the studied brain structure.
7. The frequency band limits of the amplitude maxima in the amplitude-frequency characteristics, $|G(j\omega)|$, are determined, according to which the cutoff frequencies of the digital pass-band filters are justified.
8. The stored EEG-EPogram's are filtered within the properly chosen pass bands, as described in step 7.
9. The maximal amplitudes of the filtered EPs, and the so-called enhancement factor for each EEG-EP-epoch, are evaluated.

A.2.5 Definition of the Enhancement Factor EHF

In a given experimental record of EEG-EPogram, the enhancement factor EHF is the ratio of the maximal time-locked response amplitude (max) to the rms value of the spontaneous activity just before the stimulus, with both signals (spontaneous and evoked activities) being filtered within the same frequency pass bands:

$$\mathrm{EHF} = \frac{\max}{2\sqrt{2}\,\mathrm{rms}}$$

A.2.6 Coherence

For the signal analysis, evaluation of oscillatory dynamics and coherence analysis Brainvision Analyzer Software was used. First, the FFT of each epoch with 0–800 ms duration was calculated, and then the coherence analysis was performed with a 1.25 Hz resolution.

The choosing of the time interval 800 ms following the stimulation is based on a rationale that takes care of the complex biological properties of the EEG. In engineering studies the analyzer usually prefers an analysis period >1,000 ms for the delta band. However, to optimize the time period of analysis we first performed a power spectral analysis of EEG response and found a peak around 1.5 Hz in the power spectrum of the EP and ERP responses. Furthermore, we observed that the filtered ERP in the delta frequency range is a dampened aperiodical signal, which is almost completely flattened at around 500–600 ms. An analysis of coherence for longer periods would include unexpected artifacts and/or alpha after discharges. In the theta band the second response window is found around 400 ms. These were the chain of reasoning to choose the period 0–800 ms as the optimal time period for the coherence analysis. We also extended our analysis to 1,000 ms; here similar results with minimal deviation were found. We also recommend to brain research scientist to develop controls and not to use strict formula presented in some of the engineering textbooks. The method used was the cross-spectrum/autospectrum, and the mathematical relations are described in the following:

$$\mathrm{Coh}(c_1,c_2)(f) = |\,\mathrm{CS}(c_1,c_2)(f)\,|^2 \,/(|\,\mathrm{CS}(c_1,c_1)(f)\,|\,|\,\mathrm{CS}(c_2,c_2)(f)\,|),$$

in conjunction with

$$\mathrm{CS}(c_1,c_2)(f) = \sum c_{1,i}(f)\, c_{2,i}(f)^*$$

Then, Fisher's Z transformation was used to normalize the distribution of average coherence values.

Appendix B
Chaos

In the new branch of nonlinear dynamics, new concepts, new expressions, and new methods, such as the correlation dimension, are used. The present appendix describes some of these newly introduced steps, and complements the results of Chap. 14.

B.1 New Type of Expressions

For the non-specialist in chaotic dynamics reading appropriate books is essential to obtain a deeper understanding (e.g., Abraham and Shaw 1983; Babloyantz and Destexhe 1986; Haken 1977; Schuster 1988). However, the following important definitions are given to orient the reader.

- *Phase space:* In general, phase space is identified with a topological manifold. An n-dimensional phase space is spanned by a set of n independent linear vectors. This requirement is usually sufficient. Takens (1981), for example, proposed to span a ten-dimensional phase space by $x(t),\ldots, x(t + 9\tau)$ where τ is a fixed time increment. Every instantaneous state of a system is therefore, represented by a set $(x1,\ldots, xn)$, which defines a point in the phase space. The sequence of such states observed the time scale and defines a curve in the phase space, called a trajectory.
- *Attractor:* Some dynamic systems show the tendency to turn, under various but limited conditions, to a reproducible active state and stay there, i.e., as time increases, trajectories either penetrate the entire phase-space or they converge to a lower dimensional subset (see Fig. 15.2) called the attractor. Trajectories can also converge to a sequence of attractors. In the latter case, transition from one attractor to another is called a state change or bifurcation. Attractors can be periodic, quasi-periodic, or chaotic, which is also called a strange attractor (Abraham and Shaw 1983; Freeman and Skarda 1985). Usually, the following types of attractors are distinguished (Fig. B.1).
- *Fixed point.* The simplest stable state solution. With increasing time, all trajectories in phase space tend to terminate at this point. Stable fixed points are static

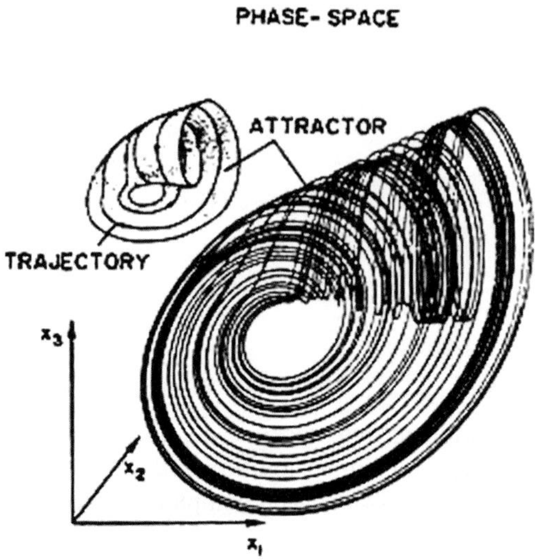

Fig. B.1 Trajectories converge to a lower dimensional subset called an attractor (modified from Abraham and Shaw 1983; Röschke and Başar 1989)

attractors (see Fig. 15.3a). A standard example is a pendulum that has come to rest after some time of oscillation because of friction.

- *Limit cycle.* A closed and recurrent trajectory in phase space. All trajectories tend to terminate in this cycle; no other closed cycle lies in its neighborhood. Without external drive, the limit cycle corresponds to a periodic stable position of the non-linear system, whose amplitude and frequency are determined by internal parameters of self-sustained oscillations. Stable limit-cycles act as periodic attractors (see Fig. 15.3b). The standard example is the attractor of a van der Pool oscillator. Limit cycles regularly occur with driven oscillators (Başar 1976, 1980).
- *Torus.* The system's trajectories move on a two-dimensional toroidal surface. Two frequencies are present, oscillations around the torus and along the torus (oscillations with two incommensurable frequencies). The trajectory never closes or covers the whole torus (see Fig. 15.3c). The trajectory on the torus is a quasi-periodic motion.
- *Strange attractor.* The manifestation of a strange attractor is its activity, which appears to be random, but which is deterministic and reproducible if the input and initial conditions can be replicated (e.g., a Lorenz attractor or Rössler attractor; see Fig. 15.3d). Because in practice they cannot be replicated, the manifestation is usually that after many trajectories, the phase plane is not evenly filled, as it would be for a random time series, but is occupied by a quasi-patterned line, never exactly repeated but clearly constrained (Fig. B.2).

One of the most important properties of a strange attractor is its sensitive dependence on initial conditions. This means that points that are arbitrarily close initially,

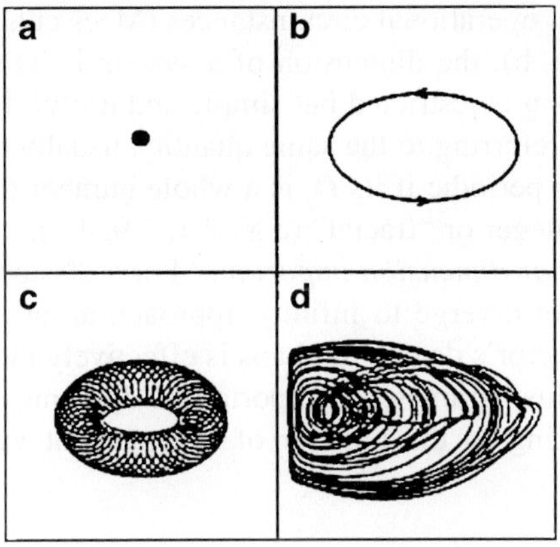

Fig. B.2 (**a**) Fixed point. (**b**) Limit cycle. (**c**) Torus. (**d**) Projection of a strange attractor (from Başar 1990)

become macroscopically separated after a sufficiently long time. In other words, similar causes do not produce similar effects.

This is an extensive statement, which seems to damage the causality principle of natural philosophy. However, by examining the properties of a strange attractor more precisely, one finds that a strange attractor may have a strong conformity, called self-similarity, which is an invariance with respect to scaling. Self-similar objects possess a fractal dimension, and fractals, are thus named because their fractal or Hausdorff dimension exceeds their topological dimension and they show strange properties, known in linear systems (see Schuster 1988).

Noise. Bullock (1976) describes the noise in the general neurophysiological approach as follows: unwanted action that interferes with desired signals. He further states:

> We should recognize the sharp difference between this dictionary definition and another, current usage that refers to a stochastic sequence (whiteness). In the first meaning, noise is determined by the state of the receiver (sleep, attention) and depends on the usefulness, regardless of the character; any unwanted sequence is regarded as noise whether it is a hiss, a whistle, or a voice. In the second meaning, noise is determined by the state of the sender (filter settings) and depends on the statistical character regardless of the use; any quasi-random sequence is regarded as noise whether it is unwanted interference or a high resolution signal. The first meaning overlooks the difficulty of knowing what may be of value to the receiver; the second overlooks the difficulty of avoiding the common English sense, as in "signal-to-noise-ratio."

B.2 The Correlation Dimension

This has become the most widely used measure to describe chaotic behavior. A valuable first step in the study of dynamic behavior, particularly when chaos is present, is measuring its dimension and investigating how the dimensionality can change under different operational circumstances (Mees et al. 1987). As expressed by Rapp et al. (1985a, b), the dimension of a system is its number of degrees of freedom. This definition is restricted but simple and useful. It is important to compare systems only by referring to the same quantity, usually the correlation dimension (D_2). A system is periodic if its D_2 is a whole number (e.g., 2.0, 3.0, 4.0) and chaotic if D_2 is non-integer or "fractal" (e.g., 2.1, 3.9, 4.5).

Why is the descriptor dimension important? For a dissipative dynamic system, trajectories that do not diverge to infinity approach an attractor. The unpredictability and so the attractor's degree of chaos is effectively measured by the parameter's "dimension." This parameter is important to dynamics because it provides a precise way of referring to the number of independent variables inherent in a motion.

B.2.1 Computation of Correlation Dimension

Here is a schematic illustration of a proposal made by Grassberger and Procaccia (1983) to compute the correlation dimension D_2. First, a phase space must be constructed.

In general, the phase-space is identified with a topological manifold. An n-dimensional phase-space is spanned by a set of n-independent embedding vectors. This requirement is usually sufficient. Following a proposal made by Takens (1981) referred to as the "time shift method" the n-dimensional vectors $\underline{x} = x(t),\ldots,x(t+(n-1)\tau)$ are constructed where τ is a fixed time increment. Every instantaneous state of a system is therefore, represented the vector $\underline{x}$, which defines a point in the phase-space.

Once the phase space is constructed, the correlation integral as a function of variable distances R is defined as

$$C(R) = \lim_{N\to\infty} \frac{1}{N^2} \sum_{i,j=1;i\neq j}^{N} \theta(R - |\underline{x}_i - \underline{x}_j|),$$

where N is the number of data points and θ is the Heavyside function. Then $C(R)$ is a measure of the probability that two arbitrary points $\underline{x}_i, \underline{x}_j$ will be separated by a distance $<R$. For this definition of $C(R)$ the following relationship holds:

$$C(R) \propto R^{D_2}$$

and then, if the amount of data and the embedding dimension are sufficiently large

$$D_2 = \lim_{R\to 0} \frac{\log C(R)}{\log R}$$

The main point is that C(R) behaves as a power of R for small R. By plotting log C(R) vs. log R D2 can be calculated from the slope of the curve. For an attractor with an unknown dimension, it is necessary to calculate C(R) for several embedding dimensions. In fact, following Takens' proposal one should choose the dimension n of the embedding phase-space at least twice the dimension of the attractor. The evaluation of the calculated correlation dimension D2 should converge towards a saturation value, i.e.,

$$\Delta D_2 / \Delta n = 0$$

for some n > no. Two examples can illustrate this relationship:

1. If the attractor is a simple curve in the phase-space (see Fig. 15.4), the number of points lying inside the circle with radius r is $N(r_0) = 8$, $N(2r_0) = 16$, $N(3r_0) = 24$, and so on. By plotting $N(r)$ vs. log r, one finds a straight line with slope $m = 1.00$. This is exactly the dimension of the simple curve, which was assumed to be an attractor (Fig. B.3).
2. Assume that the attractor represents a two-dimensional manifold in the phase-space. For every point of the attractor, the number of points lying inside a circle

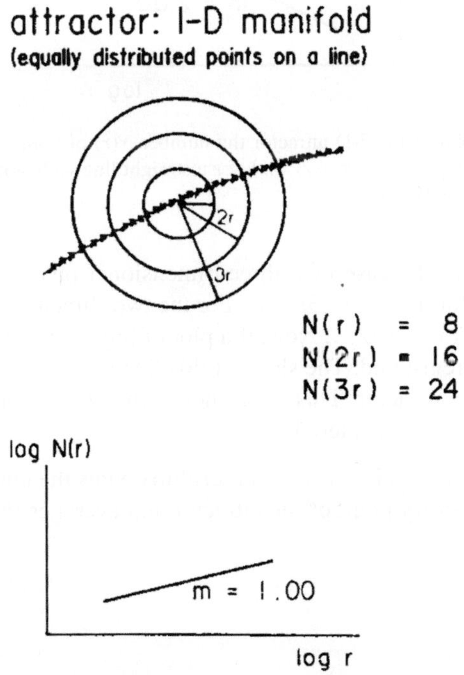

Fig. B.3 By counting the number of points N(r) lying inside a circle of radius r and plotting log N(r) vs. log r a straight line with slope m = 1.00 is obtained (from Röschke and Başar 1989)

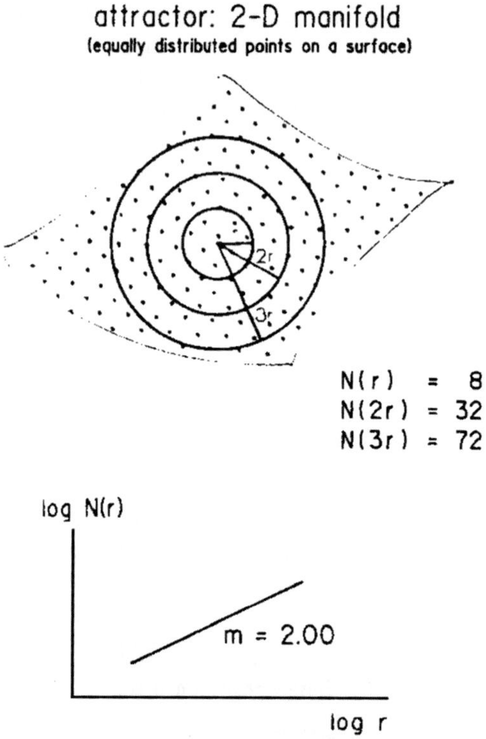

Fig. B.4 For every point of the 2-D attractor the number *N*(*r*) of points lying inside a circle of radius *r* are counted. By plotting log *N*(*r*) vs. log *r* a straight line with slope $m = 2.00$ is obtained (from Röschke and Başar 1989)

are counted (or in the case of a three-dimensional phase-space, inside a ball) which have a radius of r_o, $2r_o$, $3r_o$, etc. For the two-dimensional manifold $N(r_o) = 8$, $N(2r_o) = 32$, $N(3r_o) = 72$,… Now, if a plot of log (*r*) vs. log *N*(*r*) is performed, a straight line is registered. The slope of this line is $m = 2.00$. This is exactly the dimension of the attractor. This result never changes, even if a higher dimensional phase-space is considered.

In fact, Grassberger and Procaccia's algorithm counts the number of points lying inside the circle for every point of the attractor and averages the results (Fig. B.4).

Appendix C
Oscillatory Systems and Some Basic Experiments on Resonance in Physics and Systems Theory

C.1 Introduction

This appendix tries to understand theoretically the physical mechanisms that can play a basic role in the occurrence of resonance phenomena in general, and in the brain. Issues such as the role of dynamic patterns in biological research, theory of oscillators, and resonance phenomena in nature and the brain constitute the core materials of this appendix. The phenomena described in this appendix are of basic importance not only for the brain research scientist, but also for those scientists studying biophysical phenomena. We will try to bring together the resonance phenomena in elementary particle physics and the catastrophe models in the theory of morphogenesis, as well as the simplest harmonic oscillator.

C.2 The Harmonic Oscillator and Resonance Phenomena

The motions can be broadly categorized into two classes, according to whether the thing that is moving stays near one place or travels from one place to the other. Examples of the first class are an oscillating pendulum, a vibrating violin string, electrons vibrating in atoms, etc. Parallel examples of traveling motion are ocean waves rolling toward the beach, the electron beam of a TV tube, and a ray of light emitted from a star and detected with the eye. This and the coming section deal with the motion of a closed system that has been given an initial excitation (by some external disturbance) and is thereafter allowed to oscillate freely without further influence. Such oscillations are called free or natural oscillations (Crawford 1965). Examples of simple systems that stay in one vicinity and oscillate or vibrate about an average position are a pendulum oscillating in a plane, a mass on a spring, and an electrical LC circuit (Fig. C.1).

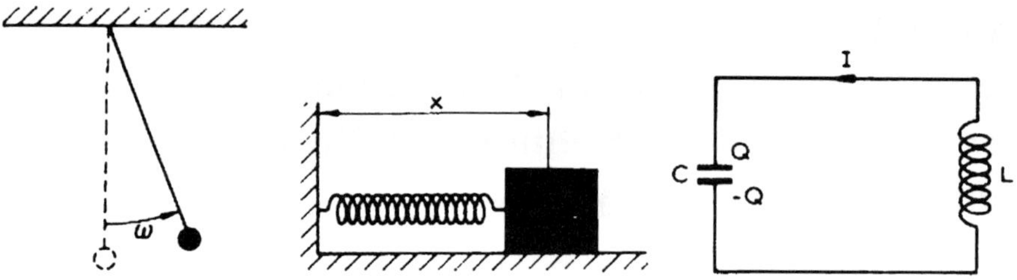

Fig. C.1 Systems with one degree of freedom (the pendulum is constrained to swing in a plane)

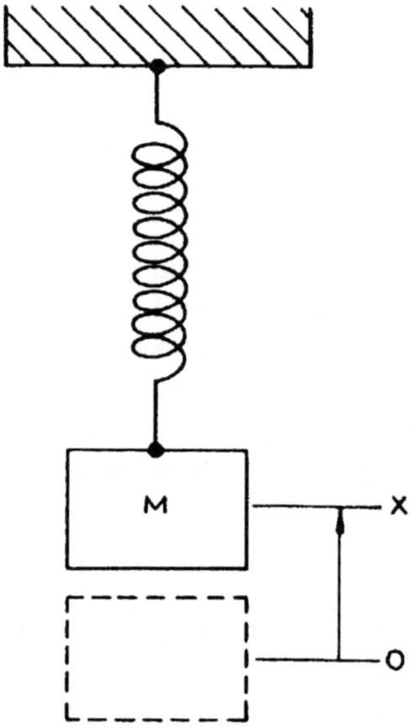

Fig. C.2 A mass on a string: a simple mechanical example of a harmonic oscillator

C.3 The Linear Harmonic Oscillator

The simplest mechanical system whose motion follows a linear differential equation with constant coefficients is a mass on a spring: First the spring stretches to balance the gravity; once it is balanced, we then discuss the vertical displacement of the mass from its equilibrium position (Fig. C.2). We shall call this upward displacement x, and we shall also suppose that the spring is perfectly linear, in which case the force pulling back when the spring is stretched is precisely proportional to the amount of stretch. That is, the force is $-kx$ (with a minus sign to remind us that it pulls back). Thus (according to dynamics law), the mass times the acceleration must equal $-kx$ (Feynman et al. 1963).

$$m(\mathrm{d}^2x/\mathrm{d}t^*) = -\,kx \quad \text{with } (k/m) = p^2, \tag{C.1}$$

$$(\mathrm{d}^2x/\mathrm{d}t^2) = -\,p^2x. \tag{C.2}$$

This is the most general equation of the harmonic oscillator. The solution of the equation is known to be:

$$x = C_1 \sin pt + C_2 \cos pt, \tag{C.3}$$

where C_1 and C_2 are constants of integration that must be evaluated from the initial conditions. That this expression is in fact a solution of the differential equation (C.2) may be verified by direct substitution.

The motion of the mass as a function of time is shown in Fig. C.3, where it is seen that the mass performs oscillations about the position of equilibrium. Because there is no energy loss in this ideal system, the oscillation continues indefinitely with the same amplitude A. The portion of the motion included between two points at which the mass has the same position, as B and C in Fig. C.3, is called one cycle of the oscillation. The time required for the completion of one cycle is called the period, \tau , of the oscillation. The number of cycles that occur in one second is called the frequency, f, of the oscillation. Because the motion is known to be harmonic, the displacement and velocity can be written:

$$x = A \sin \omega t \text{ and } v = A\omega \cos \omega t, \tag{C.4}$$

where ω denotes the angular frequency of the harmonic motion, which is equal to $2\pi f$.

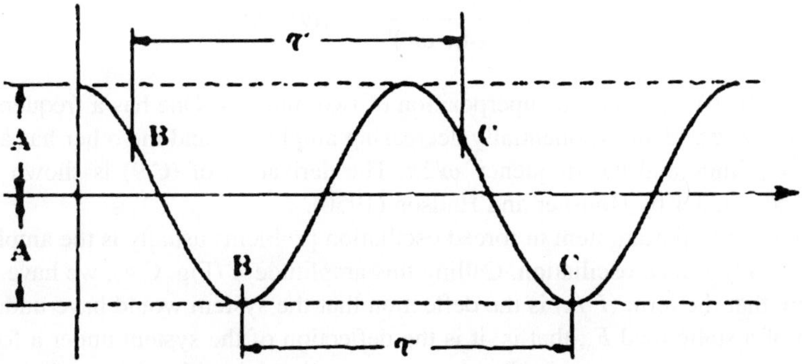

Fig. C.3 A cosine waveform

C.4 Forced Oscillations of the Linear Harmonic Oscillator: Classical Resonance

Oscillations that are maintained by an exciting force are said to be forced oscillations. First we will describe the differential equation of the motion of a simple harmonic oscillator on which a sinusoidal exciting force is acting. For the basic oscillation problem, we shall consider a system that consists of a linear restoring force, F_s, a viscous damping force, F_d, and a sinusoidal exciting force, $F_e(t)$, with angular frequency ω:

$$F_S = -kx,$$

$$F_d = -ex(x = dx/dt)$$

$$F_e t = F_o \sin \omega t.$$

Substituting these terms into the equation of motion, which dictates that the mass times the acceleration must be equal to the sum of all the forces in the same direction with acceleration, we obtain:

$$mx = -kx - cx + F_0 \sin\omega t \ (x = \mathrm{d}^2 x/\mathrm{d}t^2).$$

This equation is written in the standard form as:

$$x + 2nx + p^2 x = (F_0/m)\sin\omega t, \tag{C.5}$$

where, $(k/m) = p^2$ and $(c/m) = 2n$.

The complete solution of this differential equation is:

$$x = \mathrm{e}^{-nt}\left\{C_1 \sin(p^2 - n^2)^{1/2} t + C_2 \cos(p^2 - n^2)^{1/2} t\right\} + \frac{F_0/m}{\{(p^2 - \omega^2)^2 + 4n^2\omega^2\}^{1/2}} \sin(\omega t + \Phi). \tag{C.6}$$

Equation (C.6) represents a superposition of two motions. One has a frequency of $(p^2 - n^2)^{1/2}/2\pi$ and an exponentially decreasing amplitude, and the other has a constant amplitude and the frequency $\omega/2\pi$. The derivation of (C.9) is shown in an extended manner by Housner and Hudson (1950).

The most important item in forced oscillation problems usually is the amplitude of the steady forced oscillation. Calling this amplitude A (Fig. C.4), we have.

Note that the term (F_0/k) is the deflection that the system would have under the action of a static load F_0; that is, it is the deflection of the system under a forcing function with zero frequency. The expression on the right side of the equation thus represents a dynamic amplification or magnification factor and gives the ratio

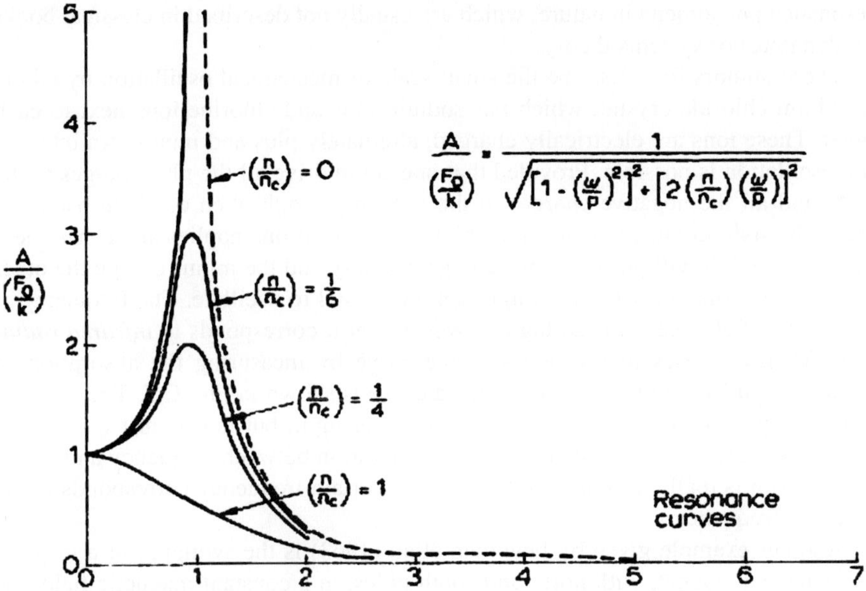

Fig. C.4 The classical resonance curves

between the dynamic and static deflections. The variation of this magnification factor with frequency ratio and damping ratio is shown in Fig. C.4. The most significant feature of Fig. C.4 is the fact that, near the frequency ratio $(\omega/p) = 1$, the magnification factor can become very large if the damping ratio is small. The infinite value indicated at $(n/n_c) = 0$ would, of course, not exist in practice, because it is impossible to reduce the damping to zero, and it would require an infinite time to reach the infinite amplitude even if the damping were zero. The occurrence of large displacements near $(\omega/p) = 1$ is called *resonance* and the frequency for which $\omega = p$ is called the *resonant frequency.*

C.5 Resonance in Nature

Although we have discussed the case of the harmonic oscillator in mechanics and its resonance formula, we would like to give some other examples of resonance phenomena. We recommend the reader who is interested in understanding the whole philosophy of the resonance phenomena to refer to Feynman et al. (1963). Feynman et al. describe, in a very elegant manner, that the resonance equation is the same in all the studied phenomena in classical physics.

There are many circumstances in nature in which something is oscillating and in which the resonance phenomenon occurs. Feynman et al. studied some of the resonance phenomena in nature, which are usually not described in classical books on dynamics or systems theory.

These authors first describe the small scale of mechanical oscillation by taking a sodium chloride crystal, which has sodium ions and chlorine ions next to each other. These ions are electrically charged, alternately plus and minus. An interesting oscillation is possible. Provided that one could drive all the plus charges to the right and all the negative charges to the left, they would then oscillate back and forth, the sodium lattice against the chlorine lattice. If one applies an electric field on the crystal, it will push the plus charge one way and the minus charge the other way! An external electric field might get the crystal to oscillate. The frequency of the electric field needed is so high, however, that it corresponds to *infrared radiation!* Then one tries to find a resonance curve by measuring the absorption of infrared light by sodium chloride. Such a curve is shown in Fig. C.5. The abscissa is not frequency, but is given in terms of wavelength, but that is just a technical matter, because for a wave there is a definite relation between frequency and wavelength; so it is really a frequency scale, and a certain frequency corresponds to the resonant frequency.

Another example given by Feynman et al. (1963) is the swinging of a magnet. If we have a magnet, with north and south poles, in a constant magnetic field, the N end of the magnet will be pulled one way and the S end the other way, and there will in general be a torque on it, so it will vibrate about its equilibrium position like a compass needle. However, the magnets we are talking about are *atoms.* These atoms have an angular momentum, the torque does not produce a simple motion in the direction of the field, but instead, a *precession.* Now, looked at from the side, any one component is swinging. One can disturb or drive that swinging and measure absorption. The curve in Fig. C.6 represents a typical such resonance curve. What has been done here is slightly different technically.

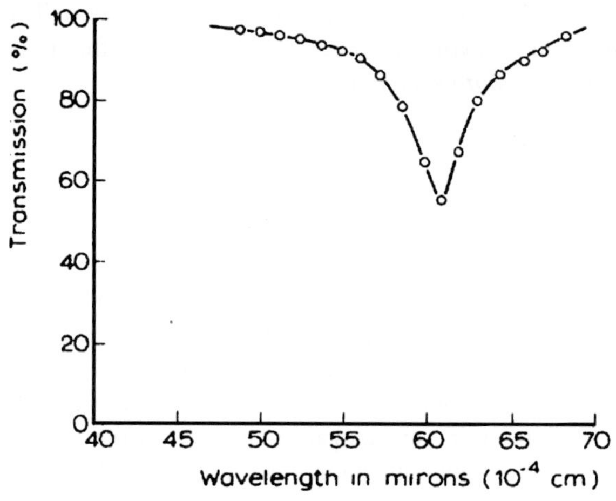

Fig. C.5 Transmission of infrared radiation through a thin (0.17 μm) sodium chloride film

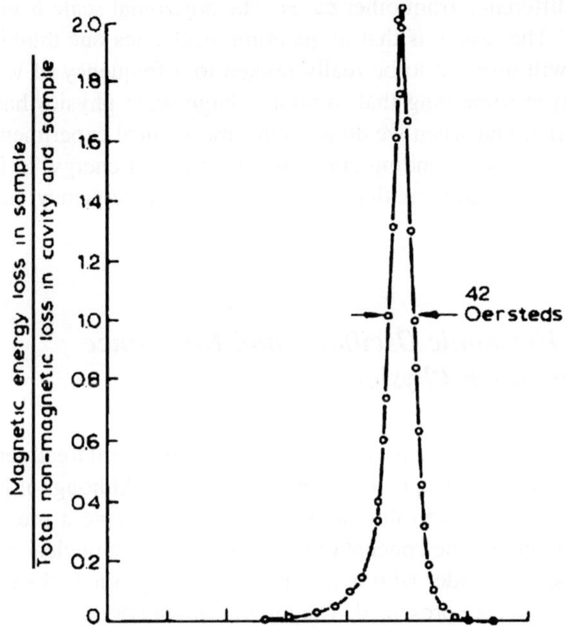

Fig. C.6 Magnetic energy loss in paramagnetic organic compound as function of applied magnetic field intensity (modified from Feynman et al. 1963)

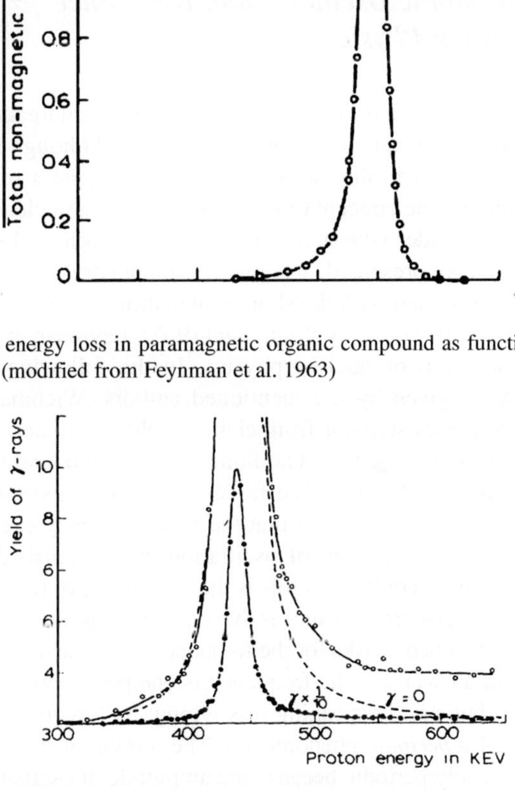

Fig. C.7 The intensity of gamma radiation from lithium as a function of the energy of the bombarding protons. The *dashed curve* is a theoretical one calculated for protons with an angular momentum 1 = 0 (modified from Feynman et al. 1963)

Feynman and coworkers go further. Their next example has to do with atomic nuclei. The motions of protons and neutrons in nuclei are oscillatory in certain ways, and this can be demonstrated by the following experiment. We bombard a lithium atom with protons, and we discover that a certain reaction producing γ-rays, actually has a very sharp maximum typical of resonance. We note in Fig. C.7, however, one difference from other cases: The horizontal scale is not a frequency, it is an *energy*! The reason is that in quantum mechanics one thinks of classically as the energy will turn out to be really related to a frequency of wave amplitude. When we analyze something that in simple large-scale physics has to do with a frequency, we find that when we do quantum-mechanical experiments with atomic matter, we get the corresponding curve as a function of energy. In fact, this curve is a demonstration of this relationship. It shows that frequency and energy have some deep interrelationship.

C.5.1 The Harmonic Oscillator and Resonance in Quantum Physics

The problem of the harmonic oscillator and its mathematical treatment presents one of the most important concepts in the modern physics. Although it is very difficult to handle such a problem in this book, we will try to give a presentation of the harmonic oscillator and the concept of the cross-section in nuclear and high energy nuclear physics. The reader who is not mathematically oriented should accept the general concepts that are presented. The mathematician or the engineer who studies this book cannot be satisfied with the short explanations. Therefore we recommend that that person study the books by Wichman (1967), Feynman et al. (1963), or any other books on quantum or nuclear physics. In the following, we will mostly use the interpretation given by the mentioned authors. Wichman starts with the example of the harmonic oscillator from classical physics to describe the phenomena inside of the atom. Consider a pendulum set in motion, and then left to swing by itself. We assume that the frictional forces (the most important of which is air resistance) are small, but not zero, so that the pendulum may execute several hundred oscillations before its energy of oscillation has diminished to $1/e$ times its original value. (The time required for this is the "mean-life of the oscillatory state.") Let the time interval between two successive outswings to the right be 1 s.

Suppose now that someone asks for the frequency of the pendulum. Without much reflection we would answer that the frequency is one per second. This is certainly a reasonable answer, but strictly speaking it is wrong: by "frequency" we understand the repetition rate of a *periodic* phenomenon. The motion of the pendulum is, however, only approximately periodic because the amplitude of oscillation diminishes as time goes on. The frequency of a *damped* harmonic motion is not precisely defined, although for all practical purposes it may be very well defined indeed.

An atom emitting radiation is analogous to a damped pendulum in some respects. The emission process does not go on forever, and this must mean that the "oscillation inside the atom" is a damped oscillation. Therefore, there is not a *precisely* defined frequency, because the oscillatory phenomenon is not strictly periodic. The electromagnetic radiation emitted by that "something that is oscillating inside the atom" is thus not monochromatic. The emitted line has a finite width (see Fig. C.8).

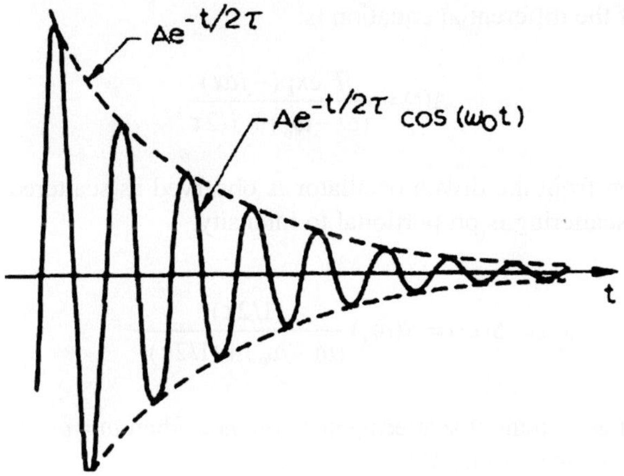

Fig. C.8 An exponentially dampled oscillatory process, showing the amplitude as a function of time. Because the process is not strictly periodic in time, it is wrong to say that the frequency of the oscillation is ω_0, because the concept of frequency refers to *aperiodic* phenomena. If the damping is not too large, it is fair to say that the frequency is *approximately* ω_0. It is intuitively clear that the smaller the damping, i.e., the smaller the decrease in amplitude for two successive maxima, the better is the frequency defined

Further, the atom by itself can be considered, just after it has been excited. The amplitude of whatever it is that oscillates inside the atom will be denoted by $A(t)$ and the time dependence of $A(t)$ will be assumed:

$$tA(t) = A\exp(-j\omega_0 t -)2\tau,$$

where A is constant. This is the time dependence of the amplitude of a damped harmonic oscillator of mean frequency ω_0, in the complex representation. The oscillator amplitude $A(t)$ given in the preceding satisfies the first order differential equation:

$$\frac{\mathrm{d}A(t)}{\mathrm{d}t} + \left(j\omega_0 + \frac{1}{2\tau}\right)A(t) = 0. \qquad \text{(C.7)}$$

This homogeneous differential equation describes the oscillator in the absence of any external influences. *Suppose now that monochromatic light, of frequency* ω, *is incident on the oscillator.* The equation must then be modified by addition of a term describing the harmonically varying applied driving force, which is F exp $(-j\omega t)$. The resulting inhomogeneous differential equation for the oscillator is then of the form:

$$\frac{\mathrm{d}A(t)}{\mathrm{d}t} + \left(j\omega_0 + \frac{1}{2\tau}\right)A(t) = F\exp(-j\omega t). \qquad \text{(C.8)}$$

The solution of the differential equation is:

$$A(t) = \frac{jF\exp(-j\omega t)}{(\omega - \omega_0) + j/2\tau}. \quad \text{(C.9)}$$

The emission from the driven oscillator is observed as scattered radiation, and the amount of scattering is proportional to intensity.
We may write:

$$S(\omega) = S(\omega_0)\frac{(1/2\tau)^2}{(\omega - \omega_0) + (1/2\tau)^2}, \quad \text{(C.10)}$$

where $S(\omega_0)$ is the amount of scattering at resonance when $\omega = \omega_0$. A schematic plot of $S(\omega)$ vs. ω is shown in Fig. C.9.

The function $S(\omega)$ expresses the intensity of response of the system under an external perturbation at the frequency ω. *This kind of resonant response is a very general phenomenon in quantum physics, and it is by no means restricted to the interaction of light with atoms.* We find the same resonant response when we study the scattering of material particles, such as protons, of a well-defined energy, from nucleus, or the scattering of pions from a proton. One might well say that a quasi-stable energy level of a quantum-mechanical system exists in *precisely* the sense that the system exhibits a resonant response, as given by (A.10), at the appropriate frequency.

In nuclear physics the resonance formula is known as the Breit–Wigner one-level resonance formula, after G. Breit and E.P. Wigner. What is meant if we sup-

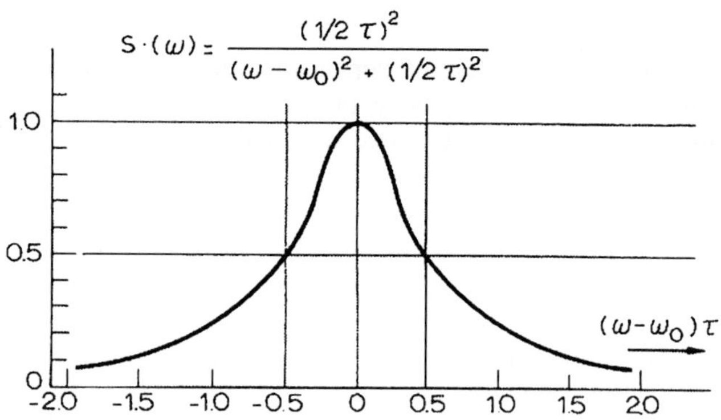

Fig. C.9 The universal resonance curve. It describes the response of any linear (or approximately linear) system to a sinusoidally varying external force in the neighborhood of a resonant frequency provided no other resonant frequency is close by. (Two bell-shaped curves play a particularly important role in physics: the resonance curve and the gaussian curve. As usually drawn they may look very similar. It must be remembered, however, that the gaussian curve falls off very rapidly outside the central region, whereas the resonance curve has a long "tail.") (modified from Wichman 1967)

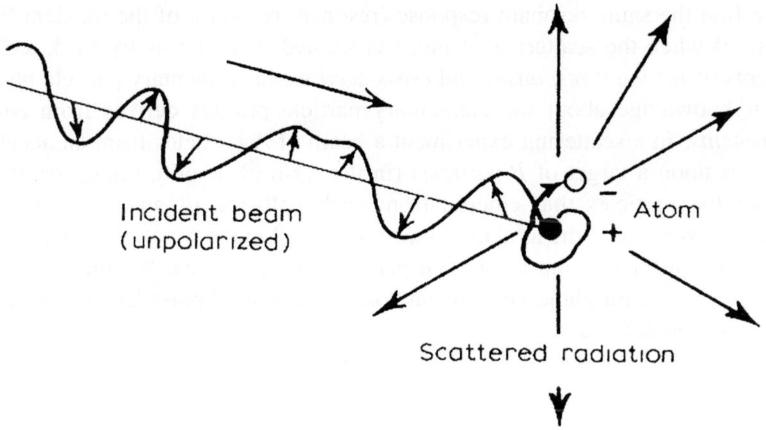

Fig. C.10 A beam of radiation falls on an atom and causes the charges (electrons) in the atom to move. The moving electrons in turn radiate in various directions (from Feynman et al. 1963)

pose that light of frequency ω is incident on the oscillator; that is, a beam of radiation falls on an atom and causes the charges (electrons) in the atom to move? The moving electrons in turn radiate in various directions (Fig. C.10). *What fraction* of the incoming light is scattered? Feynman et al. (1963) describe such a problem as follows:

Here's an idea: Say that the atom scatters a total amount of intensity that is the amount that would fall on a certain geometrical area, and we give the answer by giving that area. That answer, then, is independent of the incident intensity; it gives the ratio of the energy scattered to the energy incident per square meter. In other words the ratio

$$\frac{\text{Total energy scattered per second}}{\text{Energy incident per square meter per second}}$$

is an area. The significance of this area is that, if all the energy that impinged on that area were to be spewed in all directions, then that is the amount of energy that would be scattered by the atom.

This area is called a *cross-section for scattering:* The idea of a cross-section is used constantly, whenever some phenomenon occurs in proportion to the intensity of a beam. In such cases, one always describes the amount of the phenomenon by saying what the effective area would have to be to pick up that much of the beam. It does not mean in any way that this oscillator *actually has* such an area. If there were nothing present but a free electron shaking up and down there would be no area directly associated with it, physically. It is merely a way of expressing the answer to a certain kind of problem; it tells us what area the incident beam would have to hit in order to account for that much energy coming off.

We find the same resonant response (resonant response of the incident light to the atom) when the scattering of pions is studied. Now let us try to describe the concepts of *resonant responses* and *cross-sections* in elementary particle physics.

Our knowledge about the elementary particle physics derives from *collision experiments*. In a scattering experiment a beam of A-particles from an accelerator impinges upon a target of B-particles (in the form of a solid, liquid, or gas). We observe the particles that emerge from each collision of an A-particle with a B-particle. We denote the total cross-section as σ_T. Let the (on the average) uniform density of particles in the layer be n particles per unit area. We imagine that the target is a very thin plane layer of randomly distributed particles. The total cross section is then defined:

$$\sigma_T = \frac{p}{n},$$

where p is the probability that A-particle incident perpendicularly on the layer undergoes *some interaction* with one of B-particles that is removed from the incident beam. Wichman (1967) explains the total cross-section in terms of the following model:

> A circular disc of area σ_T is assigned to each B-particle. The discs are oriented perpendicularly to the incident beam of A-particles, and we imagine that they have the property that an A-particle that hits a disc is removed from the beam, whereas the A-particle is unaffected if it misses the discs. Consider again our thin target layer of n B-particles per unit area. The total area covered by the discs contained in a region of area F is equal to $nF\sigma_T$. This means that a fraction *nor* of the layer is "opaque" and a fraction $(1 - n\sigma_T)$ is "transparent". The probability that an *A-particle* in the incident beam is removed from the beam is accordingly $p = n\sigma_T$. The relation (6.15) can thus be interpreted in this manner, but the reader should understand that the opaque discs exist only in our imagination. The cross-section is a very convenient measure of the tendency of particles A and B to interact with each other, but it should not be thought that it refers to geometric properties of either one of the particles (Fig. C.11).

This was an example from collision experiments in particle physics. In a similar manner, one can define other kinds of cross-sections in elementary particle physics; for example, let us think that the *A-particle* can react with the B-particle to produce a C- and D-particle.

$$A + B \rightarrow C + D$$

For example $\gamma + p \rightarrow \pi^+ + \nu$, as in Fig. C.12. The reaction cross-section $\sigma_{AB\rightarrow CD}$ for this process is then defined by:

$$\sigma_{AB\rightarrow CD} = \sigma_{T P_{AB\rightarrow CD}},$$

where $P_{AB\rightarrow CD}$ is the probability that the preceding reaction takes place when an A-particle is removed from the beams through an interaction with a B-particle in the target (Wichman 1967). For such reactions in particle physics, the Breit–Wigner resonance formula can be derived through a different line of reasoning. In particle

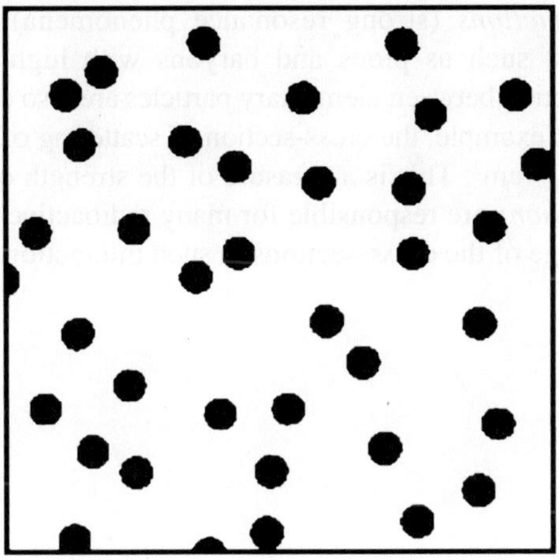

Fig. C.11 We can express the effectiveness by which the *B*-particles (in the target) remove *A*-particles from the incident beam by a total cross-section σ_T. With each *B*-particle is associated a circular disc of area σ_T such that an *A*-particle (imagined to be a point) interacts with the *B*-particle if and only if it hits the disc. The figure shows these imaginary discs for a very thin layer of *B*-particles. If there are *n* *B*-particles per unit area the total "blocked" area within a unit area will be $n\sigma_T$. The probability that an *A*-particle passes through such a layer is accordingly $(1 - n\sigma_T)$. The figure should not, of course, be taken literally. The *B*-particles are not in reality small discs or spheres (from Wichman 1967)

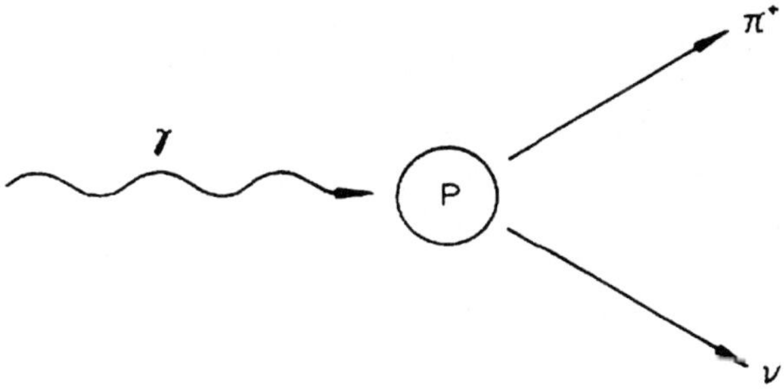

Fig. C.12 The schema of a strong interaction in high energy elementary particle physics

physics, phenomena of particle interactions play an important role; and, the resonant peaks in particle interactions are quantified with cross-sections. In other words, resonance phenomena in elementary particle physics or atomic physics are measured with the cross-section of the reactions.

The interactions in elementary particle physics are grouped into four categories.

1. Strong interactions
2. Electromagnetic interactions
3. Weak interactions
4. Graviton interaction

The *strong interactions* (strong resonance phenomena) are interactions of elementary particles; such as pions and baryons with high cross-sections. The strength of the interaction between elementary particles are also characterized in terms of cross-sections. For example, the cross-section of scattering of neutrons on protons is in the range of 10^{-25} cm^2. This is a measure of the strength of the interaction.

The *weak interactions* are responsible for many radioactive decays for example, the β-decay. The range of the cross-sections in such interactions is around 10^{-44} cm^2 (Heisenberg 1961).

References

Abraham, R. H., Shaw, C. D., 1983. Dynamics: The Geometry of Behaviour. Aerial, Santa Cruz, CA.

Adey, W. R., 1966. Neurophysiological correlates of information transaction and storage in brain tissue. In: Stellar, E., Sprague, J. M. (Eds.), Progress in Physiological Psychology. Academic Press, New York, pp. 1–43.

Adey, W. R., 1989. Cell membranes, electromagnetic fields, and intercellular communication. In: Başar, E., Bullock, T. H. (Eds.), Brain Dynamics – Progress and Perspectives. Springer, New York, pp. 26–42.

Adey, W. R., Dunlop, C.W., Hendrix, C. E., 1960. Hippocampal slow waves. Distribution and phase relationships in the course of approach learning. Arch. Neurol. 3, 74–90.

Adler, G., Brassen, S., Jajcevic, A., 2003. EEG coherence in Alzheimer's dementia. J. Neural. Transm. 110, 1051–1058.

Adrian, E. D., 1931. Potential changes in the isolated nervous system of *Dytiscus marginalis*. J. Physiol. 72, 132–151.

Adrian, E. D., 1934. The Basis of Sensation. The Action of the Sense Organs. Christophers, London.

Adrian, E. D., 1937. Synchronized reactions in the optic ganglion of *Dytiscus*. J. Physiol. 91, 66–89.

Adrian, E. D., 1941. Afferent discharges to the cerebral cortex from peripheral sense organs. J. Physiol. 100, 159–191.

Adrian, E. D., 1942. Olfactory reactions in the brain of the hedgehog. J. Physiol. 101, 459–473.

Adrian, E. D., 1951. Rhythmic discharges from the thalamus. J. Physiol. 113, 9–108.

Adrian, E. D., Matthews, B. H. C., 1934. The Berger rhythm: Potential changes from the occipital lobes in man. Brain 57, 355–385.

Aladjalova, N. A., 1957. Infra-slow rhythmic oscillations of the steady potential of the cerebral cortex. Nature 179(4567), 957–959.

Aladjalova, N. A., 1964. Slow electrical processes in the brain. Progress in Brain Research, Vol 7, Elsevier, Amsterdam.

Allers, K. A., Kreiss, D. S., Walters, J. R., 2000. Multisecond oscillations in the subthalamic nucleus: effects of apomorphine and dopamine cell lesion. Synapse 38, 38–50.

Allers, K. A., Ruskin, D.N., Bergstrom, D. A., Freeman, L. E., Ghazi, L. J., Tierney, P. L., Walters, J. R., 2002. Multisecond periodicities in basal ganglia firing rates correlate with theta bursts in transcortical and hippocampal EEG. J. Neurophysiol. 87(2), 1118–1122.

Allers, K. A., Ruskin, D. N., Bergstrom, D. A., Walters, Jr., 1999. Correlations of multisecond oscillations in firing rate in pairs of basal ganglia neurons. Soc. Neurosci. Abstr. 25, 1929.

Akiskal, H. S., Hantouche, E. G., Bourgeois, M. L., Azorin, J. M., Sechter, D., Allilaire, J. F., Chatenêt-Duchêne, L., Lancrenon, S., 2001. Toward a refined phenomenology of mania: combining clinician-assessment and self-report in the French EPIMAN study. J. Affect. Disord. 67 (1–3), 89–96.

Andreasen, N., 2005. The Creating Brain. The Neuroscience of Genius. Dana, New York.

Andersen, P., Andersson, S. A., 1968. Physiological Basis of the Alpha Rhythm. Appleton-Century-Crofts, New York.

Apostol, G., Creutzfeldt, O. D., 1974. Crosscorrelation between the activity of septal units and hippocampal EEG during arousal. Brain Res. 67, 65–75.

Babiloni, C., Cassetta, E., Binetti, G., Tombini, M., Del Percio, C., Ferreri, F., Ferri, R., Frisoni, G., Lanuzza, B., Nobili, F., Parisi, L., Rodriguez, G., Frigerio, L., Gurzì, M., Prestia, A., Vernieri, F., Eusebi, F., Rossini, P. M., 2007. Resting EEG sources correlate with attentional span in mild cognitive impairment and Alzheimer's disease. Eur. J. Neurosci. 25(12), 3742–3757.

Babiloni, C., Frisoni, G., Steriade, M., Bresciani, L., Binetti, G., Del Percio, C., Geroldi, C., Miniussi, C., Nobili, F., Rodriguez, G., Zappasodi, F., Carfagna, T., Rossini, P. M., 2006. Frontal white matter volume and delta EEG sources negatively correlate in awake subjects with mild cognitive impairment and Alzheimer's disease. Clin. Neurophysiol. 117(5), 1113–1129.

Babiloni, C., Ferri, R., Binetti, G., Vecchio, F., Frisoni, G. B., Lanuzza, B., Miniussi, C., Nobili, F., Rodriguez, G., Rundo, F., Cassarino, A., Infarinato, F., Cassetta, E., Salinari, S., Eusebi, F., Rossini, P. M., 2009. Directionality of EEG synchronization in Alzheimer's disease subjects. Neurobiol. Aging 30(1), 93–102.

Babloyantz, A., 1986. Molecules, Dynamics and Life. New York, Wiley.

Babloyantz, A., 1988. Chaotic dynamics in brain activity. In: Başar, E. (Ed.), Dynamics of Sensory and Cognitive Processing of the Brain. Springer, Berlin, pp. 196–202.

Babloyantz, A., 1989. Estimation of correlation dimensions from single and multi-channel recordings – a critical view. In: Başar, E., Bullock, T.H. (Eds), Brain Dynamics – Progress and Perspectives. Springer, Berlin, pp. 122–130.

Babloyantz, A., Destexhe, A., 1986. Low dimension chaos in an instance of epilepsy. *Proc. Natl. Acad. Sci. U.S.A.* 83, 3513–3517.

Babloyantz, A., Kaczmarek, L. K., 1979. Self-organization in biological systems with multiple cellular contacts. Bull. Math. Biol. 41, 193–201.

Babloyantz, A., Salazar, J. M., Nicolis, C., 1985. Evidence of chaotic dynamics of brain activity during the sleep cycle. Phys. Lett. A III, 152–156.

Baddeley, A. D., 1986a. Working Memory. Oxford University Press, Oxford, UK.

Baddeley, A. D., 1986b. Modularity, mass-action and memory. Quart. J. Exp. Psychol. 38, 527–533.

Baddeley, A., 1996. The fractionation of working memory. Proc. Natl. Acad. Sci. U.S.A 93, 13468–13472.

Bailey, C. H., Giustetto, M., Huang, Y., Hawkins, R. D., Kandel, E. R, 2000. Is heterosinaptic modulation essential for stabilizing Hebbian plasticity and memory? Nat. Rev. Neurosci. 1, 11–20.

Bailey, C. H., Chen, M., 1983. Morphological basis of long-term habituation and sensitization in Aplysia. Science 220, 91–93.

Balconi, M., Lucchiari, C., 2006. EEG correlates (event-related desynchronization) of emotional face elaboration: a temporal analysis. Neurosci. Lett. 392, 118–123.

Baldeweg, T., Spence, S., Hirsch, S. R., Gruzelier, J., 1998. Gamma-band electroencephalographic oscillations in a patient with somatic hallucinations. Lancet 352, 620–621.

Barlow, H. B., 1972. Single units and sensation: a neuron doctrine for perceptual psychology. Perception 1, 371–394.

Barlow, H. B., 1995. The neuron doctrine in perception. In: Gazzaniga, M. S., Bizzi, E. (Eds.), The Cognitive Neurosciences. MIT Press, Cambridge, MA.

Banaschewski, T., Brandeis, D., 2007. Annotation: what electrical brain activity tells us about brain function that other techniques cannot tell us a child psychiatric perspective. J. Child Psychol. Psychiatry 48, 415–435.

Barman, S. M., Gebber, G. L., 1993. Lateral tegmental field neuronsplay a permissive role in governing the 10-Hz rhythmin sympathetic nerve discharge. Am. J. Physiol. 265(5 Pt 2), R1006–R1013.

Barman, S. M., Orer, H. S., Gebber, G. L., 1995. A 10-Hz rhythm reflects the organization of a brainstem network that specifically governs sympathetic nerve discharge. Brain Res. 671(2), 345–350.

Barman, S. M., Kitchens, H. L., Leckow, A. B., Geber, G. L., 1997. Oct Pontine neurons are elements of the network responsible for the 10-Hz rhythm in sympathetic nerve discharge. Am. J. Physiol. 273 (4 Pt 2), H1909–H1919.

Barry, R. J., De Pascalis, V., Hodder, D., Clarke, A. R., Johnstone, S. J., 2003a. Preferred EEG brain states at stimulus onset in a fixed interstimulus interval auditory oddball task and their effects on ERP components. Int. J. Psychophysiol. 47 (3), 187–198.

Barry, R. J., Johnstone, S. J., Clarke, A. R., 2003b. A review of electrophysiology in attention-deficit/hyperactivity disorder: II. Event-related potentials. Clin. Neurophysiol. 114(2), 184–198.

Bartels, A., Zeki, S., 2000. The neural basis of romantic love. Neuroreport 11, 3829–3834.

Bartels, A., Zeki, S., 2004. The chronoarchitecture of the human brain-natural viewing conditions reveal a time-based anatomy of the brain. NeuroImage 22, 419–433.

Bartley, S. H., Bishop, G. H., 1933. The cortical response to stimulation of the optic nerve in the rabbit. Am. J. Physiol. 103, 159–172.

Barut, A., 1967. The Theory of the Scattering Matrix. Macmillan, New York.

Başar, E., 1974. Biological systems analysis and evoked potentials of the brain. T-l-T. J. Life Sci. 4, 37–58.

Başar, E., 1976. Biophysical and Physiological Systems Analysis. Addison-Wesley, Amsterdam.

Başar, E., 1980. EEG-Brain Dynamics. Relation between EEG and Brain Evoked Potentials. Elsevier, Amsterdam.

Başar, E., 1983a. Toward a physical approach to integrative physiology: I. Brain dynamics and physical causality. Am. J. Physiol. 14, R510–R533.

Başar, E., 1983b. Synergetics of neuronal populations. A survey on experiments. In: Başar, E., Flohr, H., Haken, H., Mandell, A. (Eds.), Synergetics of the Brain. Springer, Berlin, pp. 183–200.

Başar, E., 1988. EEG-dynamics and evoked potentials in sensory and cognitive processing by the brain. In: Başar, E. (Ed.), Dynamics of Sensory and Cognitive Processing by the Brain. Springer, Berlin, pp. 30–55.

Başar, E., 1990. Chaos in Brain Function. Springer, Berlin.

Başar, E., 1992. Brain natural frequencies are causal factors for resonancesand induced rhythms. In: Başar, E., Bullock, T. H. (Eds.), Induced Rhythm in the Brain. Birkhauser, Boston, pp. 425–467.

Başar, E., 1998. Brain Oscillations I: Principles and Approaches. Springer-Verlag, Heidelberg.

Başar, E., 1999. Brain Function and Oscillations: II. Integrative Brain Function. Neurophysiology and Cognitive Processes. Springer-Verlag, Heidelberg.

Başar, E., 2004. Memory and Brain Dynamics: Oscillations Integrating Attention, Perception, Learning and Memory. CRC Press, Boca Raton, FL.

Başar, E., 2006. The theory of the whole-brain-work. Int. J. Psychophysiol. 60, 133–138.

Başar, E., 2007. Brain Oscillations: Cutting Edges. Int. J. Psychophysiol. 64(1), 35–49.

Başar, E., 2008. Oscillations in "brain-body-mind" – A holistic view including the autonomous system. Brain Res. 1235, 2–11.

Başar, E., 2009. S-Matrix and Feynman space-time diagrams to quantum brain approach: an extended proposal. Neuroquantology 7(1), 30–45.

Başar, E., Bullock, T. H. (Eds.), 1992. Induced Rhythms in the Brain. Birkhäuser, Boston.

Başar, E., Güntekin, B., 2007. A breakthrough in neuroscience needs a "Nebulous Cartesian System" oscillations, quantumdynamics and chaos in the brain and vegetative system. Int. J. Psychophysiol. 64, 108–122.

Başar, E., Güntekin, B., 2008. A review of brain oscillations in cognitive disorders and the role of neurotransmitters. Brain Res. 1235, 172–193.

Başar, E., Güntekin, B., 2009. An essay on Darwin's theory and Bergson's creative evolution in the era of neuroquantology. NeuroQuantology 7(4), 609–622.

Başar, E., Eroğlu, C., 1976. Spectral analysis of spontaneous activity in smooth muscles. In: Bülbring, E., Shuba, M. F. (Eds.), Physiology of Smooth Muscle. Raven, New York, pp. 137–146.

Başar, E., Özesmi, C., 1972. The hippocampal EEG activity and a systems analytical interpretation of averaged evoked potentials of the brain. Kybernetik 12, 45–54.

Başar, E., Stampfer, H. G., 1985. Important associations among EEG-dynamics, event-related potentials, short-term memory and learning. Int. J. Neurosci. 26(3–4), 161–180.

Başar, E., Ungan, P., 1973. A component analysis and principles derived for the understanding of evoked potentials of the brain: studies in the hippocampus. Kybernetik 12, 133–140.

Başar, E., Weiss, Ch., 1968. Analyse des Frequenzganges druckinduzierter Änderungen des Strömungswiderstandes isolierter Rattennieren. Pflügers Arch. 304, 121–135.

Başar, E., Weiss, Ch., 1969. Rate sensitivity of the mechanism of pressure induced change of vascular resistance. Kybernetik 5, 241–247.

Başar, E., Weiss, C., 1981. Vasculature and Circulation. Elsevier, Amsterdam.

Başar, E., Başar-Eroğlu, C., Demiralp, T., Schürmann, M., 1993. Compound P300–40 Hz response of the human brain. Electroencephalogr. Clin. Neurophysiol. 87, 14.

Başar, E., Başar-Eroğlu, C., Karakaş, S., Schürmann, M., 1999a. Are cognitive processes manifested in event-related gamma, alpha, theta and delta oscillations in the EEG? Neurosci. Lett. 15, 165–168.

Başar, E., Başar-Eroğlu, C., Karakaş, S., Schürmann, M., 1999b. Oscillatory brain theory: A new trend in neuroscience: the role of oscillatory processes in sensory and cognitive functions. IEEE Eng. Med. Biol. 18(3), 56–66.

Başar, E., Başar-Eroğlu, C., Karakaş, S., Schürmann, M., 2000. Brain oscillations in perception and memory. Int. J. Psychophysiol. 35(2–3), 95–124.

Başar, E., Başar-Eroğlu, C., Karakaş, S., Schürmann, M., 2001a. Gamma, alpha, delta, and theta oscillations govern cognitive processes. Int. J. Psychophysiol. 39, 241–248.

Başar, E., Başar-Eroğlu, C., Karakaş, S., Schürmann, M., 2001b. Event-related oscillations in perception and memory. Int. J. Neurosci. 39(2–3), 87–156.

Başar, E., Başar-Eroğlu, C., Röschke, J., 1988. Do coherent patterns of the strange attractor EEG reflect deterministic sensory-cognitive states of the brain? In: Markus, M., Müller, M., Nicolis, G. (Eds.), From Chemical to Biological Organization. Springer, Berlin, pp. 297–306.

Başar, E., Başar-Eroğlu, C., Röschke, J., Schütt, A., 1989a. The EEG is a quasi-deterministic signal anticipating sensory-cognitive tasks. In: Başar, E., Bullock, T. H. (Eds.), Brain Dynamics. Progress and Perspectives. Springer, Berlin, pp. 43–71.

Başar, E., Başar-Eroğlu, C., Röschke, J., Schult, J., 1989b. Chaos and alpha reparation in brain function. In: Cotterill, R. (Ed.), Models of Brain Function. Cambridge University Press, Cambridge, UK, pp. 365–395.

Başar, E., Başar-Eroğlu, C., Röschke, J., Schult, J., 1990. Strange attractor EEG as sign of cognitive function. In: John, E. R., Harmony, T., Prichep, L., Valdes-Sosa, M., Valdes-Sosa, P. (Eds.), Machinery of the Mind. Birkhauser, Boston, pp. 91–114.

Başar, E., Demir, N., Gönder, A., Ungan, P., 1979a. Combined dynamics of EEG and evoked potentials I. Studies of simultaneously recorded EEG–EP-grams in the auditory pathway, reticular formation and hippocampus of the cat brain during the waking stage. Biol. Cybernet. 34, 1–19.

Başar, E., Demiralp, T., Schürmann, M., Başar-Eroğlu, C., Ademoğlu, A., 1999c. Oscillatory brain dynamics, wavelet analysis, and cognition. Brain Language 66(1), 146–183.

Başar, E., Durusan, R., Gönder, A., Ungan, P., 1979b. Combined dynamics of EEG and evoked potentials II. Studies of simultaneously recorded EEG-Programs in the auditory pathway, reticular formation and hippocampus of the cat brain during sleep. Biol. Cybernet. 34, 21–30.

Başar, E., Dössel, O., Fuchs, M., Rahn, E., Saermark, K., Schürmann, M., 1992. Evoked alpha responses from frontal-temporal areas in multichannel SQUID systems. Proceedings of IEEE Symposium on Neuroscience and Technology, Lyon, 28–33.

Başar, E., Eroğlu, C., Ungan, P., 1974. An analysis of portal vein spontaneous contractions. Pflügers Arch. 352, 135-143.

Başar, E., Flohr, H., Haken, H., Mandell, A. J. (Eds.), 1983. Synergetics of the Brain (Proceedings of the International Symposium on Synergetics at Schloss Elmau, Bavaria, May 2 –7). Springer, Berlin.

Başar, E., Gönder, A., Özesmi, C., Ungan, P., 1975a. Dynamics of brain rhythmic and evoked potentials I. Some computational methods for the analysis of electrical signals from the brain. Biol. Cybernet. 20, 137–143.

Başar, E., Gönder, A., Özesmi, C., Ungan, P., 1975b. Dynamics of brain rhythmic and evoked potentials II. Studies in the auditory pathway, reticular formation, and hippocampus during the waking stage. Biol. Cybernet. 20, 145–160.

Başar, E., Gönder, A., Özesmi, C., Ungan, P., 1975c. Dynamics of brain rhythmic and evoked potentials III. Studies in the auditory pathway, reticular formation, and hippocampus during sleep. Biol. Cybernet. 20, 161–169.

Başar, E., Gönder, A., Ungan, P., 1976a. Important relation between EEG and brain evoked potentials I. Resonance phenomena in subdural structures of the cat brain. Biol. Cybernet. 25, 27–40.

Başar, E., Gönder, A., Ungan, P., 1976b. Important relation between EEG and brain evoked potentials II. A system analysis of electrical signals from the human brain. Biol. Cybernet. 25, 41–48.

Başar, E., Gönder, A., Ungan, P., 1980. Comparative frequency analysis of single EEG-evoked potential Records. J. Biomed. Eng. 2, 9–14.

Başar, E., Güntekin, B., Öniz, A., 2006. Principles of oscillatory brain dynamics and a treatise of recognition of faces and facial expressions. Prog. Brain Res. 159, 43–63.

Başar, E., Hari, R., Lopes da Silva, F. H., Schürmann, M. (Eds.), 1997a. Brain alpha activity: new aspects and functional correlates. Int. J. Psychophysiol. 26, 1–482.

Başar, E., Özgören, M., Başar-Eroğlu, C., Karakaş, S., 2003. Superbinding: spatio-temporal oscillatory dynamics. Theory Biosci. 121, 370–385.

Başar, E., Özgören, M., Karakaş, S., 2001d. Superbinding in integrative brain function and memory. In: Proceedings of 23rd Annual International Conference of IEEE Engineering in Medicine and Biology Society, Minneapolis, MN.

Başar, E., Özgören, M., Karakaş, S., 2001c. A brain theory based on neural assemblies and superbinding. In: Reuter, H., Schwab, P., Gniech, K. D. (Eds.), Wahrnehmen und Erkennen. PABST Science Publishers, Lengerich, pp. 11–24.

Başar, E., Özgören, M., Karakaş, S., 2002. Supersynergy and superbinding in integrative brain functions. Int. J. Psychophysiol. 45 (1–2), 37–37.

Başar, E., Özgören, M., Karakaş, S., Başar-Eroğlu, C., 2004. Super-synergy in brain oscillations and the grandmother percept. Int. J. Bifurcation Chaos 14, 453–491.

Başar, E., Özgören, M., Öniz, A., Schmiedt, C., Başar-Eroğlu, C., 2007. The brain oscillations differentiate the picture of the own grandmother. Int. J. Psychophysiol. 64(1), 81–90.

Başar, E., Ruedas, G., Schwarzkopf, H. J., Weiss, C. H., 1968a. Untersuchungen des zeitlichen Verhaltens druckabhängiger Änderungen des Strömungswiderstandes im Coronargefäßsystem des Rattenherzens. Pflügers Arch. 304, 189–202.

Başar, E., Rosen, B., Başar-Eroğlu, C., Greitschus, F., 1987. The associations between 40 Hz-EEG and the middle latency response of the auditory evoked potential. Int. J. Neurosci. 33, 103–117.

Başar, E., Rahn, E., Demiralp, T., Schürmann, M., 1998. Spontaneous EEG activity controls frontal visual evoked potential amplitudes. Electroencephalogr. Clin. Neurophysiol. 108, 101–109.

Başar, E., Schmiedt-Fehr, C., Öniz, A., Başar-Eroğlu, C., 2008. Brain oscillations evoked by the face of a loved person. Brain Res. 1214, 105–115.

Başar, E., Schütt, A., Bullock, T. H., 1999. Dynamics of potentials from the brain of anamniotes (vertebrates). In: Başar, E. (Ed.), Integrative Brain Function. Neurophysiology and Cognitive Processes, Brain Function and Oscillations, Vol. II. Springer, New York, pp. 109–116.

Başar, E., Schürmann, M., Sakowitz, O., 2001c. The selectively distributed theta system: functions. Int. J. Psychophysiol. 39, 197–212.

Başar, E., Schürmann, M.; Başar-Eroğlu, C., Karakaş, S., 1997b. Alpha oscillations in brain functioning: an integrative theory. In: Başar, E., Hari, R., Lopes da Silva, F. H., Schürmann, M. (Eds.), Brain Alpha Activity: New Aspects and Functional Correlates. Special issue of the Int. J. Psychophysiol. 26, 5–29.

Başar, E., Tischner, H., Weiss, Ch., 1968b. Untersuchungen zur Dynamik druckinduzierter Änderungen des Strömungswiderstandes der autoregulierenden, isolierten Rattenniere. Pflügers Arch. 299, 191–213.

Başar, E., Yordanova, J., Kolev, V., Başar-Eroğlu, C., 1997c. Is the alpha rhythm a control parameter for brain responses? Biol. Cybernet. 76, 471–480.

Başar-Eroğlu, C., Başar, E., 1991. A compound P300–40 Hz response of the cat hippocampus. Int. J. Neurosci. 60, 227–237.

Başar-Eroğlu, C., Demiralp, T., 2001. Event-related theta oscillations: an integrative and comparative approach in the human and animal brain. Int. J. Psychophysiol. 39, 167–195.

Başar-Eroğlu, C., Başar, E., Demiralp, T., Schürmann, M., 1992. P300 response: possible psychophysiological correlates in delta and theta frequency channels. Int. J. Psychophysiol. 13, 161–179.

Başar-Eroğlu, C., Başar, E., Schmielau, F., 1991. P300 in freely moving cats with intracranial electrodes. Int. J. Neurosci. 60, 215–226.

Başar-Eroğlu, C., Başar, E., Zetler, G., 1996a. Effects of ceruletide and haloperidol on auditory-evoked potentials in the cat brain. Int. J. Neurosci. 85, 131–146.

Başar-Eroğlu, C., Brand, A., Hildebrandt, H., 2002. Brain oscillations in working memory. Int. J. Psychophysiol. 45, 36.

Başar-Eroğlu, C., Brand, A., Hildebrandt, H., Kedzior, K. K., Mathes, B., Schmiedt, C., 2007. Working memory related gamma oscillations in schizophrenia patients. Int. J. Psychophysiol. 64, 39–45.

Başar-Eroğlu, C., Demir, N., Başar, E., Weiss, Ch., 1979. Low frequency vascular dynamics. Pflügers Arch, European Journal of Physiology, Suppl. 379, R8.

Başar-Eroğlu, C., Demiralp, T., Schürmann, M., Başar, E., 2001. Topological distribution of oddball "P-300" responses. Int. J. Psychophysiol. 39, 213–220.

Başar-Eroğlu, C., Kolev, V., Ritter, B., Aksu, F., Başar, E., 1994. EEG auditory evoked potentials and evoked rhythmicities in three-year-old children. Int. J. Neurosci. 75, 239–255.

Başar-Eroğlu, C., Strüber, D., Kruse, P., Başar, E., Stadler, M., 1996b. Frontal gamma band enhancement during multistable visual perception. Int. J. Psychophysiol. 24, 113–125.

Başar-Eroğlu, C., Schmiedt-Fehr, C., Marbach, S., Brand, A., Mathes, B., 2008. Altered oscillatory alpha and theta networks in schizophrenia. Brain Res. 1235, 143–152.

Bauer, L. O., Hesselbrock, V. M., 1993. EEG, autonomic and subjective correlates of the risk for alcoholism. J. Stud. Alcohol. 54, 577–589.

Bearden, C. E., Hoffman, K. M., Cannon, T. D., 2001. The neuropsychology and neuroanatomy of bipolar affective disorder: a critical review. Bipolar Disord. 3, 106–150.

Beauregard, O. C., 1987. Bergson's Duration and Quantal Spacetime Non-Separability. In: Papanicolaou, A. C., Gunter, P. A. Y. (Eds.), Bergson and Modern Thought: Towards a Unified Science. Harwood, New York.

Beim-Graben, P., 2001. Estimating and improving the signal-to-noise ratio of time series by symbolic Dynamics. Physical Review E 64, 1–15.

Beim-Graben, P., Saddy, J. D., Schlesewsky, M., 2000. Symbolic dynamics of event-related brain potentials. Physical Review E 62, 4.

Begleiter, H., Porjesz, B., 2006. Genetics of human brain oscillations. Int. J. Psychophysiol. 60, 162–171.

Bendat, J. S., Piersol, A. G., 1968. Mesasurement and Analysis of Random Data. Wiley, New York.

Benes, F. M., Berretta, S., 2001. GABAergic interneurons: implications for understanding schizophrenia and bipolardisorder. Neuropsychopharmacology 25, 1–27.

Berger, H., 1929. Über das Elektrenkephalogramm des Menschen. I. Bericht. Archiv Fuer Psychiatrie und Nervenkrankheiten 87, 527–570.

Bergson, H., 1910. Time and Free Will: An Essay on the Immediate Data of Consciousness. Pogson, F. L. (trans.). Kessinger, Whitefish, MT.

Bergson, H., 1896. Matière et mémoire (Matter and memory). Macmillan, New York.

Bergson, H., 1907. L'évolution Créatrice. Presse Universitaires de France, Paris.

Berettini, W. H., Nurnberger Jr., J. I., Hare, T. A., Simmons-Alling, S., Gershon, E. S., Post, R. M., 1983. Reduced plasma and CSF gamma-aminotbutyric acid in affective illness: effect oflithium carrbonate. Biol. Psychiatry 18, 185–194.

Best, P. J., Ranck, J. B., 1982. Reliability of the relationship between hippocampal unit activity and sensory–behavioral events in the rat. Exp. Neurol. 75, 652–664.

Besthorn, C., Forstl, H., Geiger-Kabisch, C., Sattel, H., Gasser, T., Schreiter-Gasser, U., 1994. EEG coherence in Alzheimer's disease. Electroencephalogr. Clin. Neurophysiol. 90, 242–245.

Bhagwagar, Z., Wylezinska, M., Jezzard, P., Evans, J., Ashworth, F., Sule, A., Matthews, P. M., Cowen, P. J., 2007. Reduction in occipital cortex g-aminobutyric acid concentrations in medication-free recovered unipolar depressed and bipolarsubjects. Biol. Psychiatry 61(6), 806–812.

Bierer, L. M., Hof, P. R., Purohit, D. P., Carlin, L., Schmeidler, J., Davis, K. L., Perl, D. P., 1995. Neocortical neurofibrillary tangles correlate with dementia severity in Alzheimer's disease. Arch. Neurol. 52(1), 81–88.

Birkhoff, G. D., 1932. Probability and physical systems. Bull. Am. Math. Soc. 38, 361–379.

Bishop, G. H., 1933. Cyclic changes in excitability of the optic pathway of the rabbit. Am. J. Physiol. 103, 213–224.

Bishop, P. O., Jeremy, D., McLeod, J. G., 1953. Phenomenon of repetitive firing in lateral geniculate of cat. J. Neurophysiol. 16, 443–447.

Biswal, B., Yetkin, F. Z., Haughton, V. M., Hyde, J. S., 1995. Functional connectivity in the motor cortex of resting human brain echo-planar MRI. Magn. Reson. Med. 34, 537–541.

Bland, B. H., 1986. The physiology and pharmacology of hippocampal formation theta rhythms. Prog. Neurobiol. 26(1), 1–54.

Bland, B. H., Colom, L.V., 1993. Extrinsic and intrinsic properties underlying oscillation and synchrony in limbic cortex. Prog. Neurobiol. 41(2), 157–208.

Bleckmann, H., Tittel, G., Blübaum-Gronau, E., 1989. The lateral line system of surface-feeding Wsh: Anatomy, physiology, and Behavior. In: Coombs, S., Görner, P., Münz, H. (Eds.), The Mechanosensory Lateral Line. Neurobiology and Evolution. Springer, New York, pp. 501–526.

Brambilla, P., Perez, J., Barale, F., Schettini, G., Soares, J. C., 2003. GABAergic dysfunction in mood disorders. Mol. Psychiatry 8, 721–737.

Brazhnik, E. S., Fox, S. E., 1997. Intracellular recordings from medial septal neurons during hippocampal theta rhythm. Exp. Brain Res. 114(3), 442–453.

Brazier, M. A. B., 1968. The Electrical Activity of the Nervous System. Williams & Wilkins, Baltimore.

Bremer, F., Bonnet, V., 1950. Interprétation des réactions, rhythmiques prologuées des aires sensorielles de l écorce cérébrale. Electroencephalogr. Clin. Neurophysiol. 2, 384–400.

Brenner, C. A., Sporns, O., Lysaker, P. H., O'Donnell, B. F., 2003. EEG synchronization to modulated auditory tones in schizophrenia, schizoaffective disorder, and schizotypal personality disorder. Am. J. Psychiatr. 160, 2238–2240.

Bressler, S. L., Kelso, J. A., 2001. Cortical coordination dynamics and cognition. Trends Cogn. Sci. 1, 26–36.

Bressler, S. L., Tognoli, E., 2006. Operational principles of neurocognitive networks. Int. J. Psychophysiol. 60, 139–148.

Bonte, F. J., Harris, T. S., Hynan, L. S., Bigio, E. H., White, C. L., 2006. Tc-99m HMPAO SPECT in the differential diagnosis of the dementias with histopathologic confirmation. Clin. Nucl. Med. 31(7), 376–378.

Bowden, C. L., 2003. Valproate. Bipolar Disord. 5, 189–202.

Bowden, C. L., 2007. Brain-mind in probabilistic hyperspace. Abstract Book Page 24–25.

Bowden, C. L., 2008. Bipolar pathophysiology and development of improved treatments. Brain Res. 1235, 92–97.
Bowden, C. L., Karren, N. U., 2006. Anticonvulsants in bipolar disorder. Aust. N. Z. J. Psychiatry 40 (5), 386–393.
Bozler, E., 1947. Response of smooth muscle to stretch. Am. J. Physiol. 149, 299–301.
Böhmig, L., 1883. Beiträge zur Kenntnis des Zentralnervensystems einiger Pulmonaten Gasteropoden Helix pomatia und Limnea stagnalis Inaugurale Dissertatio, Leipzig.
Bucci, P., Mucci, A., Merlotti, E., Volpe, U., Galderisi, S., 2007. Induced gamma activity and event-related coherence in schizophrenia. Clin. EEG Neurosci. 38(2), 96–104.
Bullock, T. H., 1945. Problems in the comparative study of brain waves. Yale J. Biol. Med. 17, 657–679.
Bullock, T. H., 1974. Comparisons between vertebrates and invertebrates in nervous organization. In: Schmitt, F. O., Worden, F. G. (Eds.), The Neurosciences: Third Study Program. MIT Press, Cambridge, MA, pp. 343-431.
Bullock, T. H., 1976. In search of principles in neural integration. In: Fentress, J. C. (Ed.), Simpler Networks and Behavior. Sinauer Associates, Sunderland, MA, pp. 52–60.
Bullock, T. H., 1979. Processing of ampullary input in the brain: comparison of sensitivity and evoked responses among elasmobranchs and siluriform fishes. J. Physiol. 75, 397–467.
Bullock, T. H., 1980. Reassessment of neural connectivity and its specification. In: Pinsker, H. M., Willis, W. D. Jr. (Eds.), Information Processing in the Nervous System. Raven, New York, pp. 199–220.
Bullock, T. H., 1983. Comparative neuroscience holds promise for quiet revolutions. Science 225(4661), 473–478.
Bullock, T. H., 1984a. Physiology of the tectum mesencephali in elasmobranchs. In: Vanegas, H. (Ed.), Comparative Neurology of the Optic Tectum. Plenum, New York, pp. 47–68.
Bullock, T. H., 1984b. Ongoing compound field potentials from octopus brain are labile and vertebrate-like. Electroencephalogr. Clin. Neurophysiol. 57(5), 473–483.
Bullock, T. H., 1988a. Compound potentials of the brain, ongoing and evoked: perspectives from comparative neurology. In: Başar, E. (Ed.), Dynamics of Sensory and Cognitive Processing by the Brain. Springer Series in Brain Dynamics, vol. 1. Springer, New York, pp. 3–18.
Bullock, T. H., 1988b. The comparative neurology of expectation: stimulus acquisition and neurobiology of anticipated and unanticipated input. In: Atema, J., Popper, P. R. A. N., Tavolga, W. N. (Eds.), Sensory Biology of Aquatic Animals. Springer, New York, pp. 195–222.
Bullock, T. H., 1989b. Evolution of compound field potentials in the brain. In: Başar, E., Bullock, T. H. (Eds.), Brain Dynamics: Progress and Perspectives. Springer, Berlin, pp. 258–266.
Bullock, T. H., 1992. Introduction to induced rhythms: a widespread, heterogeneous class od oscillations. In: Başar, E., Bullock, T. H. (Eds.), Induced Rhythm in the Brain. Birkhauser, Boston, pp. 1–26.
Bullock, T. H., 2002. Biology of brain waves: natural history and evolution of an information-rich sign of activity. Istanbul.
Bullock, T. H., 2006. How do brains evolve complexity? An essay. Int. J. Psychophysiol. 60, 106–109.
Bullock, T. H., Başar, E., 1988. Comparison of ongoing compound field potentials in the brains of invertebrates and vertebrates. Brain Res. Rev. 13, 57–75.
Bullock, T. H., Corwin, J. T., 1979. Acoustic evoked activity in the brain in sharks. J. Comp. Physiol. 129, 223–234.
Bullock, T. H., Horridge, G. A., 1965. Structure and Function in the Nervous Systems of Invertebrates, vol. II. W.H. Freeman, San Francisco.
Bullock, T. H., McClune, M. C., 1989. Lateral coherence of the electrocorticogram: A new measure of brain synchrony. Electroencephalogr. Clin. Neurophysiol. 73, 479–498.
Bullock, T. H., Iragui, V. J., Alksne, J. F., 1990. Electrocorticogram coherence and correlation of amplitude modulation between electrodes both decline in millimeters in human as well as in rabbit brains. Soc. Neurosci. Abstr. 16, 1241.

Bullock, T. H., Bennett, M. V., Johnston, D., Josephson, R., Marder, E., Fields, R. D., 2005. Neuroscience. The neuron doctrine, redux. Science 310(5749), 791–793.
Bullock, T. H., McClune, M. C., Achimowicz, J. Z., Iragui-Madoz, V. J., Duckrow, R. B., Spencer, S. S., 1995a. EEG coherence has structure in the millimeter domain: subdural and hippocampal recordings from epileptic patients. Electroencephalogr. Clin. Neurophysiol. 95, 161–177.
Bullock, T. H., McClune, M. C., Achimowicz, J. Z., Iragui-Madoz, V. J., Duckrow, R. B., Spencer, S. S., 1995b. Temporal fluctuations in coherence of brain waves. Proc. Natl. Acad. Sci. U.S.A. 92, 11568–11572.
Burgess, A. P., Gruzelier, J. H., 2000. Short duration power changes in the EEG during recognition memory for words and faces. Psychophysiology 37, 596–606.
Burnstock, G., Prosser, C. L., 1960. Responses of smooth muscles to quick stretch: relation of stretch to conduction. Am. J. Physiol. 198(5), 921–925.
Byrne, J. H., Castellucci, V. F., Carew, T. J., Kandel, E. R., 1978. Stimulus-response relations and stability of mechanoreceptor and motor neurons mediating defensive gill withdrawal reflex in Aplysia. J. Neurophysiol. 41, 402–417.
Buzsaki, G., 2002. Theta oscillations in the hippocampus. Neuron 33, 325–340.
Buszaky, G., 2006. Rhythms of the Brain. Oxford University Press, New York.
Cajal, R., 1897. Advice for a Young Investigator. Swanson, N., Swanson, L. W. (trans.), 1999. MIT Press, Cambridge, MA.
Cajal, R., 1911. Histologie du système nerveux de l'homme et des vertebras. Azoulay, L. (trans.), 1972. Consejo superior de investigaciones cientificas. Instituto Ramon y Cajal, Madrid.
Callaway, E., 1983. Presidential address, 1982. The pharmacology of human information processing. Psychophysiology 20, 359–370.
Cannon, W. B., Britton, S. W., 1925. Pseudoaffective medulliadrenal secretion. Am. J. Physiol. 72, 283–294.
Capra, F., 1982. The Turning Point. Science, Society and the Rising Culture. Simon and Schuster, New York.
Capra, F., 1984. The Turning Point. Flamingo, London, p. 97.
Caro, C. G., Pedley, T. J., Schroter, R. C., Seed, W. A., 1978. The Mechanics of Circulation. Oxford University Press, Oxford, UK.
Cassidy, D. C., 1999. Werner Heisenberg und das unbestimmtheitsprinzip. Spektrum Wiss. 1, 6–13.
Castellucci, V.F., Kandel, E.R., 1974. A quantal analysis of the synaptic depression underlying habituation of the gillwithdrawal reflex in Aplysia. Proc. Natl. Acad. Sci. U.S.A. 71, 5004–5008.
Castle, E., Wessely, S., Der, G., Murray, R. M., 1991. The incidence of operationally defined schizophrenia in Camberwell 1965–84. Br. J. Psychiatry 159, 790–794.
Changeux, J.-P., 2004. The Physiology of Truth: Neuroscience and Human Knowledge. Harvard University Press, Cambridge, MA.
Chang, H. T., 1950. The repetitive discharges of corticothalamic reverberating circuit. J. Neurophysiol. 13, 235–257.
Chen, A. C., Herrmann, C. S., 2001. Perception of pain coincides with the spatial expansion of electroencephalographic dynamics in human subjects. Neurosci. Lett. 297, 183–186.
Churchland, S. P., 2002. Brain-Wise Studies in Neurophysiology. The MIT Press, Cambridge, MA.
Clark, A., 1998. Being There: Putting Brain, Body, and World Together Again. MIT Press, Cambridge, MA.
Conel, J. L., 1963. Postnatal Development of the Human Cerebral Cortex: The Cortex of the Forty-Eight-Month Infant. Harvard University Press, Cambridge, MA.
Cook, E. W. III, Miller, G. A., 1992. Digital filtering: Background and tutorial for psychophysiologists. Psychophysiology 29, 350–367.
Cooley, R. L., Montano, N., Cogliati, C., Van De Borne, P., Richenbacker, W., Oren, R., Somers, V., 1998. Evidence for a central origin of the low-frequency oscillation in RR interval variability. Circulation 98, 556–561.

Crawford, Jr., F. S., 1965. Waves. Berkeley Physics Course, Vol. 3, McGraw-Hill, New York.
Creutzfeld, O. D., 1995. Cortex Cerebri: Performance, Structural and Functional Organization of the Cortex. Oxford University Press, Oxford, UK.
Curio, G., 2000. Linking 600-Hz "spikelike" EEG/MEG wavelets ("sigma-bursts") to cellular substrates: concepts and caveats. J. Clin. Neurophysiol. 17(4), 377–396.
Curio, G., Mackert, B. M., Burghoff, M., Neumann, J., Nolte, G., Scherg, M., Marx, P., 1997. Somatotopic source arrangement of 600 Hz oscillatory magnetic fields at the human primary somatosensory hand cortex. Neurosci. Lett. 234(2–3), 131–134.
Damasio, A. R., 1994. Descartes' Error: Emotion, Reason, and the Human Brain. Grosset-Putnam, New York.
Damasio, A. R., 1997. Memory's lanes. In: Geary, J. (Ed.), Time. 149(18), 39–45.
Damasio, A. R., Damasio, H., 1994. Cortical systems for retrieval of concrete knowledge: The convergence zone framework. In: Koch, C., Davis, J. L. (Eds.), Large-Scale Neuronal Theories of the Brain, MIT Press, Cambridge, MA, pp. 61–74.
Darwin, C., 1859. On The Origin of Species by Means of Natural Selection or the Preservation of Favoured Races in the Struggle for Life. John Murray, Albemarle Street, London.
Darwin, C., 1872. The Expression of Emotion in Man and Animals. Philosophical Library, New York.
Dauwels, J., Vialatte, F., Musha, T., Cichocki, A., 2010. A comparative study of synchrony measures for the early diagnosis of Alzheimer's disease based on EEG. Neuroimage 49(1), 668–693.
De Beer, G., 1960. Darwin's notebooks on transmutation of species. Part II. Second notebook [C] (February to July 1838). Bull. Br. Mus 2, 75–118.
De Lamarck, J. B., 1809. Philosophie Zooloqiue, ou, Exposition des Considerations Relatives a L'histoire Naturelle des Animaux. 2 Vols., Paris.
De Weerd, J. A., 1981. A posteriori time-varying filtering of averaged evoked potentials. I. Introduction and conceptual basis. Biol. Cybernet. 41, 211–222.
Deiber, M. P., Missonnier, P., Bertrand, O., Gold, G., Fazio-Costa, L., Ibañez, V., Giannakopoulos, P., 2007. Distinction between perceptual and attentional processing in working memory tasks: a study of phase-locked and induced oscillatory braindynamics. J. Cogn. Neurosci. 19(1), 158–172.
Deleuze, G., 1966. Le Bergsonisme. PUF, Paris. Tomlinson, H., Habberjam, B. (Trans.), 1988. Bergsonism. Zone Books, NY.
Demiralp, T., Ademoğlu, A., Schurmann, M., Başar-Eroğlu, C., Başar, E., 1999. Detection of P300 in single trials by the wavelet transform (WT). Brain Lang. 66, 108–128.
Demiralp, T., Bayraktaroğlu, Z., Lenz, D., Junge, S., Busch, N. A., Maess, B., Ergen, M., Herrmann, C. S., 2007a. Gamma amplitudes are coupled to theta phase in human EEG during visual perception. Int. J. Psychophysiol. 64(1), 24–30.
Demiralp, T., Herrmann, C. S., Erdal, M. E., Ergenoğlu, T., Keskin, Y. H., Ergen, M., Beydağı, H., 2007b. DRD4 and DAT1 polymorphisms modulate human gamma band responses. Cereb. Cortex 17(5), 1007–1019.
Descartes, R., 1637. Discourse on Method, Optics, Geometry, and Meteorology. Oscamp, Paul, J. (trans), 2001.
Descartes, R., 1840. Discours de la méthode. Hachette, Paris.
Desimone, R., 1996. Neural mechanisms for visual memory and their role in attention. Proc. Natl. Acad. Sci. U.S.A. 26, 13494–13499.
Deutsch, D., 2003. Physics, philosophy and quantumtechnology. In: Shapiro, J. H., Hirota, O. (Eds.), Proceedings of the Sixth International Conference on Quantum Communication, Measurement and Computing. Rinton, Princeton, NJ.
Diagnostic and Statistical Manual of Mental Disorders, 4th ed, 1994. American Psychiatric Press, Washington, DC.
Dick, D. M., Jones, K., Saccone, N., Hinrichs, A., Wang, J. C., Goate, A., Bierut, L., Almasy, L., Schuckit, M., Hesselbrock, V., Tischfield, J., Foroud, T., Edenberg, H., Porjesz, B., Begleiter, H., 2006. Endophenotypes successfully lead to gene identification: results from the collaborative study on the genetics of alcoholism. Behav. Genet. 36, 112–126.

Dinse, H. R., Krüger, K., Akhavan, A. C., Spengler, F., Schöner, G., Schreiner, C. E., 1997. Low-frequency oscillations of visual, auditory and somatosensory cortical neurons evoked by sensory stimulation. Int. J. Psychophysiol. 26, 205–227.

Domhoff, G. W., 2005. Refocusing the neurocognitive approach to dreams: A critique of the Hobson versus Solms debate. Dreaming 15, 3–20.

Donchin, E., Gerbrandt, L. K., Leifer, L., Tucker, L. R., 1973. Contingent negative variations and motor response. In: McCallum, W. C., Knott, J. R. (Eds.), Event-related Slow Potentials of the Brain: Their Relations to Behavior. Proceedings of the 2nd International CNV Congress, Vancouver (1971). Elsevier, Amsterdam, pp. 187–190.

Doppelmayr, M., Klimesch, W., Sauseng, P., Hödlmoser, K., Stadler, W., Hanslmayr, S., 2005. Intelligence related differences in EEG-bandpower. Neurosci. Lett. 381, 309–313.

Doppelmayr, M., Klimesch, W., Schwaiger, J., Stadler, W., Röhm, D., 2000. The time locked theta response reflects interindividual differences in human memory performance. Neurosci. Lett. 278, 141–144.

Dougall, N. J., Bruggink, S., Ebmeier, K. P., 2004. Systematic review of the diagnostic accuracy of 99mTc-HMPAO-SPECT in dementia. Am. J. Geriatr. Psychiatry 12(6), 554–570.

Dunkin, J. J., Leuchter, A. F., Newton, T. F., Cook, I. A., 1994. Reduced EEG coherence in dementia: state or trait marker? Biol. Psychiatry 35, 870–879.

Dustman, R. E., Shearer, D. E., Emmerson, R. Y., 1993. EEG and event-related potentials in normal aging. Prog. Neurobiol. 41, 369–401.

Dvorak, I., Siska, J., 1986. On some problems encountered in the estimation of the correlation dimension of the EEG. Phys. Lett. A. 118, 63–66.

Eccles, J. C., 1973. The Understanding of the Brain. McGraw-Hill, New York.

Eccles, J. C., 1986. Do mental events cause neural events analogously to the probability fields of quantum mechanics? Proc. R. Soc. Lond. B. Biol. Sci. 227(1249), 411–428.

Echteler, S. M., Saidel, W. M., 1980. Some connections of the telcost telencephalon. Soc. Neurosci. Abstr. 6, 629.

Eckhorn, R., Bauer, R., Jordan, R., Brosch, W., Kruse, M., Munk, M., Reitboeck, H. J., 1988. Coherent oscillations: a mechanism of feature linking in the visual cortex. Biol. Cybernet. 60, 121–130.

Edelman, G. M., 1978. Group selection and phasic reentrant signaling: a theory of higher brain functions. In: Edelman, G. M., Mountcastle, V. B. (Eds.), The Mindful Brain. MIT Press, Cambridge, MA, pp. 51–100.

Edelman, G. M., 1987. Neural Darwinism. Basic Books, New York.

Eeg-Olofsson, O., 1971. The development of the EEG in normal children from age 1 to 15 years. The 14 and 6 Hz positive spike phenomenon. Neuropädiatrie. 3, 11–45.

Eichenbaum, H., 2000. A cortical-hippocampal system for declarative memory. Nat. Rev. Neurosci. 1, 41–50.

Einstein, A., 1905. On the motion – required by the molecular kinetic theory of heat – of small particles suspended in a stationary liquid. Annalen der Physik 17, 549–560.

Einstein, A., Infeld, L., 1938. Evolution of Physics. Simon and Schuster, New York.

Eggert, P., Weiss, C., 1980. Periodic microflow pattern measured with a new microflow probe within the rat kidney cortex. Pflügers Arch. 383, 223–227.

Eggert, P., Thiemann, V., Weiss, C., 1979. Periodic changes in blood flow in the in vivo rat kidney. Pflügers Arch. 382(1), 63–66.

Egner, T., Gruzelier, J. H., 2001. Learned self-regulation of EEG frequency components affects attention and event-related brain potentials in humans. Neuroreport 12(18), 4155–4159.

Ekman, P., 1992a. An argument for basic emotions. Cogn. Emotion 6, 169–200.

Ekman, P., 1992b. Are there basic emotions? Psychol. Rev. 99, 550–553.

Ekman, P., 1992c. Facial expressions of emotion: new findings, new questions. Psychol. Sci. 3(1), 34–38.

Ekman, P., Friesen, W. V., 1976. Pictures of Facial Affect. Consulting Psychologist Press, Palo Alto.

Elazar, Z., Adey, W. R., 1967. Spectral analysis of low frequency components in the electrical activity of the hippocampus during learning. Electroencephalogr. Clin. Neurophysiol. 23, 225–240.

Enoch, M. A., White, K. V., Harris, C. R., Robin, R. W., Ross, J., Rohrbaugh, J. W., Goldman, D., 1999. Association of low-voltage alpha EEG with a subtype of alcohol use disorders. Alcohol Clin. Exp. Res. 23, 1312–1319.
Eroğlu, C., 1974. An analysis of spontaneous contraction patterns of smooth mucsles in time and frequency domains. Thesis, Hacettepe University, Ankara.
Evans, E.F., 1982. Functional anatomy of the auditory system. In: Barlow, H. B., Mollon, F. D. (Eds.), The senses. Cambridge University Press, New York, pp. 251–332.
Fabiani, M., Karis, D., Donchin, E., 1990. Effects of mnemonic strategy manipulation in a Van Restorff paradigm. Electroencephalogr. Clin. Neurophysiol. 75, 22–35.
Farwell, L. A., Martinerie, J. M., Bashore, T. R., Rapp, P. E., Goddard, P. H., 1993. Optimal digital filters for long-latency components of the event-related brain potential. Psychophysiology 30, 306–315.
Fehr, T., 2002. Lokalisation langsamer Hirnaktivität bei schizophrenen Patienten mittels magnetenzephalografischer Untersuchungen und Exploration von Zusammenhängen zwischen langsamwelliger Hirnaktivität und Symptomatik. Shaker, Aachen, Germany.
Fehr, T., Wienbruch, C., Moratti, S., Rockstroh, B., Elbert, T., 2001. Statistical discrimination of controls, schizophrenics, depressives and alcoholics using local magnetoencephalographic frequency-related variables. Biomedizinische Technik 46, 242–244.
Fell, J., Dietl, T., Grunwald, T., Kurthen, M., Klaver, P., Trautner, P., Schaller, C., Elger, C. E., Fernandez, G., 2004. Neural bases of cognitive ERPs: more than phase reset. J. Cogn. Neurosci. 16, 1595–1604.
Fell, J., Hinrichs, H., Röschke, J., 1997. Time course of 40 Hz EEG activity accompanying P3 responses in an auditory oddball paradigm. Neurosci. Lett. 235, 121–124.
Fell, J., Kohling, R., Grunwald, T., Klaver, P., Dietl, T., Schaller, C., Becker, A., Elger, C. E., Fernandez, G., 2005. Phase-locking characteristics of limbic P3 responses in hippocampal sclerosis. Neuroimage 24, 980–989.
Feller, M. B., 1999. Spontaneous correlated activity in developing neural circuits. Neuron 22, 653–656.
Fellous, J. M., Sejnowski, T. J., 2000. Cholinergic induction of oscillations in the hippocampal slice in the slow (0.5–2 Hz), theta (5–12 Hz), and gamma (35–70 Hz) bands. Hippocampus 10(2), 187–197.
Ferri, C. P., Prince, M., Brayne, C., Brodaty, H., Fratiglioni, L., Ganguli, M., Hall, K., Hasegawa, K., Hendrie, H., Huang, Y., Jorm, A., Mathers, C., Menezes, P. R., Rimmer, E., Scazufca, M., 2005. Global prevalence of dementia: a Delphi consensus study. Lancet 366 (9503), 2112–2117.
Feynman, R. P., 1962. Theory of Fundamental Processes. Benjamin, New York.
Feynman, R. P., Leighton, R. B., Sands, M., 1963. The Feynman Lectures on Physics. Addison-Wesley, Reading, MA.
Flohr, H., 1991. Brain processes and phenomenal consciousness: a new and specific hypothesis. Theory Psychol. 1, 245–262.
Fessard, A., 1961. The role of neuronal networks in sensory communications within the brain. In: Rosenblith, W. A. (Ed.), Sensory Communication. MIT Press, Cambridge, MA, pp. 585–606.
Fisch, B. J., 1991. Spelmann's EEG primer (2nd ed.). Elsevier, Amsterdam.
Fischer, Y., Gahwiler, B. H., Thompson, S. M., 1999. Activation of intrinsic hippocampal theta oscillations by acetylcholine in rat septo-hippocampal co-cultures. J. Physiol. 519, 405–413.
Flippow, I. V., Williams, W. C., Krebs, A. A., Pugachev, K. S., 2007. Sound induced changes of intra-slow brain potential fluctuations in the medical geniculate nucleus and in the auditory cortex of anaesthetized rats. Brain Res. 16, 78–86.
Freeman, W. J. (Ed.), 1975. Mass Action in the Nervous System. Academic Press, New York.
Freeman, W. J., 1991. The physiology of perception. Sci. Am. 264(2), 78–85.
Freeman, W. J., 2006. A cinematographic hypothesis of cortical dynamics in perception. Int. J. Psychophysiol. 60, 149–161.
Freeman, W. J., 1999. Foreword. In: Başar, E. (Ed.), Brain Function and Oscillations: II. Integrative Brain Function. Neurophysiology and Cognitive Processes. Springer, Berlin.
Freeman, W. J., Skarda, C. A., 1985. Spatial EEG patterns, non-linear dynamics and perception: the neo-Sherringtonian view. Brain Res. Rev. 10, 147–175.

French, J. D., Vehzeano, M., Magoun, H. W., 1953. An extralemniscal sensory system in the brain. Archiv. Neurol. Psychiatry 69(4), 505–518.
Freund, T. F., Antal, M., 1988. GABA-containing neurons in the septum control inhibitory interneurons in the hippocampus. Nature 336(6195), 170–173.
Fries, P., 2005. A mechanism for cognitive dynamics: neuronal communication through neuronal coherence. Trends Cogn. Sci. 9(10), 474–480.
Frodl-Bauch, T., Bottlender, R., Hegerl, U., 1999. Neurochemical substrates and neuroanatomical generators of the event-related P300. Neuropsychobiology 40, 86–94.
Frotscher, M., Leranth, C., 1985. Cholinergic innervation of the rat hippocampus as revealed by choline acetyltransferase immunocytochemistry: a combined light and electron microscopic study. J. Comp. Neurol. 239(2), 237–246.
Fuster, J. M., 1989. The Prefrontal Cortex: Anatomy, Physiology, and Neuropsychology of the Frontal Lobe. Raven, New York.
Fuster, J. M., 1991. Role of prefrontal cortex in delay tasks: evidence from reversible lesion and unit recording in the monkey. In: Levin, H. S., Eisenberg, H. M., Benton, A. L. (Eds.), Frontal Lobe Function and Dysfunction. Oxford University Press, New York, pp. 59–71.
Fuster, J. M., 1995a. Memory in the Cerebral Cortex: An Empirical Approach to Neural Networks in the Human and Nonhuman Primate. The MIT Press, Cambridge, MA.
Fuster, J. M., 1995b. Memory in the cortex of the primate. Biol. Res. 28, 59–72.
Fuster, J. M., 1997. Network memory. Trends Neurosci. 20, 451–459.
Fuster, J. M., 2003. Cortex and Mind. Unifying Cognition. Oxford University Press, New York.
Folkow, B., 1967. Regional adjustment of intestinal blood flow. Gastroenterology. 52, 423–432.
Folkow, B., Neil, E., 1971. Circulation. Oxford University Press, London.
Foot, C., 1994. Atom Optic-A Heisenberg Microscope. Nature 371, 744–745.
Ford, J. M., 1999. Schizophrenia: the broken P300 and beyond. Psychophysiology 36, 667–682.
Ford, J. M., Krystal, J. H., Mathalon, D. H., 2007. Neural synchrony in schizophrenia: from networks to new treatments. Schizophr. Bull. 33, 848–852.
Ford, J. M., Roach, B., Hoffman, R. S., Mathalon, D. H., 2008. The dependence of P300 amplitude on gamma synchrony breaks down in schizophrenia. Brain Res. 1235, 133–142.
Ford, J. M., White, P., Lim, K. O., Pfefferbaum, A., 1994. Schizophrenics have fewer and smaller P300s: a single-trial analysis. Biol. Psychiatry 35, 96–103.
Galambos, R., 2006. Models and musings about them. Int. J. Psychophysiol. 60, 101–105.
Galambos, R. S., Makeig, P., Talmachoff, A., 1981. 40 Hz auditory potential recorded from the human scalp. Proc. Natl. Acad. Sci. U.S.A. 78, 2643–2647.
Galambos, R., Rose, J. E., Bromley, R. B., Hughes, J. R., 1952. Microelectrode studies on medial geniculate body of the cat II. Response to clicks. J. Neurophysiol. 15, 359–380.
Gallinat, J., Kunz, D., Senkowski, D., Kienast, T., Seifert, F., Schubert, F., Heinz, A., 2006. Hippocampal glutamate concentration predicts cerebral theta oscillations during cognitive processing. Psychopharmacology 187, 103–111.
Gallinat, J., Winterer, G., Herrmann, C. S., Senkowski, D., 2004. Reduced oscillatory gamma-band responses in unmedicated schizophrenic patients indicate impaired frontal network processing. Clin. Neurophysiol. 115, 1863–1874.
Ganong, W. F., 2001. Review of Medical Physiology. McGraw-Hill, New York.
Gardner, E., 1952. Fundamentals of Neurology. Saunders, Philadelphia.
Gebber, G. L., Zhong, S., Barman, S. M., 1995a. The functional significance of the 10-Hz sympathetic rhythm: a hypothesis. Clin. Exp. Hypertens. 17, 181–195.
Gebber, G. L., Zhong, S., Barman, S. M., 1995b. Synchronization of cardiac-related discharges of sympathetic nerves with inputs from widely separated spinal segments. Am. J. Physiol. 268(6 Pt 2), R1472–R1483.
Gebber, G. L., Zhong, S., Lewis, C., Barman, S. M., 1999. Differential patterns of spinal sympathetic outflow involving a 10-Hz rhythm. J. Neurophysiol. 82(2), 841–854.
Gelperin, A., 1989. Perspectives in Neural Systems and Behavior. Liss, New York, pp. 121–136.
Gerschenfeld, H. M., Ascher, P., Tauc, L., 1967. Two different excitatory transmitters acting on a single molluscan neurone. Nature (Lond.) 213, 358.

Gevins, A., 1998. The future of electroencephalography in assessing neurocognitive functioning. Electroencephalogr. Clin. Neurophysiol. 106(2), 165–172.
Gevins, A., Smith, M. E., McEvoy, L., Yu, D., 1997. High-resolution EEG mapping of cortical activation related to working memory: effects of task difficulty, type of processing, and practice. Cereb. Cortex 7, 374–385.
Givens, B., 1996. Stimulus-evoked resetting of the dentate theta rhythm: relation to working memory. Neuroreport 8, 159–163.
Gladwin, T. E., De Jong, R., 2005. Bursts of occipital theta and alpha amplitude preceding alternation and repetition trials in a task-switching experiment. Biol. Psychol. 68, 309–329.
Gleick, J., 1987. Chaos: Making a New Science. Penguin, New York.
Glickstein, M., 1993. Cerebellar function in normal movement and in motorlearning. In: Andersen, P., Hvalby, O., Paulsen, O., Hokfelt, B. (Eds.), Memory Concepts. Elsevier, Amsterdam, pp. 127–135.
Goldman-Rakic, P. S., 1996. Regional and cellular fractionation of working memory. Proc. Natl. Acad. Sci. U.S.A 93, 13473–13480
Goldman-Rakic, P. S., Friedman, H. R., 1991. The circuitry of working memory revealed by anatomy and metabolic imaging. In: Levin, H. S., Eisenberg, H. M., Benton, A. L. (Eds.), Frontal Lobe Function and Dysfunction. Oxford University Press, New York, pp. 72–91.
Goto, Y., O'Donnell, P., 2001. Synchronous activity in the hippocampus and nucleus accumbens in vivo. J. Neurosci. 21, 1–5.
Gönder, A., Başar, E., 1978. Evoked frequency stabilization in the electrical activity of the cat brain. Biol. Cybernet. 31, 193–204.
Grassberger, P., Procaccia, I., 1983. Measuring the strangeness of strange attractors. Physica D. 9, 183–208.
Gray, C. M., Singer, W., 1987. Stimulus specific neuronal oscillations in the cat cortex: a cortical function unit. Soc. Neurosci. 404, 3.
Gray, C. M., Singer, W., 1989. Stimulus-specific neuronal oscillations in orientation columns of cat visual cortex. Proc. Natl. Acad. Sci. U.S.A. 86, 1698–1702.
Greene, B., 2004. The Fabric of the Cosmos; Space, Time, and the Texture of Reality. Vintage Books, New York.
Griffith, J. S., 1971. Mathematical Neurobiology: An Introduction to the Mathematics of the Nervous System. Academic Press, New York.
Grill-Spector, K., Kushnir, T., Edelman, S., Avidan-Carmel, G., Itzchak, Y., Malach, R., 1999. Differential processing of objects under various viewing conditions in the human lateral occipital complex. Neuron 24, 187–203.
Gruzelier, J. (Ed.), 1996. New advances in EEG and cognition. Int. J. Psychophysiol. 1–187.
Guidotti, A., Auta, J., Davis, J. M., DiGiorgi Gerevini, V., Dwivedi, Y., Grayson, D. R., Impagnatiello, F., Pandey, G., Pesold, C., Sharma, R., Uzunov, D., Costa, E., 2000. Decrease in reeling and glutamic acid decarboxylase 67 (GAD67) expression in schizophrenia and bipolar disorder. Arch. Gen. Psychiatry 57, 1061–1069.
Gurtubay, I. G., Alegre, M., Labarga, A., Malanda, A., Iriarte, J., Artieda, J., 2001. Gamma band activity in an auditory oddball paradigm studiedwith the wavelet transform. Clin. Neurophysiol. 112, 1219–1228.
Guyton, A. C., 1971. Textbook of Medical Physiology. Saunders, Philadelphia.
Güntekin, B., Başar, E., 2007a. Gender differences influence brain's beta oscillatory responses in recognition of facial expressions. Neurosci. Lett. 424 (2), 94–99.
Güntekin, B., Başar, E., 2007b. Brain oscillations are highly influenced by gender differences. Int. J. Psychophysiol. 65(3), 294–299.
Güntekin, B., Başar, E., 2007c. Emotional face expressions are differentiated with brain oscillations. Int. J. Psychophysiol. 64, 91–100.
Güntekin, B., Saatçi, E., Yener, G., 2008. Decrease of evoked delta, theta and alpha coherence in Alzheimer patients during a visual oddball paradigm. Brain Res. 1235, 109–116.
Güntekin, B., Başar, E., 2009. Facial affect manifested by multiple oscillations. Int. J. Psychophysiol. 71(1), 31–36.

Güntekin, B., Başar, E., 2010. A new interpretation of P300 responses upon analysis of coherences. Cogn. Neurodyn. 4(2), 107–118.

Haenschel, C., Baldeweg, T., Croft, R. J., Whittington, M., Gruzelier, J., 2000. Gamma and beta frequency oscillations in response to novel auditory stimuli: a comparison of human electroencephalogram (EEG) data with in vitro models. Proc. Natl. Acad. Sci. U.S.A. 97(13), 7645–7650.

Haig, A. R., Gordon, E., De Pascalis, V., Meares, R. A., Bahramali, H., Harris, A., 2000. Gamma activity in schizophrenia: evidence of impaired network binding? Clin. Neurophysiol. 111, 1461–1468.

Haken, H., 1977. Synergetics. An Introduction. Springer, New York.

Haken, H., 2004. Synergetics. Introduction and Advanced Topics. Springer, New York.

Halgren, E., Stapleton, J. M., Smith, M., Altafullah, I., 1986. Generators of human scalp P3. In: Cracco, R. Q., Bodis-Wollner, I. (Eds.), Evoked Potentials. Liss, New York, pp. 269–284.

Hameroff, S., Penrose, R., 1996. Orchestral reduction of quantum coherence in brain microtubules: a model for consciousness. Math. Comput. Simul. 40, 453–480.

Hartline, P. H., 1987. Multisensory convergence. In: Adelman, G. (Ed.), Encyclopedia of Neuroscience. Birkhäuser, Boston, pp. 706–709.

Hassenstein, B., 1971. Information and Control in the Living Organism. Chapman & Hall, London.

Hawkins, R. D., Abrams, T. W., Carew, T. J., Kandel, E. R., 1983. A cellular mechanism classical conditioning in Aplysia: activity dependent amplification of presynaptic facilitation. Science 219, 400–405.

Hawking, S., 1988. A Brief History of Time: From the Big Bang to Black Holes. Bantam, New York.

Hawking, S., 2001. The Universe in a Nutshell. Bantam, New York, p. 52.

Hayek, F. A. (Ed.), 1952. The Sensory Order. University of Chicago Press, Chicago.

Hebb, D.O., 1949. The Organization of Behaviour. Wiley, New York.

Hebb, D.O., 1951. The role of neurological ideas in psychology. J. Pers. 20(1), 39–55.

Heckers, S., Stone, D., Walsh, J., Shick, J., Koul, P., Benes, F., 2002. Differential expression of glutamic acid decarboxylase 65 and 67 messenger RNA in bipolar disorder and schizophrenia. Arch. Gen. Psychiatry 59, 521–529.

Heisenberg, W., 1943. Die "beobachtbaren Grö en" in der Theorie der Elementarteilchen (The "Observable Quantities" in the Theory of Elementary Particles). Z. Phys.

Heisenberg, W., 1961. Einführung in die Theorie der Elementarteilchen (Ausarbeitung einer Vorlesung, gehalten im Sommersemester 1961). University München, München, FRG.

Herrmann, C. S., Demiralp, T., 2005. Human EEG gamma oscillations in neuropsychiatric disorders. Clin Neurophysiol. 116, 2719–2733.

Herrmann, C. S., Matthias, H. J., Engel, A. K., 2004. Cognitive functions of gamma-band activity: memory match and utilization. Trends Cogn. Sci. 8, 8347–8355.

Hiripi, L., Salanki, J., 1973. Seasonal and activity-dependent changes of serotonin levels in CNS and heart of snail (Helix pomatia). Comp. Gen. Pharmacol. 4, 285–292.

Hobson, J. A., McCarley, R., 1977. The brain as a dream state generator: an activation-synthesis hypothesis of the dream process. Am. J. Psychiatry 134, 1335–1348.

Hobson, J. A., Pace-Schott, E. F., Stickgold, R., 2000b. Dream science 2000: a response to commentaries on dreaming and the brain. Behav. Brain Sci. 23, 1019–1034.

Hogan, M. J., Swanwick, G. R., Kaiser, J., Rowan, M., Lawlor, B., 2003. Memory-related EEG power and coherence reductions in mild Alzheimer's disease. Int. J. Psychophysiol. 49, 147–163.

Hooper, H. E., 1983. In: The Hooper Visual Organization Test 1983 edition: Manual. Western Psychological Services, Los Angeles.

Housner, G. W., Hudson, D. E., 1950. Applied Mechanics Dynamics. D. Van Nostrand, Princeton, NJ.

Howard, L., Polich, J., 1985. P300 latency and memory span development. Dev. Psychol. 21, 283–289.

Huang, M., Li, Z., Ichikawa, J., Dai, J., Meltzer, H. Y., 2006. Effects of divalproex and atypical antipsycothic drugs on dopamine and actylcholine efflux in rat hippocampus and prefrontal cortex. Brain Res. 1099, 44–55.
Hughes, S. W., Crunelli, V., 2005. Thalamic mechanisms of EEG alpha rhythms and their pathological implications. Neuroscientist. 11, 357–372.
Hughes, S. W., Crunelli, V., 2007. Just a phase they're going through: The complex interaction of intrinsic high-threshold bursting and gap junctions in the generation of thalamic alpha and theta rhythms. Int. J. Psychophysiol. 64, 3–17.
Huxley, A., 1932. Brave New World. HarperCollins, London.
Iagolnitzer, D., Barut, A., 1967. The Theory of the Scattering Matrix. Macmillan, New York.
Isoglu-Alkaç, U., Başar-Eroğlu, C., Ademoğlu, A., Demiralp, T., Miener, M., Stadler, M., 2000. Alpha activity decreases during the perception of Necker cube reversals: an application ofwavelet transform. Biol. Cybernet. 82(4), 313–320.
Jacobson, E. D., 1967. The gastrointestinal circulation (Symposium). Gastroenterology 52, 332–471.
James, W., 1890. The Principles of Psychology. Holt, New York.
Jarcho, L. W., 1949. Excitability of cortical afferent systems during barbiturate anesthesia. J. Neurophysiol. 12, 447–457.
Jasper, H. H., 1958. The ten-twenty electrode system. Electroencephalogr. Clin. Neurophysiol. 10, 371–375.
Jensen, O., Tesche, C. D., 2002. Frontal theta activity in humans increases with memory load in a working memory task. Eur. J. Neurosci. 15(8), 395–399.
Jensen, O., Kaiser, J., Lachaux, J. P., 2007. Human gamma-frequency oscillations associated with attention and memory. Trends Neurosci. 30(7), 317–324.
Jeon, Y. W., Polich, J., 2003. Meta-analysis of P300 and schizophrenia: patients, paradigms, and practical implications. Psychophysiology 40(5), 684–701.
Johnson, S. L., 2005. Mania and dysregulation in goal pursuit: a review. Clin. Psychol. Rev. 25 (2), 241–262.
Jones, K. A., Porjesz, B., Chorlian, D., Rangaswamy, M., Kamarajan, C., Padmanabhapillai, A., Stimus, A., Begleiter, H., 2006. S-transform time-frequency analysis of P300 reveals deficits in individuals diagnosed with alcoholism. Clin. Neurophysiol. 117, 2128–2143.
Jung, C. G., 1951a. Aion: Researches into the Phenomenology of the Self (Collected Works Vol. 9 Part 2). Bollingen, Princeton, NJ.
Jung, C. G., 1951b. The Psychological Aspects of the Kore. Hull, R. F. C. (Trans.), 1959. In: Read, S. H., Fordham, M., Adler, G. (Eds.), The Archetypes and the Collective Unconscious (2nd ed., Vol. 9(I), pp. 182–203). Princeton University Press, Princeton, NJ.
Jung, C. G., 1959. Archetypes and the Collective Unconscious. Hull, R.F. C. (Trans.), 1964. Pantheon, New York.
Jung, R., 1963. Hirnpotentialwellen, Neuronentladungen und Gleichspannungsphänomene. In: Werner, R., Jenenser EEG-Symposium "30 Jahre Elektroenzephalographie." Volk und Gesundheit, Berlin.
Kaiya, H., Namba, M., Yoshida, H., Nakamura, S., 1982. Plasma glutamate decaroxylase activity in neuropsychiatry. Psych. Res. 6, 335–343.
Kamarajan, C., Porjesz, B., Jones, K. A., Choi, K., Chorlian, D. B., Padmanabhapillai, A., Rangaswamy, M., Stimus, A. T., Begleiter, H., 2004. The role of brain oscillations as functional correlates of cognitive systems: a study of frontal inhibitory control in alcoholism. Int. J. Psychophysiol. 51, 155–180.
Kandel, E. R., 1976. Cellular Basis of Behavior: An Introduction to Behavioral Neurobiology. W. H. Freeman, San Francisco.
Kandel, E. R., 2006. In Search of Memory: The Emergence of a New Science of Mind. W. W. Norton, New York.
Kandel, E. R., Schwartz, J. H., 1982. Molecular biology of learning: modulation of transmitter release. Science 218, 433–443.
Kant, I., 1787. Critique of Pure Reason.

Karakaş, S., Başar, E., 1998. Early gamma response is sensory in origin: a conclusion based on crosscomparison of results from multiple experimental paradigms. Int. J. Psychophysiol. 31, 13–31.

Karakaş, S., Başar, E. (Eds.), 2006. Models and theories of brain function with special emphasis on cognitive processing. Int. J. Psychophysiol. Special Issue 60(2), 186–193.

Karakaş, S., Bekci, B., Erzengin, O. U., 2003. Early gamma response in human neuroelectric activity is correlated with neuropsychological test scores. Neurosci. Lett. 340(1), 37–40.

Karakaş, S., Erzengin, O. U., Başar, E., 2000a. The genesis of human event related responses explained through the theory of oscillatory neural assemblies. Neurosci. Lett. 285, 45–48.

Karakaş, S., Erzengin, O. U., Başar, E., 2000b. A new strategy involving multiple cognitive paradigms demonstrates that ERP components are determined by the superposition of oscillatory responses. Clin. Neurophysiol. 111, 1719–1732.

Karakaş, S., Karakaş, H., Erzengin, O. U., 2002. Early sensory gamma represents the integration of bottom-up and top-down processing, Int. J. Psychophysiol. 45, 39.

Karrasch, M., Laine, M., Rinne, J. O., Rapinoja, P., Sinerva, E., Krause, C. M., 2006. Brain oscillatory responses to an auditory–verbal working memory task in mild cognitive impairment and Alzheimer's disease. Int. J. Psychophysiol. 59(2), 168–178.

Katada, A., Ozaki, H., Suzuki, H., Suhara, K., 1981. Developmental characteristics of normal and mentally retarded children's EEGs. Electroencephalogr. Clin. Neurophysiol. 52, 192–201.

Katchalsky, A., Rowland, V., Blumenthal, R. (Eds.), 1974. Dynamic patterns of brain cell assemblies. Neurosci. Res. Progr. Bull. 12, 3–87.

Kelso, J. A. S., Engstrom, D. A., 2006. The Complementary Nature. MIT Press, Cambridge, MA.

Kelly, J. P., Dodd, J., 1991. Anatomical organization of the nervous system. In: Kandel, E. R., Schwartz, J. H., Jessell, T. M. (Eds.), Principles of Neural Science. Elsevier, Amsterdam, pp. 273–282.

Kenner, T., Ono, K., 1972. Analysis of slow autooscillations of arterial flow. Pflügers Arch. 331, 347–356.

Kerkut, G. A., Lambert, J. D. C., Gayton, R. J., Loker, J. E., Walker, R. J., 1975. Mapping of nerve cells in the subesophageal ganglia of helix aspersa. Comp. Biochem. Physiol. 50A, 1–25.

Kirk, I. J., Mackay, J. C., 2003. The role of theta-range oscillations in synchronizing and integrating activity in distributed mnemonic networks. Cortex 39, 993–1008.

Kirk, I. J., McNaughton, N., 1993. Mapping the differential effects of procaine on frequency and amplitude of reticularly elicited hippocampal rhythmical slow activity. Hippocampus 3(4), 517–525.

Klimesch, W., 1996. Memory processes, brain oscillations and EEG synchronization. Int. J. Psychophysiol. 24(1–2), 61–100.

Klimesch, W., 1999. EEG alpha and theta oscillations reflect cognitive and memory performance: a review and analysis. Brain Res. Rev. 29, 169–195.

Klimesch, W., Doppelmayr, M., Pachinger, T., Ripper, B., 1997a. Brain oscillations and human memory performance: EEG correlates in the upper alpha and theta bands. Neurosci. Lett. 238, 9–12.

Klimesch, W., Doppelmayr, M., Röhm, D., Pöllhuber, D., Stadler, W., 2000a. Simultaneous desynchronization and synchronization of different alpha responses in the human electroencephalograph: a neglected paradox? Neurosci. Lett. 284, 97–100.

Klimesch, W., Doppelmayr, M., Russegger, H., Pachinger, T., Schwaiger, J., 1998. Induced alpha band power changes in thehuman EEG and attention. Neurosci. Lett. 244(2), 73–76.

Klimesch, W., Doppelmayr, M., Schimke, H., Ripper, B., 1997b. Theta synchronization and alpha desynchronization in a memory task. Psychophysiology 34, 169–176.

Klimesch, W., Doppelmayr, M., Schwaiger, J., Winkler, T., Gruber, W., 2000b. Theta oscillations and the ERP old/new effect: independent phenomena? Clin. Neurophysiol. 111, 781–793.

Klimesch, W., Freunberger, R., Sauseng, P., Gruber, W., 2008. A short review of slow phase synchronisation and memory: evidence for control processes in different memory systems? Brain Res. 1235, 31–44.

Klimesch, W., Sauseng, P., Hanslmayr, S., Gruber, W., Freunberger, R., 2007. Event-related phase reorganization may explain evoked neural dynamics. Neurosci. Biobehav. Rev. 31(7), 1003–1016.
Klimesch, W., Schack, B., Schabus, M., Doppelmayr, M., Gruber, W., Sauseng, P., 2004. Phase-locked alpha and theta oscillations generate the P1–N1 complex and are related to memory performance. Cogn. Brain Res. 19, 302–316.
Klimesch, W., Schimke, H., Schwaiger, J., 1994. Episodic and semantic memory: an analysis in the EEG theta and alpha band. Electroencephalogr. Clin. Neurophysiol. 91, 428–441.
Kocsis, B., Viana Di Prisco, G., Vertes, R. P., 2001. Theta synchronization in the limbic system: The role of Gudden's tegumental nuclei. Eur. J. Neurosci. 13, 381–388.
Kolev, V., Yordanova, J., 1997. Analysis of phase-locking is informative for studying event-related EEG activity. Biol. Cybernet. 96, 229–235.
Kolev, V., Başar-Eroğlu, C., Aksu, F., Başar, E., 1994. EEG rhythmicities evoked by visual stimuli in three-year-old children. Int. J. Neurosci. 75(3–4), 257–270.
Kolev, V., Yordanova, J., Schürmann, M., Başar, E., 1999. Event-related alpha oscillations in task processing. Clin. Neurophysiol. 110, 1784–1792.
Konopacki, J., 1998. Theta-like activity in the limbic cortex in vitro. Neurosci. Biobehav. Rev. 22 (2), 311–323.
Krause, C. M., Sillanmaki, L., Haggqvist, A., Heino, R., 2001. Test–retest consistency of the event-related desynchronization/event-related synchronization of the 4–6, 6–8, 8–10 and 10–12 Hz frequency bands during a memory task. Clin. Neurophysiol. 112, 750–757.
Kunze, H., 1917. Uber den Aufban des Centralnervensystems vonHelix pomatia L. und die Struktur seiner elemente. Zool. Anzeiger. 48, 232–240.
Kunze, H., 1921. Zur Topographie und Histologie des Centralnervensystems von Helix pomatia. L.Z. Wiss Zool. 118, 25–203.
Kupfermann, I., 1991. Localization of higher cognitive and affective functions. In: Kandel, E., Schwartz, J., Jessell, T. (Eds.), Principles of neural science. Elsevier, Amsterdam, pp. 823–838.
Kwon, J. S., O'Donnell, B. F., Wallenstein, G. V., Greene, R. W., Hirayasu, Y., Nestor, P. G., Hasselmo, M. E., Potts, G. F., Shenton, M. E., Mc Carley, R. W., 1999. Gamma frequency-range abnormalities to auditory stimulation in schizophrenia. Arch. Gen. Psychiatry 56, 1001–1005.
Lang, P. J., 1980. Behavioral treatment and bio-behavioral assessment: computer applications. In: Sidowski, J. B., Johnson, J. H., Williams, T. A. (Eds.), Technology in Mental Health Care Delivery Systems. Ablex, Norwood, NJ, pp. 119–137.
Laplace, P.S., 1878. Oeuvres complètes de Laplace. Gauthier-Villars, Paris.
Lashley, K. S. (Ed.), 1929. Brain Mechanisms and Intelligence: a Quantitative Study of Injuries to the Brain. University of Chicago Press, Chicago.
Layne, S. P., Mayer-Kressm, G., Holzfuss, J., 1986. Problems associated with dimensional analysis of electroencephalogram data. In: Mayer-Kress, G. (Ed.), Dimensions and Entropies in Chaotic Systems. Springer, New York, pp. 246–256.
Le Doux, J. E., 1999. Emotion, memory, and the brain. In: Damasio, A. (Ed.), The Sci. Am. Book of the Brain. Lyons, Guilford, CT, pp. 105–117.
Lee, K. H., Williams, L. M., Breakspear, M., Gordon, E., 2003a. Synchronous gamma activity: a review and contribution to an integrative neuroscience model of schizophrenia. Brain Res. Rev. 41, 57–78.
Lee, K. H., Williams, L. M., Haig, A., Gordon, E., 2003b. Gamma (40 Hz) phase synchronicity and symptom dimension in schizophrenia. Cogn. Neuropsychiatry 8, 57–71.
Lehrer, J., 2007. Proust Was a Neuroscientist. Mariner Books, Boston.
Leiberg, S., Lutzenberger, W., Kaiser, J., 2006. Effects of memory load on cortical oscillatory activity during auditory pattern working memory. Brain Res. 1120, 131–140.
Lenz, D., Krauel, K., Schadow, J., Baving, L., Duzel, E., Herrmann, C. S., 2008. Enhanced gamma-band activity in ADHD patients lacks correlation with memory performance found in healthy children. Brain Res. 1235, 117–132.
Leuchter, A. F., Spar, J. E., Walter, D. O., Weiner, H., 1987. Electroencephalographic spectra and coherence in the diagnosis of Alzheimer's-type and multiinfarct dementia: a pilot study. Arch. Gen. Psychiatry 44, 993–998.

Leung, L. S., Harvey, G. C., Vanderwolf, C. H., 1982. Combined video and computer analysis of the relation between the hemispheric response and behavior. Behav. Brain Res. 2, 195–200.

Levinson, A. J., Young, L. T., Fitzgerald, P. B., Daskalakis, Z. J., 2007. Cortical inhibitory dysfunction in bipolar disorder. A study using transcranial magnetic stimulation. J. Clin. Psychopharmacol. 27, 493–497.

Levitan, I. B., Kaczmarek, L. K., 2002. The Neuron: Cell and Molecular Biology. Oxford University Press, New York.

Libet, B., 1991. Control of the transition from sensory detection to sensory awarenes in man by the duration of a thalamic stimulus. Brain 114, 1731–1757.

Lieb, J., Sclabassi, R., Crandall, P., Buchness, R., 1974. Comparison of the action of diazepam and phenobarbital using EEG-derived power spectra obtained from temporal lobe epileptics. Neuropharmacology 13, 769–784.

Light, G. A., Hsu, J. L., Hsieh, M. H., Meyer-Gomes, K., Sprock, J., Swerdlow, N. R., Braff, D. L., 2006. Gamma band oscillations reveal neural network cortical coherence dysfunction in schizophrenia patients. Biol. Psychiatry 60, 1231–1240.

Llinas, R. R., 1988. The intrinsic electrophysiological properties of mammalian neurons: insights into central nervous system function. Science 242(4886), 1654–1664.

Locke, J., 1690. An Essay Concerning Human Understanding. In Four Books. Basset, London.

Locatelli, T., Cursi, M., Liberati, D., Franceschi, M., Comi, G., 1998. EEG coherence in Alzheimer's disease. Electroencephalogr. Clin. Neurophys. 106, 229–237.

Loewi, O., 1921. Über humorale Übertragbarkeit der Herznervenwirkung. I. Pflügers Arch. 189, 239–242.

Lopes da Silva, F. H., Kamphius, W., Van Neeven, J. M. A. M., Pijn, J. P. M, 1990. Cellular & network mechanism in the kindling model of epilepsy; the role of GABAergic inhibition and the emergence of strange attractors. In: John, E. R., Harmony, T., Prichep, L., Valdes-Sosa, M., Valdes-Sosa, P. (Eds.), Machinery of the Mind. Birkhauser, Boston, pp. 115–139.

Lopes da Silva, F. H., Vos, J. E., Mooibroek, J., Van Rotterdam, A., 1980. Relative contributions of intracortical and thalamo-cortical processes in the generation of alpha rhythms, revealed by partial coherence analysis. Electroencephalogr. Clin. Neurophysiol. 50(5–6), 449–456.

Lorenz, E. N., 1963. Deterrninistic nonperiodic flow. Atmos. Sci. 20, 130.

Löscher, W., 2002. Basic pharmacology of valproate: a review after 35 years of clinical use for the treatment of epilepsy. CNS Drugs 16(10), 669–694.

Lundgren, O., 1967. Studies on blood flow distribution and countercurrent exchange in the small intestine of the cat. Acta. Physiol. Scand. Suppl. 303.

Lundgren, O., Wallentin, I., 1964. Local chemical and nervous control of consecutive vascular sections in the mesenteric lymph nodes of the cat. Angiologica 1, 284–296.

Luria, A. R., 1966. Higher Cortical Functions in Man. Basic Books, New York.

Lutz, I., 1966. Die Reaktion der Gefabmuskulatur in situ auf rhythmische Dehnungsreize. Pflügers Arch. 291, R22.

Lutz, J., 1978. Autonomic vascular response in the splanchnic region. In: Bauer, R. B., Buse, R. (Eds.), The Arterial System. Dynamics, Control Theory and Regulation. Springer, New York.

MacKay, D. M., 1970. Perception and brain function. In: Schmitt, F. O. (Ed.), The Neurosciences: Second Study Program. The Rockeffeller University Press, New York, pp. 303–316.

Magill, P. J., Bolam, J. P., Bevan, M. D., 2000. Relationship of activity in the subthalamic nucleus-globus pallidus network to cortical electroencephalogram. J. Neurosci. 20, 820–833.

Makeig, S., Westerfield, M., Jung, T. P., Enghoff, S., Townsend, J., Courchesne, E., Sejnowski, T. J., 2002. Dynamic brain sources of visual evoked responses. Science 295, 690–694.

Maltseva, I., Geissler, H. G., Başar, E., 2000. Alpha oscillations as an indicator dynamic memory operations: anticipation of omitted stimuli. Int. J. Psychophysiol. 36, 185–197.

Marksteiner, J., Hinterhuber, H., Humpel, C., 2007. Cerebrospinal fluid biomarkers for diagnosis of Alzheimer's disease: beta-amyloid (1–42), tau, phospho-tau-181 and total protein. Drugs Today 43(6), 423–431.

Martindale, C., Hines, D., 1975. Creativity and cortical activation during creative, intellectual, and EEG feedback tasks. Biol. Psychol. 3, 71–80.

Martínez-Arán, A., Vieta, E., Colom, F., Torrent, C., Sanchez-Moreno, J., Reinares, M., Benabarre, A., Goikolea, J.M., Brugue, E., Daban, C., Salamero, M., 2004a. Cognitive impairment in euthymic bipolar patients: implications for clinical and functional outcome. Bipolar Disord. 6, 224–232.

Martínez-Arán, A., Vieta, E., Reinares, M., Colom, F., Torrent, C., Sánchez-Moreno, J., Benabarre, A., Goikolea, J.M., Comes, M., Salamero, M., 2004b. Cognitive function across manic or hypomanic, depressed, and euthymic states in bipolar disorder. Am. J. Psychiatry 161, 262–270.

Mathes, B., Strüber, D., Stadler, M. A., Başar-Eroğlu, C., 2006. Voluntary control of Necker cube reversals modulates the EEG delta- and gamma-band response. Neurosci. Lett. 402(1–2), 145–149.

Maxwell, J.C., 1981. Theory of Heat. Greenwood, Westport, CT.

McCarthy, G., 2000. Physiological studies of face processing in humans. In: Gazzaniga, M. S. (Ed.), The New Cognitive Neurosciences. MIT Press, Cambridge, MA, pp. 393–409.

Meador, K. J., Thompson, J. L., Loring, D. W., Murro, A. M., King, D. W., Gallagher, B. B., Lee, G. P., Smith, J. R., Flanigin, H. F., 1991. Behavioral state-specific changes in human hippocampal theta activity. Neurology 41, 869–872.

Mees, A. I., Rapp, P. E., Jennings, L. S., 1987. Singular-value decomposition and embedding dimension. Phys. Rev. A 36, 340.

Mesulam, M. M., 1990. Large scale neurocognitive networks and distributed processing for attention, language, and memory. Ann. Neurol. 28, 597–613.

Mesulam, M. M., 1994. Neurocognitive networks and selectively distributed processing. Rev. Neurol. (Paris) 150, 564–569.

Miller, E. K., 2000. The prefrontal cortex and cognitive control. Nat. Rev. Neurosci. 1, 59–65.

Miller, R., 1989. Cortico-hippocampal interplay: self organizing phase-locked loops for indexing memory. Psychobiology 17, 115–128.

Miller, R., 1991. Cortico-Hippocampal Interplay and the Representation of Contexts in the Brain. Springer, Berlin.

Miltner, W., Braun, C., Arnold, M., Witte, H., Taub, E., 1999. Coherence of gamma-band EEG activity as a basis for associative learning. Nature 397, 434–436.

Missonnier, P., Gold, G., Herrmann, F. R., Fazio-Costa, L., Michel, J. P., Deiber, M. P., Michon, A., Giannakopoulos, P., 2006. Decreased theta event-related synchronization during working memory activation is associated with progressive mild cognitive impairment. Dement. Geriatr. Cogn. Disord. 22(3), 250–259.

Mitrofanis, J., Guillery, R. W., 1993. New views of the thalamic reticular nucleus in the adult and the developing brain. Trends Neurosci. 16, 240–245.

Monod, J., 1970. Le Hasard et la Nécessité. Editions du Seuil, Paris.

Monod, J., 1971. Chance and Necessity: An Essay on the Natural Philosophy of Modern Biology. Alfred A. Knopf, New York.

Moore, J. K., Osen, K. K., 1979. The human cochlear nuclei. In: Creutzfeldt, O., Scheich, H., Schreiner, C. (Eds.), Hearing Mechanisms and Speech. Springer, Berlin, pp. 36–44.

Mountcastle, V. B., 1976. The world around us: neural command functions for selective attention. Neurosci. Res. Progr. Bull. 14, 1–47.

Mountcastle, V. B., 1992. Preface. In: Başar, E., Bullock, T. H. (Eds.), Induced Rhythms in the Brain. Birkhäuser, Boston, pp. 217–231.

Mountcastle, V. B., 1998. Perceptual Neuroscience: The Cerebral Cortex. Harvard University Press, Cambridge, MA.

Mrzljak, L., Uylings, H. B. M., Van Eden, C. G., Judaš, M., 1990. Neuronal development in human prefrontal cortex in prenatal and postnatal stages. Prog. Brain Res. 85, 185–222.

Munk, M. H., Reolfsema, P. R., König, P., Engel, A.K., Singer, W., 1996. Role of reticular activation in the modulation of intracortical synchronisation. Science 272, 271–274.

Näätänen, R., 1992. Attention and Brain Function. LEA, Hillsdale, NJ.

De Nabias, B., 1894. Recherches histologiques et organologiques sur les centres nerveux des Gasteropodes. Act. Soc. Linn., Bordeaux.

Narici, L., Pizzella, V., Romani, G. L., Torrioli, G., Traversa, R., Rossini, P. M., 1990. Evoked alpha and mu rhythms in humans: a neuromagnetic study. Brain Res. 520, 222–231.

Neuper, C., Pfurtscheller, G., 1998a. Event-related desynchronization (ERD) and synchronization (ERS) of rolandic EEG rhythms during motor behavior. Int. J. Psychophysiol. 30(1–2), 7–8.

Neuper, C., Pfurtscheller, G., 1998b. 134 ERD/ERS based brain computer interface (BCI): effects of motor imagery on sensorimotor rhythms. Int. J. Psychophysiol. 30(1–2), 53–54.

Newton, R. G., 2004. Galileo's Pendulum: From the Rhythm of Time to the Making of the Matter. Harvard University Press, Cambridge, MA, pp. 1–153.

Niedermeyer, E., 1993. The normal EEG of the waking adult. In: Niedermeyer, E., Lopes Da Silva, F. H. (Eds.), Electroencephalography: Basic Principles, Clinical Applications and Related Fields, 3rd ed. Williams & Wilkins, Baltimore, pp. 131–152.

Noe, A., Thompson, E., 2004. Are there neural correlates of consciousness? J. Conscious. Stud. 11, 3–28.

Nowakowska, C., Strong, C., Santosa, C., Wang, P., Ketter, T., 2005. Temperamental commonalities and differences in euthymic mood disorder patients, creative controls, and healthy controls. J. Affect. Disord. 85, 207–215.

Noyan, A., 1980. Fizyoloji Ders Kitabı. Anadolu Ünv. Tıp Fak. Anadolu Ünv. Yayınları (2).

Nunez, P. L., 1997. EEG coherence measures in medical and cognitive science: a general overview of experimental methods, computer algorithms, and accuracy. In: Eselt, M., Zwiener, U., Witte, H. (Eds.), Quantative and Topological EEG and MEG Analysis. Universitätsverlag Druckhaus Mayer-Jena.

O'Donnell, B. F., Hetrick, W. P., Vohs, J. L., Drishnan, P., Carroll, C. A., Shekha, A., 2004. Neural synchronization deficits to auditory stimulation in bipolar disorder. Neuroreport 15, 1369–1372.

O'Donnell, T. O., Rotzinger, S., Ulrich, M., Hanstock, C. C., Nakashima, T. T., Silverstone, P. H., 2003. Effects of chronic lithium and sodium valproate on concentrations of brain amino acids. Eur. Neuropsychopharmacol. 13, 220–227.

Obrist, W., 1976. Problems of aging. In: Remond, A. (Ed.), Handbook of Electroencephalogr. Clin. Neurophysiol. 6A, 275–290.

Oddie, S. D., Bland, B. H., Colom, L. V., Vertes, R. P., 1994. The midline posterior hypothalamic region comprises a critical part of the ascending brainstem hippocampal synchronizing pathway. Hippocampus 4(4), 454–473.

Onton, J., Delorme, A., Makeig, S., 2005. Frontal midline EEG dynamics during working memory. Neuroimage 27(2), 341–356.

Öniz, A., Başar, E., 2009. Prolongation of alpha oscillations in auditory oddball paradigm. Int. J. Psychophysiol. 71(3), 235–241.

Ösby, U., Brandt, L., Correia, N., Ekbom, A., Sparén, P., 2001. Excess mortality in bipolar and unipolar disorder in Sweden. Arch. Gen. Psychiatry 58(9), 844–850.

Özerdem, A., Güntekin, B., Tunca, Z., Başar, E., 2008. Brain oscillatory responses in patients with bipolar disorder manic episode before and after valproate treatment. Brain Res. 1235, 98–108.

Özerdem, A., Kocaaslan, S., Tunca, Z., Başar, E., 2007. Effect of Valproate on oscillatory delta frequency resoponses to visual stimuli in a group of euthymic bipolar patients in comparison to healthy controls. Biol. Psychiatry 61, 226S.

Özerdem, A., Güntekin, B., Saatçi E, Tunca, Z., Başar, E., 2010. Prog Neuropsychopharmacol Biol Psychiatry. 2010 Aug 16;34(6):861–5. Epub 2010 Apr 14.

Özgören, M., Başar-Eroğlu, C., Başar, E., 2005. Beta oscillations in face recognition. Int. J. Psychophysiol. 55, 51–59.

Papanicolaou, A. C., Gunter, P. A. Y. (Eds.), 1987. Bergson and Modern Thought: Towards a Unified Science. Harwood, New York.

Pandya, D. N., 1987. Association cortex. In: Adelman, G. (Ed.), Encyclopedia neuroscience. Vol. 2. Birkhauser, Boston, pp. 80–83.

Pandya, D. N., Van Hoesen, G. W., Mesulam, M. M., 1979. The cortical projections of the cingulate gyrus in the rhesus monkey. Anat. Rec. 193, 643–644.

Panek, R., 2004. The Invisible Century: Einstein, Freud, and the Search for Hidden Universes. Viking, New York.

Pantev, C., Makeig, S., Hoke, M., Galambos, R., Hampson, S., Gallen, C., 1991. Human auditory evoked gamma-band magnetic fields. Proc. Natl. Acad. Sci. U.S.A. 88, 8996–9000.

Pavlov, I. P., 1927. Conditioned Reflexes: An Investigation of the Physiological Activity of the Cerebral Cortex. Oxford University Press, London.

Pascal, B., 1657. De l'Esprit géométrique.

Pascal, B., 1660. Pensees. Trotter, W. F. (trans.), Eliot, T. S. (introduction), 2003. Dover, New York.

Perkel, D. H., Bullock, T. H., 1968. Neural coding. Neurosci. Res. Progr. Bull. 6, 221–348.

Perry, E., Walker, M., Grace, J., Perry, R., 1999. Acetylcholine in mind: a neurotransmitter correlate of consciousness? Trends Neurosci. 22, 273–280.

Peňáz, J., Buriánek, P., Semrád, B., 1968. Dynamic aspects of vasomotor and autoregulatory control of blood flow. In: Hudlicka, O. (Ed.), Circulation in Skeletal Muscle. Pergamon, Oxford, UK.

Penrose, R., 1989. The Emperor's New Mind: Concerning Computers, Minds and the Laws of Physics. Oxford University Press, New York.

Penttonen, M., Nurmınen, N., Mıettınen, R., Sivri, O. J., Henze, D. A., Csicsvari, J., Buzsaki, G., 1999. Ultra-slow oscillation (0.025 Hz) triggers hippocampal afterdischarges in Wistar rats. Neuroscience 94, 735–743.

Petersen, I., Eeg-Olofsson, O., 1971. The development of the electroencephalogram in normal children from the age of 1 through 15 years. Neuropädiatrie. 3, 247–304.

Petsche, H., Rappelsberger, P., Pockberger, H., 1987. EEG-Veranderungen beim Lesen. In: Weinmann, H. M. (Ed.), Zugang zum Verstandnis hoherer Hirnfunktionen durch das EEG. Zuckschwerdt, Munchen, pp. 59–74.

Petsche, H., Etlinger, S. C., 1998. EEG and Thinking: Power and Coherence Analysis of Cognitive Processes. Verlag Der Österreichischen Akademie Der Wissenscaften, Wien.

Petsche, H., Stumpf, C., Gogolak, G., 1962. The significance of the rabbit's septum as a relay station between the midbrain and the hippocampus. 1. The control of hippocampal arousal activity by the septum cells. Electroencephalogr. Clin. Neurophysiol. 14, 202–211.

Petty, F., 1995. GABA and mood disorders: a brief review and hypothesis. J. Affect. Disord. 34, 275–281.

Pfurtscheller, G., 1997. EEG event-related desynchronization (ERD) and synchronization (ERS). Electroencephalogr. Clin. Neurophysiol. 103(1), 26.

Pfurtscheller, G., 2001. Functional brain imaging based on ERD/ERS. Vision Res. 41(10–11), 1257–1260.

Pfurtscheller, G., Brunner, C., Schlogl, A., Lopes da Silva, F. H., 2006. Mu rhythm, (de)synchronization and EEG single-trial classification of different motor imagery tasks. Neuroimage 31(1), 153–159.

Pfurtscheller, G., Neuper, C., Andrew, C., Edlinger, G., 1997. Foot and hand area mu rhythms. Int. J. Psychophysiol. 26, 121–135.

Pfurtscheller, G., Neuper, C., Ramoser, H., Müller-Gerking, J., 1999. Visually guided motor activates sensorimotor areas in humans. Neurosci Lett. 269(3), 153–156.

Pikovski, A., Rosenblum, M., Kurths, J., 2001. Synchronization. Cambridge Nonlinear Science Series 12. Cambridge University Press, Cambridge, MA, pp. 1–411.

Pinsker, H., Kupfermann, I., Castellucci, V. F., Kandel, E. R., 1970. Habituation and dishabituation of the gill-withdrawal reflex in Apbeia. Science 167, 1740–1742.

Platt, C. J., Bullock, T. H., Czéh, G., Kovaevi, N., Konjevi, D. J., Gojkovi, M., 1974. Comparison of electroreceptor, mechanoreceptor and optic evoked potentials in the brain of some rays and sharks. J. Comp. Physiol. 95, 323–355.

Pratt, H., Michalewski, H. J., Barrett, G., Starr, A., 1989. Brain potentials in a memory-scanning task: I. Modality and task effects on potentials to the probes. Electroencephalogr. Clin. Neurophysiol. 72, 407–421.

Pribram, K. H., 1963. The new neurology: memory, novelty, thought and choice. In: Pribram, K. H. (Ed.), Brain and Behaviour 3-Memory Mechanisms (1969). Penguin Education, Australia.

Prigogine, I., 1980. From Being to Becoming: Time and Complexity in the Physical Sciences. Freeman, New York.

Poincaré, H., 1892. Thermodynamique. Publié par. J. Blondin, Paris, Georges Carré.

Popper, K., 1935. Logik der Forschung. Verlag von Julius Springer, Vienna.

Porjesz, B., Begleiter, H., 1996. Effects of alcohol on electrophysiological activity of the brain. In: Begleiter, H., Kissin, B. (Eds.), Alcohol and Alcoholism: The Pharmacology of Alcohol and Alcohol Dependence. Oxford University Press, New York, pp. 207–247.

Porjesz, B., Begleiter, H., 2003. Alcoholism and human electrophysiology. Alcohol Res. Health 27, 153–160.

Porjesz, B., Almasy, L., Edenberg, H. J., Wang, K., Chorlian, D. B., Foroud, T., Goate, A., Rice, J. P., O'Connor, S. J., Rohrbaugh, J., Kuperman, S., Bauer, L. O., Crowe, R. R., Schuckit, M. A., Hesselbrock, V., Conneally, P. M., Tischfield, J. A., Li, T. K., Reich, T., Begleiter, H., 2002. Linkage disequilibrium between the beta frequency of the human EEG and a GABA receptor gene locus. Proc. Natl. Acad. Sci. U.S.A. 99, 3729–3733.

Porjesz, B., Begleiter, H., Reich, T., Van Eerdewegh, P., Edenberg, H. J., Foroud, T., Goate, A., Litke, A., Chorlian, D. B., Stimus, A., Rice, J., Blangero, J., Almasy, L., Sorbell, J., Bauer, L. O., Kuperman, S., O'Connor, S. J., Rohrbaugh, J., 1998. Amplitude of visual P3 event related potential as a phenotypic marker for a predisposition to alcoholism: preliminary results from the COGA Project. Collaborative study on the genetics of alcoholism. Alcohol. Clin. Exp. Res. 22, 1317–1323.

Porjesz, B., Rangaswamy, M., Kamarajan, C., Jones, K. A., Padmanabhapillai, A., Begleiter, H., 2005. The utility of neurophysiological markers in the study of alcoholism. Clin. Neurophysiol. 116(5), 993–1018.

Porjesz, B., Rangaswamy, M., 2007. Neurophysiological endophenotypes, CNS disinhibition, and risk for alcohol dependence and related disorders. Sci. World J. 7, 131–141.

Quiroga, R. Q., Rosso, O. A., Başar, E., 1999. Wavelet entropy: a measure of order in evoked potentials. Electroencephalogr. Clin. Neurophysiol. 49, 299–303.

Quiroga, R. Q., Rosso, O. A., Başar, E., Schürmann, M, 2001a. Wavelet entropy in event-related potentials: a new method shows ordering of EEG oscillations. Biol. Cybernet. 84, 291–299.

Quiroga, R. Q., Reddy, L., Kreiman, G., Koch, C., Fried, I., 2005. Invariant visual representation by single neurons in the human brain. Nature 435(7045), 1102–1107.

Quiroga, R. Q., Sakowitz, O. W., Başar, E., Schürmann, M., 2001b. Wavelet transform in the analysis of the frequency composition of evoked potentials. Brain Res. Protocols 8, 16–24.

Rahn, E., Başar, E., 1993a. Prestimulus EEG-activity strongly influences the auditory evoked vertex responses: a new method for selective averaging. Int. J. Neurosci. 69, 207–220.

Rahn, E., Başar, E., 1993b. Enhancement of visual evoked potentials by stimulation during low prestimulus EEG stages. Int. J. Neurosci. 72, 123–136.

Rangaswamy, M., Porjesz, B., Chorlian, D. B., Wang, K., Jones, K. A., Bauer, L. O., Rohrbaugh, J., O'Connor, S. J., Kuperman, S., Reich, T., Begleiter, H., 2002. Beta power in the EEG of alcoholics. Biol. Psychiatry. 52, 831–842.

Rangaswamy, M., Porjesz, B., 2008. Human brain oscillations as effective endophenotypes: uncovering genes for cognitive (dys) function and predisposition for alcoholism spectrum disorders. Brain Res. 1235, 153–171.

Rao, S. G., Williams, G. V., Goldman-Rakic, P. S., 2000. Destruction and creation of spatial tuning by disinhibition: GABA(A) blockade of prefrontal cortical neurons engaged by working memory. J. Neurosci. 20, 485–494.

Rapp, P. E., Zimmerman, I. D., Albano, A. M., 1986. Experimental studies of chaotic neural behavior: cellular activity and electroencephalographic signals. In: Othmer, H. G. (Ed.), Nonlinear Oscillations in Biology and Chemistry, Springer, Berlin, pp. 175–805.

Rapp, P. E., Zimmerman, I. D., Albano, A. M., Deguzman, G. C., Greenbaum, N. N., 1985a. Dynamics of spontaneous neural activity in the simian motor cortex: the dimension of chaotic neurons. Phys. Lett. 110, 335–338.

Rapp, P. E., Zimmerman, I. D., Albano, A. M., deGuzman, G. C., Greenbaun, N. N., Bashore, T. R., 1985b. Experimental studies of chaotic neural behavior: cellular activity and elec- troencepha-

lographic signals. In: Othmer, H. G. (Ed.), Nonlinear Oscillations in Biology and Chemistry. Lecture Notes in Biomathematics. Springer, Berlin, pp. 175–205.
Rappelsberger, P., Pockberger, H., Petsche, H., 1982. The contribution of the cortical layers to the generation of the EEG: field potential and current source density analyses in the rabbit's visual cortex. Electroencephalogr. Clin. Neurophysiol. 53(3), 254–269.
Reichle, L. E., 2004. The Transition to Chaos in Conservative Classical Systems: Quantum Manifestations. Springer, New York.
Regan, D., 1989. Human Brain Electrophysiology: Evoked Potentials and Evoked Magnetic Fields in Science and Medicine. Elsevier, Amsterdam.
Reynolds, S. R. M., 1947. The physiologic basis of menstruation: a summary of current concepts. JAMA 135(9), 552–557.
Reynolds, S. R. M., 1963. Maternal blood flow in the uterus and placenta. In: Hamilton, W. F., Dow, P. (Eds.), Handbook of Physiology. Section 2, Circulation Vol II, pp. 1585–1618.
Rihmer, Z., Gonda, X., Rihmer, A., 2006. Creativity and mental illness. Psychiatr. Hung. 21(4), 288–294.
Rosen, R., 1969. Hierarchical organization in automata theoretic models of the central nervous system. In: Leibovic, K. N. (Ed.), Information Processing in the Nervous System. Springer, Berlin, Heidelberg, NY.
Rosso, O. A., Martin, M. T., Plastino, A., 2002. Brain electrical activity analysis using wavelet-based informational tools. Physica A 15, 587–608.
Rosso, O. A., Blanco, S., Yordanova, J., Kolev, V., Figliola, A., Schürmann, M., Başar, E., 2001. Wavelet Entropy: a new tool for analysis of short time brain electrical signals. J. Neurosci. Methods. 105, 65–75.
Rosso, S. M., Kamphorst, W., de Graaf, B., Willemsen, R., Ravid, R., Niermeijer, M. F. Spillantini, M. G., Heutink, P., van Swieten, C., 2001. Familial frontotemporal dementia with ubiquitin-positive inclusions is linked to chromosome 17q21-22. Brain 124, 1948–1957.
Röschke, J., 1986. Eine Analyse der nichtlinearen EEG-Dynamik. Diss. Georg-August-Universität, Göttingen.
Röschke, J., Başar, E., 1985. Is the EEG a simple noise or a "strange attractor"? Pflügers Arch. 405, R45.
Röschke, J., Başar, E., 1988. The EEG is not a simple noise: strange attractors in intracranial structures. In: Başar, E. (Ed.), Dynamics of Sensory and Cognitive Processing by the Brain. Springer, Berlin, pp. 203–216.
Röschke, J., Başar, E., 1989. Correlation dimensions in various parts of cat and human brain in different states In: Başar, E., Bullock, T. H. (Eds.), Brain Dynamics, Progress and Perspectives. Springer, Berlin, pp. 131–148.
Röschke, J., Başar, J., 1990. The EEG is not a simple noise: Strange attractors in intracranial structures. In: Başar, E. (Ed.), Chaos in Brain Function. Springer, Berlin.
Röschke, J., Fell, J., 1997. Spectral analysis of P300 generation in depression and schizophrenia. Neuropsychobiology 35, 108–114.
Röschke, J., Mann, K., Riemann, D., Frank, C., Fell, J., 1995. Sequential analysis of the brain's transfer properties during consecutive REM episodes. Electroencephalogr.Clin. Neurophysiol. 96, 390–397.
Ruskin, D. N., Bergstrom, D. A., Walters, J. R., 1999a. Multisecond oscillations in firing rate in the globus pallidus: synergistic modulation by D1 and D2 dopamine receptors. J. Pharmacol. Exp. Ther. 290, 1493–1501.
Ruskin, D. N., Bergstrom, D. A., Walters, J. R., 2002. Nigrostriatal lesion and dopamine agonists affect firing patterns of rodent entopeduncular nucleus neurons. J. Neurophysiol. 88(1), 487–496.
Ruskin, D. N., Bergstrom, D. A., Baek, D., Freeman, L. E., Walters, Jr., 2001a. Cocaine or selective block of dopamine transporters influences multisecond oscillations in firing rate in the globus pallidus. Neuropsychopharmacology 25, 28–40.
Ruskin, D. N., Bergstrom, D. A., Kaneoke, Y., Patel, B. N., Twery, M. J., Walters, Jr., 1999b. Multisecond oscillations in firing rate in the basal ganglia: robust modulation by dopamine receptor activation and anesthesia. J. Neurophysiol. 81, 2046–2055.

Ruskin, D. N., Bergstrom, D. A., Shenker, A., Freeman, L. E., Baek, D., Walters, Jr., 2001b. Drugs used in the treatment of attention-deficit/hyperactivity disorder affect postsynaptic firing rate and oscillation without preferential dopamine autoreceptor action. Biol. Psychiatry 49, 340–350.

Saermark, K., Lebech, J., Bak, C. K., Sabers, A., 1989. Magnetoencephalography and attractor dimension: normal subjects and epileptic patients. In: Başar, E., Bullock, . . (Eds.), Brain Dynamics. Springer, New York.

Saermark, K., Mikkelsen, K. B., Başar, E., 1992. Magnetoencephalographic evidence for induced rhythms. In: Başar, E., Bullock, T. H. (Eds.), Induced Rhythms in the Brain. Birkhäuser, Boston, pp. 129–145.

Sagawa, K., 1972. The use of control theory and systems analysis in cardiovascular dynamics. In: Bergel, D. H. (Ed.), Cardiovascular fluid dynamics. Academic Press, New York.

Sakowitz, O. W., Quiroga, R. Q., Schürmann, M., Başar, E., 2001. Bisensory stimulation increases gamma-responses over multiple cortical regions. Cognitive Brain Res. 11, 267–279.

Sakowitz, O. W., Quiroga, Q. R., Schurmann, M., Başar, E., 2005. Spatio-temporal frequency characteristics of intersensory components in audiovisually evoked potentials. Cogn. Brain Res. 23, 316–326.

Sakowitz, O. W., Schürmann, M., Başar, E., 2000. Oscillatory frontal theta responses are increased upon bisensory stimulation. Clin. Neurophysiol. 111, 884–893.

Sanquist, Th. F., Rohrbaugh, J. W., Syndulko, K., Lindsley, D. B., 1980. Electrocortical signs of levels of processing: perceptual analysis and recognition memory. Psychophysiology 17, 568–576.

Santosa, C. M., Strong, C. M., Nowakowska, C., Wang, P. W., Rennicke, C. M., Ketter, T. A., 2007. Enhanced creativity in bipolar patients: A controlled study. J. Affect. Disord. 100, 31–39.

Sato, K., 1963. On the linear model of the brain activity in electroencephalographic potentials. Folia Psychiatrica et Neurologica Japonica 17, 156–166.

Sato, K., Kitajima, H., Mimura, K., Hirota, N., Tagawa, Y., Ochi, N., 1971. Cerebral visual evoked potentials in relation to EEG. Electroencephalogr. Clin. Neurophysiol. 30, 123–128.

Sato, K., Ono, K., Chiba, G., Fukuta, K., 1977. Component activities in the autoregressive activity of physiological systems. Int. J. Neurosci. 7, 239–249.

Sauseng, P., Klimesch, W., Doppelmayr, M., Pecherstorfer, T., Freunberger, R., Hanslmayr, S., 2005b. EEG alpha synchronization and functional coupling during top-down processing in a working memory task. Hum. Brain Mapp. 26, 148–155.

Sauseng, P., Klimesch, W., Gruber,W., Doppelmayr, M., Stadler,W., Schabus, M., 2002. The interplay between theta and alpha oscillations in the human electroencephalogram reflects the transfer of information between memory systems. Neurosci. Lett. 324, 121–124.

Sauseng, P., Klimesch, W., Stadler, W., Schabus, M., Doppelmayr, M., Hanslmayr, S., Gruber, W. R., Birbaumer, N., 2005a. A shift of visual spatial attention is selectively associated with human EEG alpha activity. Eur. J. Neurosci. 22, 2917–2926.

Scheffers, M. K., Johnson, Jr. R., 1994. Recognition memory and search for attended letters: an event-related potential analysis. J. Psychophysiol. 8, 328–347.

Schmiedt, C., Brand, A., Hildebrandt, H., Başar-Eroğlu, C., 2005. Event-related theta oscillations during working memory tasks in patients with schizophrenia and healthy controls. Cogn. Brain Res. 25(3), 936–947.

Schmidtke, C. R., 1987. Bergson and a Pulsational-Wave Model of Temporality: A way to Disentangle Theories in Gerontology. In: Papanicolaou, A. C., Gunter, P. A. Y. (Eds.), Bergson and Modern Thought: Towards a Unified Science. Harwood, New York.

Schweitzer, J., 1986. Functional organization of the electroreceptive midbrain in an elasmobranch (Platyrhinoidis triseriata). J. Comp. Physiol. A. 158, 43–58.

Schuster, H. G., 1988. Deterministic Chaos second revised and enlarged edition. Physik Verlag, Weinheim.

Schürmann, M., Başar-Eroğlu, C., Başar, E., 1996. P300-delta responses are only an example of the functional interpretation of event-related brain rhythms. Psycoloquy 7(5) memory-brain.6.schurmann

Schürmann, M., Başar-Eroğlu, C., Başar, E., 1997. A possible role of evoked alpha in primary sensory processing: common properties of cat intracranial recordings and human EEG and MEG. Int. J. Psychphysiol. 26, 149–170.

Schürmann, M., Başar-Eroğlu, C., Kolev, V., Başar, E., 1995. A new metric for analyzing single-trial event-related potentials (ERPs): application to human visual P300 delta response. Neurosci. Lett. 197, 167–170.

Schürmann, M., Demiralp, T., Başar, E., Başar-Eroğlu, C., 2000. Electroencephalogram alpha (8–15 Hz), responses to visual stimuli in cat cortex, thalamus, and hippocampus: a distributed alpha network? Neurosci. Lett. 292, 175–178.

Schürmann, M., Başar-Eroğlu, C., Rahn, E., Braasch, M., Dössel, O., Fuchs, M., Başar, E., 1992. A comparative study of alpha responses in human MEG temporoparietal and occipital recordings and cat intracranial EEG recordings. Proceedings of the IEEE symposium on Neuroscience and Technology, Lyon, 132–137.

Schütt, A., Başar, E., 1992. The effects of acetylcholine, dopamine and noradrenaline on the visceral ganglion of Helix Pomatia II: stimulus evoked field potentials. Comp. Biochem. Physiol. 102C, 169–176.

Schütt, A., Başar, E., Bullock, T. H., 1992. The effects of acetylcholine, dopamine and noradrenaline on the visceral ganglion of Helix pomatia I. Ongoing compound field potentials of low frequencies. Comp. Biochem. Physiol. 102C, 159–168.

Schütt, A., Bullock, T. H., Başar, E., 1999. Dynamics of potentials from Invertebrate brains. In: Başar, E. (Ed.), Brain Function and Oscillations: II. Integrative Brain Function. Neurophysiology and Cognitive Processes, pp. 91–108.

Sheer, D. E., 1976. Focused arousal and 40 Hz EEG. In: Knight, R. M., Bakker, D. J. (Eds.), The Neuropsychology of Learning Disorders. University Park Press, Baltimore, pp. 71–87.

Sheer, D. E., 1984. Focused arousal, 40 Hz EEG, and clinical application. In: Sheer, D. E. (Ed.), Attention: Theory, Brain Function, and Clinical Applications. Academic Press, New York.

Shenoy, K. V., Kaufinan, J., McGrann, J. V., Shaw, G. L., 1993. Learning by selection in the trion model of cortical organization. Cerebral Cortex 3, 239–248.

Shepherd, G. M., 1988. Neurobiology. Oxford University Press, New York.

Sherrington, C. (Ed.), 1948. The Integrative Action of the Nervous System. Cambridge University Press, Cambridge, UK.

Silva, L. R., Amitai, Y., Connors, B. G., 1991. Intrinsic oscillations of neocortex generated by layer 5 pyramidal neurons. Science 251, 432–435.

Sim, K., Chua, T. H., Chan, Y. H., Mahendran, R., Chong, S. A., 2006. Psychiatric comorbidity in first episode schizophrenia: a 2 year, longitudinal outcome study. J. Psychiatr. Res. 40(7), 656–663.

Singer, W., 1981. Topographic organization of orientation columns in the cat visual cortex. A deoxyclucose Study. Exp. Brain Res. 44, 431–436.

Singer, W., 1989. The brain: a self-organizing system. In: Klivington, K. A. (Ed.), The Science of Mind. MIT Press, Cambridge, MA, pp. 174–179.

Skinner, B. F., 1938. The Behavior of Organisms: An Experimental Analysis. (Originally, Reprinted by the B. F. Skinner Foundation, 1991 and 1999).

Skinner, J. E., Martin, J., Landisman, C., Mommer, M., Fulton, K., Mitra, M., Burton, W., Saltzberg, B., 1989. Chaotic attractors in a model of neocortex: dimensionalities of olfactory bulb surface potentials are spatially uniform and event-related. In: Başar, E., Bullock, T. H. (Eds.), Brain Dynamics. Springer-Verlag, Berline, pp. 158–173.

Smith, H. W., 1959. The Kidney: Structure and Function in Health and Disease. Oxford University Press, New York.

Smulders, T. V., 2009. Darwin 200: special feature on brain evolution. Biol. Lett. 5(1), 105–107.

Smythe, J. W., Cristie, B. R., Colom, L. V., Lawson, V. H., Bland, B. H., 1991. Hippocampal theta field activity and theta-on/theta-off cell discharges are controlled by an ascending hypothalamo-septal pathway. J. Neurosci. 11(7), 2241–2248.

Solodovnikov, V. V., 1960. Introduction to the Statistical Dynamics of Automatic Control Systems. Dover, New York.

Sokolov, E. N., 1975. Neuronal mechanisms of the orienting reflex. In: Sokolov, E. N., Vinogradova, O. S. (Eds.), Neuronal Mechanisms of the Orienting Reflex. LEA, Hillsdale, NJ, pp. 217–238.

Sokolov, E. N., 2001. Toward new theories of brain function and brain dynamics. In: Başar, E., Schürmann, M. (Eds.), Int. J. Psychophysiol. 39, 87–89.

Solms, M., 1997. The neuropsychology of dreams: a clinico-anatomical study. LEA, Hillsdale, NJ.

Solms, M., 2000a. Dreaming and REM sleep are controlled by different brain mechanisms. Behav. Brain Sci. 23, 843–850.

Solms, M., Turnbull, O., 2002. Emotion and motivation. In: Solms, M., Turnbull, O. (Eds.), The Brain and the Inner World. Other Press, New York, pp. 105–137.

Sparks, H. V., 1964. Effect of Quick Stretch on Isolated Vascular Smooth Muscle. Circ. Res. 15(8), 254–260.

Spelman, F. A., Pinter, R. B., 1978. Low-frequency dynamics of the iliac artery of the unanesthetized baboon. Ann. Biomed. Eng. 6(3), 212–230.

Spencer, K. M., Nestor, P. G., Niznikiewicz, M. A., Salisbury, D. F., Shenton, M. E., Mc Carley, R. W., 2003. Abnormal neural synchrony in schizophrenia. J. Neurosci. 23(19), 7407–7411.

Spencer, K. M., Nestor, P. G., Perlmutter, R., Niznikiewicz, M. A., Klump, M. C., Frumin, M., Shenton, M. E., Mc Carley, R. W., 2004. Neural synchrony indexes disordered perception and cognition in schizophrenia. Proc. Natl. Acad. Sci. U.S.A. 101(49), 17288–17293.

Spencer, K. M., Niznikiewicz, M. A., Shenton, M. E., McCarley, R. W., 2007. Sensory-evoked gamma oscillations in chronic schizophrenia. Biol. Psychiatry 63(8), 744–747.

Stam, C. J., 2000. Brain dynamics in theta and alpha frequency bands and working memory performance in humans. Neurosci. Lett. 286(2), 115–118.

Stampfer, H. G., Başar, E., 1985. Does frequency analysis lead to better understanding of human event related potentials? Int. J. Neurosci. 26, 181–196.

Stapp, H., 1987. Bergson and the Unification of the Sciences. In: Papanicolaou, A. C., Gunter, P. A. Y. (Eds.), Bergson and Modern Thought: Towards a Unified Science. Harwood, New York.

Stassen, H. H., Lykken, D. T., Propping, P., Bomben, G., 1988. Genetic determination of the human EEG. Survey of recent results on twins reared together and apart. Hum. Genet. 80, 165–176.

Steriade, M., 2001. Impact of network activities on neuronal properties in corticothalamic systems. J. Neurophysiol. 86, 1–39.

Steriade, M., Corró Dossi, R., Pare, D., 1992. Mesopontine cholinergic system suppress slow rhythms and induce fast oscillations in thalamocortical circuits. In: Başar, E., Bullock, T. H. (Eds.), Induced Rhythm in the Brain. Birkhauser, Boston, pp. 251–268.

Steriade, M., Jones, E. G., Llinas, R. R., 1990. Thalamic Oscillations and Signaling. Wiley, New York.

Stewart, M., Fox, S. E., 1990. Do septal neurons pace the hippocampal theta rhythm? Trends Neurosci. 13(5), 163–168.

Strüber, D., Başar-Eroğlu, C., Hoff, E., Stadler, M., 2000. Reversalrate dependent differences in the EEG gamma-band during multistable visual perception. Int. J. Psychophysiol. 38(3), 243–252.

Sturbeck, K., 1988. Vergleichende Analyse evozierter Potentiale von Invertebraten und Vertebraten. Dissertation, Medical University Lübeck.

Stryker, M. P., 1989. Is grandmother an oscillation? Nature 338, 297–298.

Symond, M. B., Psych, B., Harris, A. W. F., Gordon, E., Williams, L. M., 2005. "Gamma synchrony" in first-episode schizophrenia: a disorder of temporal connectivity? Am. J. Psychiatry 162, 459–465.

Szilard, L., 1929. Über die Entropieverminderung in einem Thermodynamischen System bei Eingriffe Intelligenter Wesen. Zeitschrift für Physik 53, 840–960.

Takens, F., 1981. Detecting strange attractors in turbulence. In Lecture Notes in Mathematics. Dynamical Systems and Turbulence. Springer, Berlin, pp. 898, 366.

Tass, P. A., Hauptmann, C., 2007. Therapeutic modulation of synaptic connectivity with desynchronizing brain stimulation. Int. J. Psychophysiol. 64(1), 53–61.

Taylor, M. G., 1966. Use of random excitation and spectral analysis in the study of frequency dependent parameters of the cardiovascular system. Circulation Res. 18, 585–595.

Tegmark, M., 2003. Parallel universes. Not just a staple of science fiction, other universes are a direct implication of cosmological observations. Sci. Am. 288(5), 40–51.

Terzano, M. G., Parrino, L., Boselli, M., Smerieri, A., Spaggiari, M. C., 2000. CAP components and EEG synchronization in the first 3 sleep cycles. Clin. Neurophysiol. 111, 283–290.

Tesche, C. D., Karhu, J., 2000. Theta oscillations index human hippocampal activation during a working memory task. Proc. Natl. Acad. Sci. U.S.A. 97, 919–924.

Thom, R., 1975. Structural Stability and Morphogenesis (in French). Fowler, D. H. (trans.). Benjamin, New York.

Thompson, R. F., 1986. The neurobiology of learning and memory. Science 233, 941–947.

Thompson, R. F., 1990. Neural mechanisms of classical conditioning in mammals. Phil. Trans. Royal Soc. Lond. B 329, 161–170.

Tobler, I., Kopp, C., Deboer, T., Rudolph, U., 2001. Diazepam-induced changes in sleep: role of the alpha 1 GABA (A) receptor subtype. Proc. Natl. Acad. Sci. U.S.A. 98, 6464–6469.

Tononi, G., Sporns, O., Edelman, G. M., 1992. The problem of neural integration: induced rhythm and short-term correlation. In: Başar, E., Bullock, T. H. (Eds.), Induced Rhythm in the Brain. Birkhauser, Boston, pp. 363–393.

Tranel, D., Damasio, A. R., 1995. Neurobiological foundations of human memory. In: Baddeley, A. D., Wilson, B. A., Watts, F. N. (Eds.), Handbook of Memory Disorders. Wiley, New York, pp. 27–50.

Traub, R. D., Cunningham, M. O., Gloveli, T., LeBeau, F. E. N., Bibbig, A., Buhl, E. H., Whittington, M. A., 2003. GABA-enhanced collective behavior in neuronal axons underlies persistent gamma-frequency oscillations. Proc. Natl. Acad. Sci. U.S.A. 100, 11047–11052.

Traub, R.D., Whittington, M. A., Colling, S. B., Buzsaki, G., Jefferys, J. G., 1996. Analysis of gamma rhythms in the rat hippocampus in vitro and in vivo. J. Physiol. 493, 471–484.

Tümer, N., 1980. An analysis of the dynamics of active smooth muscles in the visceral organs guinea-pig uterus and stomach. Thesis, Hacettepe Uni., Ankara.

Twarog, B. M., Roeder, K. D., 1957. Pharmacological observations of the desheathed last abdominal ganglion of the cockroach. Ann. Ent. Soc. Am. 50, 231–237.

Ungan, P., Başar, E., 1976. Comparison of Wiener filtering and selective averaging of evoked potentials. Electroencephalogr. Clin. Neurophysiol. 40, 516–520.

Varela, F., Lachaux, J. P., Rodriguez, E., Martinerie, J., 2001. The brainweb: phase syncronization and large-scale integration. Nat. Rev. Neurosci. 2, 229–232.

Van Baal, G. C., de Geus, E. J., Boomsma, D. I., 1998. Genetic influences on EEG coherence in 5-year-old twins. Behav. Genet. 28, 9–19.

Van Beijsterveldt, C. E., Boomsma, D. I., 1994. Genetics of the human electroencephalogram (EEG) and event-related brain potentials (ERPs): a review. Hum. Genet. 94, 319–330.

Van Beijsterveldt, C. E., Molenaar, P. C., de Geus, E. J., Boomsma, D. I., 1998. Genetic and environmental influences on EEG coherence. Behav. Genet. 28, 443–453.

Van der Stelt, O., Belger, A., 2007. Application of electroencephalography to the study of cognitive and brain functions in schizophrenia. Schizophr. Bull. 33, 955–970.

Van der Tweel, L. H., 1961. Some problems in vision regarded with respect to linearity and frequency response. Ann. NY Acad. Sci. 89, 829–856.

Vanderwolf, C. H., Leung, L. S., 1983. Hippocampal rhythmical slow activity: a brief history and effects of entorhinal lesions and phencyclidine. In: Seifert, W. (Ed.), The Neurobiology of the Hippocampus. Academic Press, London, pp. 275–302.

Van Erp M. G., 1988. On Epilepsy: investigations on the level of nerve membrane and of the brain: Proefschrift Rijksuniversiteit, Leiden.

Vertes, R. P., 1982. Brain stem generation of the hippocampal EEG Prog. Neurobiol. 19(3), 159–186.

Vinogradova, O. S., 1995. Expression, control, and probable functional significance of the neuronal theta-rhythm. Prog. Neurobiol. 45(6), 523–583.

Vinogradova, O. S., Zolotukhina, L. I., 1972. Sensory characteristics of the neurons in the medial and lateral septal nuclei. Zh. Vyssh. Nerv. Deyat. 22, 1260–1269.

Von Helmholtz, H., 1866. Handbuch der physiologischen Optik. L. Voss, Hamburg. MacAdam, D. L., 1970. Sources of Color Science. MIT Press, Cambridge, MA.

Von Helmholtz, H., 1962. Helmholtz's treatise on physiological optics. Dover, New York.

Von Neumann, J., Burks, A. W., 1966. Theory of Self-Reproducing Automata. University of Illinois Press, Urbana, IL.

von Stein, A., Chiang, C., König, P., 2000. Top-down processing mediated by interareal synchronization. Proc. Natl. Acad. Sci. U.S.A. 97, 14748–14753.

von Stein, A., Sarnthein, J., 2000. Different frequencies for different scales of cortical integration: from local gamma to long range alpha/theta synchronization. Int. J. Psychophysiol. 38, 301–313.

Wachholder, K., 1921. Haben die rhythmischen spontankontraktionem der gefasse einen nachweisbaren einfluss auf den blutstrom? Pflügers Arch. 190, 222–229

Waldemar, G., Dubois, B., Emre, M., Georges, J., McKeith, I. G., Rossor, M., Scheltens, P., Tariska, P., Winblad, B., 2007. Recommendations for the diagnosis and management of Alzheimer's disease and other disorders associated with dementia: EFNS guideline. Eur. J. Neurol. 14(1), E1–E26.

Wastell, D. G., 1979. The application of low-pass linear filters to evoked potential data: filtering without phase distortion. Electroencephalogr. Clin. Neurophysiol. 46, 355–356.

Weiss, S., Rappelsberger, P., 2000. Long-range EEG synchronization during word encoding correlates with successful memory performance. Cogn. Brain Res. 9(3), 299–312.

Weizsäcker, C. F., 1985. Aufbau der Physik. Munich.

Whittington, M. A., Traub, R. D., Kopell, N., Ermentrout, B., Buhl, E. H., 2000. Inhibition-based rhythms: experimental and mathematical observations on network dynamics. Int. J. Psychophysiol. 38(3), 315–336.

Wichman, E. H. (1967). Quantum Physics. Berkeley Physics Course, vol. 4, McGraw-Hill, New York.

Wichmann, T., Kliem, M. A., Long, S. B., 2000. Changes in oscillatory discharge in the primate basal ganglia associated with parkinsonism and light sleep. Mov. Disord. 15(3), 37.

Wiener, N., 1948. Cybernetics or Control and Communication in the Animal and the Machine. MIT Press, Cambridge, MA.

Wiig, K. A., Heynen, A. J., Bilkey, D. K., 1994. Effects of kainic acid microinfusions on hippocampal type 2 RSA (theta). Brain Res. Bull. 33(6), 727–732.

Wilcox, R. E, Gonzales, R. A., 1995. Introduction to neurotransmitters, receptors, signal transduction, and second messengers. In: Nemeroff, C. B., Schatzberg, A. F. (Eds.), Textbook of Psychopharmacology. American Psychiatric Press, Arlington, VA, 3–29.

Winterer, G., Enoch, M. A., White, K. V., Saylan, M., Coppola, R., Goldman, D., 2003a. EEG phenotype in alcoholism: increased coherence in the depressive subtype. Acta. Psychiatr. Scand. 108, 51–60.

Winterer, G., Mahlberg, R., Smolka, M. N., Samochowiec, J., Ziller, M., Rommelspacher, H. P., Herrmann, W. M., Schmidt, L. G., Sander, T., 2003b. Association analysis of exonic variants of the GABA(B)-receptor gene and alpha electroencephalogram voltage in normal subjects and alcohol-dependent patients. Behav. Genet. 33, 7–15.

Wynn, J. K., Light, G. A., Breitmeyer, B., Nuechterlein, K. H., Green, M. F., 2005. Event-related gamma activity in schizophrenia patients during a visual backward-masking task. Am. J. Psychiatry. 162(12), 2330–2336.

Xiaohua, L., Ketter, T. A., Frye, M. A., 2002. Synaptic, intracellular, and neuroprotective mechanism of anticonvulsants: are they relevant for the treatment and course of bipolar disorder. J. Affect. Disord. 69, 1–14.

Yatham, L. N., Liddle, P. F., Shiah, I. S., Lam, R. W., 2002. PET study of [(18)F]6-fluoro-L-dopa uptake in neuroleptic- and mood-stabilizer naive first-episode nonpsychotic mania: effects of treatment with divalproex sodium. Am. J. Psychiatry 159, 768–774.

Yener, G., Güntekin, B., Başar, E., 2008. Event related delta oscillatory responses of Alzheimer patients. Eur. J. Neurol. 15(6), 540–547.

Yener, G. G., Güntekin, B., Öniz, A., Başar, E., 2007. Increased frontal phase-locking of event-related theta oscillations in Alzheimer patients treated with cholinesterase inhibitors. Int. J. Psychophysiol. 64(1), 46–52.

Yener, G., Güntekin, B., Tülay, E., Başar, E., 2009. A comparative analysis of sensory visual evoked oscillations with visual cognitive event related oscillations in Alzheimer's disease. Neurosci. Lett. 462, 193–197.

Yeragani,V. K., Cashmere, D., Miewald, J., Tancer, M., Keshavan, M. S., 2006. Decreased coherence in higher frequency ranges (beta and gamma) between central and frontal EEG in patients with schizophrenia: a preliminary report. Psychiatry Res. 141(1), 53–60.

Yoffey, J. M., Courtice, F. C., 1970. Lymphatics, lymph and the lymphomyeloid complex. Academic Press, London.

Yordanova, J., Kolev, V., 1996. Brain theta response predicts P300 latency in children. Neuroreport 7, 277–280.

Yordanova, J., Kolev, V., 1997. Developmental changes in the event-related EEG theta response and P300. Electroenceph. Clin. Neurophysiol. 104, 418–430.

Yordanova, J., Kolev, V., 1998a. Event-related alpha oscillations are functionally associated with P300 during information processing. Neuroreport. 9(14), 3159–3164.

Yordanova, J., Kolev, V., 1998b. Single-sweep analysis of the theta frequency band during an auditory oddball task. Psychophysiology 35(1), 116–126.

Yordanova, J., Rosso, O. A., Kolev, V., 2003. A transient dominance of theta event-elated brain potential component characterizes stimulus processing in an auditory oddball task. Clin. Neurophysiol. 114(3), 529–540.

Yordanova, J., Banaschewski, T., Kolev, V., Woerner, W., Rothenberger, A., 2001. Abnormal early stages of task stimulus processing in children with attention-deficit hyperactivity disorder-evidence from event-related gamma oscillations. Clin. Neurophysiol. 112(6), 1096–1108.

Yordanova, J., Kolev, V., Rosso, O. A., Schürmann, M., Sakowitz, O. W., Özgören, M., Başar, E., 2002. Wavelet entropy analysis of event-related potentials indicates modality-independent theta dominance. J. Neurosci. Methods 117, 99.

Zeeman, E. C., 1977. Catastrophe theory. Selected papers, 1972-1977. Addison-Wesley, Reading, MA.

Zeki, S., 2007. The neurobiology of love. FEBS Lett. 581, 2575–2579.

Zheng, L. L., Jiang, Z. Y., Yu, E. Y., 2007. Alpha spectral power and coherence in the patients with mild cognitive impairment during a three-level working memory task. J. Zhejiang Univ. Sci. B. 8(8), 584–592.

Zheng-yan, J., 2005. Abnormal cortical functional connections in Alzheimer's disease: analysis of inter-and intra-hemispheric EEG coherence. J. Zhejiang Univ. Sci. 6B(4), 259–264.

Zweig, S., 1932. Die Heilung durch den Geist. Fischer/Verlag, Berlin.

Author Index

A

Abraham, R.H., 299, 455, 456
Abrams, T.W., 353
Achimowicz, J.Z., 218, 220, 235, 349, 356, 411
Ademoğlu, A., 115, 157, 163, 168
Adey, W.R., 65, 123
Adler, G., 270, 294
Adrian, E.D., 108, 122, 123, 137, 200
Akhavan, A.C., 124
Akiskal, H.S., 282
Aksu, F., 223, 234
Aladjalova, N.A., 98, 187, 396, 427
Albano, A.M., 305, 310, 458
Alegre, M., 139
Alksne, J.F., 32, 349
Allers, K.A., 98, 187–191
Allilaire, J.F., 282
Almasy, L., 152, 290, 333
Altafullah, I., 124
Amitai, Y., 124
Andersen, P., 122
Andersson, S.A., 122
Andreasen, N., 385–387, 424, 427
Andrew, C., 129
Antal, M., 262
Apostol, G., 124
Arnold, M., 144, 181, 284
Artieda, J., 139
Ascher, P., 209
Ashworth, F., 262, 286
Auta, J., 262
Avidan-Carmel, G., 248
Azorin, J.M., 282

B

Babiloni, C., 267, 273
Babloyantz, A., 304, 305, 308, 310, 312, 455
Baddeley, A.D., 154, 155, 158, 159, 362
Baek, D., 98, 187, 189, 190, 396, 427
Bahramali, H., 122, 277, 280, 281
Bailey, C.H., 354, 445, 447
Bak, C.K., 127, 310
Balconi, M., 244, 252
Baldeweg, T., 113, 281, 290, 296
Başar, E., 19, 24, 26, 31, 32, 36, 41, 66–69, 71, 80, 81, 84, 85, 88, 89, 91, 95, 96, 98, 111–116, 118, 119, 122–132, 135, 137, 139–144, 151, 153–160, 162, 163, 166–170, 173, 174, 179–183, 186, 187, 194, 196, 199–201, 203, 204, 206, 208, 209, 212–214, 216, 218, 219, 223, 228–230, 232, 234, 243–249, 252, 253, 261, 263–271, 273, 280–284, 286, 287, 289, 292–295, 297, 300, 303–311, 313, 319–323, 329–333, 339, 347–349, 355–360, 362, 365, 369, 376, 389, 401, 404, 406, 410, 411, 413–415, 418, 432, 435, 448, 450, 452, 453, 456, 457, 459, 460
Başar, J., 308
Başar-Eroğlu, C., 81, 89, 98, 111, 114, 115, 118, 119, 123–131, 135, 137, 139 144, 151, 153, 155, 157, 158, 162, 163, 166–168, 170, 180–182, 199, 200, 208, 209, 223, 228–230, 234, 243, 244, 248, 249, 252, 261, 263, 264, 278–281, 284, 286, 293, 294, 297, 308, 309, 322, 331, 348, 406, 413, 415
Banaschewski, T., 31, 113, 118, 163, 181, 234, 260, 264, 297
Barale, F., 262
Barlow, H.B., 108, 180
Barman, S.M., 103, 152, 188, 189, 193, 320, 347, 396, 427
Barrett, G., 154
Barry, R.J., 166, 182
Bartels, A., 248

Bartley, S.H., 122
Barut, A., 330
Bashore, T.R., 305, 451, 452, 458
Bauer, L.O., 152, 283, 284, 333, 415
Bauer, R., 31, 111, 179, 181, 208
Baving, L., 252, 260, 290, 297
Bayraktaroğlu, Z., 163
Bearden, C.E., 281
Beauregard, O.C., 359
Becker, A., 139, 264
Begleiter, H., 32, 107, 152, 183, 260, 283, 284, 289, 290, 292, 293, 333, 406, 415, 428
Beim-Graben, P., 181
Bekci, B., 65, 131, 133–135, 137–139, 156, 160, 170, 181, 183, 333
Belger, A., 260
Benabarre, A., 286
Bendat, J.S., 449
Benes, F.M., 262, 263, 286
Bennett, M.V.,
Berettini, W.H., 262
Berger, H., 108
Bergson, H., 18, 196, 327, 346, 348, 357, 361, 389
Bergstrom, D.A., 98, 187–191, 396, 427
Berretta, S., 262, 263, 286
Bertrand, O.,
Best, P.J., 130
Besthorn, C., 270, 294, 295
Bevan, M.D., 190
Beydaği, H.,
Bhagwagar, Z., 262, 286
Bibbig, A., 264, 265, 286
Bierer, L.M.,
Bierut, L., 290
Bigio, E.H., 265
Bilkey, D.K., 262
Binetti, G., 267, 273
Birbaumer, N., 109, 162, 163
Birkhoff, G.D., 303
Bishop, G.H., 122
Bishop, P.O., 122
Biswal, B., 191
Blanco, S., 68, 69, 181, 347, 401
Bland, B.H., 63, 262
Blangero, J., 152, 333
Bleckmann, H., 213
Blübaum-Gronau, E., 213
Blumenthal, R., 29
Böhmig, L., 201
Bolam, J.P., 190
Bonnet, V., 122, 123
Bonte, F.J., 265
Boomsma, D.I., 289
Boselli, M., 190
Bourgeois, M.L., 282
Bowden, C.L., 194, 263, 281, 282, 415, 418
Bozler, E., 78
Braasch, M., 114, 125–128, 131, 142, 143, 157, 181, 284, 286
Braff, D.L., 277, 278, 280, 281, 297
Brambilla, P., 262
Brand, A., 168, 278, 279, 281, 293, 294, 297, 415
Brandeis, D., 260
Brandt, L., 282
Brassen, S., 270, 294
Braun, C., 144, 181, 284
Brayne, C., 265
Brazhnik, E.S., 262
Brazier, M.A.B., 49, 124
Breitmeyer, B., 277, 280, 281, 296
Bremer, F., 122, 123
Brenner, C.A., 280
Bresciani, L., 267, 273
Bressler, S.L., 31, 32, 112, 122, 286
Britton, S.W., 253
Brodaty, H., 265
Bromley, R.B., 32, 45, 123, 137
Brosch, W., 31, 111, 179, 181, 208
Bruggink, S., 265
Brugue, E., 286
Brunner, C., 181
Bucci, P., 279, 297
Buchness, R., 124
Buhl, E.H., 264, 265, 286, 289
Bullock, T.H., 32, 107, 109, 110, 142, 151, 169, 180, 195, 200, 201, 203, 204, 206, 209, 212–214, 216, 218–220, 235, 261, 330, 347, 349, 356, 376, 411, 457
Burgess, A.P., 113
Burghoff, M., 114
Buriánek, P., 88–90
Burks, A.W., 340, 436
Burnstock, G., 78
Burton, W.,
Busch, N.A., 163
Buszaky, G.,
Buzsaki, G., 191, 264, 265, 286
Byrne, J.H., 150

C

Cajal, R., 8, 108
Callaway, E., 290
Cannon, T.D., 281
Cannon, W.B., 253
Capra, F., 19, 300, 323, 340, 431
Carew, T.J., 150, 353
Carfagna, T., 267, 273

Carlin, L.,
Caro, C.G., 76
Carroll, C.A., 282
Cashmere, D., 277, 281, 295, 297
Cassarino, A., 267, 273
Cassetta, E., 267, 273
Cassidy, D.C., 314
Castellucci, V.F., 150, 354
Castle, E., 276
Chan, Y.H., 276
Chang, H.T., 123
Changeux, J.-P., 24, 196, 227
Chatenêt-Duchêne, L., 282
Chen, A.C., 354
Chen, M., 354
Chiang, C., 31, 112, 141, 142, 144, 286
Chiba, G., 167, 232
Choi, K., 152
Chong, S.A., 276
Chorlian, D.B., 152, 283, 284, 289, 292, 293, 333, 415
Chua, T.H., 276
Churchland, S.P., 375, 432
Cichocki, A., 273
Clark, A., 373
Clarke, A.R., 166, 182
Cogliati, C., 190
Colling, S.B., 264, 265, 286
Colom, F., 286
Colom, L.V., 262
Comes, M., 286
Comi, G., 270, 294
Conel, J.L., 226
Conneally, P.M., 152, 333
Connors, B.G., 124
Cook, E.W. III, 452
Cook, I.A., 270, 294
Cooley, R.L., 190
Coppola, R., 282, 290, 294, 295
Correia, N., 282
Corró Dossi, R., 124, 131, 179, 190
Corwin, J.T., 213
Courchesne, E., 130, 252
Courtice, F.C., 77, 78
Cowen, P.J., 262, 286
Crandall, P., 124
Crawford, F.S. Jr., 461
Creutzfeld, O.D., 55, 58
Creutzfeldt, O.D., 124
Cristie, B.R., 262
Croft, R.J., 113, 290
Crowe, R.R., 152, 333
Crunelli, V., 290
Csicsvari, J., 191
Cunningham, M.O., 264, 265, 286
Curio, G., 114
Cursi, M., 270, 294
Czéh, G., 213

D

Daban, C., 286
Dai, J., 263
Damasio, A.R., 7, 65, 166, 176, 238, 253, 254, 353
Damasio, H., 65
Darwin, C., 247, 255, 345, 402
Daskalakis, Z.J., 281, 286
Dauwels, J., 273
Davis, J.M., 262
Davis, K.L.,
De Beer, G., 345
de Geus, E.J., 289
de Graaf, B., 68, 69, 181, 347, 401
de Guzman, G.C., 305, 458
De Jong, R., 244
De Lamarck, J.B., 345
De Nabias, B., 201
De Pascalis, V., 122, 166, 182, 277, 280, 281
De Weerd, J.A., 452
Deboer, T., 289
Deiber, M.P.,
Del Percio, C., 267, 273
Deleuze, G., 17, 19, 361–363
Delorme, A., 264
Demir, N., 81, 89, 98, 112, 137, 142
Demiralp, T., 111, 114, 115, 125–128, 130, 131, 135, 137, 139, 142, 143, 157, 158, 162, 163, 166, 168, 170, 181, 199, 209, 260, 263, 264, 279, 284, 286, 348
Der, G., 276
Descartes, R., 9, 196, 346
Desimone, R., 159, 362
Destexhe, A., 455
Deutsch, D., 336
Dick, D.M., 290
Dietl, R., 139, 264
DiGiorgi Gerevini, V., 262
Dinse, H.R., 124
Dodd, J., 53, 57
Domhoff, G.W., 369
Donchin, E., 154, 160
Doppelmayr, M., 109, 113, 123, 129–131, 135, 156, 157, 162, 163, 169, 181, 252, 290, 333
Dössel, O., 114, 125–128, 131, 142, 143, 157, 180, 181, 284, 286
Dougall, N.J., 265
Drishnan, P., 282

Dubois, B., 265
Duckrow, R.B., 218, 220, 235, 349, 356, 411
Dunkin, J.J., 270, 294
Dunlop, C.W., 65
Durusan, R., 142
Dustman, R.E., 227
Duzel, E., 252, 260, 290, 297
Dvorak, I., 310
Dwivedi, Y., 262

E
Ebmeier, K.P., 265
Eccles, J.C., 65, 340
Echteler, S.M., 215
Eckhorn, R., 31, 111, 179, 181, 208
Edelman, G.M., 166, 177, 180, 182, 346, 354–356
Edelman, S., 248
Edenberg, H.J., 152, 290, 333
Edlinger, G., 129
Eeg-Olofsson, O., 228
Eggert, P., 84, 93
Egner, T., 113
Eichenbaum, H., 444, 446
Einstein, A., 30, 32, 35, 316, 399, 401, 419
Ekbom, A., 282
Ekman, P., 243, 255
Elazar, Z., 123
Elger, C.E., 139, 264
Emmerson, R.Y., 227
Emre, M., 265
Engel, A.K., 171, 181, 279, 283
Enghoff, S., 130, 252
Engstrom, D.A., 19, 112, 122
Enoch, M.A., 282, 283, 290, 294, 295
Erdal, M.E.,
Ergen, M., 163
Ergenoğlu, T.,
Ermentrout, B., 286, 289
Eroğlu, C., 80, 89, 137
Erzengin, O.U., 65, 131, 133–135, 137–139, 156, 160, 170, 181, 183, 333
Etlinger, S.C., 163, 269
Eusebi, F., 267, 273
Evans, E.F., 50
Evans, J., 262, 286

F
Fabiani, M., 154
Farwell, L.A., 451, 452
Fazio-Costa, L.,
Fehr, T., 415
Fell, J., 113, 131, 139, 264
Feller, M.B., 139, 264
Fernandez, G., 139, 264
Ferreri, F., 267, 273
Ferri, C.P., 265
Ferri, R., 267, 273
Fessard, A., 11, 110, 135
Feynman, R.P., 462, 465–468, 471
Fields, R.D.,
Figliola, A., 68, 69, 181, 347, 401
Fisch, B.J., 233
Fischer, Y., 262
Fitzgerald, P.B., 281, 286
Flanigin, H.F., 264
Flippov, I.V., 191
Flohr, H., 120, 133, 134, 435
Folkow, B., 76, 78, 99
Foot, C., 33
Ford, J.M., 260, 264, 278, 279, 281, 292, 293
Foroud, T., 152, 290, 333
Forstl, H., 270, 294, 295
Fox, S.E., 262
Franceschi, M., 270, 294
Frank, C., 113, 131
Fratiglioni, L., 265
Freeman, L.E., 98, 187–191, 396, 427
Freeman, W.J., 29–32, 103
French, J.D., 53, 355
Freunberger, R., 109, 113, 123, 129, 130, 156, 157, 162, 163, 169, 181, 252, 290, 333
Freund, T.F., 262
Fried, I., 252
Friedman, H.R., 155
Fries, P., 285, 286
Friesen, W.V., 243
Frigerio, L., 267, 273
Frisoni, G.B., 267, 273
Frodl-Bauch, T., 290
Frotscher, M., 262
Frumin, M., 278, 280, 281, 296, 297, 358
Frye, M.A., 263
Fuchs, M., 114, 125–128, 131, 142, 143, 157, 180, 181, 284, 286
Fukuta, K., 167, 232
Fulton, K.,
Fuster, J.M., 41, 62, 112, 147, 149, 154, 157, 158, 164, 165, 167, 241, 362, 369, 425, 432

G
Galambos, R., 32, 45, 123, 137, 138
Galderisi, S., 279, 297
Gallagher, B.B., 264

Gallen, C., 138
Gallinat, J., 263, 264, 280, 296
Ganguli, M., 265
Ganong, W.F., 263, 264, 280, 296
Gardner, E., 263, 264, 280, 296
Gasser, T., 270, 294, 295
Gayton, R.J., 202, 209
Gebber, G.L., 103, 152, 188, 189, 193, 320, 347, 396, 427
Geiger-Kabisch, C., 270, 294, 295
Geissler, H.G., 129, 157
Gelperin, A., 209
Georges, J., 265
Gerbrandt, L.K., 160
Geroldi, C., 267, 273
Gerschenfeld, H.M., 209
Gershon, E.S., 262
Gevins, A., 163, 264
Ghazi, L.J., 98, 187–191
Giannakopoulos, P.,
Giustetto, M., 445, 447
Givens, B., 163, 264
Gladwin, T.E., 244
Gleick, J., 303
Glickstein, M., 158
Gloveli, T., 264, 265, 286
Goate, A., 152, 290, 333
Goddard, P.H., 451, 452
Gogolak, G., 124
Goikolea, J.M., 286
Gojkovi, M., 213
Gold, G.,
Goldman, D., 282, 283, 290, 294, 295
Goldman-Rakic, P.S., 62, 155, 157, 164
Gonda, X., 282
Gönder, A., 96, 112, 114, 117, 137, 142, 167, 153
Gonzales, R.A., 63
Gordon, E., 122, 277, 279–281, 296
Goto, Y., 190
Grace, J., 290
Grassberger, P., 299, 458, 460
Gray, C.M., 108, 111, 137, 181, 208
Grayson, D.R., 262
Green, M.F., 277, 280, 281, 296
Greenbaum, N.N., 305, 458
Greene, B., 324
Greene, R.W., 277, 280, 281, 296
Greitschus, F., 157, 208, 331
Griffith, J.S., 35, 36, 66
Grill-Spector, K., 248
Gruber, W.R., 109, 113, 123, 129, 130, 156, 157, 162, 163, 169, 181, 252, 290, 333
Grunwald, T., 139, 264
Gruzelier, J., 113, 252, 281, 290, 296
Guidotti, A., 262
Guillery, R.W., 290
Güntekin, B., 19, 153, 163, 170, 196, 243–247, 252, 263–271, 273, 281–283, 286, 287, 289, 292–295, 323, 329, 333, 339, 347, 349, 355, 356, 358, 360, 365, 369, 389, 411, 413–415, 418
Gunter, P.A.Y., 358, 359
Gurtubay, I.G., 139
Gurzì, M., 267, 273
Guyton, A.C., 49, 73, 74, 186

H

Haenschel, C., 113, 290
Haggqvist, A., 129
Haig, A.R., 122, 277, 279–281
Haken, H., 26, 27, 29, 220, 299, 367, 401, 408, 435, 455
Halgren, E., 124
Hall, K., 265
Hameroff, S., 340
Hampson, S., 138
Hanslmayr, S., 109, 113, 123, 129, 130, 156, 157, 162, 163, 169, 181, 252, 290, 333
Hanstock, C.C., 263
Hantouche, E.G., 282
Hare, T.A., 262
Hari, R., 123, 124, 129, 130, 162, 181
Harris, A.W.F., 122, 277, 279–281, 296
Harris, C.R., 283, 294
Harris, T.S., 265
Hartline, P.H., 125
Harvey, G.C., 263
Hasegawa, K., 265
Hasselmo, M.E., 277, 280, 281, 296
Hassenstein, B., 26
Haughton, V.M., 191
Hauptmann, C., 333
Hawking, S., 318, 355, 356, 382, 384
Hawkins, R.D., 353, 445, 447
Hayek, F.A., 39, 159, 165, 182, 423, 428, 435
Hebb, D.O., 20, 29, 346, 352, 353, 356
Heckers, S., 262
Hegerl, U., 290
Heino, R., 129
Heisenberg, W., 1, 329, 330, 360
Hendrie, H., 265
Hendrix, C.E., 65
Henze, D.A., 191
Herrmann, C.S., 163, 171, 181, 252, 260, 263, 264, 279, 280, 290, 296, 297, 354
Herrmann, F.R.,

Herrmann, W.M., 282, 290, 294, 295
Hesselbrock, V.M., 152, 283, 290, 333
Hetrick, W.P., 282
Heutink, P., 68, 69, 181, 347, 401
Heynen, A.J., 262
Hildebrandt, H., 168, 278, 279, 281, 293, 297
Hines, D., 425
Hinrichs, A., 290
Hinrichs, H., 139, 264
Hinterhuber, H., 265
Hirayasu, Y., 277, 280, 281, 296
Hiripi, L., 209
Hirota, N., 167, 232
Hirsch, S.R., 281, 296
Hobson, J.A., 369
Hodder, D., 166, 182
Hödlmoser, K., 162, 333
Hof, P.R.,
Hoff, E., 163
Hoffman, K.M., 281
Hoffman, R.S., 260, 264, 278, 279, 281, 292, 293
Hogan, M.J.
Hoke, M., 138
Holzfuss, J., 305, 310
Hooper, H.E., 369
Horridge, G.A., 200, 201
Housner, G.W., 464
Howard, L., 154
Hsieh, M.H., 277, 278, 280, 281, 297
Hsu, J.L., 277, 278, 280, 281, 297
Huang, M., 263
Huang, Y., 265, 445, 447
Hudson, D.E., 464
Hughes, J.R., 32, 45, 123, 137
Hughes, S.W., 290, 316
Humpel, C., 265
Huxley, A., 376
Hyde, J.S., 191
Hynan, L.S., 265

I

Iagolnitzer, D., 330
Ibañez, V.,
Ichikawa, J., 263
Infarinato, F., 267, 273
Infeld, L., 30, 32, 35, 316, 399, 401, 419
Iragui, V.J., 32, 349
Iragui-Madoz, V.J., 218, 220, 235, 349, 356, 411
Iriarte, J., 139
Isoglu-Alkaç, U., 163
Itzchak, Y., 248

J

Jacobson, E.D., 99
Jajcevic, A., 270, 294
James, W., 238
Jarcho, L.W., 122
Jasper, H.H., 57
Jefferys, J.G., 264, 265, 286
Jennings, L.S., 458
Jensen, O., 163, 286
Jeon, Y.W., 279
Jeremy, D., 122
Jezzard, P., 262, 286
Jiang, Z.Y., 272
Johnson, R. Jr., 154
Johnson, S.L., 154, 282
Johnston, D.,
Johnstone, S.J., 166, 182
Jones, E.G., 124, 131, 179, 190
Jones, K.A., 152, 260, 283, 284, 289, 290, 292, 293, 415
Jordan, R., 31, 111, 179, 181, 208
Jorm, A., 265
Josephson, R.,
Judaš, M., 226
Jung, C.G., 419
Jung, T.P., 130, 252
Junge, S., 163

K

Kaczmarek, L.K., 48, 202, 312
Kaiser, J., 252, 369
Kaiya, H., 262
Kamarajan, C., 152, 260, 289, 290, 292, 293
Kamphius, W., 308
Kamphorst, W., 68, 69, 181, 347, 401
Kandel, E.R., 150, 152, 353, 354, 445, 447
Kaneoke, Y., 189, 190
Kant, I., 152
Karakaş, S., 65, 69, 118, 119, 123, 124, 129–131, 133–135, 137–139, 144, 151, 153, 155–157, 160, 162, 166, 170, 180–183, 200, 261, 264, 284, 286, 333, 406
Karhu, J., 163, 264
Karis, D., 154
Karrasch, M., 265
Karren, N.U., 282
Katada, A., 227
Katchalsky, A., 29
Kaufinan, J., 177
Kedzior, K.K., 278, 297
Kelly, J.P., 53, 57
Kelso, J.A.S., 19, 112, 122, 286

Kerkut, G.A., 202, 209
Keshavan, M.S., 277, 281, 295, 297
Keskin, Y.H.,
Ketter, T.A., 263, 282
King, D.W., 264
Kirk, I.J., 262, 264
Kitajima, H., 167, 232
Klaver, P., 139, 264
Kliem, M.A., 189
Klimesch, W., 109, 113, 123, 129–131, 135, 156, 157, 162, 163, 169, 181, 252, 290, 333
Klump, M.C., 278, 280, 281, 296, 297, 358
Kocaaslan, S., 283, 284, 292
Koch, C., 252
Kocsis, B., 143, 181
Kohling, R., 139, 264
Kolev, V., 31, 68, 69, 113, 114, 118, 123–131, 142, 143, 157, 162, 163, 167, 181, 223, 228–230, 234, 264, 284, 286, 297, 347, 401
König, P., 31, 112, 141, 142, 144, 283, 286
Konjevi, D.J., 213
Konopacki, J., 263
Kopell, N., 286, 289
Kopp, C., 289
Koul, P., 262
Kovaevi, N., 213
Kovlev, V., 31, 113, 118, 129, 163, 234
Krauel, K., 252, 260, 290, 297
Krause, C.M., 129, 265
Kreiman, G., 252
Kreiss, D.S., 189, 190
Krüger, K., 124
Kruse, M., 31, 111, 179, 181, 208
Kruse, P., 137
Krystal, J.H., 260, 264, 278, 279, 281, 292, 293
Kunze, H., 201
Kuperman, S., 152, 283, 284, 333, 415
Kupfermann, I., 55, 57, 354
Kurthen, M., 139, 264
Kurths, J., 417
Kushnir, T., 248
Kwon, J.S., 277, 280, 281, 296

L

Labarga, A., 139
Lachaux, J.P., 31, 112, 286, 374
Laine, M., 265
Lam, R.W., 262, 263
Lambert, J.D.C., 202, 209
Lancrenon, S., 282
Landisman, C.,
Lanuzza, B., 267, 273
Laplace, P.S., 355
Lashley, K.S., 109
Lawlor, B., 369
Lawson, V.H., 262
Layne, S.P., 305, 310
Le Doux, J.E., 238, 256
LeBeau, F.E.N., 264, 265, 286
Lebech, J., 127, 310
Lee, G.P., 264
Lee, K.H., 279
Lehrer, J., 378
Leiberg, S., 252
Leifer, L., 160
Lenz, D., 163, 252, 260, 290, 297
Leranth, C., 262
Leuchter, A.F., 270, 294
Leung, L.S., 263, 264
Levinson, A.J., 281, 286
Levitan, I.B., 48, 202
Lewis, C., 103, 152, 188, 193, 320, 347, 396, 427
Li, T.K., 152, 333
Li, Z., 263
Liberati, D., 270, 294
Libet, B., 170
Liddle, P.F., 262, 263
Lieb, J., 124
Light, G.A., 277, 278, 280, 281, 296, 297
Lim, K.O., 260, 264, 278, 279, 281, 292, 293
Lindsley, D.B., 154
Litke, A., 152, 333
Llinas, R.R., 124, 131, 179, 190
Locatelli, T., 270, 294
Locke, J., 8, 15–16, 20, 196, 346, 358, 366, 408, 441
Loewi, O., 42, 259, 367, 369–370, 379, 424
Loker, J.E., 202, 209
Long, S.B., 189
Lopes da Silva, F.H., 123, 124, 129, 130, 162, 181, 269, 308
Lorenz, E.N., 299, 303, 456
Loring, D.W., 264
Löscher, W., 263
Lucchiari, C., 244, 252
Lundgren, O., 77, 100
Luria, A.R., 109, 252
Lutz, I., 85
Lutz, J., 85
Lutzenberger, W., 252
Lysaker, P.H., 280

M

Mackay, D.M., 165
Mackay, J.C., 262, 264
Mackert, B.M., 114
Maess, B., 163
Magill, P.J., 190
Mahendran, R., 276
Mahlberg, R., 282, 290, 294, 295
Makeig, S., 130, 138, 252, 264
Malach, R., 248
Malanda, A., 139
Maltseva, I., 129, 157
Mandell, A.J., 435
Mann, K., 113, 131
Marbach, S., 294, 415
Marder, E.,
Marksteiner, J., 265
Martin, J.,
Martin, M.T., 181
Martindale, C., 425
Martinerie, J.M., 31, 112, 286, 374, 451, 452
Martínez-Arán, A., 286
Marx, P., 114
Mathalon, D.H., 260, 264, 278, 279, 281, 292, 293
Mathers, C., 265
Mathes, B., 163, 278, 294, 297, 415
Matthews, B.H.C., 108, 200
Matthews, P.M., 262, 286
Matthias, H.J., 489
Maxwell, J.C., 30, 43, 348, 350–353, 357, 358, 431, 433
Mayer-Kressm, G., 305, 310
McCarley, R.W., 277, 278, 280, 281, 296, 297, 358, 369
McCarthy, G., 243
McClune, M.C., 142, 218, 220, 235, 349, 356, 411
McEvoy, L., 163, 264
McGrann, J.V., 177
McKeith, I.G., 265
McLeod, J.G., 122
McNaughton, N., 262, 264
Meador, K.J., 264
Meares, R.A., 122, 277, 280, 281
Mees, A.I., 458
Meltzer, H.Y., 263
Menezes, P.R., 265
Merlotti, E., 279, 297
Mesulam, M.M., 58, 113
Mιettιnen, R., 191
Meyer-Gomes, K., 277, 278, 280, 281, 297
Michalewski, H.J., 154
Michel, J.P.,
Michon, A.,
Miener, M., 163
Miewald, J., 277, 281, 295, 297
Mikkelsen, K.B., 127, 310
Miller, E.K., 183, 446
Miller, G.A., 452
Miller, R., 123, 124, 130, 131, 170, 264
Miltner, W., 144, 181, 284
Mimura, K., 167, 232
Miniussi, C., 267, 273
Missonnier, P.,
Mitra, M.,
Mitrofanis, J., 290
Molenaar, P.C., 289
Mommer, M.,
Monod, J., 24, 29, 41–43, 350, 361, 398, 433
Montano, N., 190
Mooibroek, J., 269
Moore, J.K., 51
Moratti, S., 451, 452
Mountcastle, V.B., 64, 65, 108–109, 112, 151, 165, 169, 354, 376, 445
Mrzljak, L., 226
Mucci, A., 279, 297
Müller-Gerking, J., 181
Munk, M.H., 31, 111, 179, 181, 208, 283
Murray, R.M., 276
Murro, A.M., 264
Musha, T., 273

N

Näätänen, R., 131
Nakamura, S., 262
Nakashima, T.T., 263
Namba, M., 262
Narici, L., 167, 168, 180, 232
Neil, E., 76, 78, 99
Nestor, P.G., 277, 278, 280, 281, 296, 297, 358
Neumann, J., 114
Neuper, C., 129, 181
Newton, R.G., 11, 442
Newton, T.F., 270, 294
Nicolis, C., 305
Niedermeyer, E., 227, 228
Niermeijer, M.F., 68, 69, 181, 347, 401
Niznikiewicz, M.A., 278, 280, 281, 296, 297, 358
Nobili, F., 267, 273
Noe, A., 374
Nolte, G., 114
Nowakowska, C., 282
Noyan, A., 73
Nuechterlein, K.H., 277, 280, 281, 296

Nunez, P.L., 269
Nurminen, N., 191
Nurnberger, J.I. Jr., 262

O

Obrist, W., 227
Ochi, N., 167, 232
O'Connor, S.J., 152, 283, 284, 333, 415
Oddie, S.D., 262
O'Donnell, B.F., 277, 280–282, 296
O'Donnell, P., 190
O'Donnell, T.O., 263
Öniz, A., 153, 160, 163, 168, 243, 244, 248, 249, 252, 265–270, 273, 281, 286, 289, 292, 293, 323, 333, 413–415
Ono, K., 167, 232
Önton, J., 264
Oren, R., 190
Orer, H.S., 189
Ösby, U., 282
Osen, K.K., 51
Ozaki, H., 227
Özerdem, A., 263, 264, 281–284, 286, 287, 289, 292, 294, 295, 414, 415
Özesmi, C., 112, 123, 137, 167, 453
Özgören, M., 31, 69, 113, 118, 153, 155, 163, 166, 168, 180–182, 234, 243, 244, 252, 264, 281, 297

P

Pace-Schott, E.F., 369
Pachinger, T., 113, 123, 129, 130, 156, 157, 162, 163, 169, 181, 252, 290, 333
Padmanabhapillai, A., 152, 260, 289, 290, 292, 293
Pandya, D.N., 58
Panek, R., 341
Pantev, C., 138
Papanicolaou, A.C., 358, 359
Pare, D., 124, 131, 179, 190
Parisi, L., 267, 273
Parrino, L., 190
Pascal, B., 8, 13–14, 20–21, 326, 348, 357, 358, 366, 369, 374, 376, 377, 407, 441
Patel, B.N., 189, 190
Pavlov, I.P., 149
Pecherstorfer, T., 109, 162, 163
Pedley, T.J., 76
Penáz, J., 88–90
Penrose, R., 340, 364, 378, 379, 424
Penttonen, M., 191
Perez, J., 262
Perkel, D.H., 107, 109, 110
Perl, D.P.,
Perlmutter, R., 278, 280, 281, 296, 297, 358
Perry, E., 290
Perry, R., 290
Petersén, I., 228
Petsche, H., 124, 163, 269
Petty, F., 286
Pfefferbaum, A., 260, 264, 278, 279, 281, 292, 293
Pfurtscheller, G., 129, 181
Piersol, A.G., 449
Pijn, J.P.M, 308
Pikovski, A., 417
Pinsker, H., 354
Pinter, R.B., 86, 87
Pizzella, V., 167, 168, 180, 232
Plastino, A., 181
Platt, C.J., 213
Pockberger, H., 269
Poincaré, H., 14, 303, 364, 367, 378–379, 385, 426
Polich, J., 154, 279
Pöllhuber, D., 113, 123, 129, 130, 156, 157, 162, 163, 169, 181, 252, 290, 333
Popper, K., 24, 323
Porjesz, B., 32, 107, 152, 183, 260, 269, 283, 284, 289, 290, 333, 406, 414, 415, 428
Post, R.M., 262
Potts, G.F., 277, 280, 281, 296
Pratt, H., 154
Prestia, A., 267, 273
Pribram, K.H., 164, 165, 367
Prigogine, I., 25, 29–30, 39, 43–44, 299, 350, 353, 408, 433–435
Prince, M., 265
Procaccia, I., 299, 458, 460
Prosser, C.L., 78
Psych, B., 279, 280, 296
Purohit, D.P.,

Q

Quiroga, R.Q., 68, 69, 130, 156, 181, 252

R

Rahn, E., 114, 125–128, 130, 131, 142, 143, 157, 166, 168, 180, 181, 284, 286, 330
Ramoser, H., 181
Ranck, J.B., 130
Rangaswamy, M., 152, 260, 283, 284, 289, 290, 292, 293, 415
Rao, S.G., 62

Rapinoja, P., 265
Rapp, P.E., 305, 310, 451, 452, 458
Rappelsberger, P., 163, 269
Ravid, R., 68, 69, 181, 347, 401
Reddy, L., 252
Regan, D., 49, 448
Reich, T., 152, 283, 284, 333, 415
Reichle, L.E., 419
Reinares, M., 286
Reitboeck, H.J., 31, 111, 179, 181, 208
Rennicke, C.M., 282
Reolfsema, P.R., 283
Reynolds, S.R.M., 100, 101
Rice, J.P., 152, 333
Richenbacker, W., 190
Riemann, D., 113, 131
Rihmer, A., 282
Rihmer, Z., 282
Rimmer, E., 265
Rinne, J.O., 265
Ripper, B., 113, 123, 129, 130, 156, 157, 162, 163, 169, 181, 252, 290, 333
Ritter, B., 223, 234
Roach, B., 260, 264, 278, 279, 281, 292, 293
Robin, R.W., 283, 294
Rockstroh, B., 451, 452
Rodriguez, E., 31, 112, 286, 374
Rodriguez, G., 267, 273
Roeder, K.D., 209
Röhm, D., 113, 123, 129–131, 135, 156, 157, 162, 163, 169, 181, 252, 290, 333
Rohrbaugh, J.W., 152, 154, 283, 284, 294, 333, 415
Romani, G.L., 167, 168, 180, 232
Rommelspacher, H.P., 282, 290, 294, 295
Röschke, J., 113, 131, 139, 203, 264, 305–309, 322, 348, 456, 459, 460
Rose, J.E., 32, 45, 123, 137
Rosen, B., 157, 208, 331
Rosen, R., 36, 66, 305
Rosenblum, M., 417
Ross, J., 283, 294
Rossini, P.M., 167, 168, 180, 232, 267, 273
Rosso, O.A., 31, 68, 69, 113, 118, 163, 181, 234, 264, 297, 347, 401
Rosso, S.M., 68, 69, 181, 347, 401
Rossor, M., 265
Rothenberger, A., 31, 113, 118, 163, 181, 234, 264, 297
Rotzinger, S., 263
Rowan, M., 369
Rowland, V., 29
Rudolph, U., 289
Ruedas, G., 84, 88
Rundo, F., 267, 273
Ruskin, D.N., 98, 187–191, 396, 427
Russegger, H., 113, 129, 130, 156, 157, 162, 163, 181, 252, 290, 333

S

Saatçi, E., 263, 264, 281–284, 286, 287, 289, 292, 294, 295, 415
Sabers, A., 127, 310
Saccone, N., 290
Saddy, J.D., 181
Saermark, K., 127, 180, 310
Sagawa, K., 87
Saidel, W.M., 215
Sakowitz, O.W., 31, 113, 118, 130, 156, 163, 180, 181, 234, 252, 264, 297
Salamero, M., 286
Salanki, J., 209
Salazar, J.M., 305
Salinari, S., 267, 273
Salisbury, D.F., 278, 280, 281, 296, 297, 358
Saltzberg, B.,
Samochowiec, J., 282, 290, 294, 295
Sánchez-Moreno, J., 286
Sander, T., 282, 290, 294, 295
Sanquist, Th.F., 154
Santosa, C.M., 282
Sarnthein, J., 31, 112, 141, 142, 144, 286
Sato, K., 167, 232
Sattel, H., 270, 294, 295
Sauseng, P., 109, 113, 129, 130, 156, 157, 162, 163, 181, 252, 290, 333
Saylan, M., 282, 290, 294, 295
Scazufca, M., 265
Schabus, M., 109, 113, 129, 130, 156, 157, 162, 163, 181, 252, 290, 333
Schack, B., 113, 129, 130, 156, 157, 162, 163, 181, 252, 290, 333
Schadow, J., 252, 260, 290, 297
Schaller, C., 139, 264
Scheffers, M.K., 154
Scheltens, P., 265
Scherg, M., 114
Schettini, G., 262
Schimke, H., 113, 129, 130, 156, 157, 162, 163, 181, 252, 290, 333
Schlesewsky, M., 181
Schlogl, A., 181
Schmeidler, J.,
Schmidt, L.G., 282, 290, 294, 295
Schmidtke, C.R., 360, 363
Schmiedt, C., 163, 243, 244, 252, 278, 279, 281, 293, 297

Schmiedt-Fehr, C., 248, 249, 286, 294, 413, 415
Schmielau, F., 140, 141
Schöner, G., 124
Schreiner, C.E., 124
Schreiter-Gasser, U., 270, 294, 295
Schroter, R.C., 76
Schuckit, M.A., 152, 290, 333
Schult, J., 308, 309, 322
Schürmann, M., 31, 68, 69, 111, 113–115, 118, 119, 123–131, 135, 137, 139, 142–144, 151, 156–158, 162, 163, 166, 168, 170, 180, 181, 199, 200, 209, 234, 252, 261, 263, 264, 284, 286, 297, 347, 348, 401, 406
Schuster, H.G., 455, 457
Schütt, A., 169, 203, 204, 206, 209, 212, 214, 216, 219, 261, 347, 348
Schwaiger, J., 113, 129–131, 135, 156, 157, 162, 163, 181, 252, 290, 333
Schwartz, J., 353
Schwarzkopf, H.J., 84, 88
Schweitzer, J., 213
Sclabassi, R., 124
Sechter, D., 282
Seed, W.A., 76
Sejnowski, T.J., 130, 139, 252, 264
Semrád, B., 88–90
Shaw, C.D., 299, 455, 456
Shaw, G.L., 177
Shearer, D.E., 227
Sheer, D.E., 137, 208
Shekha, A., 282
Shenker, A., 98, 187, 189, 190, 396, 427
Shenoy, K.V., 177
Shenton, M.E., 277, 278, 280, 281, 296, 297, 358
Shepherd, G.M., 58
Sherrington, C., 37, 108, 121, 149, 180
Shiah, I.S., 262, 263
Shick, J., 262
Sillanmaki, L., 129
Silva, L.R., 124
Silverstone, P.H., 263
Sim, K., 276
Simmons-Alling, S., 262
Sinerva, E., 265
Singer, W., 31, 108, 111, 137, 179, 181, 208, 283
Siska, J., 310
Sivri, O.J., 191
Skarda, C.A., 29–32, 103
Skinner, B.F., 149
Skinner, J.E.,
Smerieri, A., 190
Smith, H.W., 76
Smith, J.R., 264
Smith, M.E., 124, 163, 264
Smolka, M.N., 282, 290, 294, 295
Smulders, T.V., 345
Smythe, J.W., 262
Soares, J.C., 262
Sokolov, E.N., 151, 180, 181, 183, 445
Solms, M., 195, 227, 237, 238, 253, 254, 369–371, 373, 413, 423
Solodovnikov, V.V., 113
Somers, V., 190
Sorbell, J., 152, 333
Spaggiari, M.C., 190
Spar, J.E., 270, 294
Sparén, P., 282
Sparks, H.V., 78
Spelman, F.A., 86, 87
Spence, S., 281, 296
Spencer, K.M., 278, 280, 281, 296, 297, 358
Spencer, S.S., 218, 220, 235, 349, 356, 411
Spengler, F., 124
Spillantini, M.G., 68, 69, 181, 347, 401
Sporns, O., 166, 182, 280, 354
Sprock, J., 277, 278, 280, 281, 297
Stadler, M.A., 137, 163
Stadler, W., 109, 113, 129–131, 135, 156, 157, 162, 163, 181, 252, 290, 333
Stam, C.J., 163
Stampfer, H.G., 129–131, 157, 159, 160, 166, 167
Stapleton, J.M., 124
Stapp, H., 359
Starr, A., 154
Steriade, M., 124, 131, 179, 190, 267, 273
Stewart, M., 262
Stickgold, R., 369
Stimus, A.T., 152, 289, 292, 293, 333
Stone, D., 262
Strong, C.M., 282
Strüber, D., 137, 163
Stryker, M.P.,
Stumpf, C., 124
Sturbeck, K., 205, 217
Suhara, K., 227
Sule, A., 262, 286
Suzuki, H., 227
Swanwick, G.R., 369
Swerdlow, N.R., 277, 278, 280, 281, 297
Symond, M.B., 279, 280, 296
Syndulko, K., 154
Szilard, L., 350

T

Tagawa, Y., 167, 232
Takens, F., 455, 458, 459
Tancer, M., 277, 281, 295, 297
Tariska, P., 265
Tass, P.A., 333
Taub, E., 144, 181, 284
Tauc, L., 209
Taylor, M.G., 81, 86
Tegmark, M., 375
Terzano, M.G., 190
Tesche, C.D., 163, 264
Thiemann, V., 84
Thom, R., 25, 28, 43–44, 299, 404, 408, 433–435
Thompson, E., 374
Thompson, J.L., 264
Thompson, R.F., 158
Tierney, P.L., 98, 187–191
Tischfield, J.A., 152, 290, 333
Tischner, H., 80
Tittel, G., 213
Tobler, I., 289
Tognoli, E., 31, 32
Tombini, M., 267, 273
Tononi, G., 166, 182, 354
Torrent, C., 286
Torrioli, G., 167, 168, 180, 232
Townsend, J., 130, 252
Tranel, D., 176, 353
Traub, R.D., 264, 265, 286, 289
Trautner, P., 139, 264
Traversa, R., 167, 168, 180, 232
Tucker, L.R., 160
Tülay, E., 265–270, 273, 289, 292, 293, 323, 333, 414, 415
Tümer, N., 83
Tunca, Z., 263, 264, 281–284, 286, 287, 289, 292, 294, 295, 414, 415
Turnbull, O., 195, 227, 237, 238, 253, 254, 369–371, 373, 413, 423
Twarog, B.M., 209
Twery, M.J., 189, 190

U

Ulrich, M., 263
Ungan, P., 80, 96, 112, 123, 132, 137, 142, 167, 404, 452, 453
Uylings, H.B.M., 226

V

Van Baal, G.C., 289
Van Beijsterveldt, C.E., 289
Van De Borne, P., 190
Van der Stelt, O., 260
Van der Tweel, L.H., 448
Van Eden, C.G., 226
Van Eerdewegh, P., 152, 333
Van Erp M.G.,
Van Hoesen, G.W., 58
Van Neeven, J.M.A.M., 308
Van Rotterdam, A., 269
van Swieten, C., 68, 69, 181, 347, 401
Vanderwolf, C.H., 263, 264
Varela, F., 31, 112, 286, 374
Vecchio, F., 267, 273
Vernieri, F., 267, 273
Vertes, R.P., 143, 181, 262
Vialatte, F., 273
Viana Di Prisco, G., 143, 181
Vieta, E., 286
Vinogradova, O.S., 124, 263
Vohs, J.L., 282
Volpe, U., 279, 297
Von Helmholtz, H., 165
Von Neumann, J., 340, 436
von Stein, A., 31, 112, 141, 142, 144, 286
Vos, J.E., 269

W

Wachholder, K., 78
Waldemar, G., 265
Walker, M., 290
Walker, R.J., 202, 209
Wallenstein, G.V., 277, 280, 281, 296
Wallentin, I., 77
Walsh, J., 262
Walter, D.O., 270, 294
Walters, J.R., 98, 187–191, 396, 427
Wang, J.C., 290
Wang, K., 152, 283, 284, 333, 415
Wang, P.W., 282
Wastell, D.G., 452
Weiner, H., 270, 294
Weiss, C.H., 71, 80, 81, 84, 88, 89, 91, 93, 95, 96, 98, 112, 186, 187, 320, 321
Weiss, S., 163
Weizsäcker, C.F., 1
Wessely, S., 276
Westerfield, M., 130, 252
White, C.L., 265
White, K.V., 282, 283, 290, 294, 295
White, P., 260, 264, 278, 279, 281, 292, 293
Whittington, M.A., 113, 264, 265, 286, 289, 290
Wichman, E.H., 468, 470, 472
Wichmann, T., 189

Wienbruch, C., 415
Wiener, N., 1–3, 8, 25–26, 39, 43–44, 299, 339, 351, 404, 408, 429, 431–436
Wiig, K.A., 262
Wilcox, R.E., 63
Willemsen, R., 68, 69, 181, 347, 401
Williams, G.V., 62
Williams, L.M., 279, 280, 296
Winblad, B., 265
Winkler, T., 113, 129, 130, 156, 157, 162, 163, 181, 252, 290, 333
Winterer, G., 32, 45, 123, 137, 282, 290, 294, 295
Witte, H., 144, 181, 284
Woerner, W., 31, 113, 118, 163, 181, 234, 264, 297
Wylezinska, M., 262, 286
Wynn, J.K., 277, 280, 281, 296

X
Xiaohua, L., 263

Y
Yatham, L.N., 262, 263
Yener, G.G., 265–270, 273, 289, 292, 293, 323, 333, 414, 415
Yeragani, V.K., 277, 281, 295, 297
Yetkin, F.Z., 191
Yoffey, J.M., 77, 78
Yordanova, J., 31, 68, 69, 113, 118, 123, 124, 129, 130, 162, 163, 167, 181, 223, 228–230, 234, 264, 297, 347, 401
Yoshida, H., 262
Young, L.T., 281, 286
Yu, D., 163, 264
Yu, E.Y., 272

Z
Zappasodi, F., 267, 273
Zeeman, E.C., 28, 404
Zeki, S., 248, 413
Zetler, G., 137
Zheng, L.L., 272
Zheng-yan, J., 271, 273, 293–295
Zhong, S., 103, 152, 188, 193, 320, 347, 396, 427
Ziller, M., 282, 290, 294, 295
Zimmerman, I.D., 305, 310, 458
Zolotukhina, L.I., 124, 263
Zweig, S., 341, 380

Subject Index

A

Acetylcholine (ACh), 61–63, 203, 206, 209, 210, 213, 261, 263, 265, 290, 369, 379, 396, 412, 414, 422, 427
  esteem, 42, 60, 259, 424
Aging, 65, 158, 233–234, 290
Aladjalova, N.A., 98
Alpha, theta, delta, beta, gamma, 111, 145, 156, 167, 200, 224, 225, 286, 319, 406
Alzheimer disease (AD), 61, 260, 263, 265–276, 284–288, 292–295, 333, 409, 412, 426
Andreasen, N., 341, 343, 385–387, 424, 426, 427
Anesthesia, 189–190, 192, 368
APLR. *See* Attention, perception, learning and remembering
Aplysia, 112, 148, 149, 197, 200–203, 217–220, 225, 324, 334–335, 346–349, 354, 356, 402, 403
Archetypes, 372, 403, 428
Arterial supply, 101
Attention, perception, learning and remembering (APLR), 149, 159–161, 164–169, 172–173, 177, 179, 182, 183
Autonomous nervous system, 71–72, 74, 320–322
Auto-oscillations, blood, 100–101

B

Balzac, H., 380
Barlow, J., 435
Başar, E., 95, 96, 98, 435
Berger, H., 37, 38, 108, 123, 304, 308
Bergson, H., 1–3, 8, 13, 16–21, 23, 24, 43, 196, 326–327, 341–342, 345–346, 348, 356–366, 376–378, 380, 386, 397, 400, 408, 424–425, 428, 431–433, 435, 437
Bipolar disorder, 213, 260, 262, 281–288, 290, 409–412, 426
Black box (the brain), 66
Blood flow distribution, cat, 99–100
Brain body, 7–8, 101–103, 105, 145, 193–194, 320–321, 329–340, 358, 396–397, 403–406, 408–409, 413–415, 417–421, 426–429
Brain-body Feynman diagrams, 332–339
Brain-body-mind incorporation, 101–103
Brain-body-mind integration, 434
Brain's S-matrix, 330, 332, 404, 424
Brain's string theory, 317–318, 339–340, 400, 417, 419–421, 424, 427
Brownian motion, 30, 32, 400–401
Burst waves, 104

C

Cajal, R.Y., 8, 36–37, 48, 108–109
Capra, F., 430
Cartesian system, 9, 10, 300, 321–323, 325–328, 355, 405, 434
  brain-body-mind integration, 434
  Descartes, 432–433
  "Maxwell demon", 433
  multidisciplinary science platform, 433
  neuroscience, 432
  shift, 436–437
  string theory, 435
Causality, 8, 15, 182–183, 323, 325, 329–331, 353, 366, 398, 407, 424, 428, 432
Cerebral cortex, 52–60, 121, 179, 193, 221, 227, 262, 346, 354, 356, 396
Churchland, S.P., 432
Cognitive, 6, 46, 120–122, 128–129, 132–141, 161–163, 265, 335–336, 348–353

Coherence, 68, 141–144, 167–169, 181–182, 188, 189, 220, 235, 269–273, 349–351, 381, 410–411, 424
Coronary system, 24, 80, 85, 88, 93, 95, 402, 403
Coupled oscillations, 418
Coupled oscillators, 188, 193–194, 417–418, 424, 426
Cranial nerves, 46, 185–187, 320, 355
Creative, 18–19, 196, 345, 361–362, 364–365, 384–385, 408, 425
Creative processes, 195, 364–365, 386
Cybernetics, 1, 3, 8, 25–26, 43, 300, 339, 395, 404, 408, 434–436

D

DA. *See* Dopamine
Darwin, C., 1, 6, 18, 19, 24–25, 42, 43, 69, 196–197, 224, 247, 255, 339, 341, 345–347, 359, 361, 365–366, 395, 398, 401–403, 408, 431, 433
Descartes, R., 7–13, 23, 431–433, 437
Dissipative structure, 25, 29–30, 44, 300, 339, 353, 395, 408, 434–436
Dopamine (DA), 61, 63, 190, 208–212, 248, 261–263, 277, 382
Dream, 33, 259, 326–327, 355, 365, 369–370, 379
Duration, 2, 12, 15, 17, 18, 111, 118, 141, 142, 144, 156, 168, 176, 181, 199, 205, 220, 234, 239, 257, 319, 326–327, 341, 359–366, 378, 380, 386, 397
Dynamic memory, 147, 154, 157, 159–164, 177, 253, 254

E

Eccles, J.C., 432
EEG brain dynamics, 432, 434
Einstein, A., 12, 13, 19, 23–25, 30–33, 35, 182–183, 259, 316–317, 341, 359, 365, 366, 379, 396–397, 399–401, 403, 419, 424, 427, 431, 437
Electroencephalogram (EEG), 31–32, 43, 45, 64–65, 103, 108, 110–112, 122–132, 147–148, 152, 160–161, 174–176, 180, 182, 187, 203–208, 227–229, 232–233, 260, 267, 289, 294–297, 301, 304–312, 320–321, 330–332, 340, 374, 376, 401, 410–411, 432
Emotion, 5, 15, 58, 237–243, 253–257, 384–385, 423–426, 428
Entropy, 65, 68, 181, 223, 302, 317, 324–325, 347–353, 355–358, 401–402, 426, 428, 432, 433
Episodic memory, 163, 241–243, 253–255, 341, 362–364, 377–378, 383–385, 425, 426, 428, 432
Event related coordination, 112
Event related delta, 249, 250, 273, 289
Event related theta, 130–131, 249–250, 266, 269, 270, 273
Evoked response, 138, 142, 168, 220, 232, 249, 262, 330
Evolution, 18–21, 24, 29, 43, 65, 151, 179, 195–197, 212–213, 218–224, 235–236, 300,, 330, 334, 339–340, 345–350, 356–357, 361–362, 376, 398, 402, 406, 408

F

Face recognition, 145, 221, 237, 245
Feature detectors, 108, 150–151, 169, 180, 181
Fessard, A., 11, 46, 66, 108, 110–111, 185–186, 199, 319–321, 420
Feynman diagrams, 329, 331–332, 359, 360, 366, 404, 407, 424, 436
Freeman, W., 103, 104
Freud, S., 13, 19, 341, 342, 371–372, 382
Frontal lobe, 55, 57–58, 60, 62, 121, 132, 155, 158, 195, 222, 225, 370
Fuster, J.M., 432

G

Galileo, 1, 8, 11–12, 18, 23, 182–183
Gamma-aminobutyric acid (GABA), 61–63, 152, 260, 262–263, 281, 283, 286, 289, 412, 422
Gebber, G.L., 103
Genetical causality, 183
Genetic disorder, 152, 260, 289–290
Geometrical mind, 348
Globally coupled oscillation, 418
Glutamate, 61–63, 263, 264, 283, 286
Gold fish, 195, 200, 215–223, 347
Grandmother, 68, 69, 155–158, 180, 182, 238–243, 256–257, 384–385

H

Haken, H., 432435
Hawking, S., 318, 355, 356, 382, 384
Hayek, F.A., 1, 39, 41, 159, 165, 182, 339, 346, 362, 408, 423, 425, 428, 432, 435

Hearth, 192
Hebb, D.O., 20, 29, 39, 41, 66, 339, 346, 347, 352, 356, 423
Heisenberg, W., 1–2, 15, 33–35, 299, 313, 314, 317, 323, 329, 331, 337, 355, 359, 360, 376, 407–408, 431, 436, 437
Helix pomatia, 195, 200, 201, 203, 208, 210, 211, 218, 261, 347, 356, 406
Hippocampus, 25, 48, 49, 53, 69, 112, 113, 115, 118, 119, 123, 124, 130–132, 139, 142, 166, 168, 179, 190, 191, 227, 232, 248, 262, 264, 286, 305, 308, 324, 356, 382, 396, 403
Hume, D., 8, 15, 17, 366, 398, 406, 408, 432

I
Intestines, 71, 78, 88, 90, 95, 98, 100, 201, 320, 402
Intuition, 3, 13–14, 16–20, 196, 255, 257, 345, 346, 363, 374, 377, 425
Intuitive mind, 13–14, 374, 377, 429
Invertebrate ganglia, 122, 141, 200, 203, 212, 218, 219, 221, 223, 225, 262, 347, 356, 409, 432

J
Jung, C.G., 19, 38, 341, 343, 371, 372, 374, 403, 419, 428

K
Kandel, E.R., 6, 202, 255, 257, 343, 346, 347, 353, 355, 371, 381, 384, 426, 432
Kant, I., 17, 358, 371
Kelso, S., 136
Kelvin, L., 433
Kidney, 24–25, 74, 84–85, 88–91, 93–95, 101, 188, 403, 411, 427

L
Law, 12, 16, 29–30, 32, 33, 35, 36, 43, 316, 317, 324, 331, 332, 350, 352, 355, 357, 382, 399, 401, 419, 431, 433
Leibniz, G., 10, 16, 18, 436
Local circulatory control, 98, 101
Locke, J., 8, 15–16, 20, 196, 346, 358, 366, 408
Loewi, O., 42, 259, 369, 370, 379, 424
Lymphatic system, 46, 73, 76–78, 95, 96, 112, 186, 187, 192, 321, 337, 427
Lymph nodes, 74, 76–78, 80, 84, 95, 186, 321, 402

M
Matching, 121, 135, 153, 154, 160, 162, 164–173, 182, 190, 285–286, 320
Maturating brain, 6, 195, 221–223, 234–236, 302, 346, 348–349, 357, 401–403, 409, 410
"Maxwell demon", 433
Maxwell, J.C., 431
Maxwell's diagram, 30, 43, 348, 350–353, 357, 362, 431, 433
Memory, 6, 17, 23, 58, 65, 112–113, 119, 147–169, 172–177, 180, 182, 240–243, 253–257, 272, 358, 361–362, 370, 377–378, 383–387, 423, 425–426
Memory longer activity, 149, 151, 167, 174–179, 257–259
Metaphysics, 16–19, 21, 33, 299, 323, 326–327, 362–363, 365, 375–376, 387, 397, 407–408
Micro-Darwinism, 403–404
Mind, 5–8, 13–16, 38, 42, 46, 101, 105, 145, 183, 193–194, 197, 212–213, 234–236, 255, 259–260, 290, 299, 329–331, 339–340, 348, 358, 371, 373–374, 377, 381–387, 396–401, 411–415, 417–418, 423–429
Monod, J., 24, 29, 41–43, 350, 361, 398, 433
Mozart, W.A., 367, 379–380, 385–386, 426
Multiple causality, 15, 225, 302, 321, 323, 325–326, 338, 340, 366, 398, 400, 435
Multiple matching, 164–170
Multiple oscillation memory, 112–113, 151, 155, 156, 176, 182
Mutual excitation, 192, 321
Myogenic-coordination, whole-bodyintegration and tuning
  brain-body-mind incorporation, 101–103
  interim synthesis, flow and vascular resistance, 95, 96
  local circulatory control, 98–101
  respiratory coordination, 103–104
  response susceptibility, peristalsis organs, 103
  system, 95–97
  ultraslow oscillations, 105

N
NE. *See* Norepinephrine
Nebulous cartesian system (NCS), 300, 317, 325–327, 329–331, 340, 355, 372, 376, 407, 429, 435
Neural coding, 107, 109–110
Neurochemical modulation, 208–213

Neuron doctrine, 37, 69, 108, 180
Neuron population, 107–108, 314–315
Neurotransmitter, 11, 45, 60–63, 152, 194, 197, 203, 209–210, 212–213, 260–263, 286, 289, 295, 321, 395, 406, 409, 413, 418, 421, 424, 426–427
Newton, Sir Isaac., 431, 436, 437
Norepinephrine (NE), 61, 63, 261, 396, 427

O

Overall frequency tuning, 396
Overall myogenic coordination, 98–103
Overall myogenic system, 71–72, 95–98, 105, 186–188, 192–193, 321
Overall tuning, 103, 129, 150–151, 179, 181, 320, 321, 323, 396, 420, 427

P

P 300, 128, 130, 134–135, 139, 154, 157, 159–160, 168, 170, 221, 264, 279, 280, 283, 286, 413
Parietal cortex, 53, 55–57
Pascal, B., 8, 13–14, 20–21, 326, 348, 357, 358, 366, 369, 374, 376, 377, 407
Pathologic brain, 145, 197, 213, 259, 290–297, 397, 401, 409
Perceptual memory, 119, 149, 151, 153–156, 158, 165
Peristalsis, 46, 78, 79, 87, 98–101, 103, 112, 187, 192–193, 321, 337, 402, 414
Persistent memory, 148–149, 152, 165, 172–177, 182, 253, 320
Phyletic memory, 20, 41, 148–153, 155, 156, 166, 169, 183, 237, 301, 320, 325, 348, 358
Physiological memory, 148, 151–158, 176, 177, 423
P 300-40 Hz, 139
Poincaré. H., 14, 303, 364, 367, 378–379, 385, 426
Prigogine, I., 25, 29–30, 39, 43–44, 299, 350, 353, 408, 433–435
Principle, 8–9, 11, 13–15, 66–69, 103, 108, 110, 132–137, 145, 157, 181–182, 185–186, 199, 302, 317, 354, 355, 371, 374, 375, 396–401, 405–407, 420, 427, 428, 432

Q

Quantum brain uncertainly principle, 314, 336
Quantum dynamics, 8, 19, 45, 300, 302, 312, 316, 321, 323, 417
Quasi-invariants, 197, 225, 295, 406, 410–413, 427

R

Ray, 33, 151, 183, 200, 213, 221, 313, 330, 347
Re-entrant signaling, 354, 355
Reflexes, 6, 148, 149, 151, 153, 183, 346
Resonance, 11–12, 64, 88, 101, 103, 129, 130, 169171, 192, 193, 205, 214, 305, 334, 340, 376, 398, 400, 404, 407, 418, 420–422, 426
Respiratory coordination, 103–104
Response susceptibility, 103, 180–181, 192, 407
brain, 167–168, 181, 227–228, 232–234
Response susceptibility, peristalsis organs, 103
Reticular formation (RF), 25, 51, 53, 112, 117, 120–121, 124, 131–133, 139, 142, 217–218, 221, 232, 262, 286, 305–307, 332
Reynolds, S.R.M., 100

S

Schizophrenia, 276–281, 292–297
Serotomine, 63, 209–210, 261
S-matrix, 40, 329–331, 333, 337, 340, 360–361, 366, 404, 407, 424, 436
Smooth muscles, 12, 25, 45–46, 71, 76, 78–80, 88–90, 93, 96, 98–101, 148, 186, 187, 320–322, 402, 432
Spinal cord, 37, 46, 53, 71, 105, 112, 148, 149, 152, 186–194, 200, 203, 302, 320, 321, 323, 346, 357, 404, 406, 417–418, 420–422
Stomach, 71, 78–79, 81–82, 95, 98–101, 103, 320, 402
String theory, 105, 194, 300, 317–318, 339–340, 395, 400, 417, 419, 424, 426–429, 435
Superposition, 31, 111–113, 122, 129, 132–137, 145, 156, 157, 166, 168, 169, 181–182, 325, 333, 375, 398, 403, 405, 415, 427
Super-synergy, 69, 179, 181–182, 421
Synchrony, 12, 30–32, 104, 111, 167, 190, 199, 203, 212, 219–221, 224, 225, 231, 233, 244, 259–260, 263, 273, 278–280, 282, 286, 289–293, 295–297, 319, 347–349, 353, 356, 366, 397, 406, 427, 432
Syncytium, 37, 45–46, 48, 419–421, 426, 427
Synergetic, 25–29, 32, 44, 300, 339, 395, 401, 408, 432, 434–436

Synergetic catastrophe theory, 25, 300, 339, 395, 408, 435
Synergetics, 432

T
Thom, R., 25, 28, 43–44, 299, 404, 408, 433–435
Tissue compartments, 99
Traveling backwards in time, 384

U
Ultraslow oscillation, 45–46, 98, 105, 187–193, 325, 405, 406, 411, 417, 418, 420–421, 427
Unconscient, 341, 343, 369–371, 373, 380, 381
Unifying concepts, 1, 419, 423
Uterus, 78, 80, 83, 95, 98, 100, 101, 103, 402

V
Valproate, 262–263, 282–284, 291–292, 295, 413
Vascular resistance, 80, 81, 85, 88–93, 95, 96
Vegetative system, 45–46, 95, 96, 112, 152, 183, 185–188, 192, 194, 212, 221, 302, 320, 321, 323, 332, 334, 340, 355, 358, 395, 396, 398, 404–406, 417–418, 420–422, 427
Von Helmholtz, H., 8, 39–41, 165, 169, 376
von Weizsäcker, C.F., 431, 432

W
Weiss, C., 95, 96
Whole-brain work, 46, 177, 179–183
Wiener, N., 1–3, 8, 25–26, 39, 43, 299, 339, 351, 404, 408, 429, 431–436
Working memory (WM), 62, 129, 149, 154, 155, 160, 163–167, 241, 264, 272, 278–279, 281, 291–295, 314, 351, 377, 411